TRENDS AND TECHNOLOGIES

Third Edition

TRENDS AND TECHNOLOGIES

Third Edition

Editors

Banwari Lal and Priyangshu M Sarma

The Energy and Resources Institute

First edition 1995
Second edition 2005
Reprint 2008
Third edition 2011
First reprint 2013

ISBN 978-81-7993-424-1

Published by
TERI Press
The Energy and Resources Institute
Darbari Seth Block
IHC Complex, Lodhi Road
New Delhi – 110 003
India

Tel. 2468 2100 or 4150 4900
Fax 2468 2144 or 2468 2145
India +91 • Delhi (0)11
E-mail teripress@teri.res.in
Website www.teriin.org

Printed in India

Foreword

Economic development and population growth have led to a major increase in production and consumption of goods and services. There has also been a substantial increase in the generation of waste, both in magnitude and variety, emanating from the expanding cycle of production and consumption. This trend has been most pronounced in urban areas, and now that over 50% of the world's population lives in towns and cities, the quantities and types of waste generated have grown rapidly. Rural areas have not been an exception to this global phenomenon, and with the rise in incomes and consumption levels in villages across the world, waste generation in these locations has also increased substantially. The overall effect has been that both in economic and ecological terms, costs associated with waste generation and management have increased manifold in recent years.

Much is spoken and written about on the need for a new set of three "R"s guiding our actions and thinking. This imperative refers to the need for human society to reduce, reuse, and recycle goods produced and consumed worldwide. As a philosophy, human society need not bring about a radical regression in lifestyles and standards of living, but essentially, every community and individual would now have to ensure that one does not consume purely for the sake of consuming, but in response to needs and aspirations that are bound by a conservation ethic. Also, our pricing systems would have to internalize the cost of externalities imposed by private entities on the global commons and society at large; we would need to ensure that there is a set of incentives and disincentives for producers and consumers to move towards the greater reuse of materials at every stage of the production/consumption cycle. Government policy through carefully designed fiscal measures can create conditions for a major shift to greater reuse and recycling of materials and products. For instance, bio-waste can be converted into feedstock for several products, and through the applications of green chemistry, a new set of chemicals and products could be produced, which are inherently efficient in the use of natural resources.

In a country like India where waste materials ranging from municipal garbage and agricultural residues are simply burned, this practice not only leads to higher levels of pollution but also results in the wastage of that material, which could be productively employed through technological innovations for producing useful products and services. A new set of technologies would need to be rapidly evolved to provide options by which waste can be converted to useful products, with consequent environmental benefits and economic gain. *Wealth from Waste* is a useful compilation of select technological interventions, which address these objectives, and which are assessed in depth for creating an understanding and application of these methods. The interested reader, therefore, would go through this volume fully understanding the usefulness of waste as a resource and the value of innovation to create wealth from waste. This book would also provide pointers to innovations and research directions, which could be undertaken in the future. The book is of value not only for academics and researchers, but also for leaders of industry, policy-makers, and the public at large.

R K Pachauri, PhD
Director-General, TERI

Preface

With rapid urbanization and increasing economic growth, witnessed especially in the developing countries, there is a consistent increase in the generation of waste, subsequently bringing the issues related to its impacts to the fore. The problem of industrial or municipal solid waste, domestic or agro-industrial waste, plastic waste or waste water effluent is serious in terms of their magnitude and potential impacts on air, water, land, and health of people managing these waste streams. Generation of waste is related directly or indirectly to the growth and development of human society. The distribution and the quantum of waste generated, however, vary from place to place, depending on the population density and consumption pattern of society. In developing countries, rapid urbanization and exponential growth in industries have led to the generation of large quantities of waste. Further, waste generation is always directly proportional to the population density. This makes the issues of generation of waste and methods for its mitigation prime concerns in developing countries.

Depending on its constituents, waste can be at times extremely "dirty", providing favourable conditions for the proliferation of disease-causing pathogens that are dangerous and hazardous in terms of toxicity, impacting the food chain, the ecosystem, and the entire environment. Thus, there is a need to come up with innovative, integrated solutions to mitigate and minimize the issues associated with waste and waste water. A method of safe disposal of waste can provide an immediate solution. However, it depends on the quantum, composition, and costs. Depending on the waste composition, nature's biological cycle can play a significant role in converting a type of waste to another benign form. However, these natural efforts are becoming futile owing to the large quantum of waste generated due to rapid growth and development. On the contrary, sometimes, the natural biological phenomena also generate some form of waste that can have a more detrimental impact on the environment. For example, the methanogenic bacteria generates tonnes of methane gas annually, which is almost three times more potent than carbon dioxide, through anaerobic digestion of the waste in landfills world over, contributing to global warming.

Thus, the concepts of integrated waste management options and innovation involve adopting alternative strategies, especially valorization of waste, so that the waste is no longer considered a by-product, but a value-added product. This will not only make the generators of the waste to comply with the waste regulations but also encourage them to invest in waste treatment strategies, thus creating a viable value addition chain for revenue generation. In the last two decades, there have been increasing efforts both from scientists and industrialists to safeguard the environment, especially from the adverse impacts of the waste. Today, various methods of treatment of solid and liquid waste have been successfully developed and implemented globally. However, the waste management practice is not sufficient to make the environment completely free from the detrimental impact of the wastes. But, with continuous improvement in the fields of science and technology, scientists have developed new technologies, using which human beings will be able to not only enhance their knowledge on disposing of the hazardous waste but also generate some alternative useful products from the same waste, which can be highly useful to society. Today, biofuels, bioenergy, bio-fertilizer, feedstock for green chemicals, and most importantly, the concept of bio-refineries have changed the outlook of waste valorization. Generation of profitable revenues from waste is no longer a concept; it is now a reality.

In this context *Wealth from Waste: trends and technologies* is an effort to put together the various options available to meet the above requirements. Articles included in this book not only aim to promote the concept and technologies, but also highlight the tremendous business potential that well-placed systems of waste management can offer.

Wealth from Waste is a must-read for policy-makers in the developing countries, governments, environmentalists, development practitioners, researchers, waste-management experts, and decision-makers in corporate organizations.

Acknowledgements

The editors would like to gratefully acknowledge all the contributing authors for timely submission of the manuscripts and also promptly addressing all the queries. A special gratitude to Ms Neena Mata, TERI, for following up with the correspondence and providing logistical support from the beginning till the finalization of the draft of this book.

We would also like to extend our sincere gratitude to Dr R K Pachauri, Director-General, TERI, who not only was the source of inspiration for bringing out this edition of *Wealth from Waste* but also provided his expert guidance throughout the reviewing and editing process. We are also very thankful to him for kindly agreeing to write the foreword for this book.

We acknowledge the effort of the entire editorial and production team of TERI press, which was instrumental in bringing out this edition.

The editors would also take this opportunity to thank the members of the microbial biotechnology group of TERI for their support and cooperation.

Contents

Biofuels

A K Agarwal

Associate Professor, Department of Mechanical Engineering
Indian Institute of Technology, Kanpur, India

Introduction

The world is presently confronted with the twin crises of fossil fuel depletion and environmental degradation. Indiscriminate extraction and generous consumption of fossil fuels have led to the reduction in underground carbon resources. The search for an alternative fuel, which promises a harmonious correlation with sustainable development, energy conservation, efficiency, and environmental preservation, has become highly imperative in the present context. To developing countries of the world, the fuels of bio-origin can provide a feasible solution to this crisis. Petrol- and diesel-driven automobiles are the major sources of greenhouse gas (GHG) emission. Scientists around the world have explored several alternative energy resources that have the potential to quench the ever-increasing energy thirst of today's population. Various biofuel-based energy resources explored include biomass and biogas energy, primary alcohols, vegetable oils, bio-diesel, and so on. These alternative energy resources are largely environment-friendly, but they need to be evaluated on a case-to-case basis for their advantages, disadvantages, and specific applications. Some of these fuels can be used directly, while others need to be formulated to bring their relevant properties close to conventional fuels. The power used in the agricultural and transportation sectors is essentially based on diesel and gasoline fuels, and it is essential that alternatives to these petroleum fuels are developed. Given the recent widespread use of petroleum-based fuels in various sectors, this study concentrates on assessing the viability of using various alternative fuels in the existing internal combustion engines.

The present energy scenario has stimulated active research interest in non-petroleum, renewable, and non-polluting fuels. The world reserves of primary energy and raw materials are, obviously, limited. According to an estimate, the reserves for coal will last for 218 years; for oil, 41 years; and for natural gas, 63 years, under a business-as-usual scenario.

Crude oil prices are stuck at near-record levels and are stabilizing at about $70 per barrel now. The variations in the energy prices are shown in Figure 1.

The enormous growth witnessed in the world population during the last few decades as well as the large-scale technical development and enhanced standard of living in the industrial nations have led to an intricately delicate situation in the field of energy supply and demand. This necessitates the need for developing and commercializing fossil fuel alternatives of bio-origin. This may well be the main reason behind the growing awareness and interest in unconventional energy sources, as various countries try to offset the oil monopoly.

Diesel fuels have an important function in the industrial economy of a country. Compared to the rest of the world, India's demand for diesel fuels is roughly six times that of gasoline. The consumption of diesel fuels in India in 1994/95 was 28.30 million tonnes, which was 43.2% of the total consumption of petroleum products. This requirement was met by importing crude petroleum as well as petroleum products. The import bill on these items was Rs 178.38 billion in 1994/95. With the expected growth rate of diesel consumption being more than 14% per annum, shrinking crude oil reserves, and limited refining capacity, India will be heavily dependent on imports of crude petroleum and petroleum products.

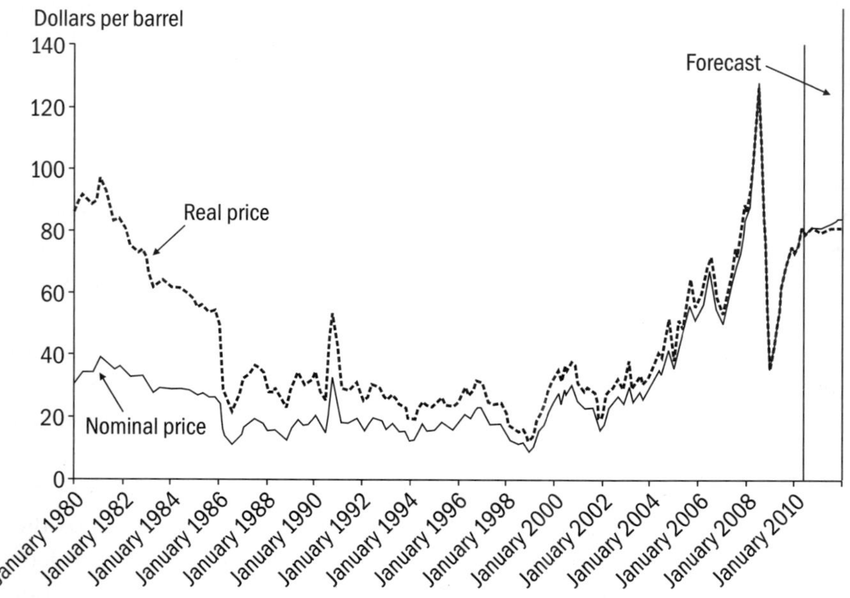

Figure 1 Crude oil prices

Source US Energy Information Administration

Environmental implications of using fossil fuels

Environmental concerns have increased significantly across the world during the past decade or so—particularly after the Earth Summit 1992. Excessive use of fossil fuels has given rise to global environmental degradation effects such as the greenhouse effect, acid rain, ozone depletion, climate change, and so on. There is a growing realization worldwide that something constructive has to be done soon to reduce the GHG emissions, or else our future is doomed. In the Kyoto Conference on global climate change, nations world over committed to reduce GHG emissions significantly.

Use of various fossil fuels such as petroleum products and coal leads to several other environmental problems such as reduction in underground carbon energy sources, significant modifications in the earth's surface layer, and subsidence of ground surface after extraction of fuels and minerals. The indiscriminate use of these fossil fuels has led to an increase in CO_2 levels in the atmosphere—from 280 parts per million (ppm) in the pre-industrial era to 360 ppm now. These CO_2 levels are still increasing, as a function of fuel burnt, leading to the greenhouse effect, acid rains, smog, and change of climate world over. These environmental implications are being felt in the day-to-day life also.

Besides, the combustion of fossil fuel has an adverse effect on human health because of increased air pollution in cities, acid rains, build-up of CO_2, changing heat balance of the earth, and so on. In fact, projections for the 30-year period between 1990 and 2020 indicate that vehicle travel, and consequently fossil fuel demand, will almost triple, and the resulting emissions will pose a serious problem. The main reason for increased pollution levels, in spite of the stringent emission norms that have been enforced, is the increased demand for energy in all sectors and, most significantly, the increased use of automobiles. Globally, the number of motor vehicles on the roads today is half a billion, about 10 times greater than their number in 1950.

Petroleum-fuelled motor vehicles are the major contributor to air pollution. For example, in the USA, 66% of carbon monoxide (CO) is emitted from vehicles alone. As can be seen in Figure 2, vehicular emissions of oxides of nitrogen (NO_x), lead, CO_2, particulates, and volatile organic compounds (VOCs) are significantly high. In spite of the plans to mandate stricter pollution control, the increased number of vehicles on the road and the associated increase in vehicle kilometres travelled will make the future task of cleaning the air very difficult, if we continue to burn fossil fuels for transportation and power at this rate.

Combustion of various fossil fuels leads to the emission of several pollutants, which are categorized as regulated and unregulated pollutants. Regulated pollutants are the ones the limits of which have been prescribed by environmental legislations, whereas unregulated pollutants are the pollutants for which no limits have been prescribed. Regulated pollutants include NO_x, CO, hydrocarbon (HC), particulates, whereas unregulated pollutants include

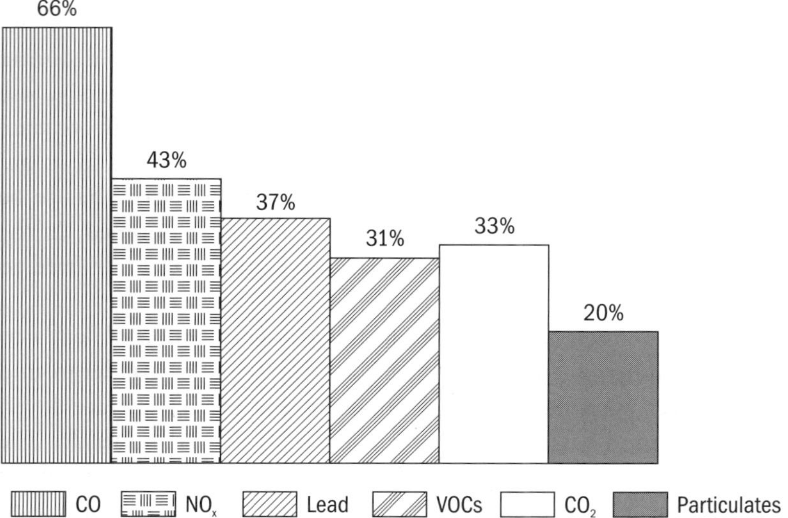

Figure 2 Percentage share of pollutants in vehicle exhaust (US, 1987)

formaldehyde, benzene, toluene, BTX (xylene), aldehydes, sulphur dioxide (SO_2), CO_2, methane, and so on.

These pollutants contribute to several effects on human health, which can be categorized as short- and long-term health effects. The short-term health effects are due to CO, NO_x, particulates, formaldehyde, and so on, while the long-term health effects are due to polycyclic aromatic hydrocarbons (PAHs), benzene, toluene, xylene, formaldehyde, and so on. CO is fatal in large doses, as it aggravates heart disorders, affects the central nervous system, and impairs the oxygen carrying capacity of the blood by forming carboxyhaemoglobin. NO_x cause irritation in the respiratory tract. HCs cause drowsiness, eye irritation, and coughing.

These pollutants also contribute to several regional and global environmental effects. Regional environmental effects, such as summer smog, are due to aldehydes, CO, NO_x, and so on. Winter smog is due to particulates. Acidification is caused by NO_x, sulphuric oxides, and so on. Several global effects such as ozone layer depletion, global warming, and so on are caused by CO_2, CO, methane, non-methane HCs, NO_x, and so on.

Alternative fuels for transportation sector

Transportation and agricultural sectors are the major consumers of fossil fuels and the biggest contributors to environmental pollution.

There are several factors that need to be taken care of before recommending any alternative fuel to be used in the existing technologies on a large scale. These factors are stated below.

- *Extent of modifications required in the existing hardware* If any alternative fuel mandates extensive modification in the existing hardware, involving huge capital, then it may be difficult to use this fuel.

- *Investment costs for developing infrastructure for processing these alternative fuels* Excessive infrastructure cost may act as a constraint for the development of the energy resource.
- *Environmental compatibility compared to conventional fuels* If the new fuel is more polluting, then it will be unacceptable for use.
- *Additional cost to the user in terms of routine maintenance, equipment wear, and lubricating oil life* Excessive additional cost will have an adverse effect on the widespread acceptance of this fuel.

Some of the alternative fuels of biological origin, such as primary alcohols and vegetable oils, are discussed in detail in this chapter.

Methanol and ethanol are two abundantly available primary alcohol, which can be used as alternative fuels that possess the potential to be produced from biomass sources. Alcohol fuels, methanol and ethanol, have physical properties and emission characteristics similar to those of petroleum fuels. Alcohol can be produced from coal, natural gas, crude oil, biomass or even organic waste.

Methanol (formula CH_3OH) is a simple compound. It does not contain sulphur or complex organic compounds. The organic emissions (ozone precursors) from methanol combustion will have lower reactivity than gasoline fuels, and hence, lower ozone-forming potential. If pure methanol is used, the emission of benzene and PAHs is very low. Methanol gives higher engine efficiency and is less flammable than gasoline, but the range of the methanol-fuelled vehicle is half that of the other vehicles because of the lower density and calorific value of methanol, thereby requiring a larger fuel tank. M100 (100% methanol) has invisible flames and is explosive in enclosed tanks. The cost of methanol is somewhat higher than the gasoline fuel. Methanol is toxic, has corrosive characteristics, and emits ozone-producing formaldehyde. Methanol poses environmental hazard in case of spill, as it is totally miscible with water.

Ethanol is similar to methanol, but is considerably cleaner, less toxic, and less corrosive. It gives greater engine efficiency. Ethanol is grain alcohol, and can be produced from agricultural crops like sugar cane, corn, and so on. Ethanol is more expensive to produce, has lower range, poses cold starting problems, and requires large harvest of these crops. Higher energy input is required in ethanol production compared to other energy crops, and it leads to environmental problems such as soil degradation.

Alcohol fuels can be successfully used as internal combustion (IC) engine fuels, either directly or by conversion to bio-diesel. The transesterification process utilizes methanol or ethanol and vegetable oils as the process inputs. This route of utilizing alcohol as a diesel engine fuel is definitely a superior route, as the toxic emissions containing aldehydes are drastically reduced. The problem of corrosion of various engine parts utilizing alcohol as fuel is also solved by way of transesterification.

Henry Ford and alcohol fuels

Ethanol has been known as a fuel for many decades. Indeed, when Henry Ford designed the Model T, he expected that ethanol, made from renewable biological materials, would be a major automobile fuel. However, gasoline emerged as the dominant transportation fuel in the early 20th century, because of the ease of operation of gasoline engines with the materials then available for engine construction and the increasing supply of cheaper petroleum from oilfield discoveries.

However, gasoline had many disadvantages as an automotive source. The "new" fuel had a lower octane rating compared to ethanol, was much more toxic (particularly when blended with tetraethyl lead and other compounds to enhance octane), was generally more dangerous, and contained toxic air pollutants. Petroleum was more likely to explode and burn accidentally, cause gum formation on storage surfaces, and lead to the formation of carbon deposits in combustion chambers of engines. Pipelines were needed for distribution from "area found" to "area needed". Petroleum was much more physically and chemically diverse than ethanol, necessitating complex refining procedures to ensure the manufacture of a consistent "gasoline" product.

Because of its lower octane rating relative to ethanol, the use of gasoline meant the use of lower compression engines and larger cooling systems.

Diesel engine technology, which developed soon after the emergence of gasoline as the dominant transportation fuel, also resulted in the generation of large quantities of pollutants.

However, despite these environmental flaws, fuels derived from petroleum have dominated automobile transportation for the past three-quarters of a century. There are two key reasons: first, the cost per kilometre of travel has been virtually the sole selection criteria; second, the large investments made by the oil and auto industries in physical capital, human skills, and technology made the entry of a new cost-competitive industry difficult.

Until very recently, environmental concerns have been largely ignored. But all of that is finally changing as consumers demand fuels, such as ethanol, which are much kinder to the natural environment and human health.

Alcohol fuels

Alcohols are derived from renewable resources such as locally grown crops and even waste products such as waste paper or grass, tree trimmings, and so on. Alcohol is a likely alternative automotive fuel, since it has properties that would allow its use in existing engines, with minor hardware modifications. Alcohols have a higher octane number than gasoline. A fuel with a higher octane number can endure a higher compression ratio before engine knocking takes place, thus giving the engine the ability to deliver more power, thus making it more powerful, efficient, and economical. Alcohol fuels burn cleaner than regular gasoline and produce less CO, HCs and NO_x. Alcohol fuel has high heat of vaporization; therefore, it reduces the peak temperature inside the cylinder, and hence, reduces NO_x emissions and increases the engine power. However, the aldehyde emission goes up significantly. Aldehydes play an important role in the formation of photochemical smog.

Bata and Roan Vermon (1989) studied the effect of adding ethanol to gasoline on the exhaust emissions from a 1978 Ford 2.3-l engine, operating under fixed conditions. They observed lower HC emissions and a 40%–50% decrease in CO concentration compared to the base fuel. On the other hand, aldehyde emissions increased by 60%–100%. Evaporative emissions also increased for ethanol-containing gasoline.

Stump, Knapp, and Racy (1996) examined tailpipe and evaporative emissions from three pre-1985 passenger motor vehicles operating on the ethanol (9 vol.%) and non-oxygenated fuel. A general reduction in HC, CO, benzene, and 1,3-butadiene emissions was observed when the ethanol fuel was used. Both formaldehyde and acetaldehyde emissions increased (almost doubled) with the ethanol blend.

Ethanol is one of the potential fuels for replacing diesel in the compression ignition (CI) engines also. The application of ethanol as a supplementary CI engine fuel may reduce environmental pollution, strengthen agricultural economy, create job opportunities, and reduce diesel fuel requirements, thereby contributing to the conservation of a major commercial energy source.

Ethanol was first suggested as an automotive fuel in the USA in the 1930s, but it was widely used only after 1970. Nowadays, ethanol is used as a fuel mainly in Brazil, and as a gasoline additive for enhancing octane number and improving combustion in the USA, Canada, and India. As gasoline price increases and pollutant emissions restrictions become more stringent, ethanol could be given more importance as a renewable fuel or gasoline additive (Rosillo Calle and Cortez 1998).

Properties

Ethanol is isomeric with dimethyl ether (DME), and both ethanol and DME can be expressed by the chemical formula C_2H_6O. The oxygen atom in ethanol possibly induces three hydrogen bonds. Although they may have the same physical formula, the thermodynamic behaviour of ethanol differs significantly from that of DME on account of the stronger molecular association via hydrogen bonds in ethanol.

Advantages of alcohol as fuel

Economic considerations

- The production of alcohol is far cheaper, simple, and eco-friendly. This way, it would be a lot cheaper than gasoline.
- Alcohol can be produced locally, thus cutting down on transportation costs.

Alcohol–gasoline blends: engine modification required

While blends of ethanol in gasoline are commonly used in vehicles designed to operate on gasoline, vehicle modification is required for alcohol fuel because its properties are different from those of gasoline (see Table 1). Ethanol has low stoichiometric air–fuel ratio and high heat of vaporization, which requires carburettor recalibration and increased heating of the air–fuel mixture to provide satisfactory drivability (Kremer, Jordim, and Maia 1996).

Brazil is the country that has the most developed technology for the alcohol-fuelled Otto cycle (four-stroke cycle) IC engine. In the early 1980s, there were more than 3.5 million alcohol-powered automobiles. In order to make alcohol engines more practical, functional, durable, and economical, engineers made several changes in the regular gasoline engines. These include the following.

- Since alcohol does not evaporate as easily as gasoline, the intake manifold had to be redesigned to provide more heating for evaporation.
- The carburettor was regulated in order to change the air/fuel proportions.
- The tin and lead coating of the fuel tank was replaced with pure tin coating.
- The fuel lines (zinc steel alloy) were changed to cadmium brass.
- The fuel filtering system was changed and re-dimensioned in order to allow greater fuel flow rates.
- In order to take advantage of the alcohol's much higher octane rating, the compression ratio was increased to about 12:1.
- The valve housings, made of cast iron, were changed to an iron–cobalt synthetic alloy. This also compensated for the lack of lubrication resulting from the absence of lead in the fuel.
- The catalytic converter's catalyst was changed from palladium and rhodium to palladium and molybdenum, leading to further reduction in the alcohol engine emissions.

Table 1 Comparison of various properties of primary alcohols with water and gasoline

Properties	Methanol	Ethanol	Water	Gasoline
Formula	CH_3OH	C_2H_5OH	H_2O	C_4–C_{12}
Molecular weight	32.04	46.07	18	100–105 (avg.)
Specific weight	6.6	6.59	8.345	6.2
Boiling point (°F)	149	173.3	212	31–421
Flash point (°F)	52	55	—	–45
Stoichiometric air/fuel ratio	6.47:1	9.01:1	—	14.96:1
BTU/Gal	56 560	75 670	—	115 400
Specific heat, BTU/lb/F	0.601	0.62	1.0	0.5
Latent heat of vaporization, BTU/lb	482	367	1000	133–145
Vapour pressure PSI @ 100°F, 1 atm	4.6	2.5	0.96	7–15

Ethanol diesel blends

Ethanol is one of the potential alternative fuels for the partial replacement of mineral diesel oil in CI engines. The application of ethanol as a supplementary CI fuel may reduce environmental pollution, strengthen agricultural economy, create job opportunities, and reduce diesel fuel requirements, thereby contributing to the conservation of a major commercial energy source (Ajav, Singh, and Bhattacharya 1999; Andrew, Derek, Arthur, *et al.* 1996).

Ajav, Singh, and Bhattacharya (1999) analysed the effect of using different blends of ethanol–diesel on engine horsepower, brake-specific fuel consumption (BSFC), brake thermal efficiency, exhaust gas temperatures, and lubricating oil temperature. The results indicate no significant power reduction in the engine operating on ethanol–diesel blends (up to 20%) at 5% level of significance. BSFC increased by up to 9% (with ethanol up to 20%) in the blends as compared to neat diesel oil. The exhaust gas temperature, lubricating oil temperatures, and exhaust emissions (CO and NO_x) were lower for operations on ethanol–diesel blends as compared to operation on diesel.

Ethanol–diesel blends up to 20% can very well be used in present day constant-speed CI engines, without any modification. The BSFC is slightly increased, as shown in Figure 3, when higher blends of ethanol are used. There is no significant difference in the power produced and the efficiency of the engine, as shown in Figure 4. Exhaust gas temperatures and lubricating oil temperatures were lower for ethanol–diesel blends compared to straight

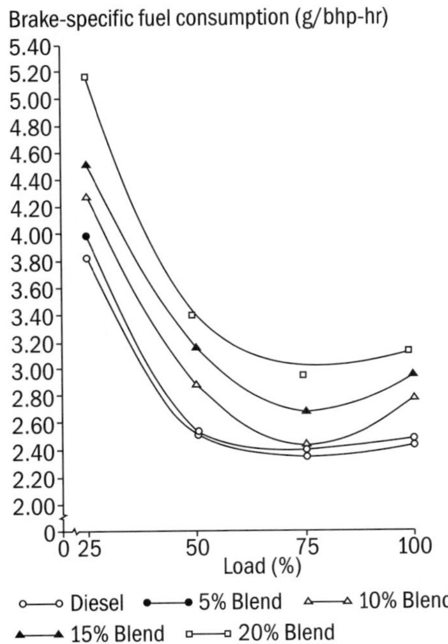

Figure 3 Brake-specific fuel consumption vs. load curve

Source Ajav, Singh, and Bhattacharya (1999)

diesel fuel, as indicated in Figure 5. The engine could be started normally during both hot and cold conditions. Up to 62% reduction in CO emission is possible with the use of ethanol–diesel blends as compared to diesel alone.

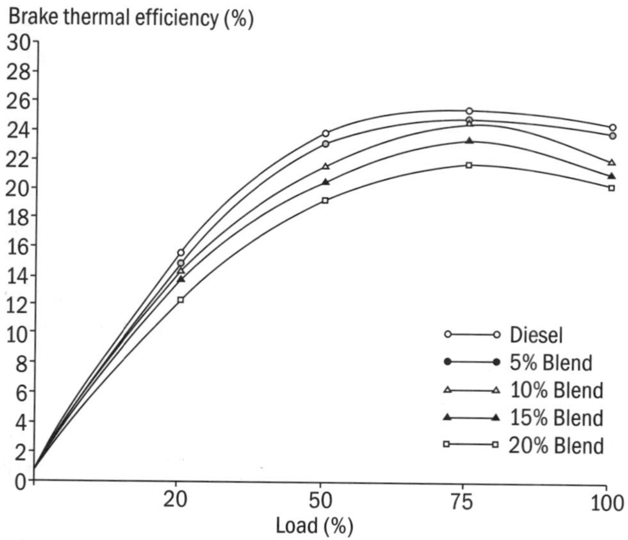

Figure 4 Brake thermal efficiency for different fuels
Source Ajav, Singh, and Bhattacharya (1999)

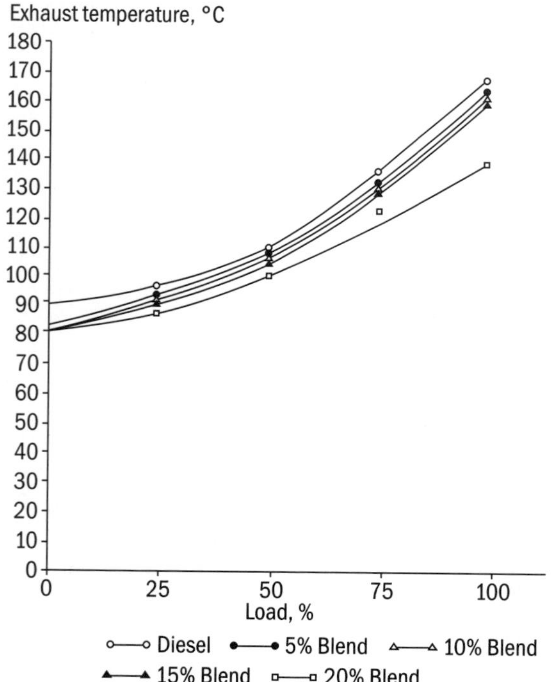

Figure 5 Engine exhaust temperature for different fuels
Source Ajav, Singh, and Bhattacharya (1999)

NO_x emissions also reduce (up to 24%) when ethanol–diesel blends are used (Ajav, Singh, and Bhattacharya 1999).

Regulated and unregulated emissions from ethanol-operated engines

Carbon dioxide

Most liquid fuels (gasoline, diesel, natural gas, propane) are derived from fossil HCs. Unlike them, ethanol is produced from renewable biological feedstock, such as agricultural crops and forestry by-products. CO_2 is released into the atmosphere when ethanol (like other fuels) is burned in an automobile engine. However, this CO_2 is recycled into organic tissues during plant growth. Only about 40% or less of the organic matter is actually removed from farm fields for ethanol production. The rest is returned to the soil as organic matter, increasing fertility and reducing soil erosion. With modern conservation farming practices, this soil organic matter will build up, representing a net removal of CO_2 from the atmosphere. An increase of only 1% in the soil organic matter means an atmospheric reduction of over 40 tonnes of CO_2 per hectare of farmland. Ethanol use in gasoline has tremendous potential for a net reduction in atmospheric CO_2 levels.

Carbon monoxide

CO, formed by the incomplete combustion of fuels, is produced most readily from petroleum fuels, which contain no oxygen in their molecular structure. Because ethanol and other "oxygenated" compounds contain oxygen, their combustion in automobile engines is more complete. The result is a substantial reduction in CO emissions. Research shows that the reductions can be up to 30%, depending on the type and age of automobile, the automobile emission system used, and the atmospheric conditions in which the automobile operates.

Because of health concerns over CO, the 1990 amendments to the US Clean Air Act mandate the use of oxygenated gasoline in many major urban centres during the winter (when atmospheric CO levels are highest) to reduce this pollution.

Other octane additives

Because of its high octane rating, adding ethanol to gasoline can lead to the reduction or removal of aromatic HCs (such as benzene) and other hazardous high-octane additives commonly used to replace tetraethyl lead in gasoline.

Ozone

Because of its potential in reducing HCs and CO in the exhaust, adding ethanol to gasoline results in an overall reduction in the ozone-forming potential of the exhausts. Ethanol has no significant effect on the emissions of nitrous oxide, another common contributor to atmospheric ozone. Adding ethanol to gasoline can potentially increase the volatility of gasoline. This potential is controlled if all ethanol-blended gasoline sold meets the volatility standards required for other types of gasoline. The US Clean Air Act allows gasohol (gasoline plus 10% ethanol) to have a higher volatility than gasoline. This results in greater VOC emissions.

Adding ethanol to gasoline does release slightly higher concentration of aldehydes during fuel combustion. Yet the resulting concentrations are extremely small and are effectively reduced by the three-way catalytic converter installed in the exhaust systems of these vehicles.

The addition of ethanol to the fuel up to 10% w/w results in an increase in the Reid vapour pressure, indicating increased evaporative emissions for ethanol fuels. Generally, benzene and toluene emissions decrease by adding ethanol to gasoline, although this beneficial effect of ethanol was lost with the operation of the catalyst. Acetic acid was detected in the exhaust gases in some cases only for the base and the 3% ethanol-blended-fuel.

A comparison of the emissions of various pollutants from different fuels is given in Table 2.

Fuel cycle emissions in heavy-duty vehicles

A study to calculate the life cycle emissions from various alternative fuels for heavy-duty vehicles also highlights the great environmental benefits accrued by using ethanol as vehicular fuel (Table 3). Beer, Grant, Brown, *et al.* (2000) confirm that the renewable fuels, bio-diesel and ethanol, are environment friendly, even when considered on a life cycle basis vis-à-vis ultra low sulphur diesel (ULSD) (Tom, Grant, William, *et al.* 2002). They have the lowest life cycle GHG emissions (in grams greenhouse gases per kilometre travelled). In fact, vehicles running on both these fuels emit larger quantities of CO_2 than those running on conventional fuels, but as most of this CO_2 is from renewable carbon stocks, that fraction is not counted towards the GHG emissions from the fuel. The other alternative gaseous fuels like liquefied petroleum gas (LPG) and compressed natural gas (CNG) come next in preference.

Table 2 Comparison of pollutants from different fuels

Fuel	CO	HC	NO_x	SO_x	PM	CO_2
Diesel	B	D	D	D	D	C
Gasoline	D	C	C	C	B	D
Alcohol	C	C*	C	A	A	A
A	B	C	C*			D
Zero or despicable	Low emissions	Medium emissions	The nature of the emissions is less hazardous for the environment than those of fossil fuels			High emissions

CO – carbon monoxide; HC – hydrocarbon; NO_x – oxides of nitrogen; SO_x – oxides of sulphur; PM – particulate matter; CO_2 – carbon dioxide

Table 3 Fuel-cycle fossil fuel greenhouse gas emissions (g/MJ) for heavy-duty vehicles in CO_2 equivalents

	Diesel	LSD	ULSD	LPG	CNG	LNG	E95 (wood)	BD20	BD100
Pre-combustion	11	12	13	11	6	9	–29	2	–41
Combustion	69	69	69	59	54	55	65	84	89
Total	80	81	82	70	60	64	36	87	48

LSD – low sulphur diesel; ULSD – ultra low sulphur diesel; LPG – liquefied petroleum gas; CNG – compressed natural gas; LNG – liquefied natural gas

Production of alcohol

Alcohols have been attracting attention worldwide. Consumer wants a cleaner fuel that can lower the risk of adverse impacts on the environment and health. Governments aim to reduce reliance on imported energy and promote domestic renewable energy programmes that could utilize domestic resources and create new economic activities. The Government of India has taken an initiative in this direction by taking a policy decision of supplying gasoline mixed with ethanol.

Though the proportion of biofuels used is relatively small compared to more traditional energy forms, the scenario is changing rapidly. When factors like developing vast agri-resources and new technologies that reduce cost, and emphasis on the environment and pollution abatement are coupled with a strong will from both the government and private entrepreneurs, the markets for biofuels will slowly but surely gain momentum.

The fuel "ethanolization" of the world alcohol industry is set to continue. If all the recently announced ethanol projects are implemented, the production of ethanol fuel worldwide could grow to 55 billion litres by 2010 against a figure of less than 20 billion litres in 2001 (Figure 6).

Automobile industry scenario in India and its impact on environment

Indian automobile industry grew rapidly after the 1980s. Growth in two-wheeler vehicles has been phenomenal as a result of increasing collaboration with foreign firms. Two-wheelers account for more than 80% of the total production of vehicles. The total number of registered vehicles in India, is now about 35 million. After a slight slowdown, the Indian automobile industry has been experiencing robust growth, which is in excess of 25% per year. After the liberalization of Indian economy in the 1990s, many of the leading multi-national companies (MNCs) have set up operations in India either individually or in collaboration with Indian partners; examples being General Motors, Ford, Mercedes, Daewoo, Hyundai, and Toyota. Today, India is also a major exporter of automobiles to destinations such as Australia, New Zealand, South East Asia, South America, Middle East, Africa, and Europe. The level of investment in the productive assets in the automobile

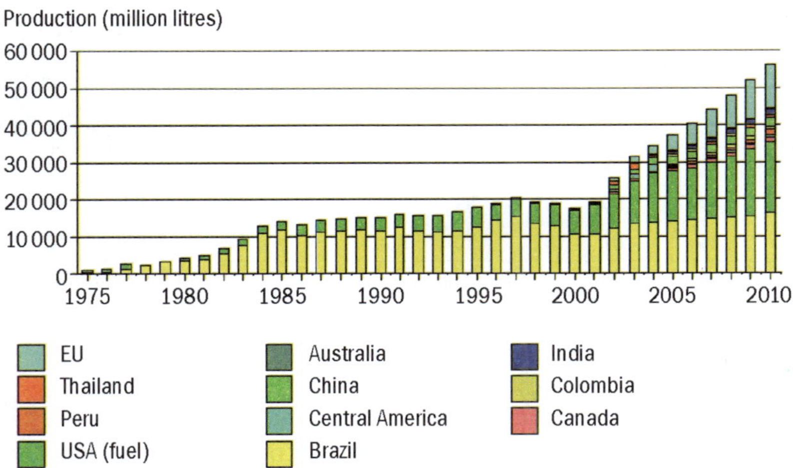

Figure 6 World ethanol production in million litres
Source Annual Report of the Ministry of Economy, Trade and Industry, Japan, 2009

industry grew from $1.2 billion in 1991 to $2.5 billion in 1995 to $11.5 billion in 2000. Vehicular emission accounts for about 30%–70% of air pollution in urban India. The pollution is often a result of a combination of antiquated automobile technology and poor fuel quality, poor maintenance of vehicles, and lack of appropriate traffic planning. Two-wheelers consume 60% of petrol in India, and they account for 70% of the total HC emission. Petrol vehicles emit more CO and HC, while diesel vehicles emit more NO_x and particulate matter (PM) (97% of the total HC emissions and 76% of the total CO emissions are from vehicles running on petrol). Vehicular emissions in Delhi outrank emissions from other sources of pollution such as thermal power plants, industries, and domestic coal burning, and account for as much as 64% of the total air pollution load in the city. Alternative fuels and technologies are likely to replace petrol and diesel. Resources include natural gas, biomass material, water, and non-fossil electricity sources, such as nuclear, hydro, and solar power. Resources can be converted into a wide range of fuels, including ethanol and methanol, methane, LPG, hydrogen (fuel cells), and electricity. LPG has been used successfully in commercial vehicles such as taxis in the USA, Canada, the Netherlands, Germany, Japan, Italy, and Thailand. About 600 000 vehicles have been converted to run on CNG in Italy, the USA, Canada, New Zealand, and Italy.

Vegetable oils as fuels

Rudolf Diesel (Figure 7) invented and modelled the diesel engine to run on a host of fuels, including coal dust suspended in water, heavy mineral oil, and vegetable oil. Diesel's first engine experiments were catastrophic failures, but by the time he showed his engine at the World Exhibition in Paris in 1900,

Figure 7 Dr Rudolf Diesel

his engine was running on 100% peanut oil. Diesel was visionary. In 1911, he stated, "The diesel engine can be fed with vegetable oils and would help considerably in the development of agriculture of the countries, which use it." In 1912, Diesel said, "The use of vegetable oils for engine fuels may seem insignificant today. But such oils may become, in course of time, as important as petroleum and the coal tar products of the present time." Since Diesel's untimely death in 1913, his engine has been modified to run on the polluting petroleum fuel, now known as "diesel". Nevertheless, his ideas on agriculture and his invention provided the foundation for a society fuelled with clean, renewable, and locally grown fuel.

In the 1930s and 1940s, vegetable oils were used as engine fuels from time to time, but usually only in emergency situations. Recently, because of the increase in crude oil prices, limited resources of fossil oil, and environmental concerns, there has been a renewed focus on vegetable oils and animal fats to develop bio-diesel. Continued and increasing use of petroleum will intensify local air pollution and magnify the global warming problems caused by CO_2. In particular cases, such as the emission of pollutants in the closed environment of underground mines, bio-diesel fuel has the potential to reduce the level of pollutants and the level of potential and probable carcinogens (Krawczyk 1996).

Higher viscosity is responsible for various undesirable combustion properties of neat vegetable oils. Four well-known techniques have been proposed to reduce the viscosity of vegetable oils: dilution, pyrolysis, micro-emulsion, and transesterification. Bio-diesel can be produced from vegetable oils for its utilization in IC engines. The advantages of using vegetable oils as fuels are as follows.

- Vegetable oils are liquid fuels from renewable sources.
- They do not over-burden the environment with emissions.
- Vegetable oils have the potential for making marginal land productive due to the property of the cultivated crops of fixing nitrogen in the soil.
- Vegetable oil production requires less energy input.
- Vegetable oils have higher energy content than other energy crops such as those producing alcohol. The heat content in vegetable oils is 90% of that of diesel, and they have a favourable output/input ratio of about 2–4:1 for un-irrigated crop production.

- The current prices of vegetable oils in the world are nearly competitive with petroleum fuel prices.
- Vegetable oil combustion has cleaner emission spectra.
- Vegetable oils require simpler processing technology.

However, they have the following limitations.
- These are not economically feasible yet.
- They need further R&D work for the development of on-farm processing technology.

Due to the rapid decline in crude oil reserves, the use of vegetable oils as engine fuel is again being promoted in many countries. Depending on the climate and soil conditions, different nations are looking into different vegetable oils for producing bio-diesel fuel. For example, soyabean oil in the USA, rapeseed and sunflower oils in Europe, palm oil in South East Asia (mainly Malaysia and Indonesia), and coconut oil in The Philippines are being considered as substitutes for diesel fuels.

Vegetable oils are available from various sources such as cultivated plants, forests, and agricultural by-products. In India, vegetable oils can be extracted from cultivated oilseeds like groundnut, rape–mustard, soyabean, sesame, niger, sunflower, safflower, castor, linseed; perennial oil-bearing materials like coconut, oil palm; derived oil-bearing materials like cotton seed, rice bran; oilseeds of forest and tree origin like sal, mahua, karanja, kusum, neem; and so on. India produces about 6.7 million tonnes of non-edible oils such as linseed, castor, karanja (*Pongamia glabra*), neem (*Azadirachta indica*), palash (*Butea monosperma*), kusum (*Schlelchera trijuga*), *Jatropha curcas*, and so on. Some of these oils produced are not being properly utilized even now, and it has been estimated that some other plant-based and forest-derived oils have a much higher production potential.

The production details for oilseeds in India are given in Table 4. The data indicates that the use of vegetable oils as the source of diesel fuels in India requires additional steps to increase the production of oilseeds. It is also

Table 4 Production of oilseeds in 2007/08

Oil	World (MT)	India (MT)
Soyabean	220.9	1.481
Rapeseed	48.4	7.000
Cotton seed	46.2	9.800
Peanut	32	1.545
Sunflower seed	27.2	1.000
Palm kernel	11.1	0.013
Copra	5.7	0.736
Total	391.5	21.575

Source US Department of Agriculture

necessary to develop new and more productive plant sources, the seeds of which have high oil content (Srivastava and Prasad 2000).

The Planning Commission of India has identified two plant species, ratanjyot (*J. curcas*) and karanja (*P. glabra*), as the potential solution to the existing energy crisis. *J. curcas* is a multi-purpose plant with many attributes and considerable potential. It is a tropical plant that can be grown in low to high rainfall areas and can be used to reclaim land, as a hedge, and/or as a commercial crop. Thus, growing it could provide employment, improve the environment, and enhance the quality of rural life. The establishment, management, and productivity of jatropha under various climatic conditions are not fully documented. The plant produces many useful products, especially the seed, from which oil can be extracted; this oil has properties similar to those of palm oil. Since it can be used in place of kerosene and diesel and as a substitute for fuelwood, it has been promoted to make rural areas self-sufficient in fuels for cooking, lighting, and motive power. Jatropha is easy to establish, grows relatively quickly, and is hardy, being drought tolerant. It is not browsed because its leaves and stems are toxic to animals, but after treatment, the seeds or seed cake can be used as animal feed. Various parts of the plant have medicinal value—its bark contains tannin, the flowers attract bees, and thus, the plant has honey production potential; its wood and fruit can be used for numerous purposes, including fuel. Of particular importance is the fact that the fruit of jatropha contains viscous oil that can be used for making soap in the cosmetics industry and as a diesel/kerosene substitute or extender. This latter use may be important when examining practical substitutes for fossil fuels to counter GHG accumulation. Also, like all trees, it fixes atmospheric carbon, stores it in wood, and assists in the build-up of soil carbon.

However, for several reasons, both technical and economic, the full potential of jatropha is yet to be realized. Jatropha is widely grown in Mexico, Nicaragua, north-eastern Thailand, and parts of India. It is now being promoted in southern Africa, Brazil, Mali, and Nepal. There are several governments, international organizations, national bodies, and non-governmental organizations (NGOs) promoting the cultivation and use of *J. curcas* and other oil-bearing plants. These include the World Bank, the International Plant Genetic Research Institute, Austrian and German Technical Assistance Programmes, the Rockefeller Foundation, Appropriate Technology International and Intermediate Technology Development Group (USA- and UK-based NGOs), the Biomass Users Network (BUN), and Plant Oil Producers Association of Zimbabwe.

An acceptable alternative fuel for engine has to fulfil the environmental and energy security needs without compromising on operating performance. Vegetable oils can be successfully used in CI engines through engine and fuel modifications. Engine modifications include dual fuelling and modification in injection system and heated fuel lines. Fuel modifications include blending of vegetable oils with diesel oil, transesterification, cracking/pyrolysis, and hydrogenation to reduce polymerization.

From amongst the large number of vegetable oils available in the world, if any specific oil needs to be adopted as a continuing energy source, it is essential that the oilseed variety having higher productivity and oil content be produced. Nevertheless, valid technologies must be developed for using vegetable oils as alternative diesel fuel that will permit crop production in an emergency situation. Vegetable oil in its raw form cannot be used in engines. It has to be converted to a more engine-friendly fuel called bio-diesel. System design approaches have taken care to ensure that these modified fuels can be utilized in the existing diesel engine without any substantial hardware modifications. It will be expensive and time consuming to incorporate even a minor design alteration in the system hardware of a large number of existing engines operating in the rural agricultural sector of the country.

In its simplest form, the carbon cycle of vegetable oil consists of fixation of carbon and release of oxygen by plants through the process of photosynthesis and then combining of oxygen and carbon to form CO_2 through the processes of combustion or respiration. It is appropriate to mention here that the CO_2 released by petroleum diesel was fixed in the atmosphere during the formative years of the earth, whereas the CO_2 released by bio-diesel gets continuously fixed by plants and may be recycled by the next generation of crops. The time taken by the carbon cycle to fix CO_2 and for its release after the combustion of bio-diesel is quite small (a few years; Figure 8) compared to the cycle time of petroleum oils (a few million years; Figure 9). It is well known that petroleum refiners are now facing new sulphur and aromatic compound specifications.

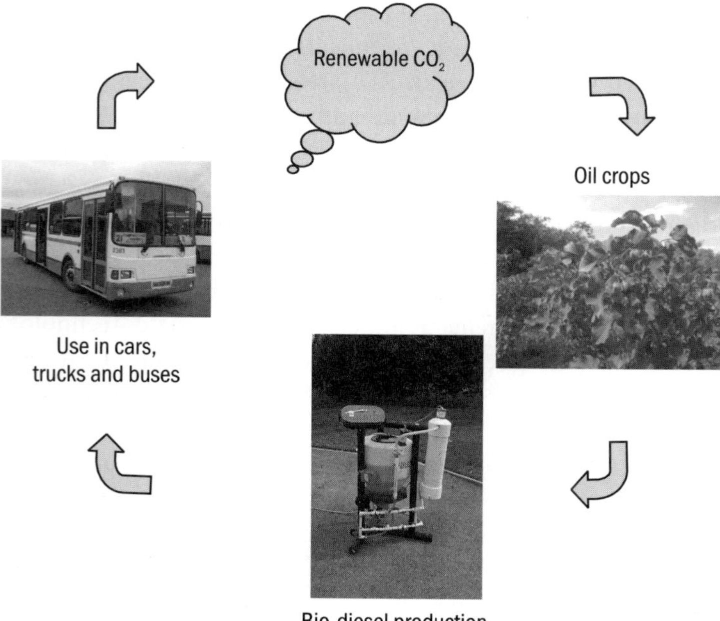

Figure 8 Bio-diesel CO_2 cycle

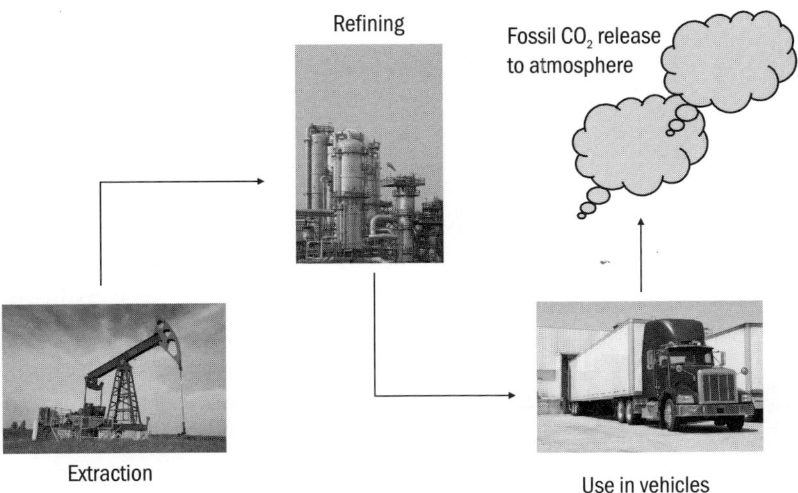

Figure 9 Petro-diesel CO_2 cycle

Since bio-diesel is a fuel made up of esters derived from oils and fats from renewable biological sources, it has been shown to emit substantially lower quantities of most of the regulated pollutants compared to petroleum diesel fuel.

Vegetable oils have energy density, cetane number, heat of vaporization, and stoichiometric air/fuel ratio comparable with mineral diesel fuel. The large molecular size of the component triglycerides results in the oil having higher viscosity compared to that of mineral diesel fuel. Viscosity affects the handling of the fuels by the pump and injector system and the shape of the fuel spray. The high jet penetration and poor atomization result in larger droplets. The fuel jet tends to be a solid stream instead of a spray of small droplets, and hence, the fuel does not get mixed with air required for burning. Larger droplets have poor combustion and thus result in loss of power and economy. In small engines, the fuel spray may even impinge upon the cylinder walls, washing away the lubricating oil film and causing the dilution of the crank case oil, leading to excessive wear of moving parts. Neat vegetable oil, due to high viscosity and low volatility, poses some problems when subjected to prolonged use in CI engines.

The petroleum diesel fuel molecules are saturated, non-branched molecules with the number of carbon atoms ranging between 12 and 18, whereas vegetable oils are the mixtures of organic compounds ranging from simple straight-chain compounds to complex structures of proteins and fat-soluble vitamins. Fats and oils are primarily water-insoluble, hydrophobic substances in the plant and animal kingdom, which are made up of 1 mole of glycerol and 3 moles of fatty acids and are commonly referred to as triglycerides. Vegetable oils are usually triglycerides, generally with a number of branched chains of different lengths, and have structure notation as shown subsequently.

$$
\begin{array}{l}
\qquad\;\; O \\
\qquad\;\; \| \\
CH_2\text{–}O\text{–}C\text{–}R^1 \\
| \qquad\;\; O \\
| \qquad\;\; \| \\
CH\text{–}O\text{–}C\text{–}R^2 \\
| \qquad\;\; O \\
| \qquad\;\; \| \\
CH_2\text{–}O\text{–}C\text{–}R^3
\end{array}
$$

where R^1, R^2, R^3 represent the HC chain of acids.

Fatty acids vary in carbon chain length and in the number of unsaturated bonds (double bonds). The structures of common fatty acids are given in Table 5, and the fatty acid composition of some vegetable oils is given in Tables 6A and B.

Vegetable oils have about 10% less heating value than diesel oil due to the O_2 content in their molecules, and the viscosity of vegetable oil is several times higher than that of the mineral diesel oil due to large molecular weight and complex chemical structure. The fuel-related properties of some of the vegetable oils are listed in Table 7.

Their high viscosity of 35–45 cSt compared to 4 cSt for diesel oil at 40°C leads to problem in pumping and spray characteristics (atomization, penetration, and so on). The inefficient mixing of oil with air contributes to incomplete combustion. The high flash point is associated with its lower volatility characteristics. This results in high carbon deposit formation,

Table 5 Chemical structure of common fatty acids

Fatty acid	Systematic name	Structure[a]	Formula
Lauric	Dodecanoic	12:0	$C_{12}H_{24}O_2$
Myristic	Tetradecanoic	14:0	$C_{14}H_{28}O_2$
Palmitic	Hexadecanoic	16:0	$C_{16}H_{32}O_2$
Stearic	Octadecanoic	18:0	$C_{18}H_{36}O_2$
Arachidic	Eicosanoic	20:0	$C_{20}H_{40}O_2$
Behenic	Docosanoic	22:0	$C_{22}H_{44}O_2$
Lignoceric	Tetracosanoic	24:0	$C_{24}H_{48}O_2$
Oleic	cis-9-Octadecenoic	18:1	$C_{18}H_{34}O_2$
Linoleic	cis-9,cis-12-Octadecadienoic	18:2	$C_{18}H_{32}O_2$
Linolenic	cis-9,cis-12,	18:3	$C_{18}H_{30}O_2$
	cis-15-Octadecatrienoic		
Erucic	cis-13-Docosenoic	22:1	$C_{22}H_{42}O_2$

[a] xx:y indicates xx carbons in the fatty acid chain with y double bonds.

Source Srivastava and Prasad (2000)

Table 6A Chemical composition of vegetable oils

	Fatty acid composition, wt.%									
Vegetable oil	14:0	16:0	18:0	20:0	22:0	24:0	18:1	22:1	18:2	18:3
Corn	0	12	2	Tr	0	0	25	0	6	Tr[a]
Cotton seed	0	28	1	0	0	0	13	0	58	0
Crambe	0	2	1	2	1	1	19	59	9	7
Linseed	0	5	2	0	0	0	20	0	18	55
Peanut	0	11	2	1	2	1	48	0	32	1
Rapeseed	0	3	1	0	0	0	64	0	22	8
Safflower	0	9	2	0	0	0	12	0	78	0
H.O. safflower	Tr	5	2	Tr	0	0	79	0	13	0
Sesame	0	13	4	0	0	0	53	0	30	0
Soyabean	0	12	3	0	0	0	23	0	55	6
Sunflower	0	6	3	0	0	0	17	0	74	0

[a] Tr = traces.

Source Srivastava and Prasad (2000)

Table 6B Chemical composition of vegetable oils

	Fatty acid composition, wt.%						
Vegetable oil	14:0	16:0	18:0	20:0	24:0	18:1	18:2
Rice bran	0.4–0.6	11.7–16.5	1.7–2.5	0.4–0.6	0.4–0.9	39.2–43.7	26.4–35.1
Sal	—	4.5–8.6	34.2–44.8	6.3–12.2	—	34.2–44.8	2.7
Mahua	—	16.0–28.2	20.0–25.1	0.0–3.3	—	41.0–51.0	8.9–13.7
Neem	0.2–0.26	13.6–16.2	14.4–24.1	0.8–3.4	—	49.1–61.9	2.3–15.8
Karanja	—	3.7–7.9	2.4–8.9	—	1.1–3.5	44.5–71.3	10.8–18.3

Source Srivastava and Prasad (2000)

injector coking, piston ring sticking, as well as lubrication oil dilution and oil degradation. The combination of high viscosity and low volatility in vegetable oils causes poor cold-start, misfire, and ignition delay. The poly-unsaturated nature of the vegetable oils causes long-term problems such as gum formation, ring sticking, and so on. Because of these problems, vegetable oils must be chemically modified to a more suitable and compatible fuel for existing engines.

Oil utilization as engine fuel

Neat vegetable oils are not suitable as fuel for diesel engines; hence, they have to be modified to bring their combustion-related properties closer to those of diesel oil. This fuel modification is mainly aimed at reducing the viscosity to get rid of flow- and combustion-related problems. Considerable efforts have been made to develop vegetable oil derivatives that approximate the properties

Table 7 Properties of vegetable oils

Vegetable oil	Kinematic viscosity[a]	Cetane no.	Heating value (MJ/kg)	Cloud point (°C)	Pour point (°C)	Flash point (°C)	Density (kg/l)	Carbon residue (Wt %)	Ash (Wt %)	Sulphur (Wt %)
Corn	34.9	37.60	39.50	–1.1	–40.0	277	0.9095	0.24	0.010	0.01
Cotton-seed	33.5	41.8	39.5	1.7	–15.0	234	0.9148	0.24	0.010	0.01
Cramble	53.6	44.6	40.5	10.0	–12.2	274	0.9044	0.23	0.050	0.01
Linseed	22.2	34.6	39.3	1.7	–15.0	241	0.9236	0.22	<0.01	0.01
Peanut	39.6	41.8	49.8	12.8	–6.7	271	0.9026	0.24	0.005	0.01
Rapeseed	37.0	37.6	39.7	–3.9	–31.7	246	0.9115	0.30	0.054	0.01
Safflower	31.3	41.3	39.5	18.3	–6.7	260	0.9144	0.25	0.006	0.01
H.O. saf-flower	41.2	49.1	39.5	–12.2	–20.6	293	0.9021	0.24	<0.001	0.02
Sesame	35.5	40.2	39.3	–3.9	–9.4	260	0.9133	0.25	<0.01	0.01
Soyabean	32.6	37.9	39.6	–3.9	–12.2	254	0.9138	0.27	<0.01	0.01
Sunflower	33.9	37.1	39.6	7.2	–15.0	274	0.9161	0.23	<0.01	0.01
Palm	39.6	42.0	—	31.0	—	267	0.9180	—	—	—
Babassu	30.3	38.0	—	20.0	—	150	0.9460	—	—	—
Tallow	—	—	40.0	—	—	201	—	6.21	—	—

[a] At 40°C.

Source Srivastava and Prasad (2000)

and performance of HC-based fuels. Vegetable oils can be used at least in four different ways.

1 Direct use and blending
2 Microemulsion
3 Pyrolysis (thermal cracking)
4 Transesterification

Direct use and blending

Caterpillar (Brazil) in 1980 used pre-combustion chamber engines with a mixture of 10% vegetable oil to maintain total power without any alterations or adjustments to the engine. At that point, it was not practical to substitute diesel fuel with 100% vegetable oil, but a blend of 20% vegetable oil and 80% diesel fuel was successful. Some short-term experiments used up to a 50/50 ratio. Pramanik (2003) found that 50% blend of jatropha oil can be used in diesel engine without any major operational difficulties, but further study is required for the long-term durability of the engine.

Direct use of vegetable oils and/or blends of the oils has generally been considered to be unsatisfactory and impractical for both direct and indirect diesel engines. High viscosity, acid composition, free fatty acid content as well as gum formation due to oxidation, polymerization during storage and combustion, carbon deposits, and lubricating oil thickening are obvious problems. The probable reasons for the problems and the potential solutions are shown in Table 8 (Ma and Hanna 1999).

Microemulsions

To solve the problem of high viscosity of vegetable oils, microemulsions with solvents such as methanol, ethanol, and 1-butanol have been studied. A microemulsion is defined as a colloidal equilibrium dispersion of optically isotropic fluid microstructures with dimension generally in the range 1–150 nm, formed spontaneously from two normally immiscible liquids. They can improve spray characteristics by explosive vaporization of the low-boiling constituents in the micelles. Short-term performance of microemulsions of aqueous ethanol in soyabean oil was nearly as good as that of No. 2 diesel, in spite of the lower cetane number and energy content (Srivastava and Prasad 2000).

Pyrolysis (thermal cracking)

Pyrolysis is the conversion of one substance into another by means of heat or by heat in the presence of a catalyst. The pyrolysed material can be vegetable oils, animal fats, natural fatty acids, and methyl esters of fatty acids. The pyrolysis of fats has been investigated for more than 100 years, especially in those areas of the world that lack deposits of petroleum. Many investigators have studied the pyrolysis of triglycerides to obtain products suitable for diesel engines. Thermal decomposition of triglycerides produces alkanes, alkenes, alkadines, aromatics, and carboxylic acids.

Table 8 Problems and potential solutions for using vegetable oils as engine fuels

Problem	Probable cause	Potential solution
Short term		
Cold weather starting	High viscosity, low cetane, and low flash point of vegetable oils	Preheat fuel prior to injection; chemically alter fuel to an ester
Plugging and gumming of filters, lines, and injectors	Natural gums (phosphatides) in vegetable oil; ash formation	Partially refine the oil to remove gums. Filter to 4 μm
Engine knocking	Very low cetane of some oils; improper injection timing	Adjust injection timing; preheat fuel prior to injection; chemically alter fuel to an ester
Long term		
Coking of injectors and carbon deposits on piston and head of engine	High viscosity of vegetable oil, incomplete combustion of fuel; poor combustion at part load	Heat fuel prior to injection; switch engine to diesel fuel when operating at part load; chemically alter the vegetable oil to an ester
Excessive engine wear	High viscosity, incomplete combustion of fuel; poor combustion at part load; possibly free fatty acids in vegetable oil; dilution of engine lubricating oil due to blow-by of vegetable oil	Heat fuel prior to injection; switch engine to diesel fuel when operating at part load; chemically alter the vegetable oil to an ester; increase lubricating oil changes; lubricating oil additives to inhibit oxidation
Failure of engine lubricating oil due to polymerization	Collection of poly-unsaturated vegetable oil blow-by in crank-case to the point where polymerization occurs	Same as above for excessive engine wear

Source Ma and Hanna (1999)

Transesterification

Vegetable oils have to undergo the process of transesterification in order to be usable in IC engines. Bio-diesel is the product of transesterification. It is biodegradable, non-toxic, and essentially free from sulphur, and it is renewable and can be produced from agriculture and plant resources. Bio-diesel is an alternative fuel that has a harmonious correlation with sustainable development, energy conservation, management, efficiency, and environmental preservation.

Transesterification is the reaction of a fat or oil with an alcohol to form esters and glycerol. Alcohol combines with the triglycerides to form glycerol and esters. A catalyst is usually used to improve the reaction rate and yield. Since the reaction is reversible, excess alcohol is required to shift the equilibrium to the product side. The alcohols that can be used in the transesterification process are methanol, ethanol, propanol, butanol, and amyl alcohol (Ma and Hanna 1999).

Methanol and ethanol are used most frequently, especially methanol because of its low cost and its physical and chemical advantages (polar and shortest chain alcohol). It can quickly react with triglycerides, and NaOH is easily dissolved in it. The reaction can be catalysed by alkalis, acids or enzymes. The alkalis include NaOH, KOH, carbonates, and the corresponding sodium and potassium alkoxides such as sodium methoxide, sodium ethoxide, sodium propoxide, and sodium butoxide. Sulphuric acid, sulphonic acids, and hydrochloric acid are usually used as acid catalysts. Lipases also can be used as biocatalysts. Alkali-catalysed transesterification is much faster than acid-catalysed transesterification and is most often used commercially (Ma and Hanna 1999; Freedman, Pryde, and Mounts 1984).

For an alkali-catalysed transesterification, the glycerides and alcohol must be substantially anhydrous because water makes the reaction partially change to saponification, which produces soap. The soap lowers the yield of esters and renders the separation of ester and glycerol and water washing difficult. The glycerol is then removed by gravity separation, and the remaining ester is mixed with hot water for the separation of the catalyst. Moisture can be removed using silica gel crystals. Ester formation eliminates almost all the problems associated with vegetable oils. The reaction mixture is stirred continuously at 70°C. Saponification reaction also takes place simultaneously along with transesterification process, but soap formation is not a major problem if water is less than 1% (Ma and Hanna 1999; Agarwal and Das 2001).

$$
\begin{array}{lllll}
CH_2-O-\overset{O}{\overset{\|}{C}}-R^1 & & CH_2-OH & & R^4-O-\overset{O}{\overset{\|}{C}}-R^1 \\
| & & | & & \\
CH-O-\overset{O}{\overset{\|}{C}}-R^2 + 3\,R^4OH & \rightleftharpoons & CH-OH & + & R^4-O-\overset{O}{\overset{\|}{C}}-R^2 \\
| & & | & & \\
CH_2-O-\overset{O}{\overset{\|}{C}}-R^3 & & CH_2-OH & & R^4-O-\overset{O}{\overset{\|}{C}}-R^3
\end{array}
$$

(Triglyceride esters) Vegetable oil — Alcohol — Glycerol — Esters

R^1, R^2, R^3, R^4 represent various alkyl group.

The process of transesterification brings about drastic change in the density of vegetable oils. The bio-diesel produced by this process is, therefore, totally miscible with diesel oil in any proportion. Viscosity of bio-diesel is very close to that of diesel oil, thus it faces no problems in the existing fuel handling system. The flash point of the bio-diesel gets lowered after esterification and the cetane number improves. Even lower concentrations of bio-diesel act as cetane number improver for bio-diesel blend. The calorific value of bio-diesel is also found to be very close to that of diesel oil. Some typical observations from the engine tests suggest that the thermal efficiency of the engine generally improves cooling losses, and exhaust gas temperature increases and smoke opacity generally gets lowered for bio-diesel blends. Possible reason may be the additional lubrication properties of the bio-diesel and the consequent reduced frictional horsepower (FHP). The energy thus saved increases thermal efficiency, cooling losses, and exhaust losses from the engine. The thermal efficiency starts reducing after a certain concentration of bio-diesel is reached. Flash point, density, pour point, cetane number, and calorific value of bio-diesel are very close to those of mineral diesel oil (Agarwal 1998; Agarwal and Das 2001).

Diesel engine can perform satisfactorily in the long term when run on bio-diesel without any hardware modifications. The optimum concentration for bio-diesel blend with improved performance is 20% bio-diesel. Increase in exhaust temperature, however, leads to increased NO_x emissions from the engine. While short-term tests are almost positive, long-term use of neat vegetable oils or their blend with diesel leads to various engine problems such as injector coking, ring sticking, injector deposits, and so on (Peterson, Wagner, and Auld 1983; Muniyappa, Brammer, and Noureddini 1996). High viscosity, low volatility, and a tendency for polymerization within the cylinder are at the root of many problems associated with the direct use of these oils as fuels. The process of transesterification yields vegetable oil ester, which has shown promise as an alternative diesel fuel as a result of improved viscosity and volatility. Several researchers have investigated different vegetable oil esters and found these to be comparable to diesel fuel (Freedman, Pryde, and Mounts 1984; Agarwal 1998; Agarwal and Das 2001; Peterson, Wagner, and Auld 1983; Muniyappa, Brammer, and Noureddini 1996; Murayama, Fujiwara, and Noto 2000).

Bio-diesel

The best way to use vegetable oil as a fuel is to convert it into bio-diesel. Bio-diesel is the name for a clean burning mono-alkyl-ester-based oxygenated fuel made from natural, renewable sources such as new and used vegetable oils and animal fats. The resulting bio-diesel is quite similar to conventional diesel fuel in its main characteristics. Bio-diesel contains no petroleum, but it is compatible with conventional diesel and can be blended in any proportion with petroleum diesel to create a bio-diesel blend. The level of blending with petroleum diesel is referred to as Bxx, where xx indicates the amount of bio-diesel in the blend (that is, B40 blend is 40% bio-diesel and 60% diesel). It can be used in CI engines with no major modification in the hardware.

A number of units are manufacturing bio-diesel worldwide. Table 9 gives the representative global production of bio-diesel.

Table 9 Global production of bio-diesel

Country	Number	Location	Total annual capacity of plants (tonnes)	Oils
Austria	11(1)	Aschach, Bruck, Pischelsdorf	56 200–60 000	Used-frying oil
Belgium	3	Feluy, Seneffe	241 000	—
Canada	1	Saskatoon		
Czechoslovakia	17(1)	Mydlovary, Olomouc	42 500–45 000	Used-frying oil
Denmark	3	Otterup, Jutland	32 000	—
France	7(1)	Rouen, Compiegne, Boussens, Peronne, Verdun, Nogent-surseine	38 100	—
Germany	8(4)	Leer, Dusseldorf, Kiel, Barby, Germunden, Thuringia	207 000	—
Hungary	17(16)	Visnye, Gyor	18 880	—
Ireland	1	—	5000	Used-frying oil
Italy	9(4)	Livomo, Cittadi, Casleto, Milano, Solbiate, Napoli, Bari, Ancona, Brescia	779 000	Sunflower oil
Nicaragua	1	—	—	Jatropha oil
Slovak Republic	10(1)	Barcelona	50 500–51 500	
Spain	1	Bilbao	500	—
Sweden	3(1)	Gothenborg, Skane, Orebro	75 000	—
Switzerland	1	Geneva	2000	—
UK	1	East Dusham	—	—
USA	4(3)	Midwest, Chicago, Quincy	190 000	Used-frying oil
Yugoslavia	2	—	5000	

Figure within parenthesis indicates the number of plants under construction.
Source Srivastava and Prasad (2000)

Properties of bio-diesel

The properties of some of the bio-diesel fuels are compared in Table 10. The characteristics of bio-diesel are close to those of diesel fuels, and therefore, bio-diesel becomes a strong candidate to replace diesel fuels, if the need arises. The conversion of triglycerides into methyl or ethyl esters through the transesterification process reduces the molecular weight to one-third that of the triglycerides and the viscosity by a factor of about eight and increases the volatility marginally. Bio-diesel has a viscosity close to that of diesel fuels. These vegetable oil esters contain 10%–11% oxygen by weight, which may lead to better combustion than HC-based diesel fuels in an engine. The cetane

Table 10 Properties of bio-diesel

	Vegetable oil methyl ester[a]					
Property	Peanut	Soyabean	Palm	Sunflower	Linseed	Tallow
Kinematic viscosity at 37.8°C	4.9	4.5	5.7	4.6	3.59[b]	
Cetane number	54	45	62	49	52	
Lower heating value (MJ/l)	33.6	33.5	33.5	33.5	35.3	
Cloud point	5	1	13	1	–	12
Pour point	–	–7	–	–	–15	9
Flash point	176	178	164	183	172	96
Density (g/ml)	0.883	0.885	0.88	0.86	0.874	
Carbon residue	–	1.74	–	–	1.83	

[a] wt%, [b] At 40°C.

number of bio-diesel is about 50. Bio-diesel has lower volumetric heating values (about 10%) than diesel fuels but has a high cetane number and flash point. The esters have cloud points and pour points that are 15–25°C higher than those of diesel fuels (Agarwal and Das 2001).

Variables affecting transesterification

The most important variables affecting the yield of bio-diesel from transesterification are as follows.

- Reaction temperature
- Molar ratio of alcohol and oil
- Catalyst
- Reaction time
- Presence of moisture and free fatty acids

Effect of reaction temperature

The rate of reaction is strongly influenced by the reaction temperature. However, given enough time, the reaction will proceed to near completion even at room temperature. Generally, the reaction is conducted close to the boiling point of methanol (60–70°C) at atmospheric pressure. The maximum yield of esters occurs at temperatures ranging from 60°C to 80°C at a molar ratio (alcohol to oil) of 6:1 (Srivastava and Prasad 2000; Ma and Hanna 1999; Freedman, Pryde, and Mounts 1984; Agarwal 1998; Agarwal, and Das 2001). Several researchers have studied the effect of temperature on the conversion of oils and fats into bio-diesel. Freedman, Pryde, and Mounts (1984) studied the transesterification of refined soyabean oil with methanol (6:1) with 1% NaOH catalyst at three different temperatures—60°C, 45°C, and 32°C. After 0.1 h, the ester yields were 94%, 87%, and 64% for 60°C, 45°C, and 32°C, respectively. After 1 h, ester formation was identical for the 60°C and 45°C reaction temperature runs and only slightly lower for the 32°C run. This shows that the temperature clearly influenced the reaction rate, and the yield of esters and transesterification can proceed satisfactorily at ambient temperatures, given enough time, in the case of alkaline catalysts.

Effect of molar ratio

Another important variable affecting the yield of ester is the molar ratio of alcohol to vegetable oil. The stoichiometry of the transesterification reaction requires 3 moles of alcohol per mole of triglyceride to yield 3 moles of fatty esters and 1 mole of glycerol. To shift the transesterification reaction to the right, it is necessary to use either excess of alcohol or remove one of the products from the reaction mixture continuously. The second option is preferred wherever feasible, since in this way, the reaction can be driven towards completion. When 100% excess methanol is used, the reaction rate is at its highest. A molar ratio of 6:1 is normally used in industrial processes to obtain methyl ester yields higher than 98% by weight. Freedman, Pryde, and Mounts (1984) studied the effect of molar ratio (from 1:1 to 6:1) on ester conversion with vegetable oils. Soyabean, sunflower, peanut, and cotton seed oils behaved similarly and achieved the highest conversions (93%–98%) at a 6:1 molar ratio. Ratios greater than 6:1 do not increase yield (already 98%–99%); however, these interfere with the separation of glycerol.

Effect of catalyst

Catalysts are classified as alkali, acid or enzymes. Alkali-catalysed transesterification is much faster than acid-catalysed reaction. However, if a vegetable oil has high free fatty acid and water content, acid-catalysed transesterification reaction is suitable. Partly because of faster esterification and partly because alkaline catalysts are less corrosive to industrial equipment than acidic catalysts, most commercial transesterification reactions are conducted with alkaline catalysts. Sodium methoxide was found to be more effective than NaOH. Sodium alkoxides are among the most efficient catalysts used for this purpose, although NaOH, due to its low cost, has attracted wide use in large-scale transesterification. The alkaline catalyst concentrations in the range of 0.5%–1% by weight lead to 94%–99% conversion of vegetable oils into esters. Further, increase in catalyst concentration does not increase the conversion, and it adds to the cost because it is necessary to remove the catalyst from the reaction medium at the end (Srivastava and Prasad 2000; Freedman, Pryde, and Mounts 1984; Agarwal 1998).

Effect of reaction time

The conversion rate increases with reaction time. Freedman, Prydes, and Mounts transesterified peanut, cotton seed, sunflower, and soyabean oils under the condition of methanol to oil ratio of 6:1 and in the presence of 0.5% sodium methoxide catalyst at 60°C. An approximate yield of 80%

was observed after 1 minute for soyabean and sunflower oils. After 1 h, the conversions were almost the same for all four oils (93%–98%) (Freedman, Pryde, and Mounts 1984). Ma and Hanna (1999) studied the effect of reaction time on transesterification of beef tallow with methanol. The reaction was very slow during the first minute due to the mixing and dispersion of methanol into beef tallow. From 1 to 5 minutes, the reaction proceeded very fast. The apparent yield of beef tallow methyl esters surged from 1% to 38%.

Effect of moisture and free fatty acids

The starting materials used for alkali-catalysed transesterification of triglycerides must meet certain specifications. The glyceride should have an acid value less than 1, and all materials should be substantially anhydrous. If the acid value is greater than 1, more NaOH is required to neutralize the free fatty acids. Water also causes soap formation and consumes the catalyst, thus reducing catalyst efficiency. The resulting soap causes an increase in viscosity and formation of gels and makes the separation of glycerol difficult. Freedman, Pryde, and Mounts found that ester yields were significantly reduced if the reactants did not meet these requirements. NaOH or sodium methoxide reacted with moisture and CO_2 in the air, which diminished their effectiveness (Freedman, Pryde, and Mounts 1984). The effects of free fatty acids and water on the transesterification of beef tallow with methanol were investigated by Ma and Hanna (1999). The results showed that the water content of beef tallow should be kept below 0.06% w/w and free fatty acid content below 0.5% w/w in order to get the best conversion. Water content was a more critical variable in the transesterification process than free fatty acids (Ma and Hanna 1999).

Economic feasibility of bio-diesel

India has rich and abundant forest resources with a wide range of plants and oilseeds. The production of these oilseeds can be stepped up many folds if the government takes the decision to use them for producing diesel fuels. The economic feasibility of bio-diesel depends on the price of the crude petroleum and the cost of transporting diesel over long distances to remote markets. It is certain that the cost of crude petroleum is bound to increase because of the increase in demand and limited supply. Further, the strict regulations on the aromatics and sulphur contents in diesel fuels will result in higher cost of production for these fuels.

The cost of producing methyl or ethyl esters from edible oils is currently much higher than that of HC-based diesel fuels. Also diverting edible oils for fuel usage will not be a wise move as the government is currently importing edible oils to meet domestic requirements. Due to the relatively high costs of vegetable oils (about three to four times the cost of diesel in India), methyl esters produced from them cannot compete economically with HC-based diesel fuels unless granted protection from the considerable tax levies applied to the latter. In the absence of tax relief, there is a need to explore alternative feedstock for the production of bio-diesel.

The cost of bio-diesel can be reduced if we consider non-edible oils and used frying oils instead of edible oils. Non-edible oils such as mahua, karanja, babassu, jatropha, neem, and so on are easily available in many parts of the world, including India, and are very cheap compared to edible oils. Most of these non-edible oils are not used to their potential and in fact are produced in surplus quantities. The potential availability of some of these non-edible oils in India is given in Table 11. Several countries, including the Netherlands, Germany, Belgium, Austria, USA, Japan, and India, discard used frying oils. With the mushrooming of fast food centres and restaurants in India, it is expected that considerable amounts of used frying oils will be generated. This oil can be used for producing bio-diesel, thus helping to reduce the cost of water treatment in the sewerage system and in the recycling of resources.

Engine performance characteristics

Bio-diesel has low heating value (10% lower than diesel) on a mass basis because of the substantial amount of oxygen present in the fuel, but at the same time, it has a higher specific gravity, 0.88, as compared to mineral diesel, which has a specific gravity of 0.85; so the overall impact is approximately 5% lower energy per unit volume. Thermal efficiency of an engine operating on bio-diesel is generally better than the one operating on diesel. Brake-specific energy consumption (BSEC) is a more reliable criterion compared to BSFC for comparing fuels with different calorific values and densities. Several experimental investigations have been carried out by researchers around the world to evaluate the engine performance of different bio-diesel oils. Masjuki, Abdulmuin, and Sii (1996) investigated preheated palm oil methyl esters (POMEs) in diesel engines. They observed that by preheating POME above the room temperature, the engine performance, especially the brake power output and exhaust emission characteristics, improved significantly. Scholl and Sorenson (1993) studied the combustion of soyabean oil methyl ester (SOME) in a direct injection diesel engine. They found that most of the relevant combustion parameters for SOME, such as ignition delay, peak pressure, and peak rise of pressure, were close to those observed for diesel oil combustion at the same load, speed, timing, and nozzle diameter (Scholl and Sorenson 1993). They also investigated the combustion and emission characteristics with SOME and diesel fuels for different injector orifice diameters. They

Table 11 Non-edible oil sources in India

Oil	Botanical name	Potential (MT)	Utilized (MT)	Per cent utilization
Rice bran	*Oryza sativa*	474 000	101 000	21
Sal	*Shorea robusta*	720 000	23 000	3
Neem	*Melia azadirachta*	400 000	20 000	6
Karanja	*Pongamia glabra*	135 000	8000	6

found that the ignition delays of the two fuels were comparable in magnitude; however, the ignition delay of SOME was found to be more sensitive than diesel-to-nozzle diameter. CO emissions from SOME were slightly lower, HC emissions reduced drastically, NO_x for two fuels were comparable, and smoke numbers for SOME were lower than those for diesel fuel. The results of their experiments are shown in Figures 10–13 (Scholl and Sorenson 1993).

Altin, Cetinkaya, and Yucesu (2001) investigated the use of sunflower oil, cotton seed oil, soyabean oil, and their methyl esters in a single-cylinder, four-stroke direct injection diesel engine. The variations of maximum engine torque values in relation to the fuel types are shown in Figure 14. The maximum torque with diesel fuel operation was 43.1 Nm at 1300 revolutions per minute (RPM). For the ease of comparison, this torque was assumed to be 100%. The observed maximum torque values of the vegetable oil fuel operations were also about 1300 RPM but were less than those of diesel fuel in all cases. The variations of maximum engine power values in relation to the fuel types are shown in Figure 15. The maximum power associated with diesel fuel operation was 7.45 kW at 1700 RPM. As before, this power was assumed to be 100%. The observed maximum power values of the vegetable oil fuel operations were also about 1700 RPM but were less than those of diesel fuel in all cases. These results may also be due to the higher viscosity and lower heating values of vegetable oils. Specific fuel consumption is one of the important parameters of an engine and is defined as the consumption per unit of power in unit time. As shown in Figure 16, the minimum specific fuel consumption values were 245 g/kWh with diesel fuel, 290 g/kWh with raw sunflower oil, and 289 g/kWh with opium poppy oil, at 1300 RPM. Specific fuel consumption values of the methyl esters were generally less than those of the raw oil fuels.

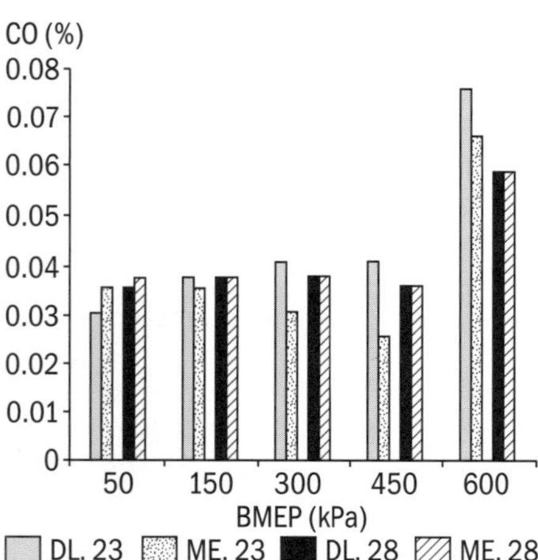

Figure 10 Carbon monoxide (CO) emissions for diesel and soyabean oil methyl ester for two nozzle diameters
Source Scholl and Sorenson (1993)

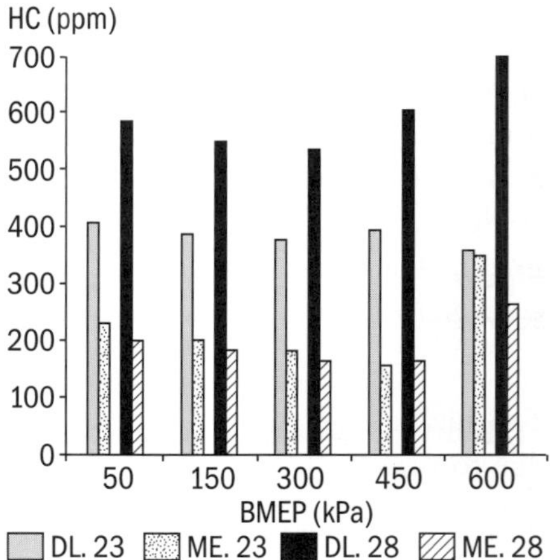

Figure 11 Hydrocarbon (HC) emissions for diesel and soyabean oil methyl ester for two nozzle diameters
Source Scholl and Sorenson (1993)

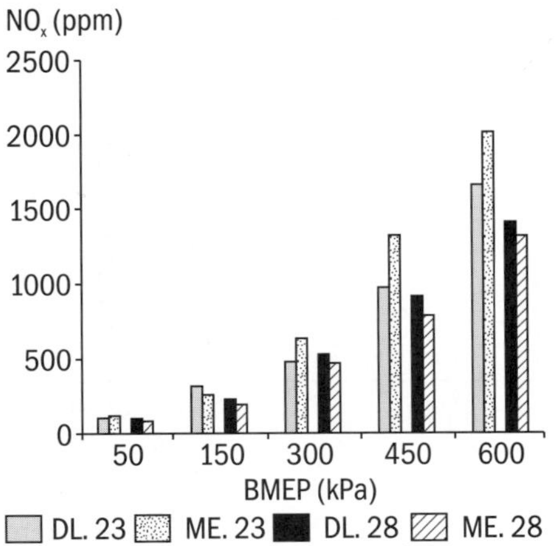

Figure 12 Oxides of nitrogen (NO_x) emissions for diesel and soyabean oil methyl ester for two nozzle diameters
Source Scholl and Sorenson (1993)

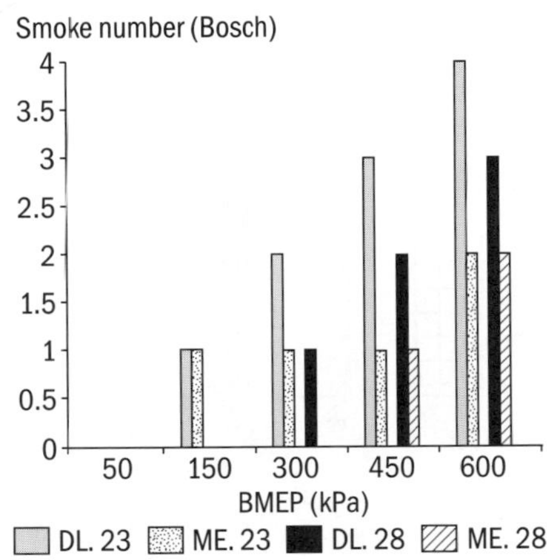

Figure 13 Smoke emissions for diesel and soyabean oil methyl ester for two nozzle diameters
Source Scholl and Sorenson (1993)

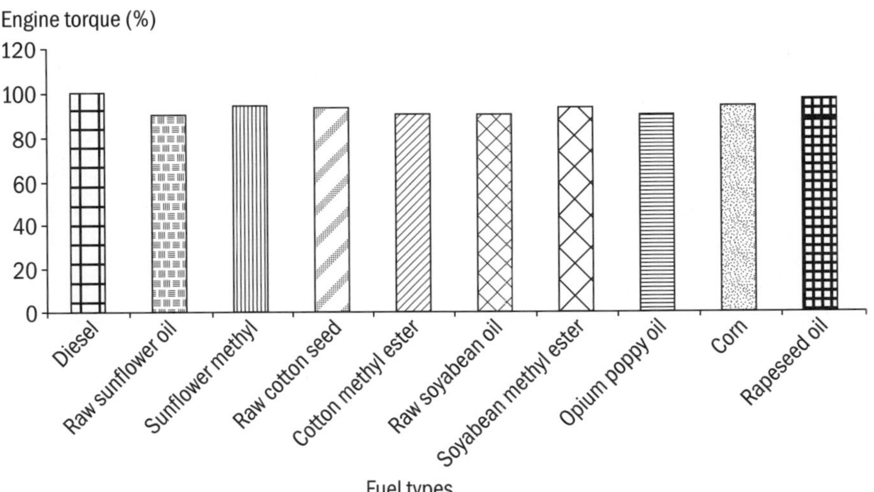

Figure 14 Variation of engine torque with respect to fuel types
Source Altin, Cetinkaya, and Yucesu (2001)

The higher specific fuel consumption values in the case of vegetable oils are due to their lower energy content. The exhaust emissions values for the above investigations are given in Figures 17–20.

Lower CO emissions can be achieved for methyl esters compared to raw vegetable oils because of the better spraying qualities of the esters. The nitrogen dioxide (NO_2) emissions are shown in Figure 19. NO_2 emission is maximum with diesel fuel. As seen from the graph, NO_2 emission with vegetable oil fuels is lower than that with diesel fuel, and the NO_2 values of the

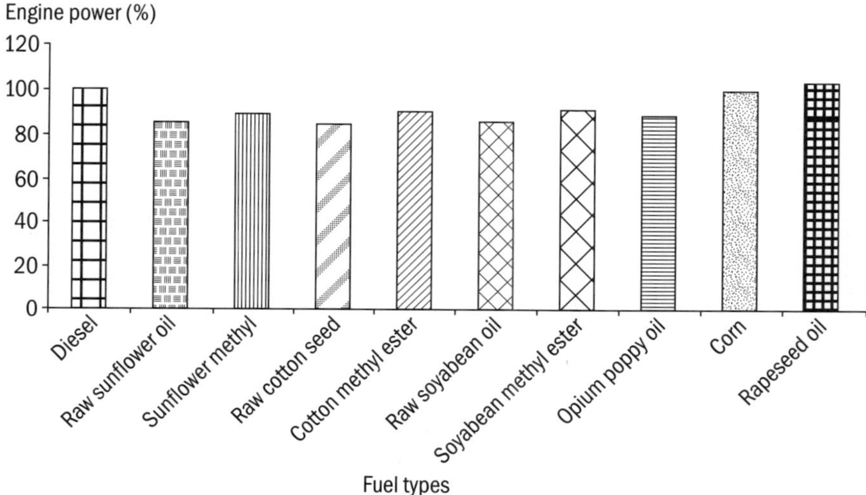

Figure 15 Variation of engine power with respect to fuel types
Source Altin, Cetinkaya, and Yucesu (2001)

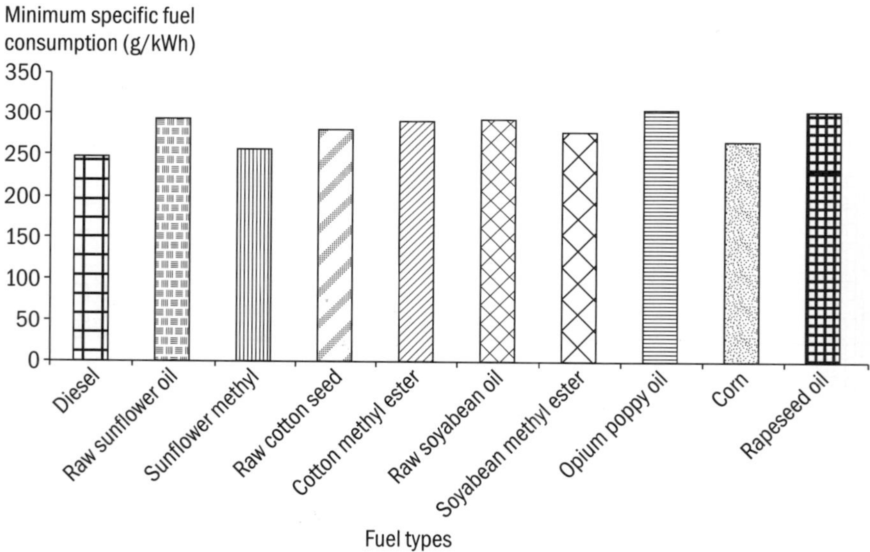

Figure 16 Variation of minimum specific fuel consumption with respect to fuel types
Source Altin, Cetinkaya, and Yucesu (2001)

methyl esters are higher than those of the raw oil fuels. The most significant factor that causes NO_2 formation is the high combustion temperatures. Since injection particles of the vegetable oils are larger than those of diesel fuel, the combustion efficiency and maximum combustion temperatures for all the vegetable oils are lower, and NO_2 emissions are less. The variations of exhaust smoke opacity in per cent in relation to the fuel type are shown in Figure 20. Smoke opacity percentages for operation with all vegetable oils are greater than those for diesel fuel. The minimum smoke opacity was with

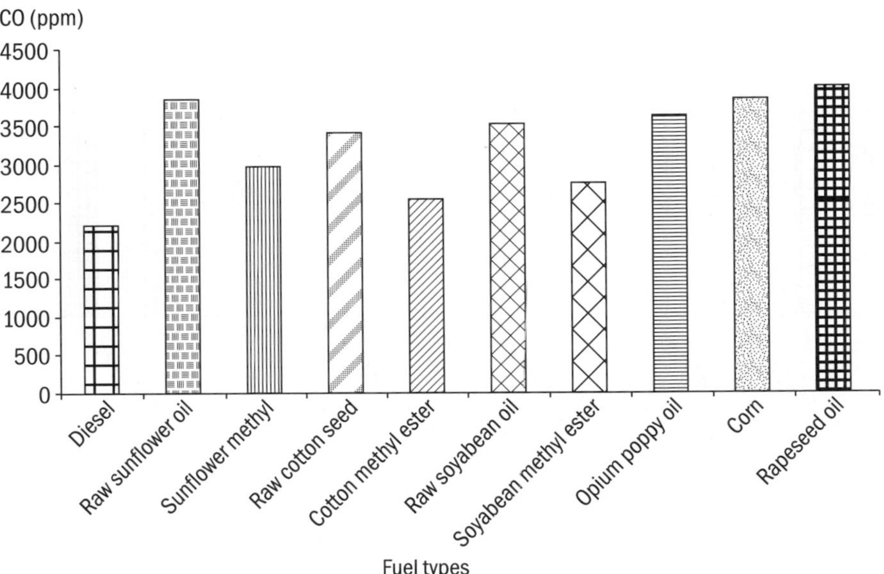

Figure 17 Variation of carbon monoxide (CO) emissions with respect to fuel types
Source Altin, Cetinkaya, and Yucesu (2001)

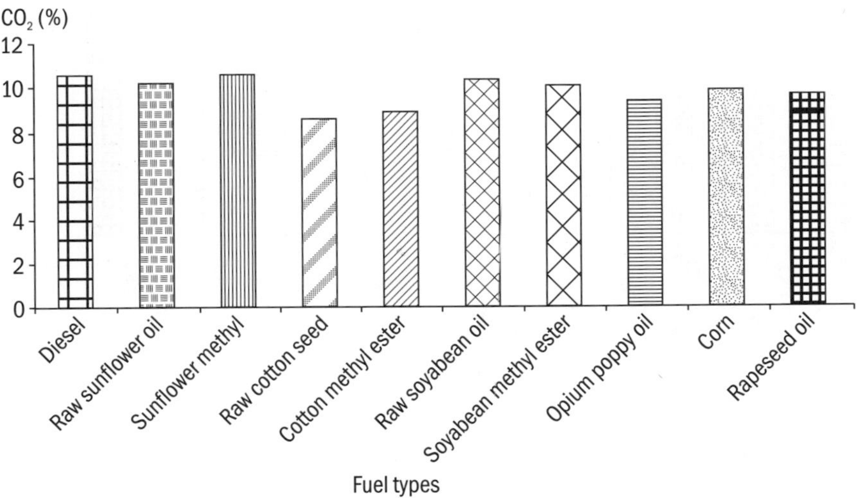

Figure 18 Variation of CO_2 emission with respect to fuel types
Source Altin, Cetinkaya, and Yucesu (2001)

diesel fuel. The opacity values of the methyl esters were somewhere between those of diesel fuel and raw oil fuels. The greater smoke opacity percentages of vegetable oil fuels are mainly due to the heavier molecules of HCs.

Murayma, Fujiwara, and Noto (2000) used waste vegetable oil esters in direct injection (DI) and indirect injection (IDI) diesel engines and found that the engine performance characteristics were almost the same with light oil.

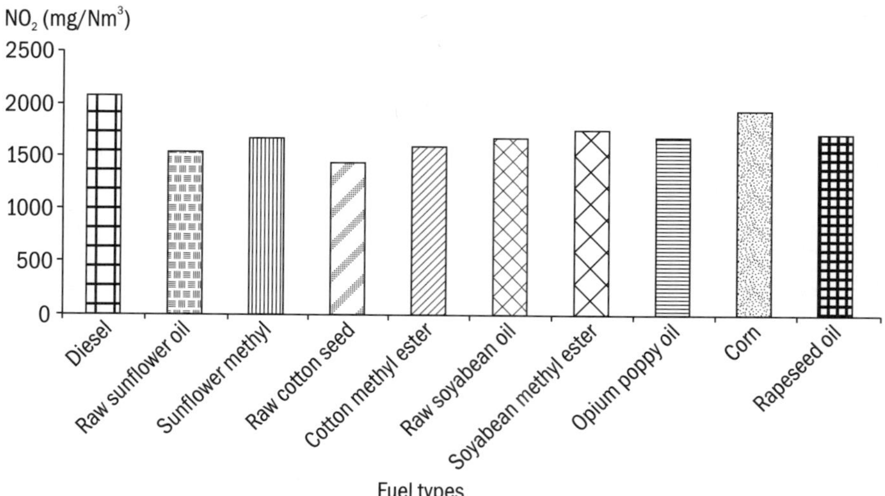

Figure 19 Variation of nitrogen dioxide (NO_2) emission with respect to fuel types
Source Altin, Cetinkaya, and Yucesu (2001)

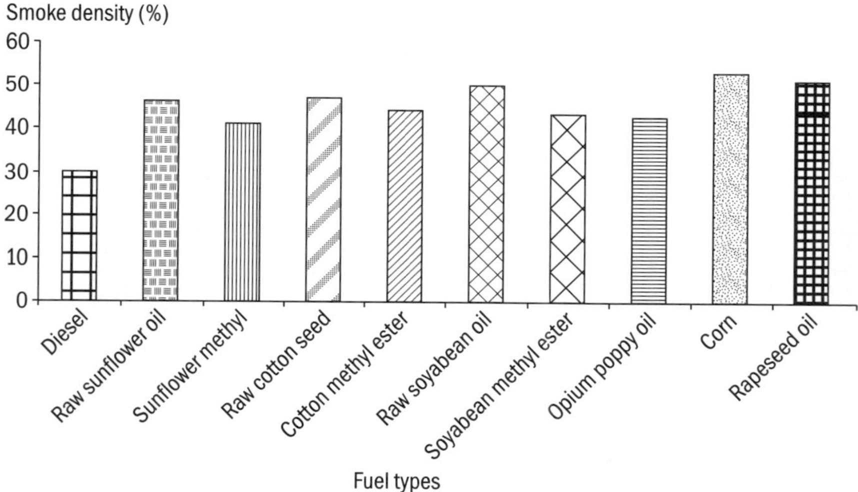

Figure 20 Variation of smoke density with respect to fuel types
Source Altin, Cetinkaya, and Yucesu (2001)

Agarwal (1998) and Agarwal and Das (2001) transesterified linseed oil and prepared linseed oil methyl ester (LOME) and carried out the engine performance experiments with different blends of bio-diesel (LOME) and diesel and compared the results with baseline data for diesel using a single-cylinder DI diesel engine. Some of the results are shown in Figures 21–26.

Higher thermal efficiencies, lower BSFC, and higher exhaust temperatures are reported for all blends of bio-diesel compared to the mineral diesel oil. The

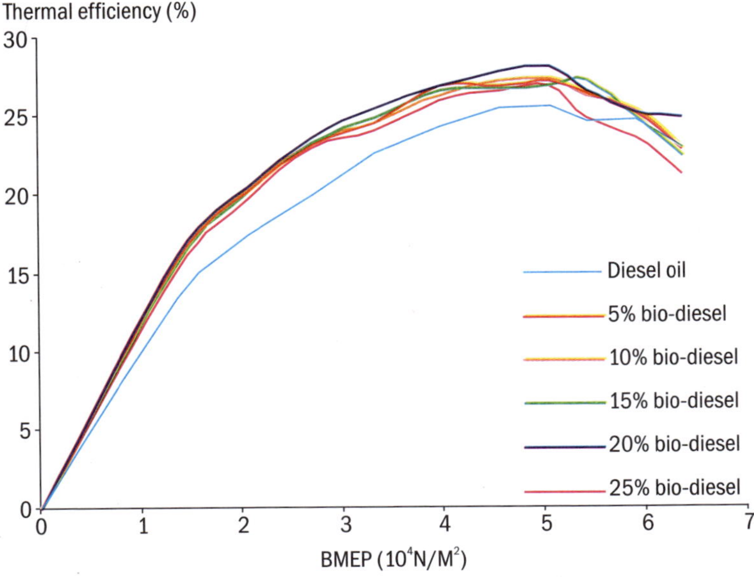

Figure 21A Comparison of thermal efficiency vs. brake mean effective pressure (BMEP) curves for lower concentrations of bio-diesel blend

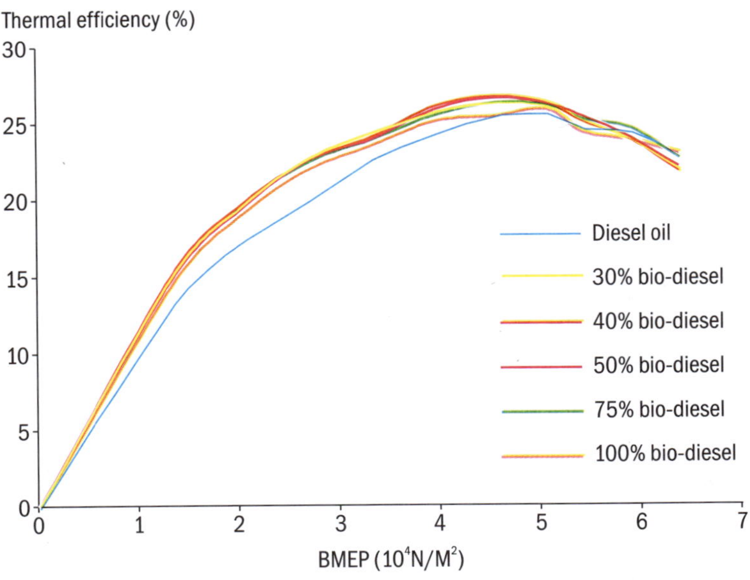

Figure 21B Comparison of thermal efficiency vs. brake mean effective pressure (BMEP) curves for higher concentrations of bio-diesel blend

concentration of NO_x increased by 5% for a B20 blend. B20 was found to be the optimum bio-diesel blend giving maximum increase in thermal efficiency, lowest BSEC, and advantage in terms of lower emissions (Agarwal and Das 2001).

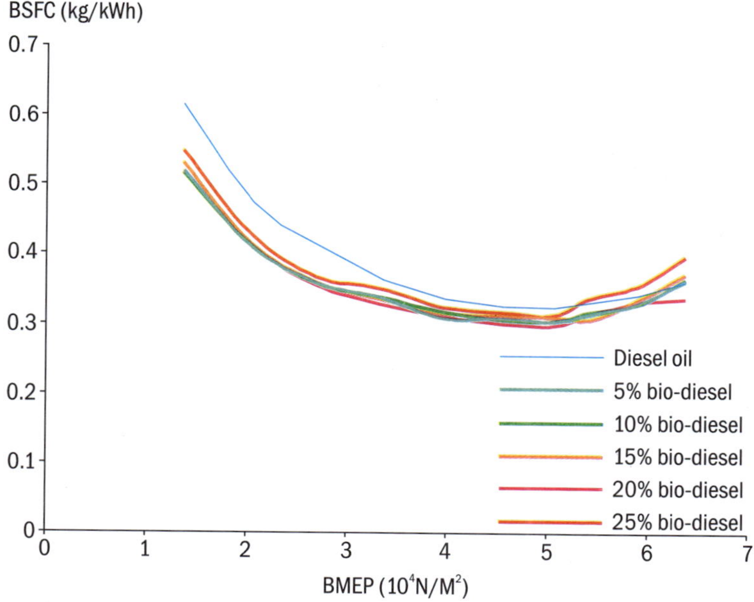

Figure 22A Comparison of brake-specific fuel consumption (BSFC) vs. brake mean effective pressure (BMEP) curves for lower concentrations of bio-diesel blend

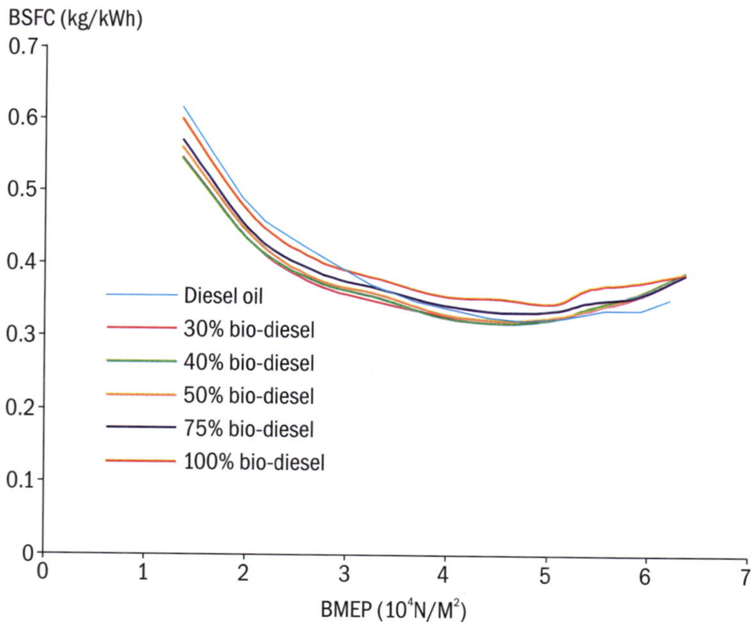

Figure 22B Comparison of brake-specific fuel consumption (BSFC) vs. brake mean effective pressure (BMEP) curves for higher concentrations of bio-diesel blend

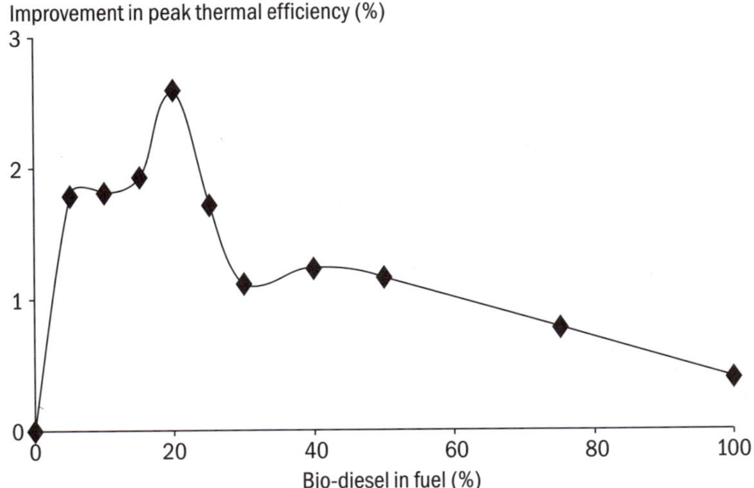

Figure 23 Improvement in peak thermal efficiency vs. concentration of bio-diesel blend

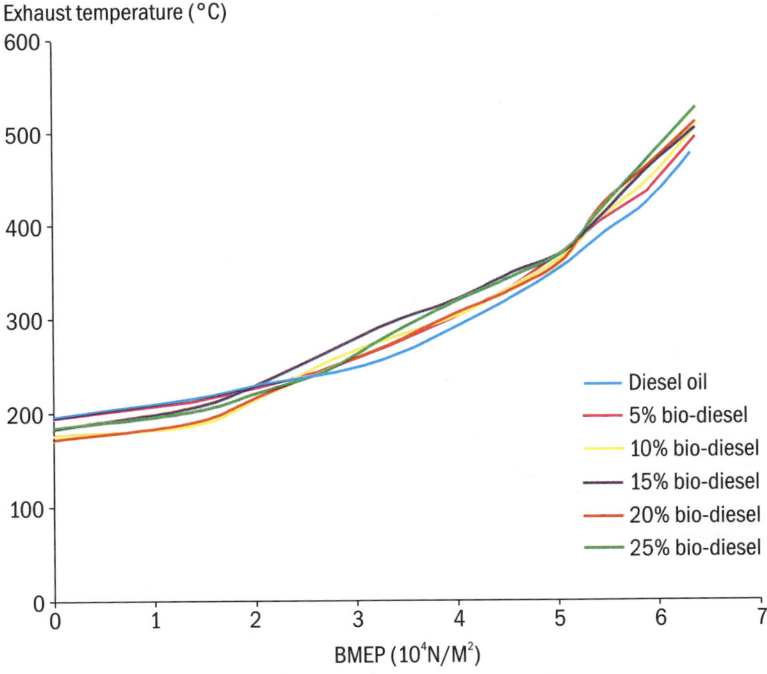

Figure 24A Comparison of exhaust temperatures vs. brake mean effective pressure (BMEP) curves for lower concentrations of bio-diesel blend

Engine emissions

Since bio-diesel is free from sulphur, less sulphate concentration and particulate reduction can be expected in the exhaust. Due to the absence of sulphur in bio-diesel, there is no risk of acid rain. The lack of aromatic HCs

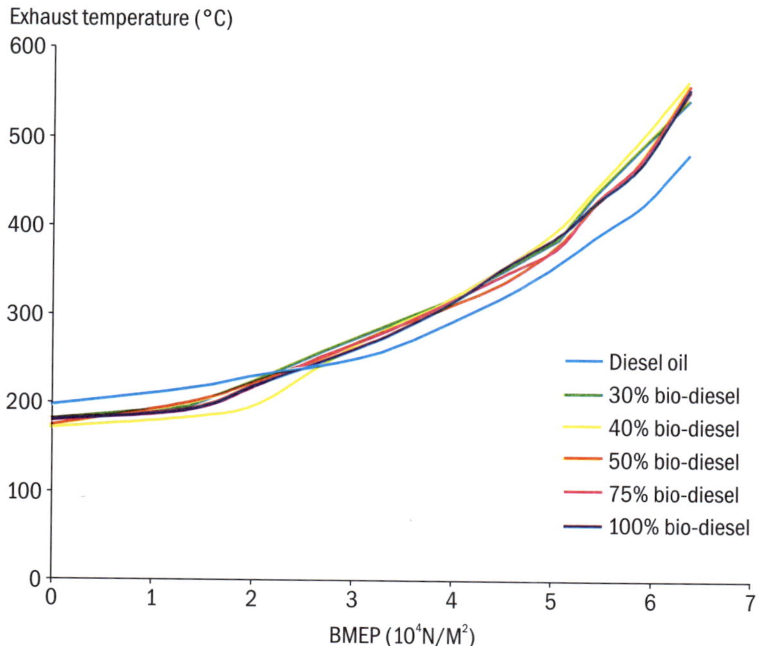

Figure 24B Comparison of smoke temperature vs. brake mean effective pressure (BMEP) for higher concentrations of bio-diesel blend

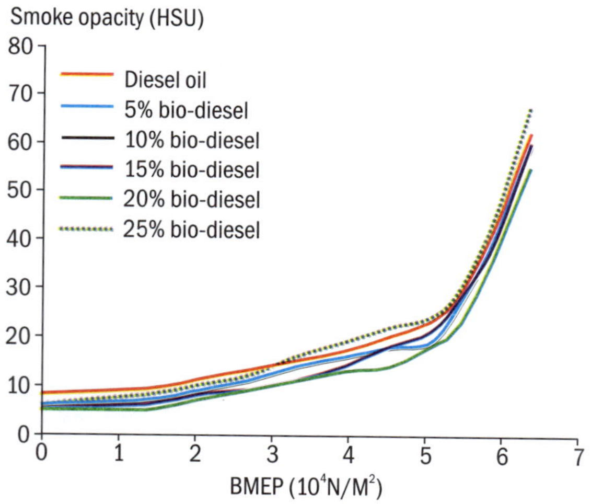

Figure 25A Comparison of smoke opacity vs. brake mean effective pressure (BMEP) curves for lower concentrations of bio-diesel blend

(benzene, toluene, and so on) in bio-diesel reduces unregulated emissions of pollutants such as ketone and benzene as well.

Breathing particulates has been found to be hazardous for human health, especially for the respiratory system. PM consists of elemental carbon (@ 31%), sulphates and moisture (@ 14%), unburnt fuel (@ 7%), unburnt lubricating oil

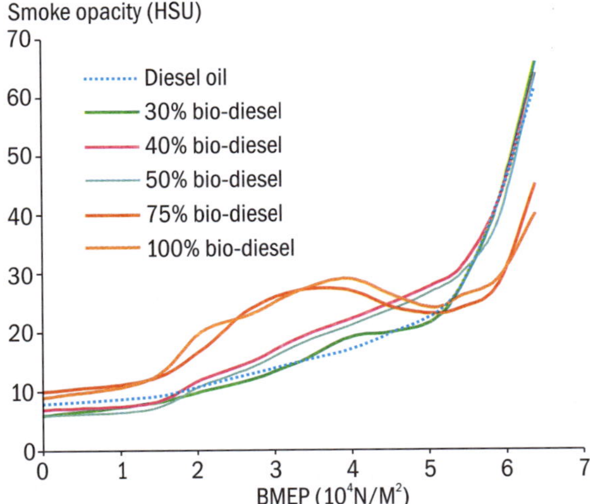

Figure 25B Comparison of smoke opacity vs. brake mean effective pressure (BMEP) curves for higher concentrations of bio-diesel blend

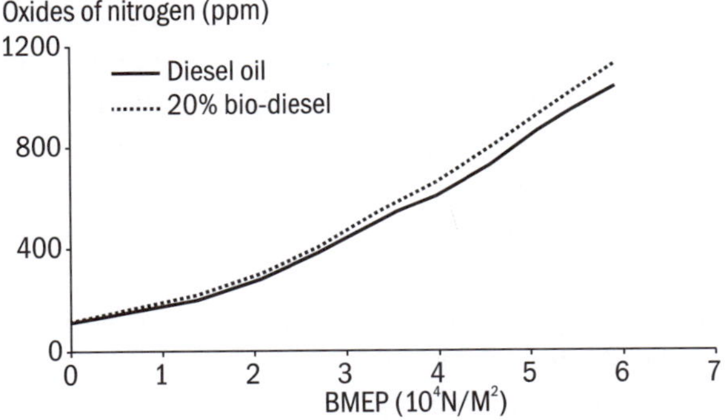

Figure 26 Concentration of oxides of nitrogen vs. brake mean effective pressure (BMEP)

(@ 40%) as well as metals and others substances (the remaining). The typical composition of PM is shown in Figure 27.

Bio-diesel is a sulphur-free (hence no sulphate formation) oxygenated fuel (hence there is more complete combustion) and causes less PM emissions. Smoke opacity is a direct measure of smoke and soot. Various studies show that smoke opacity for bio-diesel is generally lower (Agarwal 1998; Agarwal and Das 2001; Scholl and Sorenson 1993; Kalligerous, Zannikos, Stoumas, *et al.* 2004).

Several experimental investigations have been carried out on four-stroke DI diesel engines with vegetable oil methyl esters, and it has been found that HC emissions are much lower in the case of bio-diesel compared to diesel. This is due to the fact that bio-diesel is an oxygenated fuel, and therefore, more oxygen is available to burn HCs, which reduces HC emissions in the exhaust

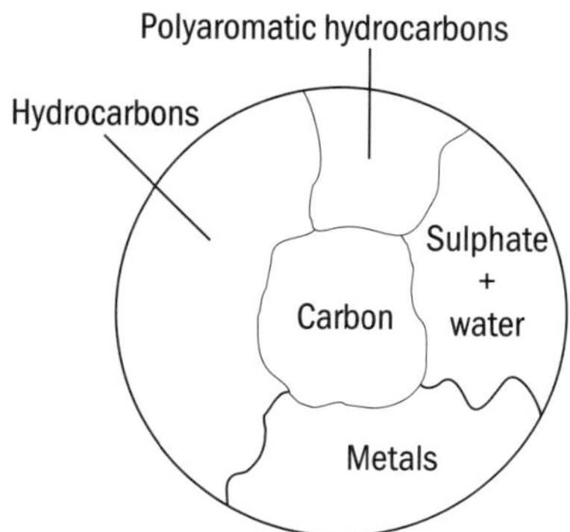

Figure 27 Typical composition of particulate matter

(Masjuki, Abdulmuin, and Sii 1996; Scholl and Sorenson 1993; Kalligerous, Zannikos, Stoumas, *et al.* 2004).

CO is a toxic combustion product, resulting from incomplete combustion of HCs. In the presence of sufficient oxygen, CO is converted to CO_2. Bio-diesel is an oxygenated fuel and results in more complete combustion; hence, CO levels are low in the exhaust. Altin, Cetinkaya, and Yucesu (2001) reported that CO emission for bio-diesel is marginally higher than that for diesel (Figure 17), while Scholl and Soresnosn (1993) contradicted by saying that CO emissions for SOME are slightly lower than those for diesel emissions (Figure 10). Kalligerous, Zannikos, Stoumas, *et al.* (2004) also reported lower CO emissions for sunflower oil.

NO_x results from the oxidation of atmospheric nitrogen. The kinetics of NO_x formation is governed by the Zeldovich mechanism, and its formation is highly dependent on temperature and availability of oxygen. Several studies indicate (Agarwal and Das 2001; Scholl and Sorenson 1993) slight increases in NO_x emissions for bio-diesel. It is quite obvious that with bio-diesel, due to improved combustion, the temperature in the combustion chamber is high, and a higher amount of oxygen is present, leading to the formation of a higher quantity of NO_x. However, bio-diesel's lower sulphur content allows the use of NO_x control technologies that cannot be otherwise used with conventional diesel. Hence, bio-diesel-based NO_x emissions can be effectively managed and eliminated. Some results on an unmodified Cummins N14 diesel engine with respect to pollutant reduction are shown in Table 12.

Bio-diesel use also shows reduction in PAHs, which are identified as carcinogen compounds, so it reduces health risk also. A 1998 bio-diesel life cycle study, jointly sponsored by the US Department of Energy and the US Department of Agriculture, concluded that bio-diesel reduces net CO_2 emissions by 78% compared to petroleum diesel. This is due to bio-diesel's closed carbon cycle. The CO_2 released into the atmosphere when bio-diesel is burned is recycled by growing plants, which are later processed into fuel. Hence, bio-diesel also helps in mitigating global warming. Peterson, Wagner,

Table 12 Bio-diesel emissions compared to conventional diesel emissions

Emission type		B100	B20
Regulated			
	Hydrocarbon	–93%	–30%
	Carbon monoxide	–50%	20%
	Particulate matter	–30%	–22%
	Oxides of nitrogen	+13%	+2%
Non-regulated			
	Sulphates	–100%	–20%
	polycyclic aromatic hydrocarbons (PAHs)	–80%	–13%
	Ozone potential of speciated hydrocarbon	–50%	–10%

and Auld (1983) also reported that CO_2 emissions are significantly lower when bio-diesel is used.

Engine tribology investigations of bio-diesel

Sliding contact between the metallic components of any mechanical system is always accompanied by wear, which results in the generation of minute particles of metal. In diesel engines, the components that are normally subjected to the wear and tear process are piston, piston ring, cylinder liner, bearing, crankshaft, tappet, and valves. In a lubrication system, wear particles remain in suspension in the oil. By analysing and examining the variations in the concentration of metallic particles in the lubricant oil after certain running duration, sufficient information on wear rate, source of element, and engine condition can be predicted.

Agarwal, Bijwe, and Das (2003a, b) have carried out detailed investigations to assess the wear of engine parts using bio-diesel fuel. They investigated the effect of long-term engine operation on engine wear for a blend of 20% LOME and diesel, both by physical dimensioning and lubricating oil analysis. Two similar new engines were completely disassembled and subjected to the dimensioning of various parts and then to long-term endurance test on 20% bio-diesel blend and diesel oil. During this test, each engine was run for 32 cycles (each of 16 h continuous running) at the rated speed. The test cycle followed is given in Table 13 (as per IS: 10000, part viii, 1980). The lubrication oil samples were collected from the engines after every 128 h for conducting oil tribological studies. After the completion of the test, both engines were disassembled for physical inspection and measuring wear of vital parts.

Carbon deposits

Deposit formations on the cylinder head, piston top, and injector tip were investigated. Figures 28A–C show the carbon deposits on cylinder head, piston top, and injector tip, respectively, for diesel-fuelled engine and Figures 29A–C show the same for bio-diesel-fuelled engine. It can be clearly noticed that the carbon deposits for bio-diesel-fuelled engines are substantially reduced. The pistons were weighed, and the amount of carbon deposits formed on the piston was found out. Deposits in the case of bio-diesel-fuelled engine were 40% less compared to diesel-fuelled engine. Carbonization of bio-diesel injector after 512 h of operation was far less than that for the diesel injector after 200 h of engine operation. This proved that the problem of carbon deposits and coking of injector tip have reduced after transesterification (Agarwal, Bijwe, and Das 2003a).

Measuring physical wear

The dimensions of the vital parts (bore, rings, different bearings, and so on) were noted before the commencement and after the completion of the long-term endurance test. Wear was measured by accurate measurements of the dimensions of various parts. It was observed that the wear of vital moving parts of the 20% bio-diesel-operated engine was about 30% lower than that of neat-diesel-operated engine (Agarwal, Bijwe, and Das 2003a, b).

Table 13 Test cycle for long-term endurance test

Load (% of rated load)	Running time (h)
100	4
50	4
110	1
No load (idling)	05
100	3
50	3.5

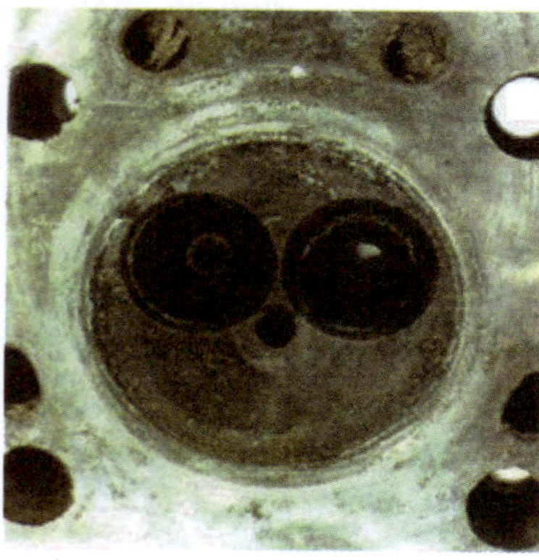

Figure 28A Carbon deposits on the cylinder head of diesel-fuelled engine after 512 hours of engine operation

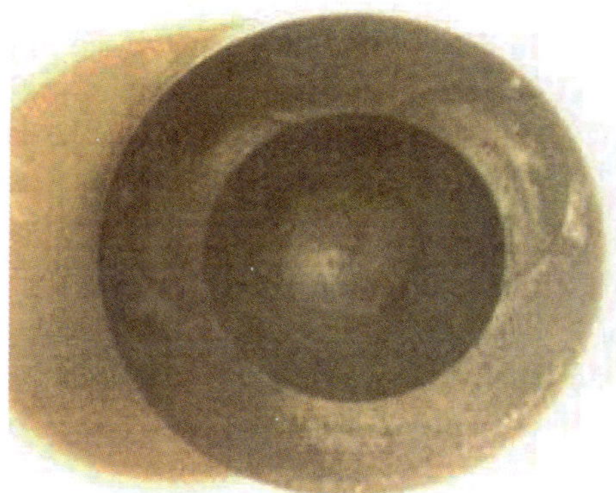

Figure 28B Carbon deposits on the piston top of diesel-fuelled engine after 512 hours of engine operation

Figure 28C Carbon deposits on the injector tip of diesel-fuelled engine after 200 hours of engine operation

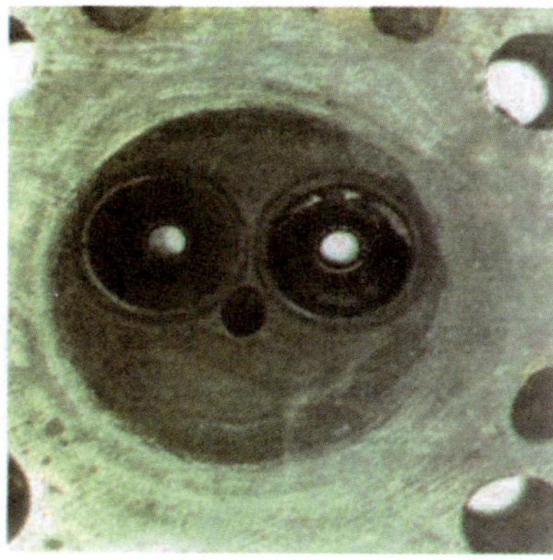

Figure 29A Carbon deposits on the cylinder head of 20% bio-diesel-fuelled engine after 512 hours of engine operation

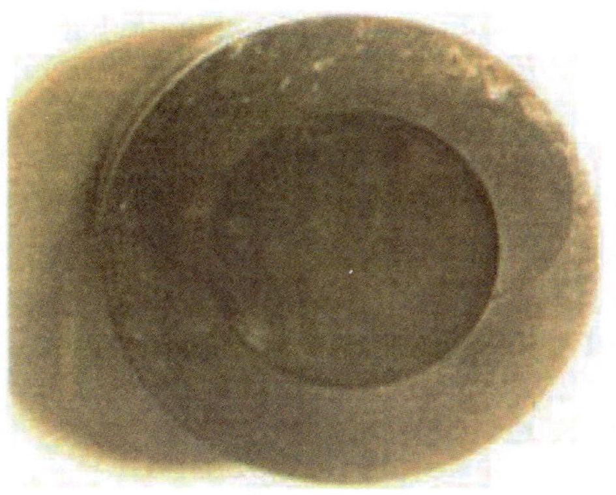

Figure 29B Carbon deposits on the piston top of 20% bio-diesel-fuelled engine after 512 hours of engine operation

Figure 29C Carbon deposits on the injector tip of 20% bio-diesel-fuelled engine after 512 hours of engine operation

Measuring wear debris through lubricating oil analysis

Oil used for the lubrication of the engine picks up the wear debris of various metals. The quantitative evaluation of wear particles present in oil gives the magnitude of engine component deterioration, whereas qualitative analysis indicates its origin.

Viscosity

Viscosity is the most important property of a lubricating oil, as it affects the wear rate of engine components. Very high viscosity lubricating oil increases the friction loss through the shearing forces of the lubricant, preventing the formation of a protective film. During engine operation at normal temperature, a small amount of fuel normally dilutes the lubricating oil, which passes on to the oil sump through blow-by losses. Engine cold-starting, rich fuel--air mixtures, and low ambient temperatures promote fuel dilution. High dilution reduces oil viscosity, pour point, and ash point, and diminishes the oil's load-carrying ability.

Change in viscosity is undesirable in the system, as it affects the lubricating efficiency of the oil. In fact, the criterion for the change of lube oil states that "change the lubricating oil if viscosity increases by 20% or more, or decreases by 10% or more." With use, the viscosity of lubricating oil may increase or decrease. Inadequate oil viscosity affects film formation and load bearing capacity, leading to excessive wear of bearings, journals, and other moving components; low oil pressure; and poor oil economy. There are two main factors responsible for the viscosity changes affecting the oil in opposite directions, as discussed subsequently.

Formation of resinous products because of oil oxidation, evaporation of lighter fractions, depletion of antiwear additives, and contamination by insolubles tends to increase the viscosity, whereas fuel dilution and shearing of viscosity index improvers tend to bring down viscosity. The extent of dominance of these mechanisms, however, differs from system to system. Hence, the net result can be reflected in either direction. If the first factor is dominating and the possibility of fuel dilution is negligible, then the viscosity value increases. Sometimes, viscosity can decrease if fuel dilution is the dominating mechanism. The viscosity of all lube oil samples were evaluated at 40°C and 100°C using a Setavis kinematic viscometer. The experimental results at these temperatures are shown in Figures 30 and 31, respectively.

An important observation was that the extent of lowering of viscosity of the lube oil is less in the case of bio-diesel-fuelled system compared to diesel-fuelled system. This may be due to the lower fuel dilution. Fuel dilution is a direct consequence of the clearance between piston rings and the cylinder liner. The more the piston ring wear, the more will be the clearance and, hence, the higher will be the fuel dilution. Since bio-diesel has inherent lubrication properties, it helps in protecting the piston rings from wearing out. The viscosity of bio-diesel helps in plugging the clearance between piston rings and the cylinder liner effectively, thus reducing blow-by losses and fuel dilution of the lubricating oil. Bio-diesel has thus proved to be more effective in protecting the moving parts of the engine.

The subtle point in this viscosity behaviour, however, cannot be neglected. As already discussed, the rate of change of viscosity is also controlled by another mechanism, that is, oil oxidation. In this case, it is also possible that the bio-diesel fuel after entering the lube oil can accelerate the rate of oxidation of the base stock, leading to slightly higher viscosity. Hence, the decrease in viscosity due to fuel dilution can get hampered. This fact was supported by Fourier transform infrared (FTIR) studies showing higher

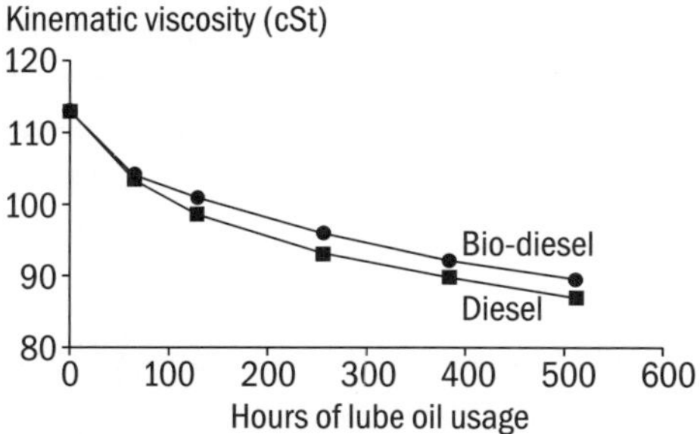

Figure 30 Kinematic viscosity at 40°C vs. hours of lube oil usage

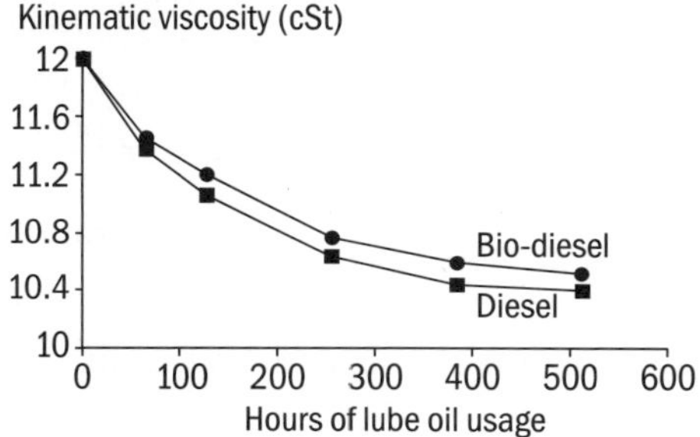

Figure 31 Kinematic viscosity at 100°C vs. hours of lube oil usage

oxidation of the base stock in case of bio-diesel-fuelled engine oil, as explained in subsequent research work (Agarwal 2003).

Ash content

The lubricating oil samples were kept in a furnace at 450°C for 4 h and then at 650°C for 2 h to produce ash. The residual ash contains the wear debris. The data on ash content in the lubricating oils for bio-diesel- and diesel-fuelled CI engines is shown in Figure 32. It was found that ash content for the 20% bio-diesel-operated engine oil was lower than that for the diesel-operated engine oil, showing that bio-diesel-fuelled engines produced a lower amount of wear debris (Agarwal 2003).

Atomic absorption spectroscopy

Concentration of the various metals present in the lubricating oil sample can be evaluated by atomic absorption spectroscopy (AAS) for quantitative and qualitative analysis. AAS works on the principle of the absorption of a certain wavelength by a particular atomic element. In AAS, a light beam of specific wavelength is projected through a pure flame (air–acetylene) onto a photo-sensor, and the radiation arriving at the photo-sensor is recorded. The fluid sample is then introduced into the flame and vaporized. The quantity of radiation arriving at the photo-sensor is reduced in proportion to the quantity of the specific element present in the sample. Results of the AAS studies carried out by Agarwal, Bijwe, and Das (2003a, b) are shown in Figures 33–39. It is obvious from all these results that the bio-diesel-fuelled engine leads to less wear of moving parts of engine in terms of the amount of metallic debris originating from the different moving parts.

Summary

Using ethanol as a fuel additive in unleaded gasoline causes an improvement in engine performance and exhaust emissions. Ethanol addition results in an improvement in brake power, brake thermal efficiency, volumetric efficiency, and fuel consumption; however, the BSFC and equivalence air–fuel ratio decrease because of the lower calorific value of the gasohol. Using an ethanol–unleaded gasoline blend leads to a significant reduction in exhaust emissions of CO and HC for all engine speeds. On the other hand, CO_2 emissions increase marginally. Ethanol diesel blends up to 20% can very well be used in present day constant speed CI engines without any hardware modification. Exhaust gas temperatures and lubricating oil temperatures were lower for ethanol diesel blends than mineral diesel fuel. The engine could be started normally under both hot and cold conditions. Significant reduction in CO and NO_x emission was observed while using ethanol diesel blends.

Bio-diesel has become a preferred choice because of its environmental benefits and the fact that it is derived from renewable resources. Researchers in various countries carried out experimental research using vegetable oils and bio-diesel as petroleum fuel substitutes. Vegetable oil methyl esters gave performance and emission characteristics comparable to those of diesel. Transesterification is a process that brings about a change in the molecular structure of the vegetable oil molecules, thus bringing down the levels of viscosity and unsaturation of vegetable oils. The viscosity of vegetable oil gets reduced drastically after transesterification. A 20% blend of bio-diesel with diesel oil improved the cetane number of diesel oil. The calorific value of bio-diesel was found to be slightly lower than petroleum diesel oil. All these tests for the characterization of bio-diesel oil demonstrated that almost all the important properties of bio-diesel are in very close agreement with the diesel oil, making it a potential candidate for applications in CI engines. A

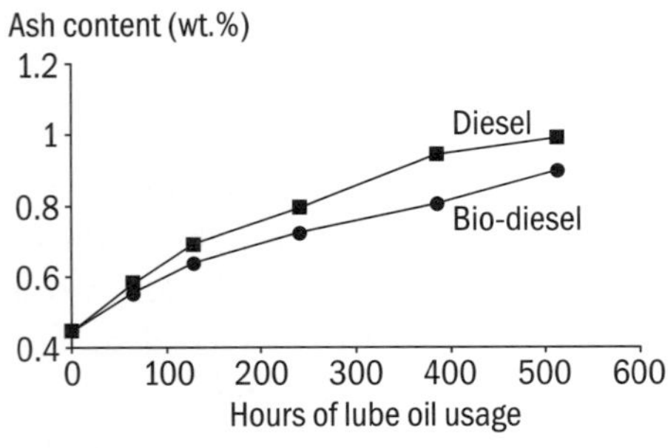

Figure 32 Ash content vs. hours of lube oil usage

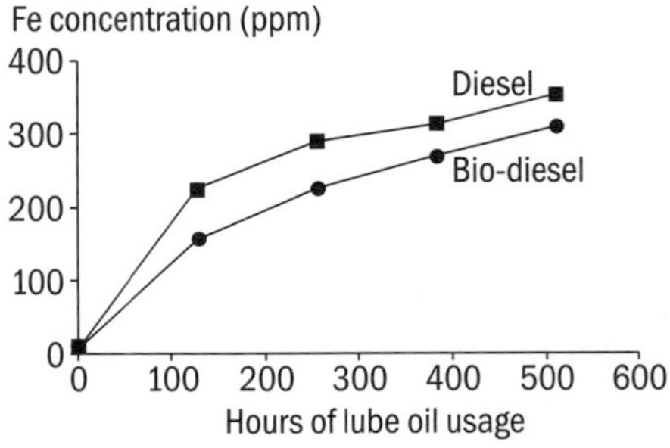

Figure 33 Iron (Fe) concentration as a function of lube oil usage

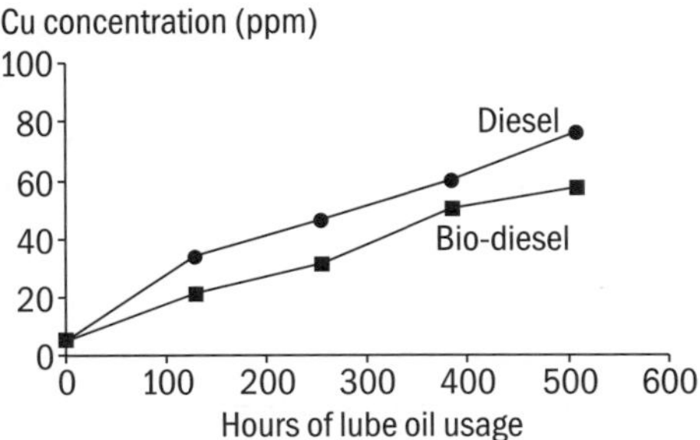

Figure 34 Copper (Cu) concentration as a function of lube oil usage

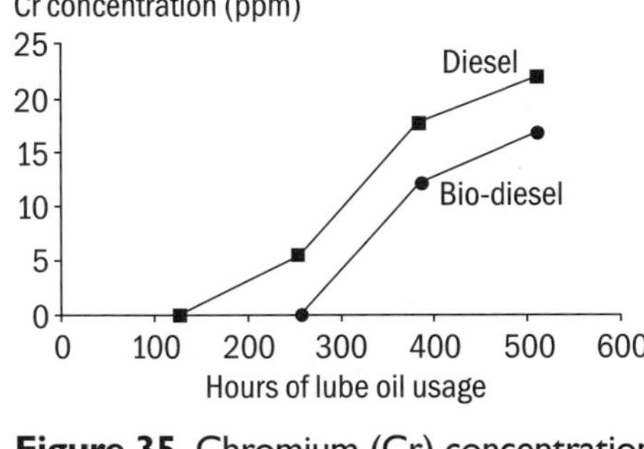

Figure 35 Chromium (Cr) concentration as a function of lube oil usage

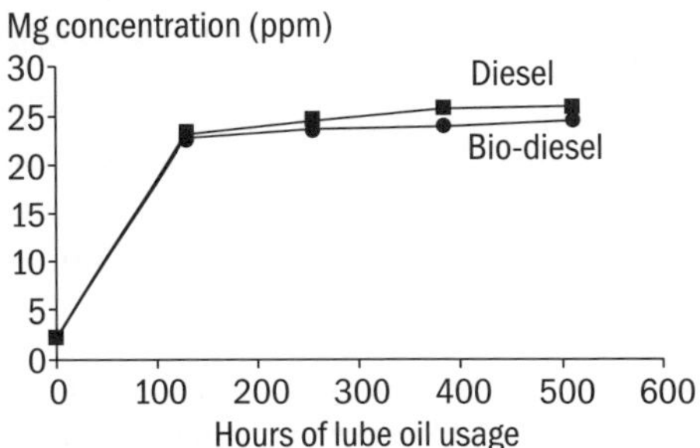

Figure 36 Magnesium (Mg) concentration as a function of lube oil usage

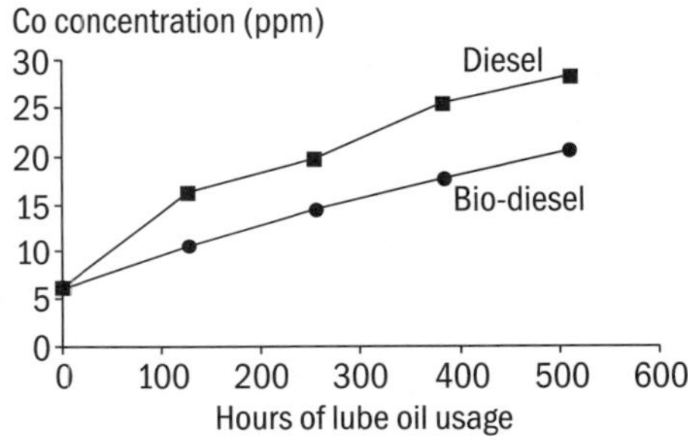

Figure 37 Cobalt (Co) concentration as a function of lube oil usage

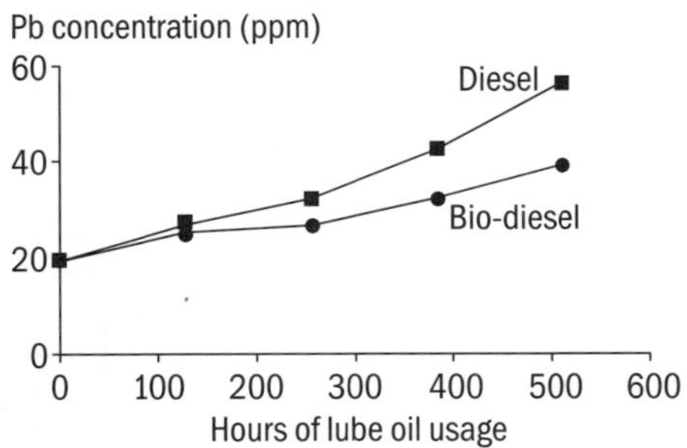

Figure 38 Lead (Pb) concentration as a function of lube oil usage

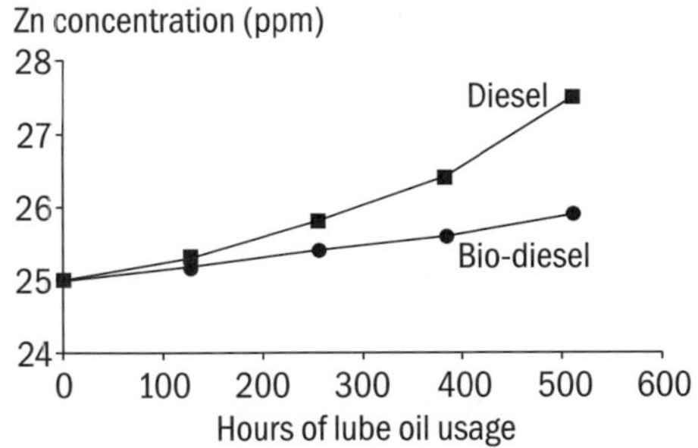

Figure 39 Zinc (Zn) concentration as a function of lube oil usage

diesel engine can perform satisfactorily on bio-diesel fuel without any engine hardware modifications. Long-term endurance test using bio-diesel proved that bio-diesel can be used for substituting mineral diesel oil in the long run. Smoke emissions reduced appreciably as a result of bio-diesel utilization in the engine. Transesterification has been found to be an effective technique to prevent some long-term problems associated with the utilization of vegetable oils such as fuel filter plugging, injector coking, formation of carbon deposits in combustion chamber, ring sticking, and contamination of lubricating oils. The carbon deposits on the piston top and injector coking substantially reduced in bio-diesel-fuelled system.

Hence, bio-diesel may be considered as diesel fuel substitute. The use of biofuels as IC engine fuels can play a vital role in helping the developed and developing countries to reduce the environmental impact of fossil fuels.

References

Agarwal A K. 1998. **Vegetable oil versus diesel fuel: development and use of biodiesel in a compression ignition engine.** *TIDE* 8(3): 191–204

Agarwal A K. 2003. **Lubrication oil tribology of a biodiesel-fuelled CI engine.** In *Proceedings of the 2003 Fall Technical Conference of the ASME Internal Combustion Engine Division*, pp. 427–442, September 2003, Erie

Agarwal A K and Das L M. 2001. **Biodiesel development and characterization for use as a fuel in compression ignition engine.** *Journal of Engineering for Gas Turbines and Power, Transaction of the ASME* **123**: 440–447

Agarwal A K, Bijwe J, and Das L M. 2003a. **Wear assessment in biodiesel fuelled compression ignition engine.** *Journal of Engineering for Gas Turbine and Power, Transactions of ASME* **125**(3): 820–826

Agarwal A K, Bijwe J, and Das L M. 2003b. **Effect of biodiesel utilisation on wear of vital parts in compression ignition engine.** *Journal of Engineering for Gas Turbine and Power, Transactions of ASME* **125**(2): 604–611

Ajav E A, Singh B, and Bhattacharya T K. 1999. **Experimental study of some performance parameters of a constant speed stationary diesel engine using ethanol–diesel blends as fuel.** *Biomass and Bioenergy* **17**(4): 357–365

Altin R, Cetinkaya S, and Yucesu H S. 2001. **The potential of using vegetable oil fuels as fuel for diesel engine.** *Energy Conversion and Management* **42**(5): 529–538

Andrew B T, Derek P M, Arthur J B, Neale G H, Ian S M, Johan J B. 1996. **Gasoline/alcohol blends: exhaust emissions, performance, and burn-rate in a multi-valve production engine.** SAE Special Publication SP-1208, SAE Paper No. 961988

Bata Reda M and Roan Vermon P. 1989. **Effect of ethanol and/or methanol in alcohol–gasoline blends on exhaust emissions.** *Journal of Engineering for Gas Turbine and Power, Transactions of ASME* **111**: 432–438

Beer T, Grant T, Brown R, Edwards J, Nelson P, Watson H, Williams D. 2000. **Life-cycle emissions analysis of alternative fuels for heavy vehicles,** pp. 1–148. CSIRO Atmospheric Research Report C/0411/1.1/F2, Australian Greenhouse Office Report

Freedman B, Pryde E H, and Mounts T L. 1984. **Variables affecting the yields of fatty esters from transesterified vegetable oils.** *Journal of American Oil Chemists Society* **61**: 1638–1643

Kalligerous S, Zannikos F, Stoumas S E, Anastopoulos G, Teas Ch., Sakellaropoulos. 2004. **An investigation of using biodiesel/marine diesel blends on the performance of a stationary diesel engine.** *Biomass and Bioenergy* **24**(2): 141–149

Krawczyk T. 1996. **Biodiesel – alternative fuel makes inroads but hurdle remains.** *INFORM* 7: 801–815

Kremer F G, Jordim J L F, and Maia D M. 1996. **Effect of alcohol composition on gasoline vehicle emissions.** SAE Special Publication SP-1208, SAE Paper No. 962094

Ma F and Hanna M A. 1999. **Biodiesel production: a review.** *Bioresource Technology* **70**: 1–15

Masjuki H, Abdulmuin M Z, and Sii H S. 1996. **Investigations on preheated palm oil methyl esters in the diesel engine.** *Proceedings of the IMechE: Part A, Journal of Power and Energy* **210**: 131–138

Muniyappa P R, Brammer S C, and Noureddini H. 1996. **Improved conversion of plant oils and animal fats into biodiesel and co-products.** *Bioresource Technology* **56**(1): 19–24

Murayama T, Fujiwara Y, and Noto T. 2000. **Evaluating waste vegetable oil as a diesel fuel.** *Proceedings of the IMechE: Part D, Journal of Automobile Engineering* **214**(2): 141–148

Peterson C L, Wagner G L, and Auld D L. 1983. **Vegetable oil substitution for diesel fuel.** *Transaction of ASAE* **26**: 322–327

Pramanik K. 2003. **Properties and use of *Jathropa curcas* oil and diesel fuel blends in compression ignition engine.** *Renewable Energy* **28**(2): 239–248

Rosillo Calle F and Cortez L A B. 1998. **Towards pro-alcohol II – a review of the Brazilian bio-ethanol program.** *Biomass and Bioenergy* **14**(2): 115–124

Scholl K W and Sorenson S C. 1993. **Combustion of soyabean oil methyl ester in a direct injection diesel engine.** *Journal of Engines* 1450–1462

Srivastava A and Prasad R. 2000. **Triglycerides-based diesel fuels.** *Renewable and Sustainable Energy Reviews* **4**(2): 111–133

Stump F, Knapp K, and Racy W. 1996. **Influence of ethanol blended fuels on the emissions from three pre-1995 light-duty passenger vehicles.** *Journal of the Air and Waste Management Association* **46**: 1149–1161.

Tom B, Grant T, Williams D, Watson H. 2002. **Fuel-cycle greenhouse gas emissions from alternative fuels in Australian heavy vehicles.** *Atmospheric Environment* **36**(4): 753–763

Bioethanol production from lignocellulosics: an overview

R C Kuhad, R Gupta, and Y P Khasa

Lignocellulose Biotechnology Laboratory, Department of Microbiology
University of Delhi South Campus, New Delhi, India

The constant increase in the consumption of fossil fuels has not only caused a global energy crisis but also increased the environmental concern. It has been predicted that annual global oil production would decrease from the current 25 billion barrels to approximately 5 billion barrels in 2050 (Campbell and Laherrere 1998). The Indian economy, like that of many other countries, depends on oil and could experience a severe crisis in future. Therefore, it becomes necessary to search for alternative resources to meet our energy demand. Ethanol is the most common renewable fuel, derived from corn grain (starch) and sugar cane (sucrose). It is expected that there will be limits to the supply of these raw materials in the years to come; therefore, lignocellulosic material is seen as an attractive feedstock for future supplies of ethanol (Chandel, Kapoor, Singh, *et al.* 2007a, b; Farrell, Plevin, Turner, *et al.* 2006; Kuhad, Singh, and Eriksson 1997; Lynd, Cushnman, Nichols, *et al.* 1991; Lynd, Wyman, Gerngross, *et al.* 1999; MacLean, Lave, Lankey, *et al.* 2000; McMillan 1997; Sheehan and Himmel 1999; Wyman 2003).

Currently, the major producers of ethanol are Brazil and the US, which account for 62% of the world ethanol production, which is derived from sugar cane juice and corn grain, respectively (Berg 2001). Brazil led world ethanol production in 2004, distilling 4 billion gallons (15 billion litres) of ethanol. The US is rapidly catching up, although the production is mainly from corn grain. China's wheat- and corn-rich provinces produced nearly 1 billion gallons of ethanol, and India churned out 500 million gallons from sugar cane molasses. France, the front-runner in the European Union's efforts to boost ethanol use, produced over 200 million gallons from sugar beets and wheat grain (Licht 2005). In Canada, ethanol fuel is produced from corn and wheat. In all, the world produced enough ethanol to displace roughly 2% of total gasoline consumption. As per the futuristic projections, about 14 billion gallons of bioethanol is expected to replace about 348 billion barrels of oil in the US by 2020.

In India, ethanol is mainly produced from sugar cane molasses. As per estimates, projected alcohol production is expected to increase in the years to come. According to Kim and Dale (2004), the total bioethanol production from crop residues is estimated to be 491 Gl/year globally. India alone has the capacity to produce 25% of that, that is, 123 Gl/year, if the entire crop residues available are used for ethanol production. Hence, to contemplate a bioethanol production plant, the lignocellulosic biomass assessment with geographical distribution and accurate information on the availability of biomass in different parts of the country is a prerequisite. With this in view, during the Ninth Plan, the Ministry of New and Renewable Energy, Government of India, had sponsored 500 *taluka*-level biomass assessment studies in 23 states to compile data on the availability of lignocellulosic biomass. As an extension of this effort, a project for the preparation of "Biomass Resource Atlas for India" has been jointly sanctioned to the Indian Institute of Science (IISc), Bengaluru, and the Regional Remote Sensing Service Centre (RRSSC), Bengaluru, which aims at the integration of the data on biomass availability obtained from *taluka*-level studies and from other reliable sources, with information on crop distribution pattern derived from geographic information system (GIS)-based maps provided by the RRSSC. A few major plant species along with their biomass types and biomass production are listed in Table 1.

Lignocellulosic materials comprise cellulose, hemicellulose, and lignin (Kuhad, Singh, and Eriksson 1997). The presence of lignin in lignocellulose limits cellulose and hemicellulose biodegradation by cellulolytic and hemicellulolytic enzymes. In order to increase the accessibility of these enzymes to the carbohydrates, the lignocellulosic biomass must be pretreated. Pretreatment can be carried out in different ways such as mechanical comminuting, steam explosion, ammonia fibre explosion, and acid or alkaline and biological pretreatments (Gupta, Sharma, and Kuhad 2009; Kumar, Barrwett, Delwiche, *et al.* 2009; McMillan 1994; Mosier, Wyman, Dale, *et al.* 2005; Palmqvist and Hahn-Hagerdal 2000a, b; Sun and Cheng 2002).

Bioconversion of various agro residues into ethanol by different yeasts has been widely studied (Abbi, Kuhad, and Singh 1996a, b; Chandel, Kapoor, Singh, *et al.* 2007b; Gupta, Sharma, and Kuhad 2009; Hahn-Hägerdal, Karhumaa, Fonseca, *et al.* 2007; Kapoor, Chandel, Kuhar, *et al.* 2007; Kapoor, Nair, and Kuhad 2008; Kuhad, Manchanda, and Singh 1999; Sun and Cheng 2002). Comparatively large number of micro-organisms and many yeast species are known to ferment hexose sugars to ethanol; however, the ability to ferment pentoses along with hexoses is limited among micro-organisms. The ideal organism for the

Table 1 Select biomass, by plant residue, data at national level (2002–04)

Species	Biomass type	Biomass (kt/year)	Total biomass (kt/year)	Species	Biomass type	Biomass (kt/year)	Total biomass (kt/year)
Arhar	Husk	606.2	5658.2	Mango	Bark	2.049	8.54
	Stalks	5052.0			Branches	2.049	
Babul	Bark	499.1	2079.0		Leaves	2.393	
	Branches	499.1			Twig	2.049	
	Leaves	581.7		Mesquite	Bark	28.58	119.08
	Twig	499.1			Branches	28.58	
Bajra	Cobs	1972.3	15810.71		Leaves	33.34	
	Husk	1793.0			Twig	28.58	
	Stalks	11953.0		Mustard	Husk	1553.2	8309.8
Bamboo	Leaves	92.41	2466.41		Stalks	6756.6	
	Stalks	2374.0		Neem	Bark	2.049	8.54
Banana	Residue	11936.5	11936.5		Branches	2.049	
Chir pine	Bark	0.484	2.007		Leaves	2.393	
	Branches	0.484			Twig	2.049	
	Leaves	0.555		Paddy	Husk	19 938.6	169 478.4
	Twig	0.484			Straw	149539.8	
Coconut	Fronds	7253.8	11734.4	*Salvadora* sp.	Bark	3.069	12.801
	Husk and pith	3166.3			Branches	3.069	
	Shell	1314.3			Leaves	3.594	
Conifers	Bark	82.14	328.56		Twig	3.069	
	Branches	82.14		Shisham	Bark	80.52	335.57
	Leaves	82.14			Branches	80.52	
	Twig	82.14			Leaves	94.01	
Cotton	Boll shells	6320.6	43202.0		Twig	80.52	
	Husk	6320.6		Sugar cane	Tops and Leaves	13211.2	13 211.2
	Stalks	30560.8		Sunflower	Stalks	1391.8	1391.8
Cotton wood	Bark	2.049	8.54	Teak	Bark	1421.2	5921.7
	Branches	2.049			Branches	1421.2	
	Leaves	2.393			Leaves	1658.1	
	Twig	2.049			Twig	1421.2	
Eucalyptus	Bark	341.1	1463.24	Tobacco	Stalks	316.6	316.6
	Branches	341.1		Turmeric	Stalks	28.7	28.7
	Leaves	398.0		Umbrella thorn Acacia	Bark	160.85	670.15
	Twig	220.2			Branches	160.85	
	Residues	162.8			Leaves	187.6	

Contd...

Table 1 Contd...

Species	Biomass type	Biomass (kt/year)	Total biomass (kt/year)	Species	Biomass type	Biomass (kt/year)	Total biomass (kt/year)
Ground-nut shell	Shell	2044.6	15675.5		Twig	160.85	
	Stalks	13630.9		Wheat	Pods	18650.4	111902.4
Jowar	Cobs	5044.8	29827.5		Stalks	93252.0	
	Husk	2017.9		White mulberry	Bark	0.959	4.007
	Stalks	17152.4			Branches	0.959	
Maize	Cobs	5612.4	28814.1		Leaves	1.13	
	Stalks	23201.7			Twig	0.959	

Source http://lab.cgpl.iisc.ernet.in/Atlas/Tables/Tables.aspx

production of ethanol from lignocellulosic hydrolysate would be the one that can utilize various sugars generated from lignocellulose hydrolysis. *Saccharomyces cerevisiae* is capable of converting only hexose sugars to ethanol. Of various xylose-fermenting yeasts, *Pichia stipitis*, *Candida shehatae*, and *Pachysolen tannophillus* are observed to convert pentose sugars to ethanol (Abbi, Kuhad, and Singh 1996a, b; Kuhad 2010). However, genetically engineered organisms are now being used for ethanol production. These organisms can greatly improve ethanol production efficiency and reduce the cost of operation (Dien, Nichols, O'Bryan, *et al.* 2000; Gonzalez, Tao, Shanmugam, *et al.* 2002; Kim, Ingram, Shanmugam, *et al.* 2007; Lawford and Rousseau 1995; Ohta, Beall, Mejia, *et al.* 1991).

This chapter describes the composition of plant residues, different strategies to pretreat the lignocellulosic material, enzymatic saccharification of cellulose and hemicellulose from residue, and ethanol fermentation from pentose and hexose sugars. Moreover, some methodologies for ethanol recovery are also discussed.

Lignocellulose

Lignocellulosic biomass is the most abundantly available and renewable natural resource on the earth. Lignocellulose comprises three main polymers: cellulose, hemicellulose, and lignin (Figure 1). The chemical properties of the components of lignocellulosics make them a substrate of choice for enormous biotechnological products (Hahn-Hagerdal, Karhumaa, Fonseca, *et al.* 2007; Kuhad, Kuhar, Kapoor, *et al.* 2007; Kuhad and Singh 1993; Kuhad, Singh, and Eriksson 1997; Sun and Cheng 2002).

Cellulose

Cellulose is a glucan polymer of D-glucopyranose units, which are linked together by β-(1→4)-glucosidic bonds. The wood cellulose has an average

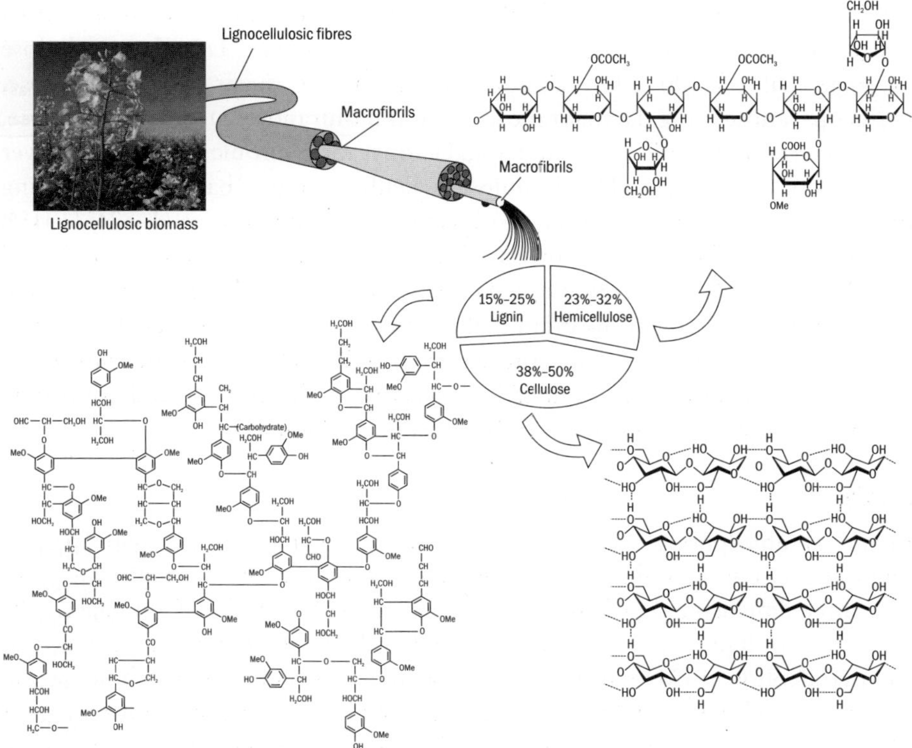

Figure 1 Structure of different components of lignocellulosic biomass

degree of polymerization (DP) of at least 9000–10 000 and possibly as high as 15 000. The average DP of 10 000 would correspond to a linear chain length of approximately 5 µm in wood. An approximate molecular weight for cellulose ranges from 10 000 Da to 150 000 Da (Goring and Timell 1962). There are several types of cellulose in wood: crystalline and non-crystalline and accessible and non-accessible. Most wood-derived cellulose is highly crystalline and may contain as much as 65% crystalline regions. The remaining portion has a lower packing density and is referred to as amorphous cellulose. Accessible and non-accessible refer to the availability of the cellulose to water, micro-organisms, and so on. The surfaces of crystalline cellulose are accessible, but the rest of the crystalline cellulose is non-accessible, whereas most of the non-crystalline cellulose is accessible, but a part of the non-crystalline cellulose is so covered with both hemicelluloses and lignin that it becomes non-accessible (Rowell, Petterson, Han, *et al.* 2005). Concepts of accessible and non-accessible cellulose are very important in moisture sorption, pulping, chemical modification, extractions, and interactions with micro-organisms.

Hemicellulose

Unlike cellulose, hemicelluloses are not chemically homogenous (Eriksson, Blanchette, and Ander 1990; Kapoor 2007; Kuhad, Singh, and Eriksson

1997). In general, the hemicellulose fraction of wood consists of a collection of polysaccharide polymers with a lower DP than cellulose (average DP of 100–200), containing mainly the sugars D-xylopyranose, D-glucopyranose, D-galactopyranose, L-arabinofuranose, D-mannopyranose, D-glucopyranosyluronic acid, and D-galactopyranosyluronic acid, with lower amounts of other sugars. Hemicelluloses usually contain a backbone consisting of one repeating sugar unit β-(1→4) linked with branch points (1→2), (1→3), and/or (1→6). They usually consist of more than one type of sugar unit and called accordingly, for example, galactoglucomanan, arabinoglucuronoxylan, arabinogalactan, glucuronoxylan, glucomannan. The hemicelluloses also contain acetyl- and methyl-substituted groups (Rowell, Petterson, Han, *et al.* 2005). Hemicelluloses are intimately associated with cellulose and contribute to the structural components of the lignocellulosic biomass.

Lignin

Lignin is amorphous and a highly complex polymer of phenylpropanoid units and is considered to be an encrusting substance in the plant biomass. The three-dimensional polymer is made up of C–O–C and C–C linkages. The precursors of lignin biosynthesis are *p*-coumaryl alcohol, coniferyl alcohol, and sinapyl alcohol. *p*-Coumaryl alcohol is a minor precursor of both softwood and hardwood lignins, whereas coniferyl alcohol is the predominant precursor of softwood lignin, and coniferyl and sinapyl alcohols are precursors of hardwood lignin (Alder 1977). Softwood lignin has a methoxyl content of approximately 15%–16%, whereas hardwood lignin has a methoxyl content of approximately 21%. Moreover, ecological factors such as age of the wood, climate, plant sustenance, and amount of sunlight also affect the chemical structure of lignin (Kuhad and Singh 1993; Kuhad, Singh, and Eriksson 1997). Lignin does not have a single repeating unit like cellulose, but instead, consists of a complex arrangement of substituted phenolic units. Lignins can be classified in several ways, but they are usually divided according to their structural elements. All wood lignins consist mainly of three basic building blocks of guaiacyl, syringyl, and *p*-hydroxyphenyl moieties, although other aromatic units also exist in many different types of woods. There is a wide variation of structures within different wood species. The lignin content of softwoods is usually in the range 18%–25%, whereas the lignin content of hardwoods varies between 25% and 35%. The phenylpropane can be substituted at α, β, or γ-positions into various combinations linked together both by ether (C–O–C) and carbon-to-carbon (C–C) linkages (Sakakibara 1991). Lignins from softwood are mainly the polymerization product of coniferyl alcohol and are called guaiacyl lignin. Hardwood lignins are mainly syringyl–guauacyl lignin, because they are a copolymer of coniferyl and sinapyl alcohols. The ratio of these two types varies in different lignins, from about 4:1 to 1:2 (Sarkanen and Ludwig 1971).

Table 2 provides a list of a few lignocellulosic residues along with their chemical composition.

Table 2 Chemical composition of various lignocellulosic residues

Substrate	Hexosans	Pentosans	Lignin
Barley wood	40	20	15
Birch wood	40	33	21
Corn cobs	42	39	14
Corn stalks	35	15	19
Corn stover	38	26	19
Cotton seed hair	80–95	5–20	0–5
Forage sorghum	34	17	16
Groundnut shells	38	36	16
Miscanthus	43	24	19
Oat straw	41	16	11
Pine	41	10	27
Rice straw	32	24	13
Rice husk	36	15	19
Rye straw	31	25	7
Saw dust	55	14	21
Sorghum straw	33	18	15
Soyabean stalks	34	25	20
Sugar cane bagasse	33	30	29
Sweet sorghum	23	14	11
Switch grass	37	29	19
Salix	41.5	22–25	25
Spruce	45	26	28
Wheat straw	30	24	18
Hardwood stem	40–55	34–40	18–25
Softwood stem	45–50	25–35	25–35
Grasses	25–40	35–50	10–30
Coastal Bermuda grass	25	35.7	6.4
Leaves	15–20	80–85	0–5
Paper	85–99	0–5	0–15
Newspaper	40–55	25–40	18–30
Waste paper from chemical pulp	60–70	10–20	5–10
Sorted refuge	60	20	20
Flax sheaves	35	24	22
Municipal solid waste	8–15	NA	24–29
Solid cattle manure	1.6–4.7	1.4–3.3	2.7–5.7
Swine waste	45	30–35	12

Bioconversion of lignocellulosic biomass

The bioconversion of lignocellulosics to ethanol consists of two main processes: hydrolysis of lignocellulosic carbohydrate to fermentable reducing sugars, and fermentation of the sugars to ethanol (Figure 2). The hydrolysis is usually catalysed by cellulase enzymes, and the fermentation is carried out by yeasts or bacteria. The factors that have been identified to affect the hydrolysis of cellulose include porosity (accessible surface area) of the waste materials, cellulose fibre crystallinity, and lignin and hemicellulose content (Alvira, Tomas-Pejo, Ballesteros, *et al.* 2010; Margeot, Hahn-Hagerdal, Edlund, *et al.* 2009; Mosier, Wyman, Dale, *et al.* 2005). The presence of lignin and hemicellulose in lignocellulosic materials makes the access of cellulase enzymes difficult, thus reducing the efficiency of the hydrolysis (Himmel, Ding, Johnsson, *et al.* 2007). Pretreatment of lignocellulosic biomass before hydrolysis can significantly improve the hydrolysis efficiency by the removal of lignin and hemicellulose, reduction of cellulose crystallinity, and increase in porosity (Kumar, Barette, Delwiche, *et al.* 2009; McMillan 1994; Mosier, Wyman, Dale, *et al.* 2005; Palmqvist and Hahn-Hagerdal 2000a, b; Sun and Cheng 2002).

Pretreatment of lignocellulosic biomass

The effect of pretreatment strategies of lignocellulosic materials has been well recognized for a long time (Alvira, Tomas-Pejo, Ballesteros, *et al.* 2010; Kumar, Barette, Delwiche, *et al.* 2009; Margeot, Hahn-Hagerdal, Edlund, *et al.* 2009; Mosier, Wyman, Dale, *et al.* 2005; Sanchez and Cardona 2008). The purpose of the pretreatment is to remove lignin and hemicellulose, reduce

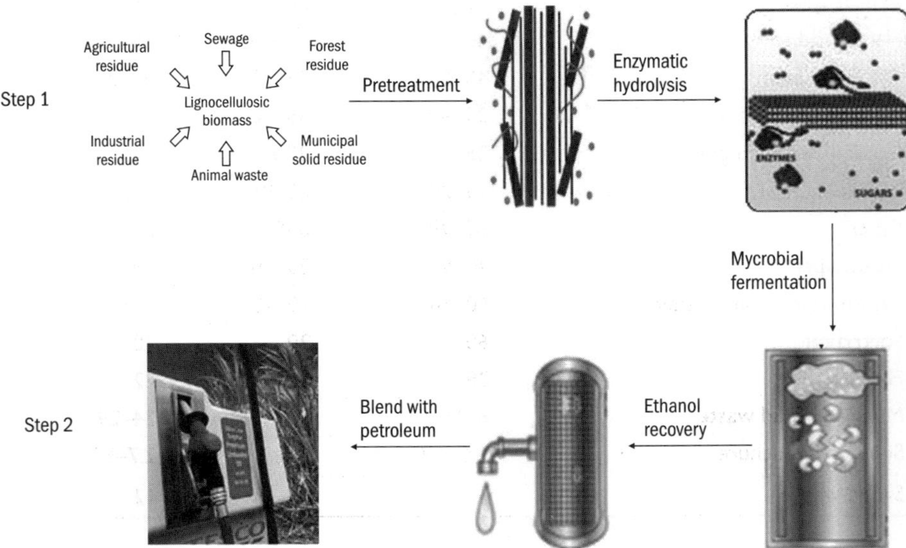

Figure 2 Schematic representation of the process of bioethanol production from lignocellulosic biomass

cellulose crystallinity, and increase the porosity of the materials. Therefore, the pretreatment strategies must meet the following requirements: (1) improving the formation of sugars or the ability to subsequently form sugars by enzymatic hydrolysis, (2) avoiding the degradation or loss of carbohydrate, (3) avoiding the formation of hydrolysis and fermentation inhibitory by-products, and (4) improving the cost-effectiveness of the process. Various physical, physico-chemical, chemical, and biological processes are used for the pretreatment of lignocellulosic materials.

Physical pretreatment

Mechanical comminution

Lignocellulosic residues can be pretreated by comminution through a combination of chipping, grinding, and milling to reduce cellulose crystallinity. This can be achieved depending on the final particle size of the material (10–30 mm after chipping and 0.2–2 mm after milling or grinding) (Sun and Cheng 2002). The milling process has been found to reduce the cellulose crystallinity and subsequently improve the digestibility of the lignocellulosic biomass efficiently. Different milling processes (ball milling, two-roll milling, hammer milling, colloid milling, and vibro energy milling) can be used to improve the enzymatic hydrolysis of lignocelullosic materials (Taherzadeh and Karimi 2008). The power requirement for mechanical comminution of agricultural materials depends on the final particle size and the waste biomass characteristics (Hendriks and Zeeman 2009; Mosier, Wyman, Dale, *et al.* 2005; Sun and Cheng 2002).

Extrusion

Extrusion process is a novel and promising physical pretreatment method for the conversion of biomass to ethanol. In extrusion, the materials are subjected to heating, mixing, and shearing, resulting in physical and chemical modifications during the passage through the extruder. Screw speed and barrel temperature are believed to disrupt the lignocellulose structure, causing defibrillation, fibrillation, and shortening of fibres, which, in turn, increases carbohydrate accessibility for enzymatic hydrolysis (Karunanithy, Muthukumarappan, and Julson 2008). The different bioreactor parameters must be taken into account to achieve the highest efficiency in the process. In recent studies, application of enzymes during extrusion process is being considered as a promising technology for ethanol production (Alvira, Tomas-Pejo, Ballesteros, *et al.* 2010).

Pulsed electric field pretreatment

Pulsed electric field (PEF) pretreatment involves the application of a short burst of high voltage to plant materials placed between two electrodes. PEF pretreatment can have serious effects on the structure of plant tissues. When a high-intensity, external electric field is applied, a critical electric potential is induced across the cell membrane, which leads to rapid electrical breakdown and local structural changes in cell membrane and cell wall. The electric field results in a significant increase in mass permeability and, in some cases, mechanical rupture of the plant tissue.

In biomass-to-fuel conversion, the biomass needs to be treated so that the cellulose in the plant fibres is exposed. Pretreatment with PEFs can facilitate this process. Using high-field strengths in the range of 5–20 kV/cm, plant cells can be significantly ruptured (Kumar, Barette, Delwiche, *et al.* 2009). By applying electric pulses with high-field strengths, PEF pretreatment can create permanent pores in the cell membrane and hence facilitate the entry of acids or enzymes that are used to break down the cellulose into its constituent sugars. In the case of the chemical modification of plant tissue, particularly in lignocellulose hydrolysis, appropriate chemicals might be needed to be transported into the tissue to aid in breakdown and digestion of the cell wall. Two advantages of PEF pretreatment are that it can be carried out at ambient conditions, and energy requirement is low because of very short pulse times (100 μs). Furthermore, the actual PEF process itself does not involve moving parts, so no complex instrument design is required. Recently, Kumar, Barette, Delwiche, *et al.* (2009) used this process for Switch grass and found considerable improvement in the enzymatic digestibility of the substrate.

Pyrolysis

Pyrolysis of lignocellulosic materials was carried out at high temperatures, above 300°C, under which the cellulose is decomposed to produce residual char and gaseous products. The dilute acid hydrolysis of residual char resulted in more than 80% conversion of cellulose to reducing sugars, comprising more than 50% glucose (Fan, Gharpuray, and Lee 1987). Moreover, the process efficiency can be enhanced under oxygen-rich conditions (Shafizadeh and Bradbury 1979) and also with the addition of some specific catalysts such as zinc chloride or sodium carbonate (Kumar, Barette, Delwiche, *et al.* 2009; Sun and Cheng 2002; Zwart, Boerrigter, and Drift 2006).

Ultrasound pretreatment

The effect of ultrasound waves on lignocellulosic biomass has been used for extracting hemicellulose, cellulose, and lignin (Sun and Tomkinson 2002). Several researchers have reported the enhanced saccharification of cellulose using ultrasonic pretreatment methods (Yachmenev, Condon, Klasson, *et al.* 2009). Higher enzymatic hydrolysis yields after ultrasound pretreatment could be due to the fact that cavitation effects caused by the introduction of ultrasound field into the enzyme–substrate suspension greatly enhance the transport of enzyme macromolecules towards the substrate surface. Furthermore, mechanical impacts, produced by the collapse of cavitation bubbles, provide an important benefit of opening up the surface of solid substrates for enzymatic action. The maximum effects of cavitation occur at 50°C, which is the optimum temperature for many enzymes (Yachmenev, Condon, Klasson, *et al.* 2009).

Chemical pretreatment

Acid hydrolysis

Mineral acids such as H_2SO_4 and HCl have been used to pretreat the lignocellulosic materials. Although concentrated mineral acids are powerful agents for cellulose hydrolysis, they are toxic, corrosive, and hazardous, and require reactors that are resistant to corrosion. Moreover, the recovery of concentrated acid is problematic enough to make the process economically feasible (Sivers and Zacchi 1995; Torget, Kim, and Lee 2000). Dilute acid hydrolysis has been successfully used for the pretreatment of lignocellulosic materials. Dilute H_2SO_4 pretreatment can lead to high reaction rates and significantly improve cellulose hydrolysis (Cara, Ruiz, Oliva, *et al.* 2008; Esteghlalian, Hashimoto, Fenske, *et al.* 1997; Rocha, Rodrigues, de Macedo, *et al.* 2009; Sun and Cheng 2002). Recently, the focus of dilute acid hydrolysis processes remained on operating under less severe conditions and achieving high yields of xylan to xylose conversion. This is necessary to achieve favourable overall process economics because of xylan, which accounts up to one-third of the total carbohydrate in many lignocellulosic materials (Gupta, Sharma, and Kuhad 2009). There are primarily two types of dilute acid pretreatment processes: high temperature, continuous-flow process for low solid load (5%–10%), and low temperature, batch process for high solid load (10%–40%). Although dilute acid pretreatment can significantly improve the cellulose hydrolysis, its cost is usually higher than some physico-chemical pretreatment processes such as steam explosion or ammonia fibre explosion/expansion (AFEX). Moreover, the neutralization of pH is necessary for the downstream enzymatic hydrolysis and fermentation processes.

Recently, a few organic acids, such as fumaric or maleic acids, are also used as alternatives to inorganic acids to enhance cellulose hydrolysis for ethanol production. These acids showed higher efficiency with less amount of furfural when compared with H_2SO_4 in terms of hydrolysis yields from wheat straw and formation of sugar degradation compounds during pretreatment (Kootstra, Beeftink, Scott, *et al.* 2009).

Alkaline hydrolysis

Alkaline hydrolysis is one of the critical methods used to pretreat the plant biomass; however, the effect of alkaline pretreatment method depends on the lignin content of the materials (Fan, Gharpuray, and Lee 1987; Kumar, Barette, Delwiche, *et al.* 2009; McIntosh and Vancov 2010; McMillan 1994; Mosier, Wyman, Dale, *et al.* 2005; Sun and Cheng 2002). The mechanism of alkaline hydrolysis is believed to be saponification of intermolecular ester bonds cross-linking xylan hemicelluloses and other components, for example, lignin and other hemicelluloses. Dilute NaOH treatment of lignocellulosic materials caused their swelling, leading to an increase in internal surface area, a decrease in the degree of polymerization and crystallinity, separation of structural linkages between lignin and carbohydrates, and disruption of the lignin structure (Carrillo, Lis, Colon, *et al.* 2005; Sun and Cheng 2002). Recently, Hu and co-workers used microwave- and radio-frequency-based dielectric heating in the alkali pretreatment of switch grass to enhance its enzymatic digestibility. In this strategy, switchgrass could be treated on a large scale at high solid loading with uniform temperature distribution (Hu, Wang, and Wen 2008; Hu and Wen 2008).

Ionic liquid pretreatment

The use of ionic liquids (ILs) as solvents for pretreatment of cellulosic biomass has recently received much attention. ILs are salts, typically composed of large organic cations and small inorganic anions, which exist as liquids at relatively low temperatures, often at room temperature. Their solvent properties can be varied by adjusting the anion and the alkyl constituents of the cation. The interesting properties of ILs include chemical and thermal stability, non-flammability, low vapour pressure, and a tendency to remain liquid in a wide range of temperatures (Hayes 2009). Because no toxic or explosive gases are released, ILs are also known as "green" solvents. Carbohydrates and lignin can be simultaneously dissolved in ILs with anion activity (for example, 1-butyl-3 methylimidazolium cation [C4mim]+) because ILs form hydrogen bonds, resulting between the non-hydrated chloride ions of the IL and the sugar hydroxyl protons in a 1:1 stoichiometry. As a result, the intricate network of non-covalent interactions among biomass polymers of cellulose, hemicellulose, and lignin is effectively disrupted while minimizing formation of degradation products. However, most of the work showing the effectiveness of ILs has been carried out using pure crystalline cellulose, and its applicability to a more complex combination of constituents in lignocellulosic biomass requires more extensive studies. The use of ILs has been shown in some lignocellulosic feedstocks such as straw (Li, Hi, and Xian 2009) or wood (Lee, Doherty, Linhardt, *et al.* 2009). The technology of ILs is under development; therefore, the commercial recovery methods have not been fully developed. In addition, techniques need to be worked out to recover hemicellulose and lignin from solutions after the extraction of cellulose (Hayes 2009).

Lime treatment

Lime pretreatment removes the lignin fraction from the polysaccharide fraction, thus making the remaining polysaccharides vulnerable to enzyme digestion (Kim and Holtzapple 2005; O'Dwyer, Zhu, Granda, *et al.* 2007). With regard to process operation, different conditions are used for different types of cellulosic materials: 100°C for 13 hours for corn stover, 150°C for 6 hours with 14 atm for poplar wood, and 100°C for 2 hours for switch grass (Mosier, Wyman, Dale, *et al.* 2005). The oxygenation of reaction mixture can greatly improve delignification, especially when treating lignin-rich woods.

The process of lime pretreatment involves slurrying the lime with water, spraying it onto the biomass, and storing the material in a pile for few weeks (Kim and Holtzapple 2005; Mosier, Wyman, Dale, *et al.* 2005). Recently, Park and co-workers developed a novel one-batch lime pretreatment process called "calcium capturing by carbonation (CaCCO)". The CaCCO process entraps the formed $CaCO_3$ in the reaction vessel throughout the conversion process, and no solid–liquid separation approach is used (Park, Shiroma, Al-Haq, *et al.* 2010).

Organosolv process

In the organosolv process, an organic or aqueous organic solvent mixture with inorganic acid catalysts (HCl or H_2SO_4) is used to break the internal lignin and hemicellulose bonds (Xu, Sun, Liu, *et al.* 2003; Zhao, Cheng, and Liu 2009). The solvents used in the process include methanol, ethanol, acetone, ethylene glycol, triethylene glycol, tetrahydrofurfuryl alcohol, and organic acids such as oxalic, acetylsalicylic, and salicylic acids (Itoh, Wada, Honda, *et al.* 2003; Kumar, Barette, Delwiche, *et al.* 2009; Sun and Cheng 2002; Xu, Sun, Liu, *et al.* 2003; Zhao, Cheng, and Liu 2009). A high yield of xylose is usually obtained with the addition of acid. Solvents used in the process are drained from the reactor, evaporated, condensed, and recycled to reduce the process cost. Removal of solvents from the system is necessary because the solvents may be inhibitory for the growth of organisms, enzymatic hydrolysis, and subsequent fermentation (Itoh, Wada, Honda, *et al.* 2003; Xu, Sun, Liu, *et al.* 2006; Zhao, Cheng, and Liu 2009).

Oxidative delignification

Pretreatment of sugar cane bagasse with hydrogen peroxide (H_2O_2) greatly enhanced its susceptibility to enzymatic hydrolysis. About 50% lignin and most of hemicellulose were solubilized by 2% H_2O_2 at 30°C within 8 hours, and 95% efficiency of glucose production from cellulose was achieved in the subsequent saccharification by cellulase at 45°C for 24 hours (Azzam 1989). Besides alkaline peroxide (Bjerre, Olesen, and Fernqvist 1996; Lissens, Klinke, Verstraete, *et al.* 2004; Martin, Thomsen, Hauggaard, *et al.* 2008), the chlorite oxidation and wet oxidation are also used as promising oxidative delignifying pretreatments. Bjerre, Olesen, and Fernqvist (1996) used wet oxidation and alkaline hydrolysis of wheat straw (20 g straw/l, 170°C, 5–10 minutes) and achieved 85% conversion yield of cellulose to glucose, whereas the sodium chlorite treatment yielded approximately 90% delignification in woody material (*Prosopis juliflora*) (Gupta, Sharma, and Kuhad 2009).

Ozonolysis

Ozone is used to degrade lignin and hemicellulose in many lignocellulosic materials. The degradation was essentially limited to lignin, and hemicellulose was slightly attacked, but cellulose was unaffected, which resulted in increased in vitro digestibility of the cellulosic substrates (Kumar, Barette, Delwiche, *et al.* 2009). According to Vidal and Molinier (1988), the ozone treatment removed 60% lignin from wheat straw, which in turn enhanced the enzymatic saccharification rate by five times. Ozonolysis pretreatment has the following advantages: (1) it effectively removes lignin, (2) it does not produce toxic residues for the downstream processes, and (3) the reactions are carried out at room temperature and pressure (García-Cubero, González-Benito, Indacoechea, *et al.* 2009). However, requirement of ozone in large quantity makes the process expensive and commercially unfeasible.

Physico-chemical pretreatment

Microwave pretreatment

Microwave-based pretreatment can be considered a physico-chemical process as both thermal and non-thermal effects are often involved. Pretreatments are carried out by immersing the biomass in dilute chemical reagents and exposing the slurry to microwave radiation for residence times ranging from 5 minutes to 20 minutes (Keshwani 2009). Preliminary experiments identified alkalis as suitable chemical reagents for microwave-based pretreatment (Zhu, Wu, Yu, *et al.* 2006). Among different alkalis used, sodium hydroxide has been found to be the most effective alkali reagent.

Ammonia fibre explosion and ammonia recycle percolation process

AFEX (explosion) is another type of physico-chemical pretreatment in which lignocellulosic materials are exposed to liquid ammonia at high temperature and pressure for certain time, and then the pressure is suddenly decreased (Lee, Jameel, and Venditti 2010; Teymouri, Laureano-Perez, Alizadeh, *et al.* 2005). The concept of AFEX is similar to steam explosion. In ammonia fibre/freeze expansion (AFEX) process, a 5%–15% ammonia solution flows through a column reactor that is packed with biomass at 1 ml/cm^2 for 14 minutes at temperature between 160°C and 180°C (Mosier, Wyman, Dale, *et al.* 2005). However, the AFEX process was not very effective for the biomass with high lignin content such as newspaper (18%–30% lignin) and aspen chips (25% lignin). Hydrolysis yield of AFEX-pretreated newspaper and aspen chips was reported as only 40% and below 50%, respectively (McMillan 1994).

Moreover, to reduce the cost and protect the environment, ammonia must be recycled after pretreatment. In an ammonia recovery process (ARP), aqueous ammonia (10–15 wt%) passes through the biomass at elevated temperatures (150–170°C), with a fluid velocity of 1 cm/min and a residence time of 14 minutes, after which the ammonia is withdrawn from the system by a pressure controller for recovery. In the ARP method, the ammonia is separated and recycled, and as the ammonia pretreatment does not produce inhibitors for the downstream biological processes, washing is not required (Galbe and Zacchi 2007; Margeot, Hahn-Hagerdal, Edlund, *et al.* 2009; Sun and Cheng 2002). Generally, AFEX and ARP processes are not differentiated in the literature, although AFEX is carried out in liquid ammonia and ARP is carried out in an aqueous ammonia solution (10%–15%). It is observed that the AFEX pretreatment simultaneously reduces lignin content and removes some hemicellulose and decrystallizes cellulose. The cost of ammonia, and especially of ammonia recovery, elevates the cost of AFEX pretreatment (Alvira, Tomas-Pejo, Ballesteros, *et al.* 2010; Kumar, Barette, Delwiche, *et al.* 2009; Mosier, Wyman, Dale, *et al.* 2005).

Autohydrolysis (steam explosion)

Steam explosion is the commonly reported method for lignocellulosic materials. The process causes hemicellulose degradation and lignin transformation due to high temperature, thus increasing the potential of cellulose hydrolysis (Boluda-Aguilar, García-Vidal, González-Castañeda, *et al.* 2010; Lee, Shi, Venditti, *et al.* 2009). Steam explosion pretreatment typically subjects lignocellulose to temperatures between 160°C and 260°C (corresponding pressures of 100–700 psia; psia = pounds per square inch absolute; atmospheric pressure is 14.5 psi) with saturated steam for a period of 10 seconds to several minutes, followed by a flashing process to explosively release the steam. This treatment results in an explosive disruption of the lignocellulose material, thus "opening up" the substrate to increase digestibility (Mosier, Wyman, Dale, *et al.* 2005). The factors that affect steam explosion pretreatment are residence time, temperature, chip size, and moisture content (Alvira, Tomas-Pejo, Ballesteros, *et al.* 2010; Duff and Murray 1996; Lee, Shi, Venditti, *et al.* 2009). The advantages of steam explosion pretreatment include low-energy requirement compared with mechanical comminution and no associated recycling or environmental costs (Kumar, Barette, Delwiche, *et al.* 2009; Sun and Cheng 2002). Limitations of steam explosion include destruction of a portion of the xylan fraction, incomplete disruption of the lignin–carbohydrate matrix, and generation of compounds that may be inhibitory to micro-organisms used in the fermentation processes (Chandel, Kapoor, Singh, *et al.* 2007b; Jurado, Prieto, Martínez-Alcalá, *et al.* 2009; Palmqvist and Hahn-Hagerdal 2000a, b).

Carbon dioxide explosion

Similar to steam and ammonia explosion pretreatment, the CO_2 explosion is also used for the pretreatment of lignocellulosic materials (Schacht, Zetzel, and Brunner 2008). It was hypothesized that CO_2 would form carbonic acid and increase the hydrolysis rate. The yields from CO_2 explosion of lignocellulosics were found to be relatively low compared with steam or ammonia explosion pretreatment (Kim and Hong 2001; Kumar, Barette, Delwiche, *et al.* 2009; Mosier, Wyman, Dale, *et al.* 2005; Zheng, Lin, and Tsao 1998). Zheng, Lin, and Tsao (1998) compared CO_2 explosion with steam and ammonia explosion and found that CO_2 explosion was more cost-effective than ammonia explosion and did not cause the formation of inhibitory compounds.

Liquid hot water pretreatment

Liquid hot water pretreatment is very similar to steam explosion, with the major difference being the explosive decompression of steam explosion pretreatment replaced by controlled cooling to keep water in the liquid phase throughout the process (Weil, Westgate, and Kohlmann 1994). This process has been shown to remove up to 80% of the hemicellulose and enhance the enzymatic digestibility of pretreated material in lignocellulosic feedstocks, such as corn stover (Mosier, Wyman, Dale, *et al.* 2005), sugar cane bagasse (Laser, Schulman, Allen, *et al.* 2002), and wheat straw (Pérez, Ballesteros, Ballesteros, *et al.* 2008). Pressured reactors are used to keep the water in the liquid state at high reaction temperatures. Termed as "uncatalysed solvolysis" by Mok and Antal (1992), various biomass samples have been pretreated with compressed liquid water. The liquid hot water pretreatment eliminates the use of expensive chemicals/catalysts to facilitate the hemicellulose depolymerization; subsequently, there is no need for neutralization or chemical recovery after the pretreatment. The resulting pretreated materials are reported to be highly amicable to the enzymatic saccharification step (Alvira, Tomas-Pejo, Ballesteros, *et al.* 2010; Lu and Mosier 2008; Mosier, Wyman, Dale, *et al.* 2005; Weil, Sarikaya, Lau, *et al.* 1998).

Control pH liquid pretreatment

An optimized and controlled pH liquid hot water pretreatment process will maximize the solubilization of hemicellulose fraction as liquid-soluble oligosaccharides while minimizing the formation of monomeric sugars. The minimization of complete hydrolysis to monosaccharides minimizes the subsequent degradation of these sugars to various aldehydes during pretreatment. By controlling the depolymerization of hemicellulose, the major xylose-containing product obtained is soluble xylan oligosaccharides (Weil, Sarikaya, Lau, *et al.* 1998). These oligosaccharides must be subsequently hydrolysed to fermentable sugars by enzymes or acids. In liquid hot water pretreatment, acetic acids and other organic acids are liberated as a result of the cleavage of O-acetyl and uronic acid substitution on hemicellulose by the action of water. These acids help catalyse further hemicellulose solubilization. They may also degrade the resulting monomeric sugars to furfural, which may have negative effects on the subsequent fermentation. With careful addition of a base or a buffer, controlled pH hydrothermolysis can maintain the pH of the liquid phase between 5.5 and 7.0 during the entire process, thus minimizing the formation of degradation products. Thus, controlled pH liquid hot water is a modified version of hot water pretreatment, which provides greater control of chemical reactions that occur during pretreatment (Lu and Mosier 2008; Weil, Sarikaya, Lau, *et al.* 1998).

Biological pretreatment

Pretreatment has become a necessity to maximize the hydrolysis of cellulosics and eventually the production of ethanol. The advantages of biological delignification of lignocellulosics over chemical and mechanical pretreatment methods include mild reaction conditions, use of non-toxic and non-corrosive chemicals, higher product yield, fewer side reactions, less energy demand, and less reactor resistance to pressure and corrosion (Kuhar, Nair, Kuhad, *et al.* 2008; Lee 1997; Sanchez 2009). In situ microbial delginification appears to be a plausible strategy to achieve improved depolymerization of hemicellulose and cellulose.

The white-rot fungi (WRF) are the most effective basidiomycetes for biological pretreatment as they degrade lignin more extensively and rapidly than any other known group of organisms (Eriksson 1993; Keller, Hamilton, and Nguyon 2003; Kuhad, Kuhar, Kapoor, *et al.* 2007; Kuhad, Singh, and Eriksson 1997). The common pattern of cellulose, hemicellulose, and lignin, but a preferential decomposition of lignin, may also occur (Eriksson 1993; Gupta, Mehta, Khasa, *et al.* 2010; Keller, Hamilton, and Nguyon 2003; Kuhad, Kuhar, Kapoor, *et al.* 2007; Kuhar, Nair, and Kuhad 2008; Kuhad, Singh, and Eriksson 1997). Some WRF have been reported to degrade lignin selectively, and this capability of select WRF can be exploited for delignification of lignocellulosics without affecting much of cellulose (Gupta, Mehta, Khasa, *et al.* 2010; Kuhar, Nair, and Kuhad 2008). Thus, select lignin-degrading WRF with comparatively low cellulose and xylanase activities could be of utmost importance for efficient delignification and eventually in the reduction of chemical and energy inputs for chemical or enzymatic hydrolysis of the substrate(s).

A few studies have been reported on the pretreatment of plant biomass with WRF for its effect on cellulose hydolysis. According to Hatakka (1983), 35% of the wheat straw is convertible to reducing sugars when pretreated with *Pleurotus ostreatus* for five weeks. Taniguchi, Suzuki, Watanabe, *et al.* (2005) also observed a similar conversion rate in rice straw pretreated with *P. ostreatus* for 60 days. Keller, Hamilton, and Nguyon (2003) observed a three- to five-fold improvement in the enzymatic cellulose digestibility in corn stover and found that biological pretreatment with *Coriolus versicolor* is effective in improving the yield of reducing sugars and also that it is feasible to produce reducing sugars from bamboo residues. Most of these fungal pretreatments have suffered because of long incubation periods. Therefore, to economize microbial pretreatment of lignocellulosics to improve the hydrolysis of carbohydrates to reducing sugars and to eventually improve ethanol yield, there is a need to test more and more basidiomycetous fungi for their ability to delignify the plant material quickly and efficiently. Recently, our group has demonstrated the potential of in situ pretreatment of *P. juliflora* with *Crenipellis* sp. RCK-1 before its chemical or enzymatic hydrolysis in increasing sugar yield and in turn producing ethanol as a biofuel (Gupta, Mehta, Khasa, *et al.* 2010; Kuhar, Nair, and Kuhad 2008).

Biological pretreatment in combination with other pretreatment technologies has also been studied (Balan, Sauca, Chundawat, *et al.* 2008; Itoh, Wada, Honda, *et al.* 2003). Itoh, Wada, Honda, *et al.* (2003) reported the production of ethanol by simultaneous saccharification and fermentation (SSF) from beech wood chips after bioorganosolvation pretreatments by ethanolysis with WRF, *Ceriporiopsis subvermispora*, *Dichomitus squalens*, *P. ostreatus*, and *C. versicolor*. The yield of ethanol obtained after pretreatment with *C. subvermispora* for eight weeks was 0.294 g/g of ethanolysis pulp (74% of theoretical) and 0.176 g/g of beech wood chips (62% of theoretical). The yield was 1.6 times higher than that obtained without the fungal treatments. The combined process enabled the separation of lignin, cellulose, and hemicellulose using only water, ethanol, and wood rot fungi. The biological pretreatments saved 15% of the electricity needed for ethanolysis. In another interesting approach, Balan, Sauca, Chundawat, *et al.* (2008) studied the effect of fungal conditioning of rice straw, followed by AFEX pretreatment and enzymatic hydrolysis. They reported that treating rice straw with WRF, *P. ostreatus*, followed by AFEX gave significantly higher glucan and xylan conversions and less severe AFEX conditions. The advantages of biological pretreatment method include low-energy requirement and mild environmental conditions (Sánchez 2009). However, the rate of hydrolysis in most biological pretreatment processes is very low.

Enzymatic hydrolysis

Enzymatic hydrolysis of cellulose is carried out by cellulase enzymes, which are highly specific. The products of the hydrolysis are usually reducing sugars, mainly glucose. The utility cost of enzymatic hydrolysis is low compared with acid or alkaline hydrolysis because enzyme hydrolysis is usually conducted at mild conditions (pH 4.8 and temperature 45–50°C), and it does not have a corrosion problem (Kuhad, Mehta, Gupta, *et al.* 2010). Both bacteria and fungi can produce cellulases for the hydrolysis of lignocellulosic materials. These micro-organisms can be aerobic or anaerobic, mesophilic or thermophilic. Bacteria belonging to genera *Clostridium, Cellulomonas, Bacillus, Thermomonospora, Ruminococcus, Bacteriodes, Erwinia, Acetovibrio, Microbispora*, and *Streptomyces* can produce cellulases, and among them, *Cellulomonas fimi* and *Thermomonospora fusca* have been studied extensively (Bisaria 1991; Duff and Murray 1996; Sun and Cheng 2002). Fungi that have been reported to produce cellulases include *Sclerotium rolfsii, P. chrysosporium* and species of *Trichoderma, Aspergillus, Schizophyllum, Fusarium*, and *Penicillium* (Duff and Murray 1996; Fan, Gharpuray, and Lee 1987; Kuhad, Manchanda, and Singh 1999; Sternberg 1976; Sun and Cheng 2002). Of all these fungal genera, *Trichoderma* has been the most extensively studied genera for cellulase production.

The cellulase system contains three major enzyme components: endoglucanase (EG; EC 3.2.1.4), cellobiohydrolase (CBH; EC 3.2.1.91), and β-glycosidase (EC 3.2.1.21). Evidence suggests that these enzymes act synergistically (Boraston, Bolam, Gilbert, *et al.* 2004; Din, Damude, Gilkes, *et al.* 1994; Gupta, Sharma, and Kuhad 2009; Teeri, Koivula, Linder 1998). The exoglucanase (CBH) acts on the ends of the cellulose chain and releases β-glucosidase as the end-product; the EG randomly attacks the internal O-glycosidic bonds, resulting in glucan chains of different lengths; and the β-glycosidases act specifically on the β-cellobiose disaccharides and produce glucose (Beguin and Aubert 1994; Kuhad, Manchanda, and Singh 1999; Kuhad, Mehta, Gupta, *et al.* 2010) (Figure 3). Structurally, cellulases typically have two separate domains: a catalytic domain (CD) and a cellulose-binding module (CBM), which are linked by a flexible linker region. The CBM is composed of approximately 35 amino acids, and the linker region is rich in serine and threonine (Divne, Stahlberg, Teeri, *et al.* 1998). The nature of the lignocellulosic substrate changes during the course of enzymatic hydrolysis (Wang, Zhang, Gao, *et al.* 2006). Initially, amorphous non-crystalline regions are attacked, because they are more accessible and easier to be hydrolysed. After the initial hydrolysis stage, as the percentage of crystalline regions in the substrate rises, the enzymatic hydrolysis rate falls rapidly. Such observed kinetic behaviour correlates to the high-crystalline nature of the bulk of cellulose fraction, and inaccessibility of the glycosidic bonds is the key rate-limiting factor for low enzymatic hydrolysis. Besides the recalcitrance of the

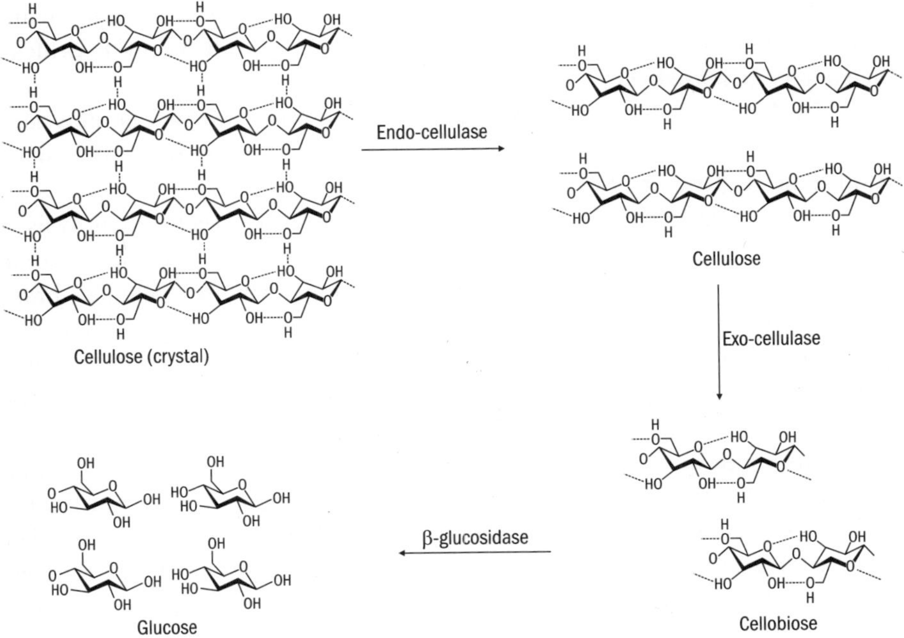

Figure 3 Enzymatic saccharification of cellulose

substrate, there are a few factors that also limit cellulase efficiency during the hydrolysis process, which include end-product inhibition, thermal deactivation of the native protein, non-specific binding to lignin (Yang and Wyman 2004), and irreversible adsorption of the enzymes to the heterogeneous substrate (Taniguchi, Suzuki, Watanabe, *et al.* 2005). Mechanistically, the reactions catalysed by cellulases are suggested to involve general acid–base catalysis by a carboxylate pair at the enzyme-active site. One residue acts as a general acid and protonates the oxygen of the O-glycosidic bond; at the same time, the other residue acts as a nucleophile. Depending on the distance between the two carboxylic groups, either inverting (~10 Å distances) or retaining (~5 Å distances) mechanisms are observed in cellulases (Withers 2001). Besides cellulases, hemicellulases are another group of polysaccharide-degrading enzymes that are specific to the hemicellulose substrate. As a heterogeneous, branched polymer, hemicellulose requires enzyme activities specific to as many as 21 different bonds (Collins, Gerdey, and Feller 2005; Polizeli, Rizzatti, Monti, *et al.* 2005; Shallom and Shoham 2003). Thus, a consortium of hemicellulases is needed for a complete breakdown of hemicellulose. Endo-1,4-xylanases are needed for hydrolysing the backbone 1,4-β-linked xylose residues; acetylxylan esterases participate in cleavage of the acetyl ester bonds; and β-D-xylosidases are used for the hydrolysis of xylan oligomers through exotype attack (Kapoor, Nair, and Kuhad 2008). Similar to cellulases, xylanases have separate catalytic domains and carbohydrate-binding modules (Divne, Stahlberg, and Teeri 1998; Teeri 1997; Teeri, Koivula, Linder, *et al.* 1998). Synergism and concerted action among xylanases enhance the effectiveness in heteropolymeric xylan hydrolysis. It has also been reported that β-xylosidases remove the short-chain oligosaccharides, minimizing the end-product inhibition of exoxylanases; thus, overall xylan hydrolysis efficiency is increased (Hu, Wang, and Wen 2008; Polizeli, Rizzatti, Monti, *et al.* 2005). Also, by adding acetylxylan esterases, acetic acids will be liberated, and a less acetylated xylan is exposed for greater accessibility to endoxylanase action (Polizeli, Rizzatti, Monti, *et al.* 2005).

To reduce the enzyme cost in the production of fuel ethanol from lignocellulosic biomass, two aspects are widely addressed: optimization of the cellulase production and development of a more efficient cellulase-based catalysis system. Additionally, protein engineering and directed evolution are powerful tools that can facilitate the development of more efficient thermophilic cellulases (Baker, McCarley, Lovettt, *et al.* 2005). Recycling and reusage of the enzymes are also a part of an attractive methodology to reduce enzymatic hydrolysis costs (Gregg, Boussaid, and Saddler 1998; Lee, Yu, and Saddler 1995; Mosier, Wyman, Dale, *et al.* 2005; Ramos, Beruil, and Saddler 1993; Singh, Kumar, and Schugerl 1991; Sun and Cheng 2002). The recovery of enzymes is largely influenced by the adsorption of the enzymes onto the substrate, especially to lignin. Another constraint in the recycling of the enzymes is enzyme inactivation. There are several strategies to recover and reuse the cellulases. The filtrate obtained after complete hydrolysis of the cellulose fraction can be concentrated by ultrafiltration to remove sugars and other small compounds that may inhibit the action of the enzymes (Tu, Chandra, and Saddler 2007). Another method for recycling enzymes is immobilization, which enables separation of the enzymes from the process flow. The principle of immobilization is to fixate the carbohydrolytic enzymes onto a solid matrix, either by adsorption or by grafting (Dourado, Bastos, Mota, *et al.* 2002; Mosier, Wyman, Dale, *et al.* 2005). The recycling techniques are mostly tested at laboratory scale. Therefore, the ability to scale up the techniques, the robustness, and feasibility still need to be demonstrated.

Fermentation

To meet the increased demand for fuel and achieve decreased air pollution and greenhouse effect, development of alternative fuels is required, which can at least partially substitute the declining petroleum resources. Ethanol fuel has almost all the advantages of liquid hydrocarbon fuels.

Many countries around the world are taking a lot of interest in ethanol as an alternative fuel, with the US and Brazil currently leading the way. In 2005, Brazil produced 4.8 million gallons of ethanol and the US produced 3.9 billion gallons; thus, these countries are the two largest ethanol producers in the world, accounting for over 90% of the total world production. There are a limited number of studies on ethanol markets, as the industry has experienced a boom only in the past few years (Gallagher, Schamel, Shapouri, *et al.* 2006). The recent increase in crude oil and gasoline prices has also opened a new market for ethanol as a fuel extender. Ethanol production in different continents is in the following ranges (billion litres/year): America (22.3), Asia (5.7), Europe (4.6), Africa (0.5), and Oceania (0.2) (Demirba 2007).

Brazil is one of the first countries to promote ethanol widely through its National Alcohol Program, which was launched in the late 1975 in response to high oil prices and declining sugar prices. According to von Lampe, Zygalsky, Nagibzadeh, *et al.* (2006), the cost of ethanol from sugar cane in Brazil at 83¢ per gallon of fuel is lower than the cost from corn in the US, at $1.09 per gallon. Besides the higher cost of production, there are additional costs in the US associated with transporting ethanol from the production locations in the Mid-west to major population areas, particularly in the coastal regions. This has led to an increase in the competitiveness of Brazilian ethanol imports despite steep tariffs in the US. Pure ethanol (100% ethanol) is used in approximately 40% of the cars in Brazil. The remaining vehicles use blends of 24% ethanol with 76% gasoline. Brazil consumes nearly 4 billion gallons of ethanol annually (Gallagher, Schamel, Shapouri, *et al.* 2006). According to a recent report of the Ministry of Mines and Energy, Brazilian ethanol production was 27 billion litres in 2009 (IBGE 2009). However, an expansion of the ethanol production up to 104 billion litres is the target till 2025 (Soccol, Vandenberghe, Medeiros, *et al.* 2010).

Ethanol is produced commercially by chemical synthesis and fermentation, and practically, majority of the first-generation ethanol is produced by the fermentation of molasses, cereal grains, and other high starch and sugar containing materials. The current molasses- and corn-starch-based technology may not be practical because molasses, cereal grains, and high starch materials as first-generation substrates are not sufficient even to meet the feed and food demand not only in our country but also in majority of other countries. The potential sources of low-cost second-generation ethanol are other renewable materials like crop residues, grasses, saw dust, wood chips, and solid waste. Especially in developed countries, considerable efforts have been focused on the microbial production of ethanol from sugars obtained from hydrolysis of lignocellulosics, whereas research in India, especially on alcohol production from lignocellulosics, has recently picked up momentum. Therefore, it is necessary to make a concerted effort to develop an economically viable process for the bioconversion of lignocellulosic feedstocks into ethanol.

At present, the process for producing ethanol from second-generation lignocellulosic feedstock is available, which follows largely the same path as the process for producing ethanol from first-generation feedstocks such as molasses, sugar beet, and other starch-rich materials. However, major differences between ethanol production from the first-generation feedstocks (molasses and starch-rich raw material) and second-generation feedstocks are as follows.

1. Pretreatment of the feedstock.
2. The release of large amounts of inorganic waste as a result of pretreatment, suggesting thereby the process is not environment-friendly.
3. Detoxification of fermentation inhibitory compounds from sugar syrups.
4. Fermentation of pentose sugars to ethanol.
5. High cost of cellulase production because of its low production level.
6. High cost of ethanol production.

Ethanol fermentation is a biological process in which sugars are fermented by micro-organisms to produce ethanol and CO_2. As compared with starch and molasses, the fermentation of lignocellulosic hydrolysates is a complex process. There are two major steps of ethanol fermentation from lignocellulosic biomass, that is, chemical pretreatment and enzymatic hydrolysis. Of these, the enzymatic hydrolysis of biomass releases the hydrolysate that contains sugars, but the acid hydrolysis yields not only pentose and hexose sugars but also a few fermentation inhibitors, such as furans and phenolics. Thus, to make the acid hydrolysates fermentable, one or a few of several detoxification methods are to be used such as overliming, ion-exchange adsorption, activated carbon adsorption, solvent extraction, or steam stripping, and enzymatic (laccase) treatments, which can improve the fermentability of an acid hydrolysate (Chandel, Kapoor, Singh, *et al.* 2007a; Mosier, Wyman, Dale, *et al.* 2005; Olsson and Hahn-Hägerdal 1996; Palmqvist and Hahn-Hagerdal 2000a, b).

A variety of micro-organisms ranging from bacteria, fungi, and yeasts are known to ferment hexose sugars (Table 3); however, the most common and efficient microbes used for hexose fermentation are *Saccharomyces cerevisiae* and *Zymomonas mobilis* (Hahn-Hagerdal, Karhumaa, Fenske, *et al.* 2007). The general requirements of an organism for ethanol production from lignocellulosic hydrolysate are that it should give a high ethanol yield and high productivity, should have high tolerance against inhibitors, should be able to ferment at low pH, and should be able to withstand high ethanol concentrations.

Various studies have been carried out using *S. cerevisiae* for the fermentation of lignocellulosic hydrolysates. An enzymatic hydrolysate of alfa-alfa when fermented with *S. cerevisiae* consumed more than 98% sugars and led to 85% fermentation efficiency with ethanol productivity of 1.3 g/l/h (Belkacemi, Tureotte, Savoic, *et al.* 1997). While as per another

Table 3 List of micro-organisms that can ferment hexose sugars to ethanol

Organism	Reference	Organism	Reference
Candida shehatae	Abbi, Kuhad, and Singh (1996a)	*Rhizopus miehei*	Millati, Edebo, and Taherzadeh (2005)
Fusaruium sporium	Mamma, Christakopoulos, Koullas, *et al.* (1995)	*Rhizomucor pusillis*	Millati, Edebo, and Taherzadeh (2005)
Kloeckera apiculata	Aguilera, Peinado, Millan, *et al.* (2006)	*Saccharomyces cerevisiae*	Kuhad, Mehta, Gupta, *et al.* (2010)
Kluyeromyces marxianus	Ballesteros, Oliva, Negro, *et al.* (2004)	*Saccharomyces bayarus*	Belloch, Orlic, Barrio, *et al.* (2008)
Mucor indicus	Abdenifar, Karimi, and Khanahmadi (2009)	*Saccharomyces paradoxus*	Belloch, Orlic, Barrio, *et al.* (2008)
Mucor hiemalis	Millati, Edebo, and Taherzadeh (2005)	*Saccharomyces kudriazevii*	Belloch, Orlic, Barrio, *et al.* (2008)
Mucor corticolous	Millati, Edebo, and Taherzadeh (2005)	*Saccharomyces cariocanus*	Belloch, Orlic, Barrio, *et al.* (2008)
Neurospora crassa	Mamma, Christakopoulos, Koullas, *et al.* (1995)	*Saccharomyces mikatae*	Belloch, Orlic, Barrio, *et al.* (2008)
Pachysolen tannophillus	Abbi, Kuhad, and Singh (1996a)	*Saccharomyces pastorianus*	Belloch, Orlic, Barrio, *et al.* (2008)
Pichia stipitis	Gupta, Sharma, and Kuhad (2009)	*Schizosaccharomyces pombe*	Hu, Bai, and An (2005)
Pichia membranifaciens	Aguilera, Peinado, Millan, *et al.* (2006)	*Terulospora del-bruecki*	Aguilera, Peinado, Millan, *et al.* (2006)
Rhizopus oryzae	Abdenifar, Karimi, and Khanahmadi (2009)	*Zymomonas mobilis*	

report, the enzymatic hydrolysate (180 g/l) of washed steam exploded oak chips when used for continuous fermentation with *S. cerevisiae*, an ethanol concentration of 77 g/l, with an ethanol productivity of 16.9 g/l/hour and ethanol yield of 0.43 g/g, was obtained (Lee, Lee, Shin, *et al.* 1999). Wang, Xu, and Zhao (2004) reported ethanol production of 41–46 g/l from various monomeric and oligomeric sugars, that is, glucose (85 g/l), fructose (91.1 g/l), and sucrose (96.6 g/l), using *S. cerevisiae*. Later on Chen, Xia, and Xue (2007) used fed-batch enzymatic saccharification strategy to achieve 110 g/l sugar concentration, and when this hydrolysate was fermented with *S. cerevisiae*, almost 95.3 g/l sugar was consumed to produce 45.7 g/l ethanol, with an ethanol yield of 94%. In another study, enzymatic hydrolysate of acid- and alkali-treated cashew apple bagasse on fermentation with *S. cerevisiae* produced 20.0 g/l and 8.2 g/l ethanol, with an ethanol productivity of 3.33 g/l/hour and 2.7 g/l/hour, respectively (Rocha, Rodrigues, de Macedo, *et al.* 2009). Recently, our group has achieved an ethanol yield of 0.48 g/g from the enzymatic hydrolysate of pretreated *P. juliflora* and *Lantana camara* containing 36.5 g/l and 37.5 g/l sugars, respectively (Gupta, Sharma, and Kuhad 2009; Kuhad, Gupta, Khasa, *et al.* 2010).

Besides the enzymatic hydrolysates, several studies have been carried out to use *S. cerevisiae* to utilize the hexose sugars present in the acid hydrolysate of lignocellulosic substrates. Brandberg, Sanandaji, Gustafsson, *et al.* (2005) fermented a non-detoxified dilute acid wood hydrolysate using *S. cerevisiae* ATCC 96581 under continuous mode (with or without cell recycling) and observed that 99% of the sugars were converted at a cell concentration of 6 g/l, leading to an ethanol concentration of 17 g/l and a productivity of 1.6 g/l/hour. To improve productivity, three different types of cell retention processes, that is, cross-flow filtration with recirculation, sedimentation, and immobilization in calcium alginate, were evaluated. It was observed that when 75% of the cells were retained by filtration, the hexose conversion increased to 94%, and the ethanol production rate was 2.3 g/l/hour. Comparable results were achieved with cell recirculation by a settler and immobilization. When the dilution rate was increased to 0.2 per hour, neither filtration nor sedimentation could prevent the washout of the cells. On the other hand, culture immobilization increased the ethanol productivity to 3.5 g/l/hour, although hexose conversion dropped (Brandberg, Karimi, Taherzadeh, *et al.* 2007).

Taherzadeh, Millati, and Niklasson (2001) used a dilute acid hydrolysate supplemented with defined mineral media for the fermentation with *S. cerevisiae* CBS 8066 immobilized in Ca-alginate beads and observed that 79%–86% of glucose was consumed with 0.45 g/g and 0.47 g/g ethanol yield. Further, to improve the ethanol productivity, the same group immobilized *S. cerevisiae* cells in different bioprocesses such as continuous stirred tank reactor (CSTR), fluidized bed bioreactor (FBBR), and combined CSTR and FBBR (Purwadi and Taherzadeh 2008). In CSTR operated at dilution rates between 0.22 per hour and 0.86 per hour, conversion of an initial glucose concentration decreased from 100% to 77%, but productivity increased from 2.1 g/l/hour to 6.6 g/l/hour. If FBBR was connected to the CSTR, glucose conversion was higher than 99%. In contrast, a single FBBR gave a higher glucose conversion of 92% at a dilution rate of 0.86 per hour and a higher ethanol productivity of 7.4 g/l/hour (Purwadi and Taherzadeh 2008).

Besides entrapment, encapsulated *S. cerevisiae* cells were also used for fermentation, as encapsulation has the advantages of checking the diffusion of nutrients as well as the cell leakage. Talebnia and Taherzadeh (2006) encapsulated *S. cerevisiae* CBS 8066 cells and observed that ethanol productivity increased with increasing dilution rate from 1.1 g/l/hour to 4.2 g/l/hour, and cell viability was high under all conditions ranging between 77% and 90%. The flocculating *S. cerevisiae* strain was also reported for the fermentation of acid hydrolysates (Purwadi, Brandberg, and Taherzadeh 2007; Tang, An, Liu, *et al.* 2006). As an additional approach, SSF has also been used for improved ethanol production. In the SSF process, the stages are virtually the same as in separate hydrolysis and fermentation systems, except that both are performed in the same reactor. It has been shown that SSF reduces the processing time, which, in turn, leads to an increase in the production of ethanol (Alfani, Gallifuoco, Saporosi, *et al.* 2000; Ballesteros, Ballesteros, Cabañas, *et al.* 1991; Soderstrom, Galbe, and Zacchi 2005; Wyman 1994). During the SSF process, the liberated glucose is quickly converted to ethanol by fermenting micro-organisms, the inhibition of cellulase by the reaction end-products is reduced, and this single fermenter process eliminates a portion of the investment cost by reducing the number of fermenters required for start-up. In addition, the SSF process reduces the contamination risk due to the presence of ethanol in the medium (Banat, Nigam, Singh, *et al.* 1998; McMillan, Newman, Templeton, *et al.* 1999; Ohgren, Bura, Lesnicki, *et al.* 2007; Soderstrom, Galbe, and Zacchi 2005). However, the performance of the SSF process is limited by the different optimum temperatures for enzymatic saccharification and microbial fermentation. The optimum temperature for cellulases is usually in the range of 40–50°C, whereas the optimum temperature of ethanol production for the most common ethanologenic yeast, *S. cerevisiae*, is between 30°C and 37°C (Suryawati, Willkins, Bellmer, *et al.* 2008). Hence, the co-existence of two process conditions for microbe and enzyme hinders the efficacy of the process.

In order to overcome this notable discrepancy, thermo-tolerant yeast strains have been used in high temperature SSF processes (Suryawati, Wilkins, Bellmer, *et al.* 2008). In such cases, the performance of SSF with a thermo-tolerant yeast at temperatures closer to the optimum temperatures for cellulase activity could enhance saccharification, reduce the operation time, and reduce cellulase dosage (Suryawati, Wilkins, Bellmer, *et al.* 2008). Szczdodrak and Targonski (1988) tested a total of 58 yeast strains belonging to 12 different genera and capable of growing and fermenting sugars at temperatures of 40–46°C. They selected several strains belonging to the genera *Saccharomyces*,

Kluyveromyces, and *Fabospora* in view of their capacity to ferment glucose, galactose, and mannose at 40°C, 43°C, and 46°C, respectively. Ballesteros, Ballesteros, Cabañas, *et al.* (1991) studied different treatments to improve the thermotolerance of some species belonging to the genera *Saccharomyces* and *Kluyveromyces,* and the best results were obtained with *Kluyveromyces marxianus* LG. Bollók, Réczey, Zacchi (2000) also used *Kluveromyces* strain in SSF experiments of softwood, and almost 70% ethanol conversion was obtained. Harikrishna, Reddy, and Chowdary (2001) have reported final ethanol concentrations of 2.0%–2.5% (w/v) in 72-hour SSF of lignocellulosic waste with thermo-tolerant yeast at 10% (w/v) initial substrate concentration. Ballesteros, Oliva, Negro, *et al.* (2004) used the mutated *K. marxianus* strain CECT 10875 and achieved an SSF yield of 50%–72% in 72–82 hours using various lignocellulosic feedstocks (poplar and eucalyptus, *Sorghum* sp. bagasse, wheat straw, and *Brassica carinata* residue). Furthermore, Ohgren, Bura, Lesnicki, *et al.* (2007) compared the separate hydrolysis and fermentation (SHF) and SSF in steam-exploded corn stover and observed that SSF gave a 13% higher overall ethanol yield than SHF. Recently, a respiratory-deficient mutant of the thermo-tolerant yeast *Candida glabrata* (Cgrd1) was subjected to ethanol production by high temperature SSF under aerobic conditions to achieve maximum ethanol (17.0 g/l) within 48 hours at 66.6% of its theoretical yield and with 0.35 g/l hour productivity (Watanabe, Nakamura, and Shima 2010).

In order to economize the ethanol production, both pentose and hexose sugars must be converted to ethanol, but even the most promising fermenting microbes do not ferment pentoses. The yeast species identified so far for the pentose fermentation are *Candida shehatae*, *Pichia stipitis*, and *Pachysolen tannophilus* (Abbi, Singh, and Kuhad 1996a, b; Hahn-Hagerdal, Karhumaa, Fenske, *et al.* 2007; Mosier, Wyman, Dale, *et al.* 2005; Palmqvist and Hahn-Hägerdal 2000a; Talbenia, Pourbafrani, Lundin, *et al.* 2008). Out of 20 yeast strains screened by Kuhad and co-workers (2010), all the 20 strains were able to metabolize the pentose sugars, whereas only 10 of them were positive for C5 fermentation, and of these 10 yeasts, *Pichia, Pachysolen*, and *Candida* sp. were found to efficiently convert the pentose sugars to ethanol (Kuhad 2010). Some other micro-organisms that can ferment pentose sugars are *Clostridium* sp., *Klebsciella* sp., *Lactobacillus* sp., *Aeromonas hydrophila, Rhizopus orizae, Fusarium oxysporum*, and *Neurospora crassa* (Chaillou, Pouwells, and Postma 1999; Dien, Cotta, and Jeffries 2003; El Kanouni, Zerdani, Zaafa, *et al.* 1998; Hahn-Hagerdal, Karhumaa, Fenske, *et al.* 2007; Millati, Edebo, and Taherzadeh 2005; Ruiz, Romero, Moya, *et al.* 2007; Sreenath, Koegel, Modes, *et al.* 1999; Vasudevan, Isak, and Ricardo 2007) (Table 4). To date, several studies have been carried out on the fermentation of xylose-rich hydrolysates from various lignocellulosic materials, and a summary of recent studies is shown in Table 5.

Several strategies have been used for the enhanced conversion of pentose sugars into ethanol. Grootjen, Jansen, Vanderlans, *et al.* (1991) applied a

Table 4 List of micro-organisms that can ferment pentose sugars

Bacteria	Reference	Fungi and yeasts	Reference
Clostridium acetobutylicum	El Kanouni, Zerdani, Zaafa, *et al.* (1998)	*Candida shehatae*	Abbi, Kuhad, and Singh (1996a)
Clostridium thermosellum	Herrero and Gomez (1980)	*Candida boidinii*	Vandeska, Amartey, Kuzmanova, *et al.* (1996)
Escherichia coli	Yomano, York, and Ingram (1998)	*Pichia stipitis*	Gupta, Sharma, and Kuhad (2009)
Klebsciella oxytocea	Ingram, Aldrich, Borges, *et al.* (1999)	*Pachysolen tannophilus*	Schneider, Wang, Chan, *et al.* (1981)
Lactobacillus pentoaceticus	Chaillou, Pouwells, and Postma (1999)	*Rhizopus orizae*	Millati, Edebo, and Taherzadeh (2005)
Lactobacillus casei	Roukas and Kotzekidou (1998)	*Mucor corticolous*	Millati, Edebo, and Taherzadeh (2005)
Lactobacillus xylosus	Sreenath, Koegel, Moldes, *et al.* (1999)	*Mucor hiemalis*	Millati, Edebo, and Taherzadeh (2005)
Lactobacills pentosus	Chaillou, Pouwells, and Postma (1999)	*Mucor indicus*	Millati, Edebo, and Taherzadeh (2005)
Lactobacillus plantanum	Sreenath, Koegel, Moldes, *et al.* (1999)	*Fusarium oxysporum*	Jeffries and Jin (2004)
Bacteriodes polygramatis	Patel (1984)	*Neurospora crassa*	Deshpande, Keskar, Mishra, *et al.* (1986)
Bacillus macerans	Dien, Cotta, and Jeffries (2003)	*Paecilomyces* sp. NF1	Mountfort and Rhodes (1991)
Bacillus polymyxa	Singh and Mishra (1993)		
Aeromonas hydrophila	Singh and Mishra (1993)		
Thermoanaerobacter ethanolicus	Singh and Mishra (1993)		

three reactor system with *P. stipitis* in which they adjusted the residence time such that glucose was depleted in the last reactor and optimal pentose conversion took place. In contrast, when free and Ca-alginate cells of *C. shehatae* NCIM 3501 were used to ferment the acid hydrolysate of rice straw, ethanol yield of 0.42 g/g and 0.5 g/g was observed, respectively. To further improve the productivity, cell recycling was applied, and an ethanol yield of 0.47 g/g was obtained (Abbi, Kuhad, and Singh 1996a). Interestingly, a long-term adaptation strategy for *P. stipitis* was also tested by using a non-detoxified dilute acid hydrolysate of wheat straw, which in turn resulted in an improved ethanol yield (Nigam 2001a, b, c). De Bari, Cuna, Nanna *et al.* (2004) reported the fermentation of detoxified

Table 5 Summary of the recent work on lignocellulose hydrolysate fermentation using non-recombinant strains of micro-organisms

Substrate	Organism	Sugar (g/l)	Ethanol (g/l)	Ethanol yield (g/g)	Reference
Corn cob	*Pichia stipitis*	30	10.4	0.34	Saracoglu and Arslan (2000)
Cashew apple bagasse	*Saccharomyces cerevisiae*	50	20	0.4	Rocha, Rodrigues, de Macedo, *et al.* (2009)
Cassava waste	*Saccharomyces cerevisiae*	38	8.73	0.23	Raman and Pothiraj (2008)
Corn stover	*Pichia stipitis*	40	15.92	0.4	Zhu, Sathitsuksanoh, Vinzant, *et al.* (2009)
Corn stover	*Pichia stipitis*	60	25	0.42	Agbogbo, Coward-Kelly, Torry-Smith, *et al.* (2006)
Corn stover	*Pichia stipitis*	40	15	0.37	Agbogbo and Wenger (2007)
Newspaper	*Saccharomyces cerevisiae*	14.64	5.64	0.39	Kuhad, Mehta, Gupta, *et al.* (2010)
Lantana camara	*Saccharomyces cerevisiae*	36.5	17	0.48	Kuhad, Gupta, Khasa, *et al.* (2010)
Poplar	*Pichia stipitis*	39	12	0.31	Fenske, Hashimoto, and Penner (1998)
Prosopis juliflora	*Pichia stipitis*	18	7.1	0.39	Gupta, Sharma, and Kuhad (2009)
Prosopis juliflora	*Saccharomyces cerevisiae*	40	18	0.45	Gupta, Sharma, and Kuhad (2009)
Red oak spent sulfite liquor	*Pichia stipitis*	49	20.2	0.41	Nigam (2001b)
Red oak wood chips	*Pichia stipitis*	36	14.5	0.4	Nigam (2001a)
Rice straw	*Pichia stipitis*	15	6	0.4	Moniruzzaman (1995)
Rice straw	*Candida shehatae*	20	9	0.45	Abbi, Kuhad, and Singh (1996a)
Rice straw	*Pichia stipitis*	33	14.9	0.45	Huang, Lin, Guo, *et al.* (2009)
Secondary fibre fines	*Saccharomyces cerevisiae*	5.4	2.12	0.39	Jefferies and Schartman (1999)
Steam exploded alkali-treated bagasse	*Saccharomyces cerevisiae*	32	14.8	0.45	Lignocellulose Biotechnology Laboratory
Steam exploded alkali-treated rice straw	*Saccharomyces cerevisiae*	36	14.8	0.41	Lignocellulose Biotechnology Laboratory

Contd...

Table 5 Contd...

Substrate	Organism	Sugar (g/l)	Ethanol (g/l)	Ethanol yield (g/g)	Reference
Steam exploded bagasse	*Saccharomyces cerevisiae*	25	10	0.4	Lignocellulose Biotechnology Laboratory
Sugar cane bagasse	*Candida shehatae*	30	8.67	0.29	Chandel, Kapoor, Singh, *et al.* (2007a)
Sugar cane bagasse	*Pachysolen tannophilus*	63.5	19	0.34	Cheng, Cai, Zhang, *et al.* (2008)
Sugar cane bagasse	*Saccharomyces cerevisiae*	79	30	0.38	Vásquez, da Silva, de Souza, *et al.* (2007)
Sugar cane bagasse k-21	*Saccharomyces cerevisiae*	45	19	0.42	Lignocellulose Biotechnology Laboratory
Sugar maple	*Pichia stipitis*	35	12.4	0.35	Stoutenberg, Perrotta, Amidon, *et al.* (2008)
Sunflower seed hull	*Pichia stipitis*	34	11	0.32	Okur and Saracoglu (2008)
Sunflower stalks	*Saccharomyces cerevisiae*	40	17.2	0.43	Sharma, Kalra, and Kocher (2004)
Switch grass	*Pichia stipitis*	39	14	0.36	Fenske, Hashimoto, and Penner (1998)
Wheat straw	*Pichia stipitis*	52	22.3	0.43	Nigam (2001c)

hydrolysate (45 g/l of glucose and 12 g/l of xylose) with *P. stipitis,* and *S. cerevisiae* immobilized in calcium alginate resulted in almost 80% of the theoretical ethanol yield. Furthermore, Kuhad and co-workers used various detoxification strategies for the acid hydrolysate of sugar cane bagasse and obtained 0.22–0.48 g/g ethanol yield (Chandel, Kapoor, Singh, *et al.* 2007a). Recently, *P. stipitis* NCIM 3498 was used to ferment the detoxified acid hydrolysates of two lignocellulosic feedstocks *(P. juliflora* and *L. camara*), which resulted in an ethanol yield ranging 60%–70% of the theoretical yields (Gupta, Sharma, and Kuhad 2009; Kuhad, Gupta, Khasa, *et al.* 2010).

Although the pentose fermenting yeast can ferment pentose sugars, the commercial exploitation of these microbes for ethanol production is limited because of their low ethanol tolerance, slow rates of fermentation, difficulty in controlling the rate of oxygen supply, sensitivity to inhibitors generated during pretreatment, and hydrolysis of lignocellulosic substrates (Hahn-Hagerdal, Karhumaa, Fenske, *et al.* 2007; Kumar, Barette, Delwiche, *et al.* 2009). In mixed sugar fermentation, the pentose uptake is competitively inhibited by hexoses, and thus, the pentose fermentation is only possible at very low glucose concentrations. Therefore, there is an urge to develop a recombinant microbe that can efficiently utilize both the sugars simultaneously with higher ethanol yield and productivity. In last one decade, genetic engineering of micro-organisms used in ethanol production has shown significant progress (Jeffries and Jin 2004). Besides *S. cerevisiae*, micro-organisms such as *Zymomonas mobilis* and *Escherichia coli* have been targeted through metabolic engineering for ethanol production from lignocellulosic biomass (Jeffries 2006; Karhumaa, Hahn-Hägerdal, and Gorwa-Grauslund 2005; Karhumaa, Fromanger, Hahn-Hägerdal, *et al.* 2007; Liu and Hu 2010; Matushika, Watanabe, Kodaki, *et al.* 2008; Runquist, Hahn-Hägerdal, and Bettiga, *et al.* 2009; Yang, Tschaplinski, Engle, *et al.* 2009). Few potent recombinant microbes are listed in Table 6.

Table 6 Genetically engineered micro-organisms capable of fermenting xylose sugars

Strain	Sugar used (g/l)	Ethanol production (g/l)	Ethanol yield (g/g)	Ethanol productivity (g/l/h)	Reference
Escherichia coli FBR5	15 A:30 X: 30 G	34.0	90	0.92	Dien, Nichols, O'Bryan, *et al.* (2000)
Escherichia coli LY01	140 X	63.2	88	0.66	Yomano, York, and Ingram (1998)
Klebsiella oxytoca M5A1	100 X	46	95	0.96	Ohta, Beall, Mejia, *et al.* (1991)
Klebsiella oxytoca P2	20 A: 40 X: 20 G	34.2	84	0.35	Bothast, Saha, Flosenzier, *et al.* (1994)
Zymomonas mobilis AX101	20 A: 40 G: 20 X	42	84	0.61	Mohagheghi, Evans, Chou, *et al.* (2002)
Zymomonas mobilis CP4	25 X	11.0	86	0.57	Zhang, Eddy, Deanda, *et al.* (1995)
Zymomonas mobilis ZM4	65 G:65 X	62	90	1.29	Joachimsthal and Rogers (2000)
Zymomonas mobilis ATCC 39767	30 G:30 X:20 A	33.5	82–84	0.82–0.65	Chou, Zhang, Mohagheghi, *et al.* (1997)
Saccharomyces cerevisiae F12	50 G:50 X	26	52	NA	Sonderegger, Jeppsson, Hahn-Hägerdal, *et al.* (2004)
Saccharomyces cerevisiae BH42	50 G:50 X	28	56	NA	Sonderegger, Jeppsson, Hahn-Hägerdal, *et al.* (2004)

Contd...

Table 6 Contd....

Strain	Sugar used (g/l)	Ethanol production (g/l)	Ethanol yield (g/g)	Ethanol productivity (g/l/h)	Reference
Saccharomyces cerevisiae A4	50 G:50 X	19	38	NA	Zaldivar, Borges, Johansson, *et al.* (2002)
Saccharomyces cerevisiae 1400	50 X	1.5	6	NA	Ho, Chen, and Brainard (1998)
Saccharomyces cerevisiae PRD1	21.7	1.6	14	0.07	Kotter and Ciriacy (1993)
Saccharomyces cerevisiae TJ1	50 X	2.7	10.6	0.02	Tantirungkij, Nakashima, Seki, *et al.* (1993)
Saccharomyces cerevisiae RBW 202-AFX	20 X	8.4	84	NA	Kuyper, Winkler, Dijken, *et al.* (2004)
Saccharomyces cerevisiae RWB 217	20 X	8.6	86	NA	Kuyper, Hartog, Toirkens, *et al.* (2005)
Saccharomyces cerevisiae H158	80 X	17.6	44	NA	Johansson, Christensson, Hobley, *et al.* (2001)
Saccharomyces cerevisiae H 2673	50 X	11.5	46	NA	Verho, Londesborough, Penttila, *et al.* (2003)
Saccharomyces cerevisiae ZU-10	80 X	30.2	75.6	0.50	Zhao and Xia (2010)
Saccharomyces cerevisiae LEK 122	20 X	2.5	24.5	0.025	Liu and Hu (2010)
Saccharomyces cerevisiae TMB 3001	10 X	4.4	88	0.061	Träff-Bjerre, Jeppsson, Hahn-Hägerdal, *et al.* (2004)
Saccharomyces cerevisiae TMB 3008	50 X	19	76	0.27	Jeppsson, Johansson, Hahn-Hägerdal, *et al.* (2002)
Saccharomyces cerevisiae TMB 3030	50 X	14	56	NA	Jeppsson, Träff, Johansson, *et al.* (2003)
Saccharomyces cerevisiae TMB 3066	50 X	21.5	86	0.073	Karhumaa, Fromanger, Hahn-Hägerdal, *et al.* (2007)

Contd...

Table 6 Contd....

Strain	Sugar used (g/l)	Ethanol production (g/l)	Ethanol yield (g/g)	Ethanol productivity (g/l/h)	Reference
Saccharomyces cerevisiae TMB 3120	10 X	4.6	92	0.064	Träff-Bjerre, Jeppsson, Hahn-Hägerdal, *et al.* (2004)
Saccharomyces cerevisiae TMB 3251	50 X	17	68	0.24	Jeppsson, Johansson, Hahn-Hägerdal, *et al.* (2002)
Saccharomyces cerevisiae TMB 3267	10 X	4.1	82	0.057	Träff-Bjerre, Jeppsson, Hahn-Hägerdal, *et al.* (2004)
Saccharomyces cerevisiae TMB 3271	50 X	15.5	62	0.28	Jeppsson, Bengtsson, Franke, *et al.* (2006)
Saccharomyces cerevisiae TMB 3400	8 X	1.44	36	NA	Wahlbom, Otero, van Zyl, *et al.* (2003)
Saccharomyces cerevisiae CMB JHV	20 X	6.4	68	NA	Vleet, Jeffries, and Olsson (2008)
Saccharomyces cerevisiae ADAP8	20 X	8.6	86	0.07	Madhavan, Tamalampudi, Ushida, *et al.* (2009)
Saccharomyces cerevisiae MA-R4	45 X	16.65	74	0.5	Matsushika, Inoue, Murakami, *et al.* (2009)
Pichia stipitis FPL UC7	80 X	30.4	76	0.41	Shi, Davis, Sherman, *et al.* (1999)
Pichia stipitis FPL-shi 21	80 X	38.4	96	0.43	Shi, Davis, Sherman, *et al.* (1999)
Pichia stipitis FPL-shi 31	80 X	24.8	62	0.15	Shi, Cruz, Sherman, *et al.* (2002)

Genetically engineered *Saccharomyces cerevisiae*

Because *S. cerevisiae* is able to metabolize the xylose isomer, xylulose, the initial effort of genetic engineering was to insert the metabolic pathway of *P. stipitis* in its genome for efficient xylose fermentation. The recombinant *S. cerevisiae* strains expressing the *P. stipitis* xylose reductase (XR) gene are able to metabolize xylose; however, most of the xylose was converted to xylitol (Amore, Kotter, Kuster, *et al.* 1991; Hallborn, Walfridsson, Airaksinen, *et al.* 1991; Takuma, Nakashima, Tantirungkij, *et al.* 1991). Ho, Chen, and Brainard (1998) were the first researchers to successfully produce a recombinant strain of *S. cerevisiae* capable of effective xylose fermentation and cofermentation of mixed sugars (xylose and glucose). The recombinant plasmids with XR and xylitol dehydrogenase (XDH) genes from *P. stipitis* and xylulokinase (XKS) gene from *S. cerevisiae* have been transformed into *Saccharomyces* sp. for the co-fermentation of glucose and xylose. Toivari, Aristidou, Ruohonen, *et al.* (2001) also reported successful xylose to ethanol production through overexpression of the endogenous XKS gene in *S. cerevisiae* XR/PsXDH integrated strains. Furthermore, *S. cerevisiae* TMB 3001 was constructed by integrating the *P. stipitis* gene xylose isomerase (xyl A) and the endogenous XKS. The recombinant strain thus developed showed aerobic growth on 2% xylose as the sole carbon source (Eliasson, Christensson, Wahlbom, *et al.* 2000). Heterologous expression of *P. stipitis* xylose reductase (PsXR) and xylitol dehydronase (PsXDH) in *S. cerevisiae* allowed its growth on xylose but with xylitol as a primary product. Recently, it has been reported that high activities of both XR and XDH are essential for increased ethanol formation and decreased xylitol formation from xylose (Karhumaa, Fromanger, Hahn-Hägerdal, *et al.* 2007). Runquist, Hahn-Hägerdal, and Bettiga (2009) used an *S. cerevisiae* strain carrying a mutated XR with improved substrate uptake and higher ethanol productivity, which showed the ability to grow anaerobically on xylose as sole carbon source. Recently, through chemical mutagenesis and adaptation, an *S. cerevisiae* strain was developed that can use xylose under oxygen-limited conditions (Liu and Hu 2010).

Genetically engineered *Zymomonas mobilis*

The *Zymomonas* is the only micro-organism that metabolizes glucose anaerobically using the Entner–Doudoroff (ED) pathway as opposed to the Embden-Meyerhof (EM) or glycolytic pathway (Matthew, Ashley, Brian, *et al.* 2005). The ED pathway yields only half as much adenosine triphosphate (ATP) per mole of glucose as the EM pathway. As a consequence, *Zymomonas* produces less biomass than yeast, and more carbon is funnelled to fermentation products. To enable the xylose utilization, four genes from *E. coli*, xylose isomerase (xyl A), xylulokinase (xyl B), transaldolase (tal), and transketolase (tkt A), were inserted into *Z. mobilis*. The transformed *Z. mobilis* CP4 showed enzymatic activity for all four enzymes and grew on xylose as the sole carbon source, attaining 86% ethanol yield (Zhang, Eddy, Deanda, *et al.* 1995). Strains with chromosomal integration of above genes were developed to achieve a stable strain for ethanol production (Zhang, Eddy, Deanda, *et al.* 1995). Similarly for arabinose utilization, Deanda, Zhang, Eddy, *et al.* (1996) incorporated five genes of arabinose utilization from *E. coli*, L-arabinose isomerase (ara A), L-ribulokinase (ara B), L-ribulose-5 phosphate-4-epimerase (ara D), tal B, and transketolase (tkt A), in *Z. mobilis* ATCC 39767. A number of other improvements have also been made to *Z. mobilis* strains, and the newest strain *Z. mobilis* AX101 ferments both arabinose and xylose and carries seven necessary recombinant genes as part of chromosomal

DNA (Lawford and Rocesseau 2002; Mohagheghi, Evans, Chou, *et al.* 2002). Later on, Kalnenieks (2006) reviewed the physiology of *Z. mobilis*, identified a number of unanswered questions about its metabolism, and determined that the structural and physiological roles of *Z. mobilis* chain remain to be elucidated. Recently, transcriptomic and metabolomic profiles of *Z. mobilis* ZM4 under aerobic and anaerobic fermentations have been elucidated using microarray, high-performance liquid chromatography, and gas chromatography–mass spectrometry (GC–MS) analysis (Yang, Tschaplinski, Engle, *et al.* 2009).

Genetically engineered *Escherichia coli*

The construction of *Escherichia coli* strains to selectively produce ethanol was one of the first successful applications of metabolic engineering. The *E. coli* has several advantages as a biocatalyst for ethanol production, including the ability to ferment a wide spectrum of sugars and no requirement of complex growth factors. Ingram's group has carried out an extensive work on the development of efficient recombinant *E. coli* strain for ethanol production. They eliminated the dependence of the host on alcohol dehydrogenase (ADH) activity by combining adh B (ADH) and pdc (pyruvate decarboxylase gene) of *Z. mobilis* to form pet operon (Beall, Ohta, and Ingram 1991; Ingram, Conway, Clark, *et al.* 1987). Further, to improve this strain, several measures have been taken (Dien, Nichols, O'Bryan, *et al.* 2000; Gonzalez, Tao, Shanmugam, *et al.* 2002; Kim, Ingram, and Shanmugam 2007; Lawford and Rousseau 1995; Ohta, Beall, Mejia, *et al.* 1991). Current work on modified *E. coli* strain KO11 is focused on formulating an inexpensive culture medium and improving inhibitor and ethanol tolerance (Kim, Ingram, and Shanmugam 2007).

Ethanol recovery: distillation and dehydration

Under ideal conditions, an ethanol and water mixture can be separated based on their difference in volatility. Because ethanol is more volatile than water (ethanol vaporizes at 78°C, whereas water vaporizes at 100°C), on heating, the ratio of ethanol to water in the vapour phase will become higher than that in the liquid phase. Therefore, in an ideal distillation column separation, the overhead product will be mainly ethanol, and water will be the main bottom product. An azeotropic mixture of ethanol (95.6%) and water (4.4%) will be formed upon completion of the distillation operation, which is determined by the difference in the boiling points between water and ethanol (Fair 2001). Because the ethanol–water mixture from fermentation is far from being ideal, the actual ethanol recovery process is a multistage and highly integrated process (Wankat 1988).

There are several dehydration processes to remove water from an azeotropic ethanol/water mixture. The first process is azeotropic distillation by adding benzene or cyclohexane to the mixture. When these components are added to the mixture, a heterogeneous azeotropic mixture in vapour–liquid–liquid equilibrium is formed, which when distilled produces anhydrous ethanol at the bottom of the column and a vapour mixture of water and cyclohexane/benzene. Extractive distillation consists of adding a ternary component, which will increase the ethanol's relative volatility. When the ternary mixture is distilled, it will produce anhydrous ethanol on the top stream of the column. The third method uses molecular sieves to remove water from ethanol. In this process, the ethanol vapour pressure passes through a bed of molecular sieve beads. The bead's pores are sized to allow absorption of water while excluding ethanol.

In a lignocellulosic fuel process, upon completion of fermentation, the liquor consists of a mixed stream of solids and liquids. Distillation and molecular sieve dehydration operations have to be used to recover a pure ethanol product of fuel grade (>99.5%). The product from the rectification column is an azeotropic mixture at an optimal point, which cannot be further improved by distillation operations. Therefore, effluent from the rectification column is fed to a dehydration separation unit, typically a vapour phase molecular sieve adsorption unit operation. Molecular sieves are crystalline metal aluminosilicates (zeolites) with a three-dimensional porous structure of silica and tetrahedral alumina (Kresge and Dhingra 2004). Zeolite materials can strongly and preferentially adsorb water from vapour mixtures; thus, they are able to remove the remaining 4.4% water content in the azeotropic mixture from the rectification column. Therefore, minimization of total energy input is a critical requirement for the economic design of an ethanol distillation/dehydration system. In the past 40 years, the installation of multistage, pressure distillation systems has reduced energy consumption by 50% compared with the earlier all-atmospheric pressure systems (Madson 2003). In modern molecular sieve dehydration systems, "pressure swing adsorption" is used to remove the water content, wherein a relatively high pressure is applied at the water removal stage and a relatively low pressure is applied at the desiccant regeneration stage. Therefore, operation temperature can be kept almost constant, and the heat of adsorption can be effectively stored and further supplied to the regeneration stage (Swain 2003). An alternative to molecular sieve material is corn grits (Ladisch and Dyck 1979). Corn grits can selectively remove water from an azeotropic mixture and are advantageous in that the materials are biorenewable, of low cost, and easily disposable. However, a major drawback of the corn grits is their mechanical stability over a long period of time (Beery and Ladisch 2001).

Economic evaluation of cellulosic ethanol production

Crop residues, forest agro-industrial waste, and municipal waste represent potential alternative sources for ethanol production worldwide. However, their enormous potential lacks endorsement, appreciation, and, of course, there are difficulties in processing technologies for their conversion into ethanol. As far as the techno-economic evaluation of lignocellulosic ethanol is concerned, the first detailed technical reports found in the literature from the US date back to the mid-1980s. Especially in 1987, the US National Renewable Energy Laboratory (NREL) received several technical reports delivered by subcontractors (Badger Engineers, Inc. 1987).

The second economical study by Aden, Ibsen, Jechura, *et al.* (2002) used the same framework with the following main differences: the feedstock is corn stover; the size of the plant is 69.3 million gallons of ethanol per year; and the on-site production of enzyme is replaced by purchased enzymes. According to the study, the lignocellulose may be converted to hexoses and pentoses by dilute acid hydrolysis efficiently at a cost of approximately $0.20 per gallon of ethanol. Enzymatic hydrolysis by lignocellulolytic enzymes is a mild treatment and has great potential for improvement; however, current technology using saccharifying enzymes is still expensive (~$0.3–0.5 per gallon ethanol) (Aden, Ibsen, Jechura, *et al.* 2002). Over the past decade, much effort has been devoted to cellulase production cost, which has increased by more than 10-fold in the range of 10–20 cents per gallon (¢/gal) of ethanol produced. According to the US Department of Energy analysis, if the enzyme cost can be brought to less than 10 ¢/gal of ethanol, the cost of ethanol production could drop as low as $1.07/gallon (in 2002 dollars), and the US Department of Energy aims to achieve this goal by 2012 (NREL Report 2007).

From 2002, the context of techno-economic evaluation of lignocellulosic ethanol has changed with the launch of the Biomass Program of the US Department of Energy (DOE). Since 2007, the design of this programme has acquired a clear strategic goal with the aim of reducing the use of gasoline by 20% by 2017 and producing 35×10^9 l of renewable and alternative fuels in 2017. Concerning the R&D in lignocellulosic bioethanol, a "multi-year program plan" is released and updated every two years, including, so far, 2005 (US DOE 2005), 2007 (US DOE 2007), and 2009 (US DOE 2009).

As an estimation, the biochemical ethanol is based on the reference scenario by the Energy Information Administration (EIA 2009), which forecasts the wholesale price of gasoline in 2012 at $2.62/gal of gasoline (US$ of 2007). Assuming a conversion factor of 0.67 gallon of gasoline per gallon of ethanol, the projected cost of ethanol is set at $1.76/gal of ethanol (US dollar of 2007). However, the ethanol cost projection of the *n*th ethanol plant is at $1.49/gal of ethanol (US dollar of 2007).

Besides the techno-economic evaluations undertaken in the US, significant contributions are made by the European research institutions, mainly in Sweden, the Netherlands, and Denmark. In the REFUEL project (2006–08), funded by the European Commission under the Intelligent Energy Europe Programme, seven EU institutes have analysed the prospects of biofuels in terms of resource potential, costs, and impacts of different biofuels, including lignocellulosic ethanol. The data for bioethanol production from cellulosic materials based on the enzymatic hydrolysis pathway is obtained from the Energy Research Centre of the Netherlands (Kuijvenhoven 2006) and the

Copernicus Institute for Sustainable Development and Innovation of Utrecht University (Hamelinck 2004). The economic evaluation (Londo, Lensink, Deurwaarder, *et al.* 2008) is based on constant € for 2002 and results in a net production cost (including the sales of electricity as a by-product) of €0.62/l in 2010 (forest wood as feedstock, production capacity of 100 000 tonnes of ethanol per year), €0.59/l in 2020 (200 000 tonnes of ethanol per year), and €0.50/l in 2030 (400 000 tonnes of ethanol per year), given the learning curve based on expected global production and number of plants; for details, readers may refer to the work of Londo, Lensink, Deurwaarder, *et al.* (2008).

Similarly, Wright and Brown (2007) evaluated the economics of advanced biochemical process for producing bioethanol from plant fibres consisting of four major operational steps: pretreatment of plant fibres to make cellulose accessible to enzymes; enzymatic hydrolysis of cellulose to glucose; fermentation of simple sugars to produce ethanol; and distillation to purify ethanol. On the basis of current evaluation, the assumed capital costs and operating costs for bioethanol production from cellulosic substrates are estimated to be \$4.03–\$5.60 per gallon of ethanol and \$1.34–\$1.69 per gallon of ethanol, respectively. Anticipating further improvements in the bioconversion technologies, projected capital costs for future plants are estimated to be \$3.33–4.44 per gallon of ethanol, with operating costs dropping to \$0.40–0.89 per gallon of ethanol (Hamelinck, van Hooijdonk, and Faaij 2005; Wright and Brown 2007).

The SSF technology is an alternative perspective that holds promise for cost reduction and higher hydrolytic efficiency due to the presence of an optimal enzyme complex produced by host-specific fungi in single or mixed culture. Genetic improvement of such host-specific fungi targeted for SSF application has greater potential than genetically improved fungi used in separated hydrolysis and fermentation. As an estimate, NREL produced SSF enzymes at a cost of 17 ¢/gal, which they propose to bring down to \$0.10 per gallon of ethanol (Aden, Ibsen, Jechura, *et al.* 2002). The updates on the technology model are provided by Aden (2008), Humbird and Aden (2009), and Aden and Foust (2009).

The major work so far has been carried out using corn stover, sugar cane bagasse, and sweet sorghum. While the work on bioconversion of other potential alternative resources has not picked up momentum, countries like India, Canada, US, Brazil, Nigeria, China, France, and Iran have long-term plans to exploit various crop residues, forest trees, municipal waste, and pulp and paper waste for ethanol production. Thus, the economy of the process may slightly vary with the country and substrate type.

Process integration and development

Increasing production capacity to commercial scale can only be done with confidence when a process is shown to be robust at an intermediate, pilot scale. An ideal pilot plant needs to be fully integrated, and should be able to evaluate the complete system (for example, enzymes and yeasts) while having sufficient flexibility to investigate alternative process configurations and test opinions for better heat integration and recycling of process streams. Recently, there has been a significant effort for building pilot facilities, some of which are listed in Table 7. From a process scale-up perspective, it is clear that the challenges lie not only in finding the most efficient organisms for the conversion of cellulose to ethanol but also in making intelligent use of the entire feedstock and exploiting all the opportunities for process integration.

Currently, it is uncertain as to which competing technology pathways will become dominant, but it is clear that the viability of cellulosic ethanol production process depends on the feedstock cost and digestibility, conversion cost and efficiency, and product revenues (Margeot, Hahn-Hagerdal, Edlund,

Table 7 List of commercial cellulosic ethanol plants

Company	Location	Feedstock
Abengoa Bioenergy	Hugoton, KS	Wheat straw
Alico	Florida	Wood, citrus waste
Bioengineering Resources	Arkansas	Urban green waste, wood chips
Bioethanol Japan	Osaka	Wood construction waste
Biogasol	Lyngby	Hay, grass, manure, fibres, paper
BlueFire Ethanol	Irvine, CA	Multiple sources
China Resources Alcohol Corporation	ZhaoDong City	Corn stover
Choren	Choren	Cellulosic biomass
Clear Fuels Technology	Hawai	Sugar cane bagasse
Colusa Biomass Energy Corporation	Sacramento, CA	Waste rice straw
Coskata	Warrenville, IL	Biomass, agricultural, and municipal waste
DuPont Danisco Cellulosic Ethanol (DDCE)	Vonore, TN	Corn cobs, switch grass
Flambeau River Biorefinery	Wisconsis	Spent pulping liquor
Fulcrum BioEnergy	Reno, NV	Municipal solid waste
Gulf Coast Energy	Mossy Head, FL	Wood waste
Iogen, Otawa	Otawa	Wheat straw, barley straw, and rice straw
KL Energy Corporation	Upton, WY	Wood
Lignol Innovation	Burnaby	Wood chips, corn stover, and switch grass
Mascoma	Lansing, MI	Wood
POET LLC	Emmetsburg, IA	Corn cobs
Range Fuels	Treutlen County, GA	Wood waste

Contd...

Table 7 Contd...

Company	Location	Feedstock
SunOpta	Little Falls, MN	Wood chips
UPM	Finland	Paper
US Envirofuels	Highlands County, FL	Sweet sorghum
Verenium	Cambridge, Massachusetts	Sugar cane bagasse and wood
Xethanol	Auburndale, FL	Citrus peels
POET LLC	South Dakota	Cellulosic biomass

Source http://en.wikipedia.org/wiki/Cellulosic_ethanol#Cellulosic_ethanol_commercialization; Bartacek, Zabranska, and Lens, *et al.* (2007); Antizar-Ladislao and Turrion-Gomez (2008)

et al. 2009). For further improvement in the bioethanol production process, following interventions in the process can be made.

- Pretreatment of lignocellulosic biomass using low cost and less polluting chemicals for improving sugar yield.
- Detoxification of sugar hydrolysates using low cost chemicals like activated charcoal, lime, or enzymes (laccase).
- Production of cost-effective and thermo-stable cellulose using native or recombinant micro-organisms.
- Saccharification of pretreated biomass using novel cocktail of enzymes (cellulase, xylanase, feruloyl esterase, and laccase).
- Fermentation of pentose sugars.
- Developing glucose and xylose co-fermenting recombinant micro-organisms.
- Ethanol recovery from fermentation broth and its distillation to desired purity.
- Economization of complete process by utilizing different by-products such as lignin and leftover fermentation broth.

Conclusion

In the last decade, the research on ethanol production from lignocellulosic materials has progressed considerably. Pretreatments and enzymatic hydrolysis using certain crop residues have been optimized, and the ethanol production processes at bench scale are ready to be taken up to pilot scale, and eventually, some of them to commercial scale. There is still a need to improve the production of cellulases, sugars, and ethanol-tolerant microbial strains. As majority of research programmes are concentrating on C6 fermentation, there is a need to give priority to research on C5 fermentation, otherwise there will remain a major gap in economic bioconversion of lignocellulosics into ethanol.

References

Abbi M, Kuhad R C, and Singh A. 1996a. **Bioconversion of pentose sugars to ethanol by free and immobilized cells of *Candida shehatae* (NCL-3501): fermentation behaviour.** *Process Biochemistry* **31**: 555–560

Abbi M, Kuhad R C, and Singh A. 1996b. **Fermentation of xylose and rice straw hydrolysate to ethanol by *Candida shehatae* NCL-3501.** *Journal of Industrial Microbiology* **17**: 20–23

Abdenifar S, Karimi K, and Khanahmadi M. 2009. **Thanol production by *Mucor indicus* and *Rhizopus oryzae* from rice straw by separate hydrolysis and fermentation.** *Biomass and Bioenergy*. doi: 10.1016/j.biombioe.2009.01.003

Aden A. 2008. ***Biochemical Production of Ethanol from Corn Stover: 2007 State of Technology Model.*** Golden: National Renewable Energy Laboratory (NREL). [NREL Report TP-510-43205]

Aden A and Foust T. 2009. **Technoeconomic analysis of the dilute sulfuric acid and enzymatic hydrolysis process for the conversion of corn stover to ethanol.** *Cellulose* **16**: 535–545

Aden A M R, Ibsen K, Jechura J, Neeves K, Sheehan J, Wallace B. 2002. ***Lignocellulosic Biomass to Ethanol Process Design and Economics Utilizing Co-current Dilute Acid Prehydrolysis and Enzymatic Hydrolysis for Corn Stover.*** Golden: National Renewable Energy Laboratory (NREL). pp. 36–39 [NREL/TP-510-32438]

Agbogbo F A and Wenger K S. 2007. **Production of ethanol from corn stover hemicellulose hydrolyzate using *Pichia stipitis*.** *Journal of Industrial Microbiology and Biotechnology* **34**: 723–727

Agbogbo F A, Coward-Kelly G, Torry-Smith M, Wenger K S. 2006. **Fermentation of glucose/xylose mixtures using *Pichia stipitis*.** *Process Biochemistry* **41**: 2333–2336

Aguilera F, Peinado R A, Millan C, Ortega J M, Mauricio M. 2006. **Relationship between ethanol tolerance, H^+-ATPase activity and the lipid composition of the plasma membrane in different wine yeast strains.** *International Journal of Food Microbiology* **110**: 34–42

Alder E. 1977. **Lignin chemistry: past, present and future.** *Wood Science and Technology* **11**: 169–218

Alfani A, Gallifuoco F, Saporosi A, Cantarella M. 2000. **Comparison of SHF and SSF process for the bioconversion of steam-exploded wheat straw.** *Journal of Industrial Microbiology and Biotechnology* **25**: 84–192

Alvira P, Tomás-Pejó E, Ballesteros M, Negro M J. 2010. **Pretreatment technologies for an efficient bioethanol production process based on enzymatic hydrolysis: a review.** *Bioresource Technology* **101**: 4851–4861

Antizar-Ladisiao B and Turrion-Gomez J L. 2008. **Second-generation biofuels and local bioenergy systems.** *Biofuels Bioproduction and Biorefinery* **2**: 455–469

Amore R, Kötter P, Kuster C, Ciriacy M, Hollenberg C P. 1991. **Cloning and expression in *Saccharomyces cerevisiae* of the NAD(P)H-dependent xylose reductase-encoding gene (XYL1) from the xylose-assimilating yeast *Pichia stipitis*.** *Gene* **109**: 89–97

Azzam A M. 1989. **Pretreatment of cane bagasse with alkaline hydrogen peroxide for enzymatic hydrolysis of cellulose and ethanol fermentation.** *Journal of Environmental Science and Health B* **24**: 421–433

Badger Engineers, Inc. 1987. ***Economic Feasibility of an Acid-Hydrolysis-Based Ethanol Plant.*** Cambridge, MA: Badger Engineers, Inc. [Subcontract Report SERI-STR-231-3142, Prepared under Subcontract ZX-3-03096-2]

Baker J O, McCarley J R, Lovettt R, Yu C H, Adney W S, Rignall T R, Vinzant T B, Decker S R, Sakon J, Himmel M E. 2005. **Catalytically enhanced endocellulase Cel5A from *Acidothermus cellulolyticus*.** *Applied Biochemistry and Biotechnology* **121**: 129–148

Balan V, Souca L D C, Chundawat S S, Vismeh R, Jones A D, Dale B E. 2008. **Mushroom spent straw: a potential substrate for an ethanol-based biorefinery.** *Journal of Industrial Microbiology and Biotechnology* **35**: 293–301

Ballesteros I, Ballesteros M, Cabañas A, Carrasco J, Martin C, Negro J M, *et al.* 1991. **Selection of thermotolerant yeasts for simultaneous saccharification and fermentation process (SSF) of cellulose to ethanol.** *Applied Biochemistry and Biotechnology* **28–29**: 307–315

Ballesteros M, Oliva J M, Negro M J, Manzanares P, Ballesteros I. 2004. **Ethanol from lignocellulosic materials by a simultaneous saccharificationand fermentation process (SSF) with *Kluyveromyces marxianus* CECT 10875.** *Process Biochemistry* **39**: 1843–1848

Banat I M, Nigam P, Singh D, Marchant P, McHale A P. 1998. **Ethanol production at elevated temperatures and alcohol concentrations. Part I: yeasts in general.** *World Journal of Microbiology and Biotechnology* **14**: 809–821

Bartacek J, Zabranska J, and Lens P N L. 2007. **Developments and constrains in fermentative hydrogen production.** *Biofuels Bioproduction and Biorefinery* **1**: 201–214

Beall D S, Ohta K, and Ingram L O. 1991. **Parametric studies of ethanol-production from xylose and other sugars by recombinant *Escherichia coli*.** *Biotechnology and Bioengineering* **38**: 296–303

Beery K E and Ladisch M R. 2001. **Adsorption of water from liquid-phase ethanol–water mixtures at room temperature using starch-based adsorbents.** *Industrial Engineering and Chemical Research* **40**: 2112–2115

Beguin P and Aubert J P. 1994. **The biological degradation of cellulose.** *FEMS Microbiology Review* **13**: 25–58

Belkacemi K, Tureotte G, Savoic P, Chornet E. 1997. **Ethanol production from enzymatic hydrolysate of cellulosic fines and hemicellulosic-rich liquors derived from aqueous/steam fractionation of forages.** *Industrial Engineering and Chemical Research* **36**: 4572–4580

Belloch C, Orlic S, Barrio E, Querol A. 2008. **Fermentative stress adaptation of hybrids within the *Saccharomyces* sensu stricto complex.** *International Journal of Food Microbiology* **122**: 188–195

Berg C. 2001. **World ethanol production 2001.** Details available at <http://www.distill.com/world_ethanol_production.htm>

Bisaria V S. 1991. **Bioprocessing of agro-residues to glucose and chemicals.** In *Bioconversion of Waste Materials to Industrial Products,* pp. 210–213, edited by A M Martin. London: Elsevier

Bjerre A B, Olesen A B, and Fernqvist T. 1996. **Pretreatment of wheat straw using combined wet oxidation and alkaline hydrolysis resulting in convertible cellulose and hemicellulose.** *Biotechnology and Bioengineering* **49**: 568–577

Bollók M, Réczey K, and Zacchi G. 2000. **Simultaneous saccharification and fermentation of steam-pretreated spruce to ethanol.** *Applied Biochemistry and Biotechnology* **84–86**: 69–80

Boluda-Aguilar M, García-Vidal L, González-Castañeda F D P, López-Gómez A. 2010. **Mandarin peel wastes pretreatment with steam explosion for bioethanol production.** *Bioresource Technology* **101**: 3506–3513

Boraston A B, Bolam D N, Gilbert H J, Davies G J. 2004. **Carbohydrate-binding modules: fine-tuning polysaccharide recognition.** *Biochemistry Journal* **382**: 769–781

Bothast R J, Saha B C, Flosenzier V A, Ingram L O. 1994. **Fermentation of L-arabinose, D-xylose, and D-glucose by ethanologenic recombinant *Klebsiella oxytoca* strain P2.** *Biotechnology Letters* **16**: 401–406

Brandberg T, Karimi K, Taherzadeh M J, Franzen C J, Grustafsson L. 2007. **Continuous fermentation of wheat-supplemented lignocellulose hydrolysate with different types of cell retention.** *Biotechnology and Bioengineering* **98**: 80–90

Brandberg T, Sanandaji N, Gustafsson L, Franzen C J. 2005. **Continuous fermentation of undetoxified dilute acid lignocellulose hydrolysate by *Saccharomyces cerevisiae* ATCC 96581 using cell recirculation.** *Biotechnology Progress* **21**: 1093–1101

Campbell C J and Laherrere J H. 1998. **The end of cheap oil.** *Science* 3: 78–83

Cara C, Ruiz E, Oliva J M, Sáez F, Castro E. 2008. **Conversion of olive tree biomass into fermentable sugars by dilute acid: re-treatment and enzymatic saccharification.** *Bioresource Technology* **99**: 1869–1876

Carrillo F, Lis M J, Colom X, Lopez-Mesas M, Valldeperas J. 2005. **Effect of alkali elluloseent on ellulose hydrolysis of wheat straw: kinetic study.** *Process Biochemistry* **40**: 3360–3364

Chaillou S, Pouwels P H, and Postma P W. 1999. **Transport of D-xylose in *Lactobacillus pentosus, Lactobacillus casei,* and *Lactobacillus plantarum:* evidence for a mechanism of facilitated diffusion via the phosphoenolpyruvate:mannose phosphotransferase system.** *Journal of Bacteriology* **181**: 4768–4773

Chandel A K, Kapoor R K, Singh A, Kuhad R C. 2007a. **Detoxification of sugarcane bagasse hydrolysate improves ethanol production by *Candida shehatae* NCIM 3501.** *Bioresource Technology* **98**: 1947–1950

Chandel A K, Kapoor R K, Singh A, Narasu M L, Viswadewan V, Kumaran S G S, Rudravaram R, Rao L V, Tripathi K K, Lal B, Kuhad R C. 2007b. **Economic evaluation and environmental benefits of biofuels: an Indian perspective.** *International Journal of Global Energy Issues* **28**: 357–381

Chen M, Xia L, and Xue P. 2007. **Enzymatic hydrolysis of corncob and ethanol production from cellulosic hydrolysate.** *International Biodeterioration and Biodegradation* **59**: 85–89

Cheng K-K, Cai B-Y, Zhang J-A, Ling H-Z, Zhou Y-J, Ge J-P, Xu J-M. 2008. **Sugarcane bagasse hemicellulose hydrolysate for ethanol production by acid recovery process.** *Biochemical Engineering Journal* **38**: 105–109

Chou Y-C, Zhang M, Mohagheghi A, Evans K, Finkelstein M. 1997. **Construction and evaluation of a xylose/arabinose fermenting strain of *Zymomonas mobilis*.** In *Abstracts in 19th Symposium on Biotechnology for Fuels and Chemicals.* [19th Symposium on Biotechnology for Fuels and Chemicals, Colorado, Springs, May 4–8]

Collins T, Gerday C, and Feller G. 2005. **Xylanases, xylanase families and extremophilic xylanases.** *FEMS Microbiology Reviews* **29**: 3–23

Deshpande V, Keskar S, Mishra C, Rao M. 1986. **Direct conversion of cellulose/ hemicellulose to ethanol by *Nuerospora crassa*.** *Enzyme and Micorbial Technology* **8**: 149–152

Deanda K, Zhang M, Eddy C, Picataggio S. 1996. **Development of an arabinose-fermenting *Zymomonas mobilis* strain by metabolic pathway engineering.** *Applied and Environmental Microbiology* **62**: 4465–4470

De Bari I, Cuna D, Nanna F, Braccio G. 2004. **Ethanol production in immobilized-cell bioreactors from mixed sugar syrups and enzymatic hydrolysates of steam-exploded biomass.** *Applied Biochemistry and Biotechnology* **113–116**: 539–557

Demirba A. 2007. **Fractionation and analysis of supercritical fluid extracts from lignite.** *Energy Sources* **24**: 817–823

Dien B S, Nichols N N, O'Bryan P J, Bothast R J. 2000. **Development of new ethanologenic *Escherichia coli* strains for fermentation of lignocellulosic biomass.** *Applied Biochemistry and Biotechnology* **84**: 181–196

Dien B S, Cotta M A, and Jeffries T W. 2003. **Bacteria engineered for fuel ethanol production: current status.** *Applied Microbiology and Biotechnology* **63**: 258–266

Din N, Damude H G, Gilkes N R, Miller R C, Warren R A J, Kilburn D G. 1994. **C-1-C-X revisited—intramolecular synergism in a cellulose.** *Proceedings in National Academy of Science* **91**: 11383–11387

Divne C, Stahlberg J, Teeri T T, Jones T A. 1998. **High-resolution crystal structures reveal how a cellulose chain is bound in the 50 angstrom long tunnel of cellobiohydrolase I from *Trichoderma reesei*.** *Journal of Molecular Biology* **275**: 309–325

Dourado F, Bastos M, Mota M, Gama F M. 2002. **Studies on the properties of celluclast/Eudragit L-100 conzugate.** *Journal of Biotechnology* **99**: 121–131

Duff S J B and Murray W D. 1996. **Bioconversion of forest products industry waste cellulosics to fuel ethanol: a review.** *Bioresource Technology* **55**: 1–33

EIA (Energy Information Administration). 2009. ***Annual Energy Outlook 2009.*** Washington: EIA, Office of Integrated Analysis and Forecasting. US Department of Energy

El Kanouni A, Zerdani I, Zaafa S, Znassni M, Loutfi M, Boudouma M. 1998. **The improvement of glucose/xylose fermentation by *Clostridium acetobutylicum* using calcium carbonate.** *World Journal of Microbiology and Biotechnology* **14**: 431–425

Eliasson A, Christensson C, Wahlbom C F, Hahn-Hägerdal B. 2000. **Anaerobic xylose fermentation by recombinant *Saccharomyces cerevisiae* carrying XYL1, XYL2, and XKS1 in mineral medium chemostat cultures.** *Applied and Environmental Microbiology* **66**: 3381–3386

Eriksson K-E L. 1993. **Lignin biodegradation and practical utilization.** *Journal of Biotechnology* **30**: 149–158

Eriksson K L, Blanchette R A, and Ander P. 1990. ***Microbial and Enzymatic Degradation of Wood and Wood Components.*** Berlin: Springer

Esteghlalian A, Hashimoto A G, Fenske J J, Penner M H. 1997. **Modeling and optimization of the dilute-sulfuric-acid pretreatment of corn stover, poplar and switchhgrass.** *Bioresource Technology* **59**: 129–136

Fair J R. 2001. **Distillation.** *Kirk-Othmer Encyclopedia of Chemical Technology* **8**: 739–785

Fan L T, Gharpuray M M, and Lee Y-H. 1987. ***Cellulose Hydrolysis Biotechnology Monographs.*** Berlin: Springer

Farrell A E, Plevin R J, Turner B T, Jones A D, O'Hare M, Kammen D M. 2006. **Ethanol can contribute to energy and environmental goals.** *Science* **311**: 306–508

Fenske J J, Hashimoto A, and Penner M H. 1998. **Relative fermentability of lignocellulosic dilute-acid prehydrolysates—application of a *Pichia stipitis*-based toxicity assay.** *Applied Biochemistry and Biotechnology* **73**: 145–157

Galbe M and Zacchi G. 2007. **Pretreatment of lignocellulosic materials for efficient bioethanol production.** *Advance Biochemical Engineerin/Biotechnology* **108**: 41–65

Gallagher P, Schamel G, Shapouri B, Brubaker H. 2006. **The international competitiveness of the U.S. corn-ethanol industry: a comparison with sugar-ethanol processing in Brazil.** *Agribusiness* **22**: 109–134

García-Cubero M T, González-Benito G, Indacoechea I, Coca M, Bolado S. 2009. **Effect of ozonolysis pretreatment on enzymatic digestibility of wheat and rye straw.** Bioresource Technology **100**: 1608–1613

Gonzalez R, Tao H, Shanmugam K T, York S W, Ingram L O. 2002. **Global gene expression differences associated with changes in glycolytic flux and growth rate in *Escherichia coli* during the fermentation of glucose and xylose.** *Biotechnology Progress* **18**: 6–20

Goring D A I and Timell T E. 1962. **Molecular weight of native celluloses.** *Tappi* **45**: 454–460

Gregg D J, Boussaid A, and Saddler J N. 1998. **Techno-economic evaluations of a generic wood to ethanol process: effect of increased cellulose yields and enzyme recycle.** *Bioresource Technology* **63**: 7–12

Grootjen D R J, Jansen M L, Vanderlans R, Luyben K. 1991. **Reactors in series for the complete conversion of glucose xylose mixtures by *Pichia stipitis* and *Saccharomyces cerevisiae*.** *Enzyme and Microbial Technology* **13**: 828–833

Gupta R, Sharma K K, and Kuhad R C. 2009. **Separate hydrolysis and fermentation (SHF) of *Prosopis juliflora*, a woody substrate, for the production of cellulosic ethanol by *Saccharomyces cerevisiae* and *Pichia stipitis*-NCIM 3498.** *Bioresource Technology* **100**: 1214–1220

Gupta R, Mehta G, Khasa Y P, Kuhad R C. 2010. **Fungal delignification of lignocellulosic biomass improves the saccharification of cellulosics.** *Biodegradation* (Submitted)

Hahn-Hägerdal B, Karhumaa K, Fonseca C, Spencer-Martins I, Gorwa-Grauslund M F. 2007. **Towards industrial pentose fermenting yeast strains.** *Applied Microbiology and Biotechnology* **74**: 937–953

Hallborn M, Walfridsson M, Airaksinen U, Ojamo H, Hahn-Hägerdal B, Penttila M. 1991. **Xylitol production by recombinant *Saccharomyces cerevisiae*.** *Biotechnology Journal* **9**: 1090–1095

Hamelinck C N. 2004. **Outlook for advanced biofuels.** Department of Science, Technology and Society and the Copernicus Institute for Sustainable Development and Innovation of Utrecht University. [Doctoral thesis]

Hamelinck C N, van Hooijdonk G, and Faaij A O C. 2005. **Ethanol from lignocellulosic biomass: techno-economic performance in short-, middle-, and long-term.** *Biomass and Bioenergy* **28**: 384–410

Harikrishna S, Reddy J T, and Chowdary G V. 2001. **Simultaneous saccharification and fermentation of lignocellulosic wastes to ethanol using thermotolerant yeast.** *Bioresource Technology* 77: 193–196

Hatakka A I. 1983. **Pretreatment of wheat straw by white-rot fungi for enzymatic saccharification of cellulose.** *Applied Microbiology and Biotechnology* **18**: 350–357

Hayes D J. 2009. **An examination of biorefining processes, catalysts and challenges.** *Catalysis Today* **145**: 138–151

Hendriks A T W M and Zeeman G. 2009. **Pretreatments to enhance the digestibility of lignocellulosic biomass.** *Bioresource Technology* **100**: 10–18

Herrero A and Gomez R F. 1980. **Development of ethanol tolerance in *Clostridium thermocellum*: effect of growth temperature.** *Applied and Environmental Microbiology* **40**: 571–577

Himmel M E, Ding S, Johnson D K, Adney W S, Nimlos M R, Brady J W, Foust T D. 2007. **Biomass recalcitrance: engineering plants and enzymes for biofuels production.** *Science* **315**: 804–807

Ho N W Y, Chen Z, and Brainard A P. 1998. **Genetically engineered *Saccharomyces* yeast capable of effective co-fermentation of glucose and xylose.** *Applied and Environmental Microbiology* **64**: 1852–1859

Hu Z and Wen Z. 2008. **Enhancing enzymatic digestibility of switch grass by microwave assisted alkali pre-treatment.** *Biochemical Engineering Journal* **38**: 369–378

Hu C K, Bai F W, and An L J. 2005. **Effect of flocculence of a self flocculating yeast on its tolerance to ethanol and the mechanism.** *Chinese Journal of Biotechnology* **21**: 123–128

Hu Z, Wang Y, and Wen Z. 2008. **Alakli (NaOH) re-treatment of switchgrass by radiofrequency bound dielectric heating.** *Applied Biochemistry and Biotechnology* **148**: 71–81

Huang C-F, Lin T-H, Guo G-L, Hwang W-S. 2009. **Enhanced ethanol production by fermentation of rice straw hydrolysate without detoxification using a newly adapted strain of *Pichia stipitis*.** *Bioresource Technology* **100**: 3914–3920

Humbird D and Aden A. 2009. ***Biochemical Production of Ethanol from Corn Stover: 2008 State of Technology Model.*** Golden: National Renewable Energy Laboratory (NREL). [NREL Report TP-510-46214]

Ingram L O, Conway T, Clark D P, Sewell G W, Preston J F. 1987. **Genetic engineering of ethanol production in *Escherichia coli*.** *Applied and Environmental Microbiology* **53**: 2420–2425

Ingram L O, Aldrich H C, Borges A C, Causey T B, Martinez A, Morales F, Saleh A, Underwood S A, Yomano I P, York S W, Zaldivar J, Zhou S. 1999. **Enteric bacterial catalysts for fuel ethanol production.** *Biotechnology Progress* **15**: 855–866

Itoh H, Wada M, Honda Y, Kuwahara M, Watanabe T. 2003. **Bioorganosolve pretreatments for simultaneous saccharification and fermentation of beech wood by ethanolysis and white rot fungi.** *Journal of Biotechnology* **103**: 273–280

Jeffries T W. 2006. **Engineering yeasts for xylose metabolism.** *Current Opinion in Biotechnology* **17**: 320–326

Jeffries T W and Jin Y-S. 2004. **Ethanol and thermo-tolerance in the bioconversion of xylose by yeasts.** *Advances in Applied Microbiology* **47**: 221–266

Jeffries T W and Schartman R. 1999. **Bioconversion of secondary fiber fines to ethanol using counter-current enzymatic saccharification and co-fermentation.** *Applied Biochemistry and Biotechnology* **77–79**: 435–444

Jeppsson M, Bengtsson O, Franke K, Lee H, Hahn-Hägerdal B, Gorwa-Grauslund M F. 2006. **The expression of a *Pichia stipitis* xylose reductase mutant with higher K_m for NADPH increases ethanol production from xylose in recombinant *Saccharomyces cerevisiae*.** *Biotechnology and Bioengineering* **93**: 665–673

Jeppsson M, Johansson B, Hahn-Hägerdal B, Gorwa-Grauslund M F. 2002. **Reduced oxidative pentose phosphate pathway flux in recombinant xylose-utilizing *Saccharomyces cerevisiae* strains improves the ethanol yield from xylose.** *Applied and Environmental Microbiology* **68**: 1604–1609

Jeppsson M, Träff K, Johansson B, Hahn-Hägerdal B, Gorwa-Grauslund M F. 2003. **Effect of enhanced xylose reductase activity on xylose consumption and product distribution in xylose-fermenting recombinant *Saccharomyces cerevisiae*.** *FEMS Yeast Research* **3**: 167–175

Joachimsthal E L and Rogers P L. 2000. **Characterization of a high-productivity recombinant strain of *Zymomonas mobilis* for ethanol production from glucose/ xylose mixtures.** *Applied Biochemistry and Biotechnology* **84–86**: 343–356

Johansson B, Christensson C, Hobley T, Hahn-Hägerdal B. 2001. **Xylulokinase overexpression in two strains of *Saccharomyces cerevisiae* also expressing xylose reductase and xylitol dehydrogenase and its effect on fermentation of xylose and lignocellulosic hydrolysate.** *Applied and Environmental Microbiology* **67**: 4249–4255

Jurado M, Prieto A, Martínez-Alcalá Á, Martínez Á T, Martínez M J. 2009. **Laccase detoxification of steam-exploded wheat straw for second generation bioethanol.** *Bioresource Technology* **100**: 6378–6384

Kapoor M K. 2007. **Alkalothermostable endo-β-1-4-xylanase from *Bacillus pumilus* strain MK001 and its application in bleaching of paper pulp.** Delhi: University of Delhi. [Doctoral thesis]

Kapoor R K, Chandel A K, Kuhar S, Gupta R, Kuhad R C. 2007. **Bioethanol from crop residues, production forecasting and economics: an Indian perspective** In *Lignocellulose Biotechnology: future prospects* New Delhi: I K International Publication

Kapoor M, Nair L M, and Kuhad R C. 2008. **Cost-effective xylanase production from free and immobilized *Bacillus pumilus* strain *MK001* and its application in saccharification of *Prosopis juliflora*.** *Biochemical Engineering Journal* **38**: 88–97

Karhumaa K, Hahn-Hägerdal B, and Gorwa-Grauslund M F. 2005. **Investigation of limiting metabolic steps in the utilization of xylose by recombinant *Saccharomyces cerevisiae* using metabolic engineering.** *Yeast* **22**: 359–368

Karhumaa K, Fromanger R, Hahn-Hägerdal B, Gorwa-Grauslund M F. 2007. **High activity of xylose reductase and xylitol dehydrogenase improves xylose fermentation by recombinant *Saccharomyces cerevisiae*.** *Applied Microbiology and Biotechnology* **73**: 1039–1046

Karunanithy C, Muthukumarappan K, and Julson J L. 2008. **Influence of high shear bioreactor parameters on carbohydrate release from different biomasses.** In American Society of Agricultural and Biological Engineers Annual International Meeting. St. Joseph: American Society of Agricultural and Biological Engineers. [ASABE 084114. ASABE, St. Joseph, Michigan]

Keller F A, Hamillton T E, and Nguyon Q A. 2003. **Microbial pretreatment of biomass potential for reducing severity of thermo-chemical biomass pre-treatment.** *Applied Biochemistry and Biotechnology* **105**: 27–41

Keshwani D R. 2009. **Microwave pretreatment of switchgrass for bioethanol production.** Raleigh: North Carolina State University. [Doctoral thesis]

Kim S and Dale E B. 2004. **Global potential bioethanol production from wasted crops and crop residues.** *Biomass and Bioenergy* **26**: 361–375

Kim S and Holtzapple M T. 2005. **Lime re-treatment and enzymatic hydrolysis of corn stover.** *Bioresource Technology* **96**: 1994–2006

Kim K and Hong J. 2001. **Supercritical CO_2 pretreatment of lignocellulose enhances enzymatic cellulose hydrolysis.** *Bioresource Technology* **77**: 139–144

Kim Y, Ingram I O, and Shanmugam K T. 2007. **Construction of an *Escherichia coli k-12* mutant for homoethanologenic fermentation of glucose or xylose without foreign genes.** *Applied and Environmental Microbiology* **73**: 1766–1771

Kootstra A M J, Beeftink H H, Scott E L, Sanders J P M. 2009. **Comparison of dilute mineral and organic acid pretreatment for enzymatic hydrolysis of wheat straw.** *Biochemical Engineering Journal* **46**: 126–131

Kotter P and Ciriacy M. 1993. **Xylose fermentation by *Saccharomyces cerevisiae.*** *Applied Microbiology and Biotechnology* **38**: 776–783

Kresge C T and Dhingra S S. 2004. **Molecular sieves.** *Kirk-Othmer Encyclopedia of Chemical Technology* **16**: 811–853

Kuhad R C. 2010. ***Process Development for the Production of Ethanol from Lignocellulosic Biomass.*** Delhi: Department of Biotechnology. [A Technical Report for Department of Biotechnology, Government of India]

Kuhad R C and Singh A. 1993. **Lignocellulose biotechnology: current and future prospects.** *Current Reviews in Biotechnology* **13**: 151–172

Kuhad R C, Singh A, and Eriksson K-E L. 1997. **Microorganisms and enzymes involved in the degradation of plant fiber cell wall.** *Advances in Biochemical Engineering and Biotechnology* **57**: 47–125

Kuhad R C, Manchanda M, and Singh A. 1999. **Hydrolytic potential of celluloytic enzymes from a mutant strain of *Fusarium oxysporum.*** *Bioprocess Engineering* **20**: 133–135

Kuhad R C, Kuhar S, Kapoor M, Sharma K K, Singh A. 2007. **Lignocellulolytic microorganisms, their enzymes and possible biotechnologies based on lignocellulolytic microorganisms and their enzymes.** In *Lignocellulose Biotechnology: future prospects*, pp. 3–32, edited by R C Kuhad and A Singh. New Delhi: IK International

Kuhad R C, Mehta G, Gupta R, Sharma K K. 2010. **Fed batch enzymatic saccharification of newspaper cellulosics improves the sugar content in the hydrolysates and eventually the ethanol fermentation by *Saccharomyces cerevisiae.*** *Biomass and Bioenergy.* doi: 10.1016/j.biombioe.2010.03.009

Kuhad R C, Gupta R, Khasa Y P, Singh A. 2010. **Bioethanol production from *Lantana camara* (red sage): pretreatment, saccharification and fermentation.** *Bioresource Technology* (submitted)

Kuhar S, Nair L M, and Kuhad R C. 2008. **Pretreatment of lignocellulosic material with fungi capable of higher lignin degradation and lower carbohydrate degradation improves substrate acid hydrolysis and the eventual conversion to ethanol.** *Canadian Journal of Microbiology* **54**: 305–313

Kuijvenhoven J. 2006. **Process design for an integrated lignocellulose to bioethanol production plant.** Paper presented at Netherlands Process Technology Symposium, October 25, 2006

Kumar P, Barrett D M, Delwiche M J, Stroeve P. 2009. **Methods for pretreatment of lignocellulosic biomass for efficient hydrolysis and biofuel production.** *Indstrial and Engineering Chemical Research* **48**: 3713–3729

Kuyper M, Hartog M M, Toirkens M J, Almering M J, Winkler A A, Dijken J P V. 2005. **Metabolic engineering of a xylose–isomerase-expressing *Saccharomyces cerevisiae* strain for rapid anaerobic xylose fermentation.** *FEMS Yeast Research* **5**: 399–409

Kuyper M, Winkler A A, Dijken J P V, Pronk J T. 2004. **Minimal metabolic engineering of *Saccharomyces cerevisiae* for efficient anaerobic xylose fermentation: a proof of principle.** *FEMS Yeast Research* **4**: 655–664

Ladisch M R and Dyck K. 1979. **Dehydration of ethanol—new approach gives positive energy-balance.** *Science* **205**: 898–900

Laser M, Schulman D, Allen S G, Lichwa J, Antal M J, Lynd L R. 2002. **A comparison of liquid hot water and steam pretreatments of sugarcane baggase for conversion to ethanol.** *Bioresource Technology* **81**: 33–44

Lawford H G and Rousseau J D. 1995. **Loss of ethanologenicity in *Escherichia coli recombinants* pLOI297 and KO11 during growth in the absence of antibiotics.** *Biotechnology Letters* **17**: 751–756

Lawford H G and Rousseau J D. 2002. **Performance testing of *Zymomonas mobilis* metabolically engineered for cofermentation of glucose, xylose and arabinose.** *Applied Biochemistry and Biotechnology* **98–100**: 429–448

Lee J. 1997. **Biological conversion of lignocellulosic biomass to ethanol.** *Journal of Biotechnology* **56**: 1–24

Lee D, Yu A H C, and Saddler J N. 1995. **Evaluation of cellulase recycling strategies for the hydrolysis of lignocellulosic substrates.** *Biotechnology and Bioengineering* **45**: 328–336

Lee J M, Jameel H, and Venditti R A. 2010. **A comparison of the autohydrolysis and ammonia fiber explosion (AFEX) pretreatments on the subsequent enzymatic hydrolysis of coastal Bermuda grass.** *Bioresource Technology* **101**: 5449–5458

Lee W G, Lee J S, Shin C S, Park S C, Chang H N, Chang Y K. 1999. **Ethanol production using concentrated oak wood hydrolysates and methods to detoxify.** *Applied Biochemistry and Biotechnology* **78**: 547–559

Lee S H, Doherty T V, Linhardt R J, Dordick J S. 2009. **Ionic liquid-mediated selective extraction of lignin from wood leading to enhanced enzymatic cellulose hydrolysis.** *Biotechnology Bioengineering* **102**: 1368–1376

Lee J M, Shi J, Venditti R A, Jameel H. 2009. **Autohydrolysis pretreatment of coastal Bermuda grass for increased enzyme hydrolysis.** *Bioresource Technology* **100**: 6434–6441

Li Q, He Y C, Xian M, Jun G, Xu X, Yang J M, Li L Z. 2009. **Improving enzymatic hydrolysis of wheat straw using ionic liquid 1-ethyl-3-methyl imidazolium diethyl phosphate pre-treatment.** *Bioresource Technology* **100**: 3570–3575

Licht F O. 2005. **World ethanol and biofuels.** *Report* **3**: 15

Lissens G, Klinke H, Verstraete W, Ahring B, Thomsen A B. 2004. **Wet oxidation treatment of organic household waste enriched with wheat straw for simultaneous saccharification and fermentation into ethanol.** *Environmental Technology* **25**: 647–655

Liu E and Hu Y. 2010. **Construction of a xylose-fermenting *Saccharomyces cerevisiae* strain by combined approaches of genetic engineering, chemical mutagenesis and evolutionary adaptation.** *Biochemical Engineering Journal* **48**: 204–210

Londo H M, Lensink S M, Deurwaarder E P, Wakker A, de Wit M P, Junginger H M, Könighofer K, Jungmeier G. 2008. **Biofuels development in the EU27+ until 2030: full-chain cost assessment and implications of policy options.** REFUEL Project, WP4 Final Report

Lu Y and Mosier N S. 2008. **Current technologies for fuel ethanol production from lignocellulosic plant biomass.** In *Genetic Improvement of Bioenergy Crops*, pp. 161–182, edited by W. Vermerris. Berlin: Springer

Lynd L R, Cushman J H, Nichols R J, Wyman C E. 1991. **Fuel ethanol from cellulosic biomass.** *Science* **251**: 1318–1323

Lynd L R, Wyman C E, and Gerngross T U. 1999. **Biocommodity engineering.** *Biotechnology Progress* **15**: 777–793

MacLean H L, Lave L B, Lankey R, Joshi S. 2000. **A life-cycle comparison of alternative automobile fuels.** *Journal of Air and Waste Management Association* **50**: 1769–1779

Madhavan A, Tamalampudi S, Ushida K, Kanai D, Katahira S, Srivastava A, Fukuda H, Bisaria V S, Kondo A. 2009. **Xylose isomerase from polycentric fungus *Orpinomyces:* gene sequencing, cloning and expression in *Saccharomyces cerevisiae* for bioconversion of xylose to ethanol.** *Applied Microbiology and Biotechnology* **82**: 1067–1078

Madson P W 2003. **Ethanol distillation: the fundamentals.** In *The Alcohol*. England: Nottingham University Press, University of Nottingham. 446 pp.

Mamma D, Christakopoulos P, Koullas D, Kekos D, Macris B J, Koukios E. 1995. **An alternative approach to the bioconversion of sweet sorghum carbohydrates to ethanol.** *Biomass and Bioenergy* **8**: 99–103

Margeot A, Hahn-Hagerdal B, Edlund M, Slade R, Monot F. 2009. **New improvements for lignocellulosic ethanol.** *Current Opinion in Biotechnology* **20**: 372–380

Martín C, Thomsen M H, Hauggaard H, Thomsem A B. 2008. **Wet oxidation re-treatment, enzymatic hydrolysis and simultaneous saccharification and fermentation of clover-ryegrass mixtures.** *Bioresource Technology* **99**: 8777–8782

Matthew H, Ashley O, Brian K, Alisa E, Benjamin J S. 2005. **Wine making 101.** Details available at <http://www.arches.uga.edu/<matthaas/strains.htm>

Matushika A, Watanabe S, Kodaki T, Makino K, Sawayama S. 2008. **Bioethanol production from xylose by recombinant *Saccharomyces cerevisiae* expressing xylose reductase, NADP$^+$-dependent xylitol dehydrogenase and xylulokinase.** *Journal of Bioscience and Bioengineering* **105**: 296–299

Matsushika A, Inoue H, Murakami K, Takimura O, Sawayama S. 2009. **Bioethanol production performance of five recombinant strains of laboratory and industrial xylose-fermenting *Saccharomyces cerevisiae*.** *Bioresource Technology* **100**: 2392–2398

McIntosh S and Vancov T. 2010. **Enhanced enzyme saccharification of *Sorghum pre-trea* straw using dilute alkali pre-treatment.** *Bioresource Technology*. doi: 10.1016/j.biortech.2010.03.116

McMillan J D. 1994. **Pretreatment of lignocellulosic biomass.** *ACS Symposium Series* **566**: 292–324

McMillan J D. 1997. **Bioethanol production: status and prospects.** *Renewable Energy* **10**: 295–302

McMillan J D, Newman M M, Templeton D W, Mohagheghi A. 1999. **Simultaneous saccharification and cofermentation of dilute-acid pretreated yellow poplar hardwood to ethanol using xylose fermenting *Zymomonas mobilis*.** *Applied Biochemistry and Biotechnology* **77/79**: 649–655

Millati R, Edebo L and Taherzadeh M J. 2005. **Performance of *Rhizopus, Rhizomucor and Mucor* in ethanol production from glucose, xylose and wood hydrolysates.** *Enzyme and Microbial Technology* **36**: 294–300

Mohagheghi A, Evans K, Chou Y C, Zhang M. 2002. **Cofermentation of glucose, xylose, and arabinose by genomic DNA-integrated xylose/arabinose fermenting strain of *Zymomonas mobilis* AX101.** *Applied Biochemistry and Biotechnology* **98–100**: 885–898

Mok W S L and Antal M J. 1992. **Uncatalyzed solvolysis of whole biomass hemicellulose by hot compressed liquid water.** *Industrial and Engineering Chemical Research* **31**: 1157–1161

Moniruzzaman M. 1995. **Alcohol fermentation of enzymatic hydrolysate of exploded rice straw by *Pichia stipitis*.** *World Journal of Microbiology and Biotechnology* **11**: 646–646

Mosier N S, Wyman C, Dale B, Elander R, Lee Y Y, Holtzapple M, Ladisch M R. 2005. **Features of promising technologies for pretreatment of lignocellulosic biomass.** *Bioresource Technology* **96**: 673–686

Mountfort D O and Rhodes L L. 1991. **Anaerobic growth and fermentation characteristics of *Paecilomyces lilacinus* isolated from *Mullet* gut.** *Applied and Environmental Microbiology* **57**: 1963–1968

Nigam J N. 2001a. **Development of xylose-fermenting yeast *Pichia stipitis* for ethanol production through adaptation on hardwood hemicellulose acid prehydrolysate.** *Journal of Applied Microbiology* **90**: 208–215

Nigam J N. 2001b. **Ethanol production from hardwood spent sulfite liquor using an adapted strain of *Pichia stipitis*.** *Journal of Industrial Microbiology and Biotechnology* **26**: 145–150

Nigam J N. 2001c. **Ethanol production from wheat straw hemicellulose hydrolysate by *Pichia stipitis*.** *Journal of Biotechnology* **87**: 17–27

NREL (National Renewable Energy Laboratory). 2007. ***NREL Report on Research Advances in Cellulosic Ethanol.*** Golden: NREL. [NREL/BR-510-40742 March 2007]

O'Dwyer J P, Zhu L, Granda C B, Holtzapple M T. 2007. **Enzymatic hydrolysis of lime-pretreated corn stover and investigation of the HCH-1 model: inhibition pattern, degree of inhibition, validity of simplified HCH-1 model.** *Bioresource Technology* **98**: 2969–2977

Ohta K, Beall D S, Mejia J P, Shanmugam K T, Ingram L O. 1991. **Genetic improvement of *Escherichia coli* for ethanol production: chromosomal integration of *Zymomonas mobilis* genes encoding pyruvate decarboxylase and alcohol dehydrogenase II.** *Applied and Environmental Microbiology* **57**: 893–900

Ohgren K, Bura R, Lesnicki G, Saddler J, Zacchi G. 2007. **A comparison between simultaneous saccharification and fermentation and separate hydrolysis and fermentation using steam-pretreated corn stover.** *Process Biochemistry* **42**: 834–839

Okur M and Saraçoglu N. 2008. **Fermentation of sunflower seed hull hydrolysate to ethanol by *Pichia stipitis*.** *Bioresource Technology* **99**: 2162–2169

Olsson L and Hahn-Hagerdal B. 1996. **Fermentationof lignocellulosic hydrolysates for ethanol production.** *Enzyme and Microbial Technology* **18**: 312–331

Palmqvist E and Hahn-Hägerdal B. 2000a. **Fermentation of lignocellulosic hydrolysates. I: inhibition and detoxification.** *Bioresource Technology* **74**: 17–24

Palmqvist E and Hahn-Hägerdal B. 2000b. **Fermentation of lignocellulosic hydrolysates. II: inhibition and detoxification.** *Bioresource Technology* **74**: 25–33

Park J, Shiroma R, Al-Haq M I, Zhang Y, Ike M, Arai-Sanoh Y, Ida A, Kondo M, Ken T. 2010. **A novel lime re-treatment for subsequent bioethanol production from rice straw—calcium capturing by carbonation (CaCCO) process.** *Bioresource Technology*. doi: 10.1016/j.biortech.2010.03.098

Patel G B. 1984. **Ethanol production during o-xylose, L-arabinose, and D-ribose fermentation by *Bacteroides polypragmatus*.** *Applied Microbiology Biotechnology* **20**: 111–117

Pérez J A, Ballesteros I, Ballesteros M, Sáez F, Negro M J, Manzanares P. 2008. **Optimizing liquid hot water pretreatment conditions to enhance sugar recovery from wheat straw for fuel-ethanol production.** *Fuel* **87**: 3640–3647

Polizeli M L T M, Rizzatti A C S, Monti R, Terenzi H F, Jorge J A, Amorim D S. 2005. **Xylanases from fungi: properties and industrial applications.** *Applied Microbiology and Biotechnology* **67**: 577–591

Purwadi R and Taherzadeh M J. 2008. **The performance of serial bioreactors in rapid continuous production of ethanol from dilute-acid hydrolyzates using immobilized cells.** *Bioresource Technology* **99**: 2226–2233

Purwadi R, Brandberg T, and Taherzadeh M J. 2007. **A possible industrial solution to ferment lignocellulosic hydrolyzate to ethanol: continuous cultivation with flocculating yeast.** *International Journal of Molecular Sciences* **8**: 920–932

Raman N and Pothiraj C. 2008. **Screening of *Zymomonas mobilis* and *Saccharomyces cerevisiae* strains for ethanol production from cassava waste.** *Rasayan Journal of Chemistry* **1**: 537–541

Ramos L P, Beruil C, and Saddler J N. 1993. **The use of enzyme recycling and the influence of sugar accumulation on the cellulose hydrolysis by *Trichoderma* cellulases.** *Enzyme and Microbial Technology* **15**: 19–25

Rocha M V P, Rodrigues T H, de Macedo G R, Gonçalves L R. 2009. **Enzymatic hydrolysis and fermentation of pretreated cashew apple bagasse with alkali and diluted ulphuric acid for bioethanol production.** *Applied Biochemistry and Biotechnology* **155**: 407–417

Roukas T and Kotzekidou P. 1998. **Lactic acid production from deproteinized whey by mixed cultures of free and coimmobilized *Lactobacillus casei* and *Lactococcus lactis* cells using fed-batch culture.** *Enzyme and Microbial Technology* **22**: 199–204

Rowell R M, Pettersen R, Han J S, Rowell J S, Tshabalala M A. 2005. **Cell wall chemistry.** In *Handbook of Wood Chemistry and Wood Composites*, pp. 35–62, edited by R M Rowell. Florida: CRC Press

Ruiz E, Romero I, Moya M, Sánchez S, Bravo V, Castro E. 2007. **Sugar fermentation by *Fusarium oxysporum* to produce ethanol** *World Journal of Microbiology and Biotechnology* **23**: 259–267

Runquist D, Hahn-Hägerdal B, and Bettiga M. 2009. **Increased expression of the oxidative pentose phosphate pathway and gluconegenesis in anaerobically growing xylose-utilizing *Sacharomyces cerevisiae*.** *Microbial Cell Factories* doi: 10.1186/1475-2859-8-49

Sakakibara A. 1991. **Chemistry of lignin.** In *Wood and Cellulosic Chemistry*, pp. 113–175, edited by D N-S Hon and N Shiraishi. New York: Marcel Dekker

Sanchez C. 2009. **Lignocellulosic residues: biodegradation and bioconversion by fungi.** *Biotechnology Advances* **27**: 185–194

Sanchez Ó J and Cardona C A. 2008. **Trends in biotechnological production of fuel ethanol from different feedstocks.** *Bioresource Technology* **99**: 5270–5295

Saraçoglu N and Arslan Y. 2000. **Comparison of different pretreatments in ethanol fermentation using corn stover hemicellulosic hydrolyzate with *Pichia stipitis* and *Candida shehatae*.** *Biotechnology Letters* **22**: 855–858

Sarkanen K V and Ludwig C H. 1971. ***Lignins: occurrence, formation, structure and reactions.*** New York: Wiley

Schacht C, Zetzl C, and Brunner G. 2008. **From plant materials to ethanol by means of supercritical fluid technology.** *Journal of Supercritical Fluids* **46**: 299–321

Schneider H, Wang P Y, Chan Y K, Maleszka R. 1981. **Conversion of n-xylose into ethanol by the yeast *Pachysolen tannophilus*.** *Biotechnology Letters* **3**: 89–92

Shafizadeh F and Bradbury A G W. 1979. **Thermal degradation of cellulose in air and nitrogen at low temperatures.** *Journal of Applied Polymer Science* **23**: 1431–1442

Shallom D and Shoham Y. 2003. **Microbial hemicellulases.** *Current Opinion in Microbiology* **6**: 219–228

Sharma S K, Kalra K L, and Kocher G S. 2004. **Fermentation of enzymatic hydrolysate of sunflower hulls for ethanol production and its scale up.** *Biomass and Bioenergy* **27**: 399–402

Sheehan J and Himmel M. 1999. **Enzymes, energy, and the environment: a strategic perspective on the U.S. Department of Energy's research and development activities for bioethanol.** *Biotechnology Progress* **15**: 817–827

Shi N Q, Davis B, Sherman F, Cruz J, Jeffries T W. 1999. **Disruption of the cytochrome c gene in xylose-utilizing yeast *Pichia stipitis* leads to higher ethanol production.** *Yeast* **15**: 1021–1030

Shi N Q, Cruz J, Sherman F, Jeffries T W. 2002. **SHAM-sensitive alternative respiration in the xylose-metabolizing yeast *Pichia stipitis*.** *Yeast* **19**: 1203–1220

Singh A and Mishra P. 1993. **Microbial production of ethanol.** In *Microbial Pentose Utilization: current application in biotechnology (Progress in Industrial Microbiology)*, Vol. 33, pp. 147–196

Singh A, Kumar P K R, and Schugerl K. 1991. **Adsorption and reuse of cellulase during saccharification of cellulosic materials.** *Journal of Biotechnology* **19**: 205–212

Sivers M V and Zacchi G. 1995. **A techno-economical comparison of three processes for the production of ethanol from pine.** *Bioresource Technology* **51**: 43–52

Soccol C R, Vandenberghe L P S, Medeiros A B P, Karp S G, Buckeridge M, Ramos L P, Pitarelo A P, Ferreira-Leitão V, Gottschalk L M F, Ferrara M A, Bon E P S, de Moraes L M P, Araújo J A, Torres F A G. 2010. **Bioethanol from lignocelluloses: status and perspectives in Brazil.** *Bioresource Technology* **101**: 4820–4825

Soderstrom J, Galbe M, and Zacchi G. 2005. **Separate versus simultaneous saccharification and fermentation of two-step steam pretreated softwood for ethanol production.** *Journal of Wood Chemistry and Technology* **25**: 187–202

Sonderegger M, Jeppsson M, Hahn-Hägerdal B, Sauer U. 2004. **Molecular basis for anaerobic growth of *Saccharomyces cerevisiae* on xylose, investigated by global gene expression and metabolic flux analysis.** *Applied and Environmental Microbiology* **70**: 2307–2317

Sreenath H K, Koegel R G, Moldes A B, Jeffries T W, Straub R J. 1999. **Enzymic saccharification of alfalfa fibre after liquid hot water pre-treatment.** *Process Biochemistry* **35**: 33–41

Sternberg D. 1976. **Production of cellulase by *Trichoderma*.** *Biotechnology and Bioengineering Symposium* 35–53

Stoutenberg R M, Perrotta J A, Amidon T E, Nakas J P. 2008. **Ethanol production from a membrane purified hemicellulosic hydrolysate derived from Sugar maple by *Pichia stipitis* NRRL Y-7124.** *Bioresources* **3**: 1349–1358

Sun Y and Cheng J. 2002. **Hydrolysis of lignocellulosic materials for ethanol production: a review.** *Bioresource Technology* **83**: 1–11

Sun R C and Tomkinson R C. 2002. **Characterization of hemicelluloses obtained by classical and ultrasonically assisted extractions from wheat straw.** Carbohydrate Polymers **50**: 263–271

Suryawati L, Wilkins M R, Bellmer D D, Huhnke R L, Maness N O, Banat I M. 2008. **Simultaneous saccharification and fermentation of Kanlow switchgrass pretreated by hydrothermolysis using *Kluyveromyces marxianus* IMB4.** *Biotechnology and Bioengineering* **101**: 894–902

Swain R L B. 2003. **Development and operation of the molecular sieve: an industry standard.** In *The Alcohol*, 446 pp. Nottingham: Nottingham University Press

Szczdodrak J and Targonski Z. 1988. **Selection of thermotolerant yeast strains for simultaneous saccharification and fermentation of cellulose.** *Biotechnology and Bioengineering* **31**: 300–303

Taherzadeh M J, Millati R, and Niklasson C. 2001. **Continuous cultivation of dilute-acid hydrolyzates to ethanol by immobilized *Saccharomyces cerevisiae*.** *Applied Biochemistry and Biotechnology* **95**: 45–57

Taherzadeh M J and Karimi K. 2008. **Pretreatment of lignocellulosic wastes to improve ethanol and biogas production: a review.** *International Journal of Molecular Sciences* 9: 1621–1651

Takuma S, Nakashima N, Tantirungkij M, Kinoshita S, Okada H, Seki T. 1991. **Isolation of xylose reductase gene of *Pichia stipitis* and its expression in *Saccharomyces cerevisiae*.** *Applied and Environmental Microbiology* **28–29**: 327–340

Talebnia F and Taherzadeh M J. 2006. **In situ detoxification and continuous cultivation of dilute-acid hydrolyzate to ethanol by encapsulated *Saccharomyces cerevisiae*.** *Journal of Biotechnology* **125**: 377–384

Talebnia F, Pourbafrani M, Lundin M, Taherzadeh M J. 2008. **Optimization study of citrus wastes saccharification by dilute acid hydrolysis.** *BioResources* 3: 108–122

Tang Y, An M, Liu K, Nagai S, Shigematsu T, Morimura S, Kida K. 2006. **Ethanol production from acid hydrolysate of wood biomass using the flocculating yeast *Saccharomyces cerevisiae* strain KF-7.** *Process Biochemistry* **41**: 909–914

Taniguchi M, Suzuki H, Watanabe D, Sakai K, Hoshino K, Tanaka T. 2005. **Evaluation of pretreatment with *Pleurotus ostreatus* for enzymatic hydrolysis of rice straw.** *Journal of Bioscience and Bioengineering* **100**: 637–643

Tantirungkij M, Nakashima N, Seki T, Yoshida T. 1993. **Construction of xylose-assimilating *Saccharomyces cerevisiae*.** *Journal of Fermentation and Bioengineering* **75**: 83–88

Teeri T T. 1997. **Crystalline cellulose degradation: new insight into the function of cellobiohydrolases.** *Trends in Biotechnology* **15**: 160–167

Teeri T T, Koivula A, Linder M, Wohlfahrt G, Divne C, Jones T A. 1998. **Trichoderma reesei cellobiohydrolases: why so efficient on crystalline cellulose?** *Biochemical Society Transactions* **26**: 173–178

Teymouri F, Laureano-Perez L, Alizadeh H, Dale B E. 2005. **Optimization of the ammonia fiber explosion (AFEX) treatment parameters for enzymatic hydrolysis of corn stover.** *Bioresource Technology* **96**: 2014–2018

Torget R W, Kim J S, and Lee Y Y. 2000. **Fundamental aspects of dilute acid hydrolysis/fractionation kinetics of hardwood carbohydrates I. Cellulose hydrolysis.** *Industrial Engineering and Chemical Research* **39**: 2817–2825

Toivari M H, Aristidou A, Ruohonen L, Penttila M. 2001. **Conversion of xylose to ethanol by recombinant *Saccharomyces cerevisiae*: importance of xylulokinase (XKS1) and oxygen availability.** *Metabolic Engineering* **3**: 236–249

Träff-Bjerre K L, Jeppsson M, Hahn-Hägerdal B, Gorwa-Grauslund M F. 2004. **Endogenous NADPH-dependent aldose reductase activity influences product formation during xylose consumption in recombinant *Saccharomyces cerevisiae*.** *Yeast* **21**: 141–150

Tu M B, Chandra R P, and Saddler J N. 2007. **Evaluating the distribution of cellulases and the recycling of free cellulases during the hydrolysis of lignocellulosic substrates.** *Biotechnology Progress* **23**: 398–406

US DOE (Department of Energy). 2005. ***Multi-Year Program Plan 2007–2012. Biomass and Biorefinery System R&D. Biomass Program, Energy Efficiency and Renewable Energy.*** Washington, DC: US DOE

US DOE (Department of Energy). 2007. ***Biomass Multi-Year Program Plan. Office of the Biomass Program, Energy Efficiency and Renewable Energy.*** Washington, DC: US DOE

US DOE (Department of Energy). 2009. ***Biomass Multi-Year Program Plan (MYPP). Office of the Biomass Program, Energy Efficiency and Renewable Energy.*** Washington, DC: US DOE

Vandeska E, Amartey S, Kuzmanova S, Jeffries T W. 1996. **Fed-batch culture for xylitol production by *Candida boidinii*.** *Process Biochemistry* **31**: 265–270

Vasudevan T G, Isak S P, and Ricardo R C O. 2007. **Molecular cloning and functional expression of a novel *Neurospora crassa* xylose reductase in Saccharomyces cerevisiae in the development of a xylose fermenting strain.** *Annals in Microbiology* 57: 223–231

Vásquez M P, da Silva J N C, de Souza M B, Pereira N. 2007. **Enzymatic hydrolysis optimization to ethanol production by simultaneous saccharification and fermentation.** *Applied Biochemistry and Biotechnology* **137–140**: 141–153

Verho R, Londesborough J, Penttila M, Richard P. 2003. **Engineering redox cofactor regeneration for improved pentose fermentation in *Saccharomyces cerevisiae.*** *Applied and Environmental Microbiology* **69**: 5892–5897

Vidal P F and Molinier J. 1988. **Ozonolysis of lignin—improvement of in vitro digestibility of poplar sawdust.** *Biomass* **16**: 1–17

Vleet J H V, Jeffries T W, and Olsson L. 2008. **Deleting the paranitrophenyl phosphatase (pNPPase), PHO13, in recombinant *Saccharomyces cerevisiae* improves growth and ethanol production on D-xylose.** *Metabolic Engineering* **10**: 360–369

von Lampe F, Zygalsky I, Nagibzadeh H, Hinrichsen G. 2006. **Influence of cation modification on the environmental degradation of high temperature superconducting Y–Ba–Cu–O films.** *Physica C: Superconductivity* **436**: 123–127

Wang D, Xu Y, and Zhao G. 2004. **Fermentation kinetics of different sugars by apple wine yeast *Saccharomyces cerevisiae.*** *Journal of Institute of Brewing* **110**: 340–346

Wang L S, Zhang Y Z, Gao P J, Shi D X, Liu H W, Gao H J. 2006. **Changes in the structural properties and rate of hydrolysis of cotton fibers during extended enzymatic hydrolysis.** *Biotechnology and Bioengineering* **93**: 443–456

Wahlbom F, Otero R R C, van Zyl W H, Hahn-Hägerdal B, Jonsson L J. 2003. **Molecular analysis of a *Saccharomyces cerevisiae* mutant with improved ability to utilize xylose shows enhanced expression of proteins involved in transport, initial xylose metabolism, and the pentose phosphate pathway.** *Applied and Environmental Microbiology* **69**: 740–746

Wankat P C. 1988. **Equilibrium staged separations.** In *Separations in Chemical Engineering*. New York: Elsevier. 707 pp.

Watanabe I, Nakamura T, and Shima J. 2010. **Strategy for simultaneous saccharification and fermentation using a respiratory-deficient mutant of *Candida glabrata* for bioethanol production.** *Journal of Bioscience and Bioengineering*. doi: 10.1016/j.jbiosc.2010.02.003

Weil J, Westgate P, Kohlmann K, Ladisch M R. 1994. **Cellulose pretreatments of lignocellulosic substrates.** *Enzyme and Microbial Technology* **16**: 1002–1004

Weil J R, Sarikaya A, Rau S L, Goetz J, Ladisch C M, Brewer M, Hendrickson R, Ladisch M R. 1998. **Pretreatment of corn fiber by pressure cooking in water.** *Applied Biochemistry and Biotechnology* **73**: 1–17

Withers S G. 2001. **Mechanisms of glycosyl transferases and hydrolases.** *Carbohydrate Polymers* **44**: 325–337

Wright M M and Brown R C. 2007. **Comparative economics of biorefineries based on the biochemical and thermochemical platforms.** *Biofuels Bioproduction and Biorefinery* **1**: 49–56

Wyman CE. 1994. **Ethanol from lignocellulosic biomass: technology, economics and opportunities.** *Bioresource Technology* **50**: 3–16

Wyman C E. 2003. **Potential synergies and challenges in refining cellulosic biomass to fuels, chemicals, and power.** *Biotechnology Progress* **19**: 254–262

Xu F, Sun J X, Liu C F, Sun R C. 2003. **Comparative study of alkali- and acidic organic solvent-soluble hemicellulosic polysaccharides from sugarcane bagasse.** *Carbohydrate Research* **341**: 253–261

Yachmenev V, Condon B, Klasson T, Lambert A. 2009. **Acceleration of the enzymatic hydrolysis of corn stover and sugar cane bagasse celluloses by low intensity uniform ultrasound.** *Journal of Biobased Materials and Bioenergy* **3**: 25–31

Yang B and Wyman C E. 2004. **Effect of xylan and lignin removal by batch and flow-through pretreatment on the enzymatic digestibility of corn stover cellulose.** *Biotechnology and Bioengineering* **86**: 88–95

Yang S, Tschaplinski T J, Engle N L, Carroll S L, Martin S L, Davison B H, Palumbo1 A V, Rodriguez Jr M, Brown S D. 2009. **Transcriptomic and metabolomic profiling of *Zymomonas mobilis* during aerobic and anaerobic fermentations.** *BioMed Central Genomics* doi: 10.1186/1471-2164-10-34

Yomano L P, York S W, and Ingram L O. 1998. **Isolation and characterization of ethanol-tolerant mutants of *Escherichia coli* KO11 for fuel ethanol production.** *Journal of Industrial Microbiology and Biotechnology* **20**: 132–138

Zaldivar J, Borges A, Johansson B, Smits H P, Villas-Boas S G, Nielsen J, Olsson L. 2002. **Fermentation performance and intracellular metabolite patterns in laboratory and industrial xylose-fermenting *Saccharomyces cerevisiae*.** *Applied Microbiology and Biotechnology* **59**: 436–442

Zhang M, Eddy C, Deanda K, Finkelstein M, Picataggio S. 1995. **Metabolic engineering of a pentose metabolism pathway in ethanologenic *Zymomonas mobilis*.** *Science* **267**: 240–243

Zhao X, Cheng K and Liu D. 2009. **Organosolv pretreatment of lignocellulosic biomass for enzymatic hydrolysis.** *Applied Microbiology and Biotechnology* **82**: 815–827

Zhao J and Xia L. 2010. **Ethanol production from corn stover hemicellulosic hydrolysate using immobilized recombinant yeast cells.** *Biochemical Engineering Journal* **49**: 28–32

Zheng Y Z, Lin H M, and Tsao G T. 1998. **Pretreatment for cellulose hydrolysis by carbon dioxide explosion.** *Biotechnology Progress* **14**: 890–896

Zhu Z, Sathitsuksanoh N, Vinzant T, Schell D J, McMillan J D, Zhang Y-H. 2009. **Comparative study of corn stover pretreated by dilute acid and cellulose solvent-based lignocellulose fractionation: enzymatic hydrolysis, supramolecular structure, and substrate accessibility.** *Biotechnology and Bioengineering* **103**: 715–724

Zhu S, Wu Y, Yu Z, Wang C, Yu F, Jin S, Ding Y, Chi R, Liao J, Zhang Y. 2006. **Comparison of three microwave/chemical pretreatment processes for enzymatic hydrolysis of rice straw.** *Biosystems Engineering* **93**: 279–283

Zwart R W R, Boerrigter H, and Drift A. 2006. **The impact of biomass pretreatment on the feasibility of overseas biomass conversion to Fisher-Tropsch products.** *Energy and Fuels* **20**: 2192–2197

Bioethanol: a status review on metabolic pathway modification of certain ethanologenic bacteria through genetic engineering

B Lal, S Patle, J Kurian

The Energy and Resources Institute, New Delhi, India

and

Ramanjaneya V R Mula

Post Doctoral Fellow, Torrey Pines Institute for Molecular Studies, Florida, USA

The world's energy supply is mainly dependent on non-renewable, crude-oil-derived (fossil) liquid fuels, of which almost 90% are consumed for energy generation and transportation. The problem of rapidly increasing population has caused many developing countries to expand their industrial base, resulting in increased energy demands. The world's energy needs have been estimated to be about 10^{21} joules per year, which equals a continuous average per capita consumption of about 4–6 kW (Tripetchkul, Hillary, and Ishizaki 1998). It is inevitable that fossil fuels such as oil, coal, and natural gas will be exhausted with time. Hence, there is a need to explore the possibilities of using alternative energy sources, which are as efficient as oil—ethanol fermentation is one such option (Nomuro, Bin, and Nakao 2002).

A few major drawbacks of non-renewable fuels are that the emissions of carbon dioxide (CO_2), carbon monoxide (CO), nitrogen oxide, and sulphur-containing residues from the combustion of these fuels cause global warming. Moreover, the dependence on imported petroleum for use as vehicle fuel is a fiscal burden and a concern for the national energy security. Studies show that vehicular emissions cause 40%–65% of urban air pollution, which deteriorates the air quality. The development of alternative transportation fuels derived from biomass resources, such as plant material and biodegradable waste, may contribute to reduced CO emissions and smog-forming compounds that cause air pollution. The other group of fuels comprises liquefied natural gas (LNG), compressed natural gas (CNG), liquefied petroleum gas (LPG), ethanol, methanol, and bio-diesel (Table 1). Other important biofuels under development are hydrogen and butanol. Among these, ethanol is the oldest of the few alternative energy fuels; fermentation of biomass converts it into pure ethanol, a useful energy source.

Ethanol produced from plant and microbial biomass is known as bioethanol: a carbon-neutral compound. The traditional process of ethanol production is through the fermentation of sugars with a species of yeast called *Saccharomyces cerevisiae*. However, the changing needs, energy demands, and technological advances to overcome the general limitations in yeast-based ethanologenic fermentations have led to a search for different methods using a broad range of substrates and novel organisms, indigenous or genetically modified. New technologies are being developed that convert the fibrous portion of the plant material into bioethanol. These feedstock materials are abundant and inexpensive.

Bioethanol: world scenario

World bioethanol production was 17.25 billion litres in 2000, which increased to over 46 billion litres in 2007. Global demand for ethanol fuel could exceed 125 billion litres by 2020 because of the new government programmes in the US, Asia, and Europe (Balat and Balat 2009). The US government's Energy Independence and Security Act, 2007, made the addition of a minimum volume of renewable fuel into the gasoline sold in the country mandatory and targeted to produce 21 billion gallons of biofuels from cellulosic feedstocks by 2022. The blend used in the US is

Table 1 Basic characteristics of a few alternative fuels

Characteristics	Compressed natural gas	Ethanol (E85)	Liquefied natural gas	Liquefied petroleum gas	Methanol
Chemical structure	CH_4	CH_3CH_2OH	CH_4	C_3H_8	CH_3OH
Primary composition	Methane	Denatured ethanol and gasoline	Cryogenically cooled methanol	Propane	Methanol and gasoline
Main fuel source	Underground reserves	Corn grains, agricultural waste	Underground reserves	By-product of petroleum refining or natural gas processing	Natural gas, coal, or woody biomass
Energy content per gallon BTU (British thermal unit)	29 000	80 460	73 500	84 500	63 350
Energy ratio	3.94:1	1.42:1	1.55:1	1.36:1	1.75:1
Compared to gasoline	25% at 300 psi	70%	66%	74%	57%
Status	Gas	Liquid	Liquid	Liquid	Liquid

10% anhydrous ethanol; it is primarily produced from corn and is blended at a 10% concentration with petrol (Graham, Lichtenberg, Roningen, *et al.* 1995; Mauguin, Bonfils, and Nacfaire 1994). Recently enacted Farm Bill, 2008, provides a monetary subsidy of $1.01 per gallon of cellulosic ethanol produced (Dwivedi, Alavalapati, and Lal 2009). In Europe, the rise in international oil prices and the growing concern over global warming drove the ethanol production, and trade increased, and in 2007, more than 1.7 billion litres of ethanol was produced from wheat and sugar beets (Farinelli, Carter, Lin, *et al.* 2009). The more sophisticated conversion technologies (enzymatically catalysed options) are, however, likely to benefit from further research and development (R&D), mainly in the form of costs becoming at par with those of the fossil fuels. Nevertheless, there is an attractive technology potential for both bio-diesel and bioethanol to attain 12% of the European market share by 2020, which equals 45 million tonnes (MT) of the oil equivalent.

In Brazil, the total ethanol production was 27 billion litres in 2009, which is estimated to increase to 105 billion litres by 2025. Most of the flex fuel vehicles in the country are running on 20%–25% ethanol blend with gasoline. The energy derived from sugar cane constitutes 13.5% of the total energy matrix of the country (Soccol, de Souza Vandenberghe, Medeiros, *et al.* 2010). However, no initiatives have been undertaken to improve ethanol production from the level maintained since 1988—this may lead to a critical fuel shortage by 2015. The problems of bioethanol production at a cheaper price and the increase in price of refined sugar in the international market persist. The total stock of cars and light vehicles was estimated to increase from 12 million to 14 million by 2000, and to 29 million by 2015 (Graham, Lichtenberg, Roningen, *et al.* 1995; Mauguin, Bonfils, and Nacfaire 1994).

Bioenergy

Energy derived from biomass resources is called bioenergy. Bioenergy releases less net CO_2 into the atmosphere and contributes to the mitigation of global warming through pollution control.

Biomass used for ethanol production comes from plants. Plants use solar energy to convert the atmospheric CO_2 into biomass (for example, leaves, roots, stems, and seeds), composed mainly of cellulose and lignin. Some biomass integrates with soil, where its carbon may be sequestered for hundreds of years. The harvested portion above ground can be used for the same purpose as fossil feedstock, that is, for fuel, electricity, and materials. When biomass is used, carbon is transformed back into CO_2 and taken up again by a new generation of trees and plants. Consequently, the net carbon emission from using the energy generated by biomass feedstock is potentially close to zero, especially as the carbon is also sequestered in soil. Thus, the production of more biomass feedstock can reduce the net carbon emissions, decrease climatic impact, and enhance the nation's energy security.

Examples of biomass that can be used for bioethanol production are lignocellulosic biomass, simple sugars, starch, forestry waste, agricultural residues, household garbage and paper products, and food processing and other industrial waste. Search for novel micro-organisms for potential industrial use, which perform in a much better way than the traditional yeast, is in the offing. Biotechnological tools have enhanced the scope to incorporate genetically desirable traits for maximizing performance of the microbe and increasing its substrate range from just monomers to polymers. Pathway interventions, mutational engineering, and mutagenesis have helped in developing strains that perform optimally in any given situation.

Biotechnological approach for increasing ethanol production

The revolution that crept into the present-day scientific arena after the launch of the Human Genome Project now promises far-reaching benefits. We have, in hand, the complete deoxyribonucleic acid (DNA) sequence or nearly complete DNA sequence of these genomes from different organisms—from microbes to plants to humans. There are more than 1000 complete microbial genomes available, and the number is increasing exponentially (Uchiyama, Higuchi, and Kawai 2010). The new genomic data and high-throughput technologies can be used for studying the proteins encoded by the genome and exploring the amazingly diverse natural capabilities in microbes. These can be used to achieve India's energy mission to enhance energy production and thereby reduce pollution. As these are all clean technologies, they will help mitigate the global climatic problems.

Genome research has opened up avenues to solve many biological, industrial, and environmental problems. It encompasses areas that identify the proteins important for all critical life functions, the gene regulatory networking that has the remote switch, and the microbial ecological functions and their interactions with potential implications on the environment-related issues. Biotechnological research should be focused at comprehending the processes of microbes and microbial systems in their native environments, and industrial applications in the present context in order to provide solutions to the problems of energy, global climate change, and environmental clean-up.

Ethanol fermentation

Ethanol, the most widely used biofuel, is produced using biomass in a process similar to that of brewing beer. Currently, most of the ethanol is obtained from molasses and starch. Although the potential of utilizing energy crops for making available starch to produce ethanol exists, ethanol production from agricultural waste that is rich in lignocelluloses is a more challenging option. Biotechnological options to use agricultural waste in the so-called simultaneous saccharification and fermentation (used in the conversion of corn starch) have to be explored.

Biomass is a renewable resource that is virtually inexhaustible and is a potential feedstock for alternative fuel production. Worldwide, biomass ranks fourth as an energy resource, meeting approximately 14% of the world's energy needs. In developing countries, biomass accounts for approximately 35% of the energy used, and in the rural areas of these countries, it is often the only accessible and affordable source of energy (Hall, Rosillo-Calle, and de Groot 1992; McGowan 1991). Dedicated energy crops, grown on our nation's wastelands, could possibly represent one of the largest biomass sources. Advances in plant sciences and process technologies promise to revolutionize the production of energy and other products from biomass.

Bottlenecks in conversion of lignocelluloses to ethanol

In the process of conversion of lignocelluloses to ethanol, the capital and process costs may well exceed those of the feedstock. Although compositions vary, the lignocellulosic material comprises 50% cellulose, 25% hemicellulose, and 25% extractable lignin. For bioconversion, the carbohydrate portion must be solubilized, whereas the lignin and residues are used to provide energy for ethanol production (Ingram and Doran 1995).

The glycosidic linkages in hemicelluloses are readily hydrolysed by dilute acid at an elevated temperature to yield a syrup containing xylose and arabinose for agricultural by-products and hardwoods or xylose, arabinose, and glucose for softwoods (Puls 1993). Saccharification of cellulose is a complex and poorly understood process, in which three enzymes—exoglucanases, endoglucanases, and glucosidases—that cleave cellulose into cellobiose and then into glucose are involved (Enari 1993; Eriksson, Blanchette, and Ander 1990). The products of saccharification induce non-competitive and non-productive inhibition of the enzyme system.

In the first case, wherein xylose, a pentose, becomes a non-amenable substrate for the efficient conversion of the same into ethanol, a microbial system is required, which can use the xylose or pentoses as efficiently as it uses glucose, fructose, and sucrose. For avoiding the end-product inhibition, simultaneous saccharification and fermentation were developed by the Gulf Oil Company (Blotkamp, Takagi, Pemberton, *et al.* 1978) by using *S. cerevisiae*. However, a few known organisms such as *Zymomonas mobilis* and certain yeasts such as *Candida shehatae* and *Pichia stipitis* are yet to be used to make bioethanol production a more profitable and viable technology.

Micro-organisms and ethanologenic processes

Micro-organisms are the largest reservoir of genetic and biochemical diversity on the earth. They evolved about 3.7 billion years ago, colonizing virtually every environment, often thriving under extremes of nutrient concentration, pH, salinity, pressure, and temperature. Genomic information is a tremendous resource for examining the function of genes and their potential applications in the industry and environment.

Zymomonas mobilis, the real workhorse for future ethanologenic processes, needs identification and characterization of molecular machines of the ethanologenic pathway, which form multiprotein complexes that execute cellular functions and control ethanol production. The yeast *S. cerevisiae* with its sequenced genome and proteome analysis gives amazing information on the different metabolic pathways and the intricately interacting proteins and their regulation. Apart from that, genetic engineering is being applied in inducing mutation in select organisms to enhance their capabilities for using them in the conversion of biomass to simpler carbon compounds and then fermenting these into ethanol in an efficient way. Development of ethanologenic *Escherichia coli* and *Klebsiella oxytoca* (Wood and Ingram 1992) is an advancement towards developing an efficient ethanologenic process to use biomass.

Background and strategy

Proteins, the chemically active components of all coding sequences of any genome, are highly dynamic entities, which carry out the primary functions of a cell in response to intracellular requirements and extracellular stimuli. Proteins are mostly organized into multiprotein complexes that carry out functions for metabolic activity, communication, growth, and structure. Hence, the need is to systematically identify, characterize, and explore these vital molecules of life. This will provide a knowledge base for linking proteome dynamics and architecture to cellular and organismic function under certain situations. The importance of post-transcriptional and post-translational regulation is recognized by emphasizing on the direct knowledge of proteins and their higher order associations in the metabolic pathways, in contrast to the less direct inferences about proteome composition drawn from the ribonucleic acid (RNA) measurements of the genes encoding the pathway enzymes. A key challenge is to learn how multiprotein machines act to regulate or execute the cell functions. An important step towards understanding the assembly, stability, and function of protein machines is to characterize protein–protein proximity and interfaces within complexes and between interacting complexes in a high-throughput manner.

The gene regulatory networks govern the facts as to which of the genes are to be expressed or switched on in a cell at any given time, how much product is made from each one, and the cell's responses to diverse environmental stimuli and intracellular signals. Gene regulatory networks are central to understanding and manipulating cells, and critical to any biotechnological exploitation of living beings, particularly microbes.

It is essential to discover the architecture, dynamics, and function of the gene regulatory networks in the target micro-organisms and construct their computational models in order to pinpoint the potential sites for modification or modulation. A comparative network structure and function analysis of the micro-organisms may give an insight into the fundamental issues with regard to the functioning of a particular pathway enzyme towards the formation of a commercially important product or function. The basic question here is how such complex living functions in a single-cell microbe are regulated so tightly. The answer may be found in the regulatory network architecture and complexity. The *cis*-acting regulatory apparatus implication, of which the gene regulatory networks are a critical part, is said to be at the interface between evolution and development (Yuh, Bolouri, and Davidson 1998).

An intriguing extension of this idea is that even subtle changes in the networks might result in an apparent variability in functionally important differences between different micro-organisms with respect to a single gene function or the function of a cluster. Tracking down the evolution of the so-called gene regulatory networks rather than individual genes will give rise to profound opportunities towards exploiting them for engineering better performing microbes for industrial and environmental applications.

A long-term goal is the mapping of microbial gene regulatory circuitries to construct maps for specific sub-genomic networks located across multiple species with a desirable single function of interest like ethanol and hydrogen production, and biodegradation of polychlorinated biphenyls.

The major focus will be to identify *cis*-acting regulatory sequences in the genome and the regulatory proteins that interact with them. Integral to this effort is the task of relating the regulatory apparatus to the groups of target genes regulated and to whatever is known about their function, which is possible through the use of comparative genomics to initiate large-scale identification and analysis of the genetic regulatory networking component.

Vigorous application of a comprehensive genome-wide approach to network mapping in select microbes has the potential to yield the first complete dissection of the regulatory networks that run a living cell (Laub, McAdams, Feldblyum, *et al.* 2000). The approach can be as follows.

- Extract regulatory elements, including operon and regulatory sequences, using sequence-level comparative genomics and decipher how these influence a pathway.
- Simulate regulatory networks using both non-dynamical models of regulatory capabilities and dynamical models of regulatory kinetics to study the potential applicability of the strategy in order to optimize the performance of the pathway under any given situation.
- Predict the behaviour of modified or redesigned gene regulatory networks in order to assess the feasibility of the modified pathway for a given situation and to identify the critical points for an enhanced understanding.
- Optimize the gene expression regulation in the pathway for better performance of the modified system in a potential micro-organism.

Genetic regulatory networks act as analogue biochemical computers to specify the identity and the level of expression of a single target gene or groups of target genes. Central to this computation is the DNA recognition and binding sequences with which the transcription factors are associated. Every gene or open reading frame that transcribes for a protein product has its own *cis*-acting sequence elements on the genome. When active transcription factors associate with the *cis*-elements of their cognate target genes, they can function to specifically repress (down-regulate) or induce (up-regulate) the synthesis of the corresponding RNA and the protein through translational process. The resulting cellular readouts are changes in the structure, metabolic capacity or behaviour of the cell, mediated by new expressions of up-regulated proteins and elimination of down-regulated proteins.

A bacterium may respond to a shortage of its preferred energy source or building-up of a metabolic product that is produced beyond limit, which the organism tolerates, or a by-product or precursor for the end-product production up-regulated, which also happens in *Z. mobilis* at acetaldehyde and alcohol production points in the Entner–Doudoroff (ED) pathway. The enhanced expression of genes for pyruvate decarboxylase (PDC) and alcohol dehydrogenase (ADH) (Ingram and Doran 1995), the protein products of which function in a biochemical pathway, allows for a different and more abundant source of energy. Some genes are regulated by a single input mechanism.

Biotechnological interventions

Biotechnological interventions have a greater role to play in areas that include production of energy, sequestering excess atmospheric carbon affecting global climate, and cleaning up environments contaminated by various hazardous chemicals and chemical processes. Each of these missions has a basis and possibly holds a solution in the biological world.

A profound understanding of genes, protein complexes, and their regulation can have a significant impact on the increase in the biomass feedstock, the amount and composition of which are controlled by genes. Achieving such knowledge requires a fusion of experimental biology and advanced computing. Understanding the plant genes and their regulation will ultimately enable manipulation of carbon storage and thus plant productivity. These advances will also allow control over the allocation of carbon to the aboveground components that can be harvested for energy or belowground components that can sequester carbon in soil. Perhaps, just as important, feedstock can be tailored to produce the substrate desired. For example, if ethanol is the desired product, lignin content could be reduced in the feedstock and plant cellulose, and starch content could be enhanced. People have been manipulating plant genomes since agriculture began, but the new biotechnological capabilities will allow a far greater efficiency and precision in these tasks.

Enhancing biological conversion to useful products

Advances in genetic engineering and molecular biology will lay a foundation for designing ways to exploit biological abilities of various organisms to work at an optimal level, for example, in converting biomass into a vast array of chemicals and other products. Although people have been using microbes such as yeast for about a millennia to make the bread rise or ferment sugars into alcohol, the process of identifying and harnessing the enormous array of capabilities in other microbes for the conversion of renewables to useful end-products is now gaining momentum.

The understanding gained across protein complexes that are important in microbes and microbial communities will help to carefully design strategies to enhance the microbial conversion capabilities. This enhanced conversion of substances will enable the production of a more desired product at a lower operational cost.

It may even be possible for engineered enzymes to degrade complex carbon substrates into simpler compounds and simultaneously convert them into ethanol. Plants can be created using engineered enzymes with regulatory switches to function at a desired point of time. Microbes that could ferment sugars at high temperatures or high ethanol concentrations, and a wide substrate range would vastly improve efficiency of the bioethanol process. These are just a few examples of how the advancing field of biotechnology can help in the bioenergy research aimed at evolving a feasible technology. The ability to manipulate and harness the potentialities of select microbes and their genes will also create novel and efficient bioprocesses that can make the processes more economical and feasible.

The future energy crisis can only be mitigated through environmentally sustainable and secure alternative energy sources. The application of biotechnology based on an understanding gained through such basic research will make genes, their regulation, their protein products, and their interactions part of an important pathway to achieve this goal of reducing reliance on fossil fuels in the future.

Potential and prospects

Incentives available across the globe for investments in ethanol research and production are lucrative enough to make ethanol available at a cheaper price. However, this may be a burden, economically, for many countries. Hence, there is a need to translate research efforts into more focused goal of ethanol production from cheap and feasible agricultural waste rather than from corn, sugar cane molasses, sugar beet, and so on, which in fact contribute greatly to inflating the cost of ethanol production apart from the cost of enzymes to be added for hydrolysis of the complex carbohydrates.

The decentralized nature of liquid biofuel production will require a minimum amount of manpower per litre or tonne of fuel produced. These plants are likely to create opportunities for new and additional jobs, especially in the marginal rural areas. Taking sweet sorghum as an example—cultivation of sorghum on 85-hectare (ha) land will be enough to create one full-time job per year. This is based on the assumption that it will take 14 hours/ha to produce and harvest the crop and 20 hours/ha for the extraction of sugar juice (assuming a yield of 20 tonnes/ha of lignocellulosic material and 10 tonnes/ha of sugar). If approximately 7 MT of ethanol is required across the European Union (EU) in the long term, 2000 direct crop production jobs and 7000 direct industrial jobs will be generated from the production of just 1 MT of bioethanol from wheat. If the market demand for bioethanol is realized, 63 000 potential jobs across the EU could materialize. This market depends, however, on the financial support from national governments and local authorities to promote the uptake of bioethanol. In the US, growth of a new bioethanol motor fuel plant would stimulate the economy with the addition of 132 000 jobs by 2010 and 275 000 new jobs by 2020.

Understanding microbial function at cellular level

The simplest microbe possesses a complex chemical and physical set-up of processes. Development of a model of the desired microbial cell and its inherent functions at the cellular level, qualitatively or quantitatively, can have a significant value, as it will help in predicting the cellular behaviour under different conditions (Arkin, Ross, and McAdams 1998; Edwards and Palsson 2000).

Present status of microbiological improvement

Although several micro-organisms, including *Clostridium* sp., have been considered as ethanologenic microbes, the yeast *S. cerevisiae* and the facultative bacterium *Z. mobilis* have been the better candidates for the industrial alcohol production. The characteristics essential for an industrially suitable micro-organism are given in Table 2.

The feedstock typically accounts for more than one-third of the production costs, thus maximizing the ethanol yield. A high ethanol yield means using those strains of bacteria that can produce fewer by-products and metabolize all major sugars, which typically include glucose, xylose, arabinose, galactose, and mannose (Wiselogel, Tyson, and Johnson 1996). Ethanologenic bacteria

Table 2 Important traits for ethanol production

Trait	Requirement
Ethanol yield	>90% of theoretical
Primary composition	Methane
Ethanol tolerance	>40 g/l
Ethanol productivity	>1 g/l/h
Robust grower and simple growth requirements	Inexpensive medium formulation
Ability to grow in undiluted hydrolysates	Resistance inhibitors
Culture growth conditions retard contaminants	Acidic pH or high temperature

Source Bothast, Nichols, and Dien (1999); Zaldivar and Olsson (2001)

that currently show maximum promise for industrial exploitation are *E. coli*, *K. oxytoca*, and *Z. mobilis*.

Recombinant *E. coli*

The first successful application of genetic modification to selectively produce ethanol using *E. coli* was carried out by Ingram, Conway, Clark, *et al.* (1987). Although it was suitable in that it had many required traits (as outlined in Table 2), its narrow pH range (6.0–8.0), quality of being less hardy than yeast, public concern over its use, and biomass utilization problems hampered its exploitation.

In *E. coli*, conversion of glucose to ethanol through the activity of pyruvate formate is an unbalanced process as only one nicotinamide adenine dinucleotide (NADH), H+ is generated from one pyruvate molecule, while two NADH, H+ are required for producing ethanol from each pyruvate molecule. However, the bacterium balances this through the production of acetic and succinic acids. Yeast and *Z. mobilis* are homo-ethanologenic as they directly form ethanol from pyruvate in the presence of the PDC and ADH. Initial attempts to incorporate the PDC into *E. coli* could not be successful as the ADH activity was not up to the level required (Dien, Cotta, and Jeffries 2003).

Figure 1 gives the typical fermentation products made using K12 *E. coli*.

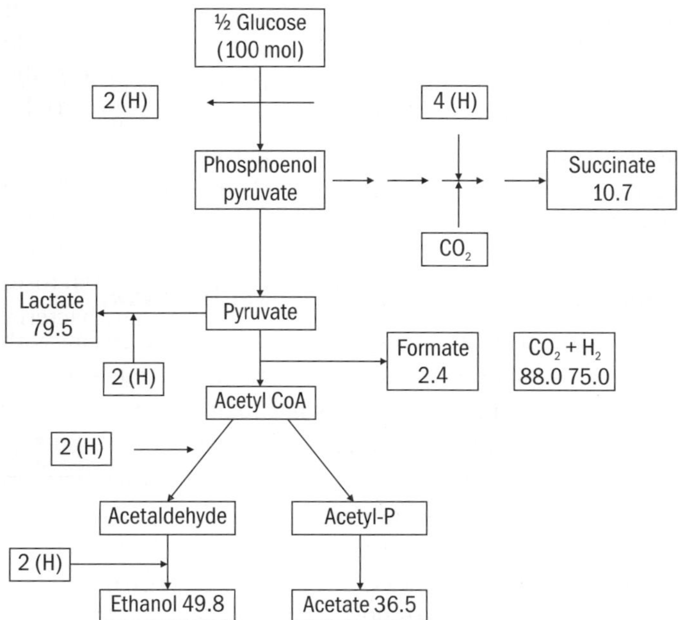

Figure 1 Typical fermentation products made by a K12 *Escherichia coli* fermenting glucose

Note Products are in moles produced per 100 moles of fermented glucose. The fermentation products accounted for 91% of the substrate used.

Source Gottschalk (1986); Ohta, Beall, Mejia, *et al.* (1991b)

An attempt was made to clone the PDC and ADH II genes from *Z. mobilis*, and put the clones under the control of *lac* promoter in a plasmid named PET operon, and *E. coli* KO11 was transformed (Ingram, Conway, Clark, *et al.* 1987). The PDC and ADH genes from *Z. mobilis* (assembled as PET or PET operon) enable *E. coli* to produce ethanol with high yield and rate (Ingram, Conway, Clark, *et al.* 1987). The results were so encouraging that different hosts were tried, and it showed that the host of the PET operon ATCC11303 (PLO1297) exhibited an ethanol tolerance of 53–56 g/l and a productivity of 0.72 g/l of ethanol per hour when xylose was supplied as a carbon source at 80 g/l (Alterthum and Ingram 1989; Beall, Ohta, and Ingram 1991). An attempt to stabilize the recombinant genes in the host through chromosomal integration of the PDC and ADH II was done (Ohta, Beall, Mejia, *et al.* 1991a, b) to evaluate the KO11 strain performance on xylose, and a global gene expression analysis by microarrays was carried out. The results showed an over-expression of the genes involved in xylose metabolism (Tao, Gonzalez, Martinez, *et al.* 2001). Performance of the strain KO11 was also analysed on the hemicellulose hydrolysates from Pinus wood, sugar cane bagasse, and corn stove (Asghari, Bothast, Doran, *et al.* 1996; Barbosa, Beck, Fein, *et al.* 1992). Later, adaptation studies by Yamano, York, and Ingram (1998); Zaldivar and Ingram (1999); and Zaldivar, Martinez, and Ingram (2000) resulted in the strain LY01 with a tolerance for organic acids, hydroxymethyl furfural, aldehydes, and ethanol. Many efforts were made by the Ingram Group of Florida Centre for Renewable Chemicals and Fuels at the University of Florida to optimize the nutritional requirements, and further work by Martinez, York, Yomano, *et al.* (1999); Underwood, Buszko, Shanmugam, *et al.* (2002); and Underwood, Zhou, Causey, *et al.* (2002) indicated that an addition of corn steep liquor, yeast autolysate, acetic acid, and acetaldehyde would increase productivity. Figure 2 gives the process of transforming *E. coli* with PET operon, converting all glucose into ethanol.

Integration of arabinose and xylose metabolism in *Z. mobilis*

The Gram negative micro-organism *Z. mobilis* is a unique bacterium with a variety of characteristics that make it a suitable candidate for ethanol production. Because of its inherent features, *Z. mobilis* has many advantages over *S. cerevisiae* (Rogers, Lee, Skotnicki, *et al.* 1982; Sprenger 1996).

The salient features of *Z. mobilis* for its preferred utilization in biotechnology of ethanol production are as follows.

- It has a shorter fermentation time (300%–400% faster than yeast).
- It gives a higher ethanol yield (92%–94% vs. 88%–90% for yeast).
- It converts sugar mixtures to ethanol with 90%–95% efficiency.
- It has higher ethanol tolerance.
- It has higher specific ethanol productivity (2.5 times).
- It produces less biomass, and more carbon is funnelled into the end-products.

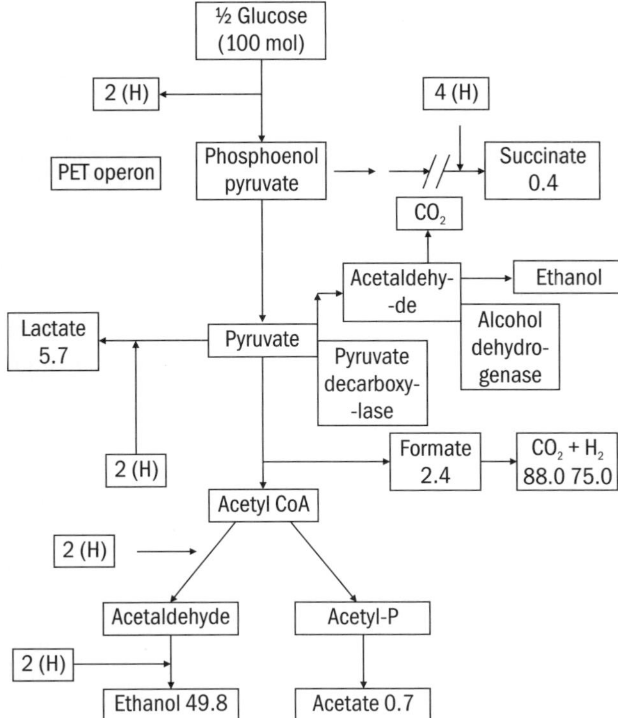

Figure 2 Transforming *Escherichia coli* with the PET operon converts almost all glucose into ethanol; the strain (KO11) also carries a mutation that blocks succinate production

Source Gottschalk (1986); Ohta, Beall, Mejia, *et al.* (1991b)

- Important enzymes of the ethanologenic pathway are over-expressed and are constitutive.
- Some strains have minimal nutritional requirements like pantothenate and biotin.
- It does not require controlled addition of oxygen throughout the process.

The general pathway, which is unique to *Z. mobilis*, is the ED pathway as opposed to the EMP pathway or the glycolytic pathway (Figure 3).

This has encouraged researchers to exploit the ability of *Z. mobilis* to use sucrose, glucose, and fructose by further manipulating the ED pathway. Attempts are also being made to construct mutants that can tolerate a higher ethanol concentration in the medium, higher substrate concentration, and higher temperature, and can increase the substrate range from hexoses such as glucose, fructose, and sucrose to pentoses such as xylose, arabinose, and also galactose and mannose. This is done only in order to overcome problems associated with *Z. mobilis*, namely, its inability to convert complex carbohydrate polymers such as cellulose, hemicellulose, and starch into ethanol, resulting in the production of by-products such as sorbitol, acetoin, glycerol, and acetic acid; formation of extracellular levan polymer; and

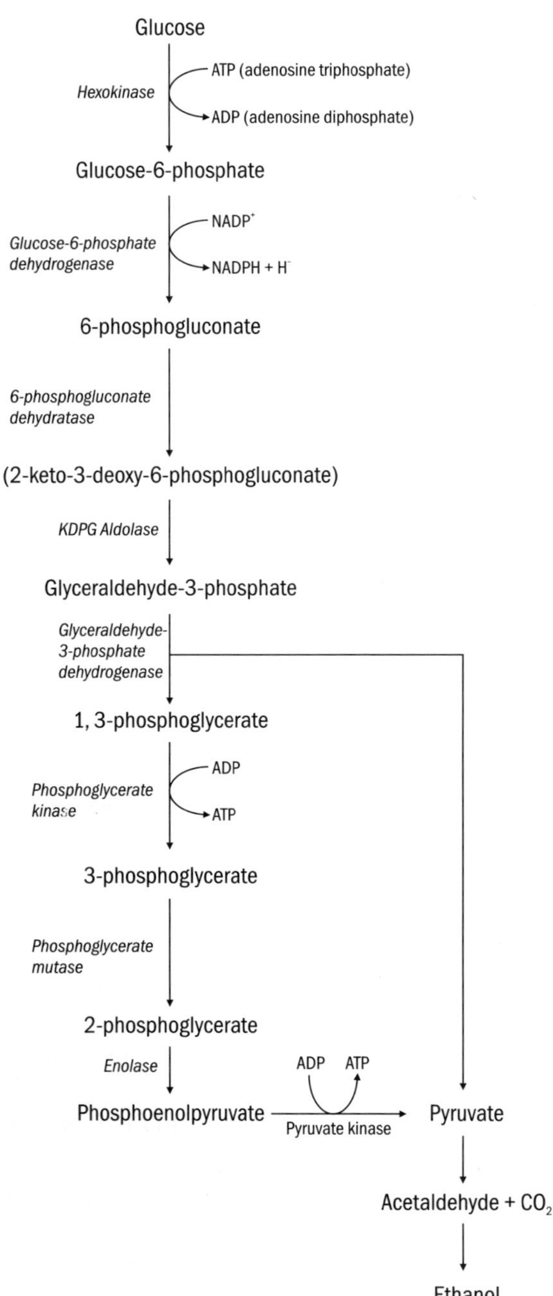

Figure 3 The unique Entner–Doudoroff (ED) pathway operating in *Zymomonas mobilis*

temperature tolerance (Dien, Cotta, and Jeffries 2003; Gunasekaran and Raj 2001).

To circumvent these problems, genetic manipulation of *Z. mobilis* has been a good option for broadening its utilizable substrate range, that is, genes encoding several hydrolytic enzymes from the related bacterial species have been cloned and transferred into *Z. mobilis*. Interestingly, a PET operon was constructed by combining the PDC and ADH II genes of *Z. mobilis* and was transferred to other bacterial strains to make them ethanologenic novel strains. Mutants of *Z. mobilis* with improved fermentation characteristics and without by-product formation have been obtained through classical mutation and selection approaches. In addition to ethanol, *Z. mobilis* has also been metabolically engineered to produce L-alanine and L-lactic acid. Genes encoding for β-carotene synthesis have also been cloned and successfully expressed in *Z. mobilis* to enrich the fermented nutrients of farm animals. Several applications of levan in food and pharmaceutical industries provide an opportunity to exploit *Z. mobilis* for its large-scale production. Merits of *Z. mobilis* offer a tremendous potential for using this organism in industrial production of various fermentation products (Gunasekaran and Raj 2001). *Z. mobilis* has been recognized as a viable alternative to yeast for ethanol production from biomass because of its high conversion yield, high ethanol tolerance, resistance to pretreated hardwood hydrolysate, and fermentation capability at a lower pH.

In previous research of *Zymomonas* metabolic engineering (Zhang, Eddy, Deanda, *et al.* 1995), the first recombinant strain of *Z. mobilis* was engineered to ferment xylose (Figure 4) through the introduction of four genes from

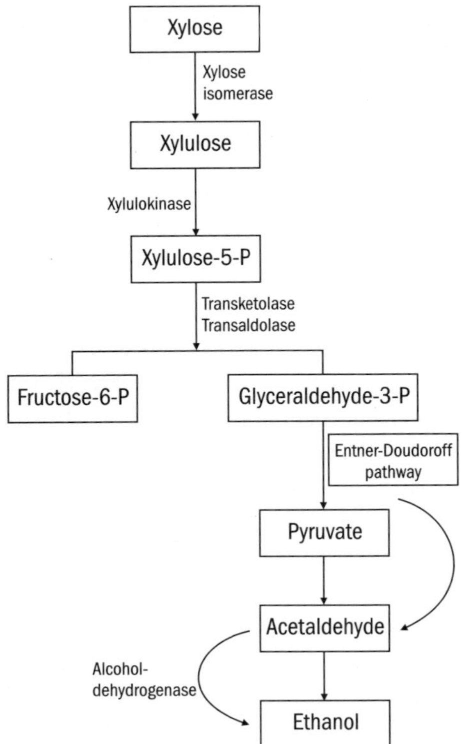

Figure 4 The modified pathway in *Zymomonas mobilis* for xylose fermentation

E. coli, namely, xylose isomerase (xylA), xylulose kinase (xylB), transketolase (tktA), and transaldolase (talB) under enolase or glyceraldehyde-3-P dehydrogenase constitutive promoter to enable the wild-type *Z. mobilis* to use xylose through xylA and xylB to xylulose-5-P and then through the tktB and talB to the ED pathway intermediates (Zhang, Eddy, Deanda, *et al.* 1995).

In spite of the inherent constraint of the available active sugar transport system, the recombinant *Z. mobilis* could ferment xylose as well as glucose. However, on xylose, the ethanol yield was at the expense of 86% of the substrate (Parker, Barnell, Snoep, *et al.* 1995). Later, Deanda, Zhang, Eddy, *et al.* (1996) used the same strategy to clone five genes of arabinose metabolism from *E. coli*, namely, L-arabinose isomerase (ara A), L-ribulose kinase (ara B), L-ribulose-5-P-4-epimerase (ara D), tktA, and talB. The first three enzymes converted arabinose to ribulose-5-P and the last two into the products of the ED pathway to get into ethanol. Although 98% theoretical conversion of arabinose was achieved with the strain ATCC39676 with the recombinant genes present, the rate of fermentation was not up to the level expected due to the low affinity of the glucose permease system for arabinose (Parker, Barnell, Snoep, *et al.* 1995).

A highly potential *Z. mobilis* strain ZM4 ATCC 31821 was transformed by pZB5, which resulted in an improved strain with a much higher ethanol tolerance on glucose and xylose mixtures (Joachimsthal and Rogers 2000). In a medium containing 65 g/l each of xylose and glucose, the two strains CP4 and ZM4 are reported to have produced 52 g/l and 62 g/l of ethanol in 60 hours and 48 hours, respectively. With increased concentration of carbon sources (75 g/l each xylose and glucose), the ethanol yield only increased to 67 g/l, and the fermentation process came to a standstill as the organism could not tolerate more than 67 g/l ethanol concentration. This was half the normal concentration of ethanol in the medium tolerated by the wild strains of *Z. mobilis* (Joachimsthal and Rogers 2000).

Other Gram negative bacteria

Although *K. oxytoca* and *Erwina chrysanthermi* transformed with PET operon did not perform better than *Z. mobilis* getting transformed with the same operon, their performance in converting cellulose into ethanol was good, particularly in case of *K. oxytoca* (Wood and Ingram 1992). The strain *K. oxytoca* utilizes cellobiose and cellotriose and, thus, is a good choice for ethanol fermentation with cellulose, as the inhibitory effects of certain cellulase preparations can be avoided. The PET operon on low and high copy number plasmids has had an influence on the ethanol production, with the low copy number plasmid pLO1555 giving higher ethanol production. However, in both cases, the ethanol yield was 92%–98% of the theoretical values (Ohta, Beall, Mejia, *et al.* 1991b). The best strains of the transformants converted 10% glucose and 10% cellobiose into 44–45 g/l of ethanol within 48 hours. Different substrates were tried to assess the efficiency of the strain, namely, mixed office paper (Brooks and Ingram 1995), sugar cane bagasse (Doran, Aldrich, and Ingram 1994), corn fibre (Moniruzzaman, Dien, Ferrer, *et al.* 1996), and sugar beet pulp (Doran, Cripe, Sutton, *et al.* 2000). Integrating cellulase components such as extracellular endoglucanase can reduce the ethanol production costs (Dien, Cotta, and Jeffries 2003). Two extracellular endoglucanase genes (cel Z and cel Y) from *E. chrysanthemi* were cloned into the recombinant *K. oxytoca* strain P2 along with auxiliary transporter gene (out) on a plasmid pCPP2006. The recombinant strain was named SZ21 (Zhou, Davis, and Ingram 2001). Although its cellulase activity was 99% less than the commercial preparations of cellulase, the ethanol productivity was 7%–16% higher than the parental strain when commercial cellulase was added. When a comparative study was done, in which galacturonic acid rich sugar beet pulp was fermented, KO11 produced significantly higher quantities of ethanol due to *E. coli* KO11 affinity for the substrate. Dien, Hospell, Wyckoff, *et al.* (1998) developed a novel hexose and pentose, utilizing the ethanologenic *E. coli* strain FBR 3 by incorporating the plasmid pLO1297. An ethanol yield of 4.38%–4.66% (w/v) with 90%–91% theoretical conversion in 70–80 hours was achieved.

At the National Chemical Laboratory, Pune, India, by utilizing a protoplast fusion technology, a hybrid microbe with *Z. mobilis* and *S. cerevisiae* was developed (Gokhale, Puntambekar, and Deobagkar 1993).

Other studies on processes and organisms

With the rapidly increasing demand for ethanol, an efficient ethanol production process and cheap substrates are needed. Lignocellulosic biomass is an abundant renewable resource, which can serve as a substrate for ethanol. The lignocellulosic structure is more resistant to decay by organisms and is not perishable like soluble sugar and starch. The simplest method of ethanol production is fermentation and distillation of sugar cane molasses and sugar beet. These materials contain high amounts of sugar, available for direct conversion to ethanol. Crops containing sugars and starch are usually expensive to produce and are perishable. Ethanol from these crops is affordable only if they are waste by-products of other processes. The availability of corn stover as a sustainable feedstock for bioethanol production was studied by Kadam and McMillan (2003). According to them, 60–80 million dry tonnes per year of corn stover should be available to fermentative routes. To achieve an ethanol production potential of 11 billion litres (3 billion gallons) per year, about 40% of the harvestable corn stover is needed. This amount should be available as long as the diversion of corn stover to non-ethanol fermentative products remains limited.

The ethanol production process from crops such as sugar cane and corn is well established; however, a process with cheaper substrate utilization is needed. The complex polysaccharides may be broken down into sugars by either acid treatment at various temperatures or by enzymatic treatment.

It has been known for over 100 years that acids act as catalysts to hydrolyse cellulose and hemicellulose into simple sugars. Acid treatment includes mild acid hydrolysis and concentrated acid hydrolysis.

The juice of rotten or discarded pineapples and the waste material generated during the production of pineapple juice served as low-cost substrates for ethanol production by *Z. mobilis*. *Z. mobilis* ATTC 10988 produced 59 g/l ethanol in undiluted pineapple juice without nutritional supplementation and without regulation of the pH, whereas 42.5 g/l ethanol was obtained using a 125 g/l sucrose medium supplemented with 10 g/l yeast extract and mineral salts. These results suggest that pineapple juice and the waste material can serve as useful low-cost substrates for ethanol production by *Z. mobilis* without the use of expensive organic nitrogen complex supplements such as yeast extract, and without the regulation of the pH during cultivation, leading to a reduction in production costs (Tanaka, Hillary, and Ishizaki 1999).

In another study, cellulosic-pyrolysate-containing levo-glucosan was chemically hydrolysed, and a maximum glucose yield of 17.35% was obtained through hydrolysis with 2 mol/l H_2SO_4 at 121°C for 20 minutes. The total initial glucose level was maintained at 41.9 g/l by diluting the hydrolysate. Then, neutralization methods were used, including either the addition of solid $Ca(OH)_2$ (to pH 6.0 or 10.4) or its combination with the absorbents. The neutralization and diatomite-shaking method gave the hydrolysate, which was completely fermented by *S. cerevisiae* and *Pichia* sp. YZ-1. A maximum ethanol yield of 0.45 g/g glucose was obtained by *S. cerevisiae* (Yu and Zhang 2002).

Another substrate, liquefied cassava starch, was used for ethanol production by co-immobilized cells of *Z. mobilis* and *S. diastaticus*. The co-immobilized cells produced 46.7 g/l ethanol from 150 g/l liquefied cassava starch, whereas the immobilized cells of yeast *S. diastaticus* alone produced 37.5 g/l ethanol. Thus, co-immobilized cells of *S. diastaticus* and *Z. mobilis* produced a high ethanol concentration as compared with the immobilized cells of *S. diastaticus* during batch fermentation of liquefied cassava starch (Amutha and Gunasekaran 2001).

Pretreatment of sunflower with *Trichoderma reesei* Rut-C 30 cellulase resulted in 57.8% saccharification (Sharma, Kalra, and Grewal 2002). Enzyme hydrolysate concentrated to 40 g/l, reducing sugars fermented with *Z. mobilis* under optimum fermentation conditions like an incubation time of 24 hours, pH 5.0, temperature 30°C, and inoculum size 3% v/v showed a maximum ethanol yield of 0.444 g/g. Ethanol production scaled up in a 1-l and 15-l fermenter under optimum conditions, revealing a maximum ethanol yield of 0.439 g/g and 0.437 g/g, respectively.

One more substrate, whey, a by-product generated in large quantities from cheese production, was used for ethanol fuel. Whey poses a major waste-disposal problem due to its high biochemical oxygen demand (BOD). When dumped untreated, whey ends up in rivers and oceans, consuming the oxygen needed to support aquatic life. Pre-disposal treatment of whey to reduce its BOD to the Environmental Protection Agency (EPA)-approved levels is extremely expensive. This study presents an economically feasible solution to the whey disposal problem (Ahmed and Morris 1991).

Sree, Sridhar, Rao, *et al.* (1998) evolved a novel solid substrate fermentation system to produce fuel ethanol from sweet sorghum and sweet potato using a thermo-tolerant yeast *S. cerevisiae* strain, VS3, and a local isolate of amylolytic *Bacillus* species, VB9. The maximum amount of ethanol produced in co-culture with a mixed substrate was 5 g/100 g of substrate at 37°C and 3.5 g/100 g of substrate at 42°C. Alcohol fermentation was also carried out using the mash of dried sweet potato with its dregs as substrate (Yu, Zhang, Zheng, *et al.* 1994).

For direct and efficient ethanol production from cellulosic materials, a novel cellulose-degrading yeast strain was developed by genetically modifying two cellulolytic enzymes on the surface of *S. cerevisiae*. This could grow on a synthetic medium containing glucan as the sole source of carbon and could directly ferment 45 g of glucan per litre to produce 16.5 g of ethanol per litre within 50 hours. Thus, 0.48 g of ethanol was produced per gram of carbohydrate used, which corresponded to 93.3% of the theoretical yield. This result indicates that efficient and simultaneous saccharification and fermentation of cellulose to ethanol were carried out by recombinant yeast cells displaying cellulolytic enzymes (Fujita, Takahashi, Ueda, *et al.* 2002).

There are micro-organisms such as *Clostridium phytofermentans* sp. nov., isolated by the researchers of the University of Massachusetts, showing combined hydrolysis and fermentation of cellulose to ethanol, acetate, CO_2, and hydrogen (Warnick, Methe, and Leschine 2002). This shows the requirement, to explore in the biosphere, for micro-organisms capable of converting the biomass into multiple products.

The fermentation process is a traditional, well-known method throughout the world, as per various studies. Extensive research has demonstrated that alcoholic fermentation of xylose, the major hemicellulosic-derived sugar, improves the economics of biomass conversion, as 70% of the raw material can be expected to convert into ethanol. One important requirement is an efficient micro-organism, which is able to ferment a variety of sugars (pentoses and hexoses) and tolerate the stress conditions. Bacterial and yeast strains have been constructed through metabolic engineering, which features traits that are advantageous for ethanol production using lignocellulose sugars. After several rounds of modifications and evaluations, three main microbial platforms—*S. cerevisiae*, *Z. mobilis*, and *E. coli*—have emerged and have performed well in pilot studies. Although efforts are going on to further enhance their properties, an improvement of the fermentation process is just one of the several factors that need to be fully optimized and integrated in order to generate a competitive lignocellulose-based ethanol production plant (Zaldivar, Nielsen, and Olsson 2001).

There are several bottlenecks in the alcoholic fermentation process and they must be overcome to achieve a high and competitive performance of bioethanol production by the yeast *S. cerevisiae*. Alfenore, Jouve, Guillouet, *et al.* (2002) described a nutritional strategy that allowed *S. cerevisiae* to produce a final ethanol titre of 19% (v/v) ethanol in 45 hours in a fed-batch culture at 30°C. This performance was achieved by implementing exponential feeding of vitamins throughout the fermentation process. A maximum instantaneous productivity of 9.5 g/l/hour was reached in the best fermentation. These performances resulted from improvements in growth, ethanol production rate, and concentration of viable cells in response to the nutritional strategy.

The ethanol yield gets reduced with molasses, and the most probable reason for this may be the impurities present in raw molasses that are inhibitory to yeast growth as well as to fermentation. For that, 22 different yeast strains were studied in isolation from the natural resources. The highest yield of ethanol by a strain from molasses and grapes, respectively, was 7.8% and 8.2% (with 16% dextrose medium) and 7.4% and 8.0% (with molasses medium). The optimum temperature for maximum ethanol production was 30–35°C, whereas the optimum pH was 5–5.5. Any variation in these conditions severely affected the ethanol production, possibly due to excessive enzyme degradation and loss of activity at a high temperature, reduced activity of glycolytic/fermentative enzymes at a low temperature, and inactivation of enzymes due to pH variation (Bajaj, Yousuf, and Thakur 2001). In another study, Jirku (1999) used immobilized *S. cerevisiae* cells and found that the maximum fermentation capacity of the system was at 30°C and was relatively pH-sensitive. A packed column reactor was used to test the biocatalyst's operational sensitivity to key fermentation variables. Results of this study as well as characteristics of the polymer, prepared by an epoxy resin and diaminopolyethylene oxide polymerization, establish the suitability of this method for ethanol production.

A complete conversion of glucose and xylose mixture was obtained by a respiratory deficient mutant of *S. diastaticus* co-cultivated with *Pichia stipitis* in continuous culture (Delgenes, Laplace, Moletta, *et al.* 1996). In another work, *S. cerevisiae* was genetically modified with the genes for pentose utilization from different sources, including fungi. The modified yeast showed production of ethanol from L-arabinose and D-xylose with limited production of by-products such as arabitol and xylitol (Bettiga, Bengtsson, Hahn-Hägerdal, *et al.* 2009). Anaerobic bacterium *Thermoanaerobium brockii* was also used for ethanol fermentation. The study showed that it followed the EMP pathway and converted glucose into ethanol, acetic acid, carbonic acid, and lactic acid. This species has actively fermented saccharides and starch into ethanol (Lamed and Zeikus 1980).

A comparison of the rates of growth and ethanol production was done by studying 11 different strains of *Zymomonas*, which revealed a wide range of characteristics (Skotnicki, Lee, Tribe, *et al.* 1981). From these studies, it was found that some strains were more tolerant to high sugar or ethanol concentrations and high incubation temperatures than others. Some strains were unable to use sucrose; some were able to produce large amounts of levan; and one strain grew well on the carbon source, but there was no production of levan. One strain, CP4, was found to be superior in all respects to most other strains and was finally chosen as a starting strain to be genetically modified for enhanced ethanol production.

Mixing has an important role in fermentation. The influence of mixing (from 100–110 revolutions per minute) on the performance of *Z. mobilis* anaerobic continuous culture was studied. It was found that the biomass yield and ethanol productivity were improved at higher stirring intensities along with a decrease in the by-product formation. Vigorous mixing led to a better coupling between catabolism and anabolism (Toma, Kalnenieks, Berzins, *et al.* 2002).

In the fermentation process, ADH and PDC are the well-studied key enzymes. According to one study, it was found that the enzyme ADH (isolated from *Z. mobilis*) is activated by ferrous ions but not by zinc ions after inactivation with the metal-complexing agents; cobalt ions also reactivated to a lesser extent. This study suggested that in this species, ADH naturally contains iron. Kinetic studies on the iron-treated enzyme indicate an alcohol-activation phenomenon that may have physiological relevance in overcoming product inhibition during fermentation (Scopes 1983).

Tamarit, Cabiscol, Aguilar, *et al.* (1997) studied the tolerance of the metal binding of ADH in *Z. mobilis* endowed with two isoenzymes—a zinc containing enzyme (ADH I) and an iron-containing enzyme (ADH II). The activity of ADH I remained fully conserved, whereas the ADH II activity decayed when the culture conditions were shifted from anaerobiosis to aerobiosis. This differential response was due to the metal present on each isoenzyme because pure preparations of the ADH I are resistant to oxidative inactivation and preparation of zinc-containing ADH II, obtained by incubation of pure ADH II with $ZnCl_2$, which showed no modification of the target for oxidative damage (His277-containing peptide). It was consistently found that the activity of the zinc-containing ADH II, once submitted to oxidative treatment, was fully restored when iron was reintroduced into the enzyme structure. These results indicate that zinc bound to these proteins plays an important role in the protection of their active centres against oxidative damage and may have relevant biochemical and physiological consequences in this species. After that, three enzymes, glucokinase (EC 2.7.1.2), fructokinase (EC 2.7.1.4), and glucose-6-phosphate dehydrogenase (EC 1.1.1.49), were isolated in high yield from the extracts of *Z. mobilis* (Scopes, Testolin, Stoter, *et al.* 1985). The principal steps in the isolation procedures involved the use of select dye–ligand adsorbent columns, with affinity elution of two of the three enzymes. Neale, Scopes, Wettenhall, *et al.* (1986) found that PDC is the key enzyme, which directs the production of CO_2 and ethanol as the end-products of sugar fermentation by *Z. mobilis*. They studied isolation, properties, and genetic expression of this enzyme in *E. coli*, which were purified to homogeneity using dye–ligand and ion exchange chromatography. The genomic fragment encoding the enzyme expressed high levels of PDC in *E. coli*.

Osman, Conway, Bonetti, *et al.* (1987) analysed the rate at which *Z. mobilis* converts glucose into ethanol and CO_2, and found that in the ED pathway, glycolytic flux did not remain constant during batch fermentation of 20% glucose. The initial rate of conversion resulted in high ethanol accumulation in the surrounding broth. As a result, glycolytic activity was repressed due to the inhibitory effects of ethanol, and the activity could be reversed immediately by ethanol removal. Fermentation activity enhanced (58 mol of CO_2 evolved per milligram of cell protein per hour) after accumulation of 1.1% ethanol in 18 hours and declined to one-half of this rate after 6.2% accumulation of ethanol in 30 hours. They also identified three possible reasons for this behaviour of *Z. mobilis*, namely, reduced number of nucleotides, reduction in internal pH from 6.3 to 5.3, and reduced specific activity of pyruvate kinase and glyceraldehyde-3-phosphate dehydrogenase of the glycolytic pathway. Loos, Kramer, Sahm, *et al.* (1994) studied the basis of osmotolerance in *Z. mobilis* when grown in a media containing high concentrations of glucose or other sugars. They described sorbitol as a compatible solute in bacteria, which enhances the growth of *Z. mobilis* at higher glucose concentrations, exceeding 0.83 M (15%). Sorbitol can protect proteins during dehydration by osmotic or thermal stress and is, therefore, used to preserve proteins during storage. The sorbitol transport system followed the Michaelis–Menten kinetics, with an apparent K_m of 34 mM and a V_{max} of 11.2 nmol/min/mg (dry mass). Sorbitol was produced by the cells themselves and was accumulated when grown on sucrose (1 M or 36%) by the action of the periplasmic enzyme glucose–fructose oxidoreductase, which converts glucose and fructose into gluconolactone and then into sorbitol. Thus, *Z. mobilis* can form and accumulate the compatible solute sorbitol from a natural carbon source, sucrose, in order to overcome the osmotic stress in a high-sugar media. No other major compatible solute (betaine, proline, glutamate or trehalose) was detected.

In further studies, Deng and Coleman (1999) introduced new genes into a cyanobacterium in order to create a novel pathway for fixed carbon utilization, which results in the synthesis of ethanol. The coding sequence of the PDC and ADH II from the bacterium *Z. mobilis* was cloned into the shuttle vector pCB4 and was then used to transform the cyanobacterium *Synechococcus* sp. strain PCC7942. The PDC and ADH genes were expressed at high levels, as demonstrated by Western blotting and enzyme activity analyses. The transformed cyanobacterium synthesized ethanol, which diffused from cells into the culture medium. As cyanobacteria have simple growth requirements and use light, CO_2, and inorganic elements efficiently, production of ethanol by cyanobacteria is a potential system for bioconversion of solar energy and CO_2 into a valuable resource.

An important study was carried out in which transcript levels of several enzymes of glycolytic and alcoholic fermentation pathways were examined in the shoot and root tissues of rice seedlings subjected to different abiotic stresses by northern blot analysis using homologous DNA probes. This work included ADH, PDC, phosphate isomerase, aldolase, glyceraldehyde phosphate dehydrogenase, enolase, glucose phosphate isomerase, and pyruvate kinase enzymes. Cold, desiccation, salt (NaCl), high temperature, and oxygen-deprivation stresses are used to see their effect on pathway regulatory enzymes. Evidently, glycolytic and alcoholic fermentation enzymes have a sufficient degree of flexibility to adjust to the increased energy demand and supply of intermediates for acclimatization to the stress conditions.

Sprenger (1996) discussed the carbon flux and its regulation, and how it branches into anabolic pathways, together with the recent approaches to broaden the substrate range of the bacterium. Sucrose, glucose, and fructose are degraded by *Z. mobilis* via an anaerobic version of the ED pathway. Sucrose splits extracellularly into glucose and fructose (or levan). The two sugars are transported into the cell via facilitated diffusion (uniport). A periplasmic enzyme glucose–fructose oxidoreductase provides the novel compatible solute sorbitol to counteract the detrimental osmotic stress.

Metabolic engineering of *Z. mobilis* strains was tried to maximize the ethanol production from mixtures of hexose and pentose sugars through the application of metabolic flux control techniques (Kompala, Ramkrishna, Jansen, *et al.* 1986). The successful engineering of ethanol-fermenting *Z. mobilis* to broaden its substrate utilization range from only the hexoses, glucose, and fructose, to include the additional pentose sugars, xylose, and arabinose found in the renewable biomass feedstock and agricultural residues was done. Pentose fermentation was accomplished by introducing two gene encodings for key enzymes in the xylose assimilation pathway, three gene encodings for key enzymes in the arabinose assimilation pathway, and two gene encodings for key enzymes in the pentose phosphate pathway into the recombinant ethanol producer *Z. mobilis*.

Conclusion

Ethanol is a low-carbon fuel with a tremendous potential to mitigate greenhouse gas emissions as it reduces the CO emissions. It promotes rural employment and offers solutions to the problems of waste disposal. Hence, ethanol produced by inexpensive sources holds out the promise of serving as an important renewable transportation fuel. The use of biomass crops as a primary ethanol feedstock is an income generation option for the rural people. As per a report by Minnesota Department of Energy and Economic Development, $1 spent on petroleum energy generates 34 cents of additional economic activity, whereas the same amount spent on biomass generates an additional $1.5 of local economic activity. Although there are limitations with respect to substrate preparation, micro-organisms, and the process to make ethanol production from renewable resources an economically feasible entity, research is being focused on evolving microbes that can convert the complex lignocellulose material into simpler metabolizable sugars and then into ethanol, overcoming the feedback inhibition of the enzymes by the hydrolysed products. The problems of inhibition by fermentation products and carrying out fermentation at higher temperatures are still to be addressed. However, technological advancements in genomics and functional genomic areas can well give a solution.

References

Ahmed I and Morris D. 1991. ***Alcohol Fuels from Whey: novel commercial uses for a waste product.*** Institute for Local Self-reliance

Alfenore S, Jouve C M, Guillouet S E, Uribelarrea J L, Goma G, Benbadis L. 2002. **Improving ethanol production and viability of *Saccharomyces cerevisiae* by a vitamin feeding strategy during fed-batch process.** *Applied Microbiology Biotechnology* **60**: 67–72

Alterthum F and Ingram L O. 1989. **Efficient ethanol-production from glucose, lactose, and xylose by recombinant *Escherichia coli*.** *Applied and Environmental Microbiology* **55**: 1943–1948

Amutha R and Gunasekaran P. 2001. **Production of ethanol from liquefied cassava starch using co-immobilized cells of *Zymomonas mobilis* and *Saccharomyces diastaticus*.** *Journal of Biosciences and Bioengineering* **92**(6): 560–564

Arkin A, Ross J, and McAdams H H. 1998. **Stochastic kinetic analyses of developmental pathway bifurcation in phase infected *Escherichia coli* cell.** *Genetics* **149**: 1633–1648

Asghari A, Bothast R J, Doran J B, Ingram L O. 1996. **Ethanol production from hemicellulose hydrolysates of agricultural residues using genetically engineered *Escherichia coli* strain KO11.** *Indian Journal of Microbiology* **16**: 42–47

Bajaj B K, Yousuf S, and Thakur R L. 2001. **Selection and characterization of yeasts for desirable fermentation characteristics.** *Indian Journal of Microbiology* **41**: 107–110

Balat M and Balat H. 2009. **Recent trends in global production and utilization of bio-ethanol fuel.** *Applied Energy* **86**(11): 2273–2282

Barbosa M D S, Beck M J, Fein J E, Potts D, Ingram L O. 1992. **Efficient fermentation of *Pinus* sp. acid hydrolysates by an ethanologenic strain of *Escherichia coli*.** *Applied and Environmental Microbiology* **58**: 1382–1384

Beall D S, Ohta K, and Ingram L O. 1991. **Parametric studies of ethanol production from xylose and other sugars by recombinant *Escherichia coli*.** *Biotechnology and Bioengineering* **38**: 296–303

Bettiga M, Bengtsson O, Hahn-Hägerdal B, Gorwa-Grauslund M F. 2009. **Arabinose and xylose fermentation by recombinant *Saccharomyces cerevisiae* expressing a fungal pentose utilization pathway.** *Microbial Cell Factories* **8**: 40

Blotkamp P J, Takagi M, Pemberton M S, Emert G H. 1978. **Enzymatic hydrolysis of cellulose and simultaneous fermentation to alcohol.** *Canadian Journal of Microbiology* **23**: 139–147

Bothast R J, Nichols N N, and Dien B S. 1999. **Fermentations with new recombinant organisms.** *Biotechnology Progress* **15**: 867–875

Brooks T A and Ingram L O. 1995. **Conversion of mixed waste office paper to ethanol by genetically engineered *Klebsiella oxytoka* strain P2.** *Biotechnology Progress* **11**: 619–625

Deanda K, Zhang M, Eddy C, Picataggio S. 1996. **Development of an arabinose-fermenting *Zymomonas mobilis* strain by metabolic pathway engineering.** *Applied and Environmental Microbiology* **62**: 4465–4470

Delgenes J P, Laplace J M, Moletta R, Navarro J M. 1996. **Comparative study of separated fermentations and co-fermentation process to produce ethanol from hardwood derived hydrolysate.** *Biomass and Bioenergy* **11**(4): 353–360

Deng M D and Coleman J R. 1999. **Ethanol synthesis by genetic engineering in Cynobacteria.** *Applied and Environmental Microbiology* **65**(2): 523–528

Dien B S, Cotta M A, and Jeffries T W. 2003. **Bacteria engineered for fuel ethanol production: current status.** *Applied Microbiology and Biotechnology* **63**(3): 258–266

Dien B S, Hospell R B, Wyckoff H A, Bothast R J. 1998. **Fermentation of hexose and pentose sugars using a novel ethanologenic. *Escherichia coli* strain.** *Enzyme and Microbial Technology* **23**: 366–371

Doran J B, Aldrich H C, and Ingram L O. 1994. **Saccharification and fermentation of sugar cane bagasse by *Klebsiella oxytoca* P2 containing chromosomally integrated genes encoding the *Zymomonas mobilis* ethanol pathway.** *Biotechnology and Bioengineering* **44**: 240–247

Doran J B, Cripe J, Sutton M, Foster B. 2000. **Fermentations of pectin-rich biomass with recombinant bacteria to produce fuel ethanol.** *Applied Biochemistry and Biotechnology* **84–86**: 141–152

Dwivedi P, Alavalapati J R R, and Lal P. 2009. **Cellulosic ethanol production in the United States: conversion technologies, current production status, economics, and emerging developments.** *Energy for Sustainable Development* **13**(3): 174–182.

Edwards S and Palsson B O. 2000. **The *Escherichia coli* MG1655 in silico metabolic genotype: its definition, characteristics and capabilities.** *Proceedings of National Academy of Sciences of the United States of America* **97**: 5528–5533

Enari T M. 1993. **Microbial cellulases.** In *Microbial Enzymes and Biotechnology*, pp. 183–223, edited by W M Forgart. New York: Applied Science Publishers

Eriksson K E, Blanchette R A, and Ander P. 1990. ***Microbial and Enzymatic Degradation of Wood and Wood Components.*** Berlin, Germany: Springer-Verlag

Farinelli B, Carter C A, Lin C-Y C, Sumner D A. 2009. **Import demand for Brazilian ethanol: a cross-country analysis.** *Journal of Cleaner Production* **17**(1): S9–S17

Fujita Y, Takahashi S, Ueda M, Tanaka A, Okada H, Morikawa Y, Kawaguchi T, Arai M, Fukuda H, Kondo A. 2002. **Direct and efficient production of ethanol from cellulosic material with a yeast strain displaying cellulolytic enzymes.** *Applied and Environmental Microbiology* **68**(10): 5136–5141

Gokhale D V, Puntambekar U S, and Deobagkar D N. 1993. ***Intergeneric Protoplast Fusion in Bacteria: a review. Biotechnology Advances II,*** pp. 199–217. Elsevier

Gottschalk G. 1986. ***Bacterial Metabolism,*** 2nd edn. New York: Springer-Verlag. 237 pp

Graham R L, Lichtenberg E, Roningen V O, Shapouri H, Walsh M E. 1995. **The Economics of Biomass Production in the United States.** In *Proceedings of the Second Conference of the Americas*, NRELICP-2008098 (DE95009230)

Gunasekaran P and Raj C K. 2001. **Ethanol fermentation technology: *Zymomonas mobilis.*** *Journal of Bacteriology* **179** (4): 1102–1104. Details available at <http://ces.iisc.renet.in/curscinew/july10/articles14htm>

Hall D O, Rosillo-Calle E, and de Groot P. 1992. **Biomass energy: lessons from case studies in developing countries.** *Energy Policy*: 62–73

Ingram L O and Doran J B. 1995. **Conversion of cellulosic materials to ethanol.** *FEMS Microbiology Reviews* **16**: 235–241

Ingram L O, Conway T, Clark D P, Sewell G W, Preston J F. 1987. **Genetic engineering of ethanol production in *Escherichia coli.*** *Applied and Environmental Microbiology* **53**: 2420–2425

Jirku V. 1999. **Whole cell immobilization as a means of enhancing ethanol tolerance.** *Journal of Industrial Microbiology and Biotechnology* **22**: 147–151

Joachimsthal E L and Rogers P L. 2000. **Characterization of a high productivity recombinant strain of *Zymomonas mobilis* for ethanol production from glucose/ xylose mixtures.** *Applied Biochemistry and Biotechnology* **84–86**: 343–356

Kadam K L and McMillan J D. 2003. **Availability of corn stover as a sustainable feedstock for bioethanol production.** *Bioresources Technology* 88(1): 17–25

Kompala D S, Ramkrishna D, Jansen N B, Tsao G T. 1986. **Investigation of bacterial growth on mixed substrates. Experimental evaluation of cybernetic models.** *Biotechnology and Bioengineering* **28**: 1044–1055

Lamed R and Zeikus J G. 1980. **Glucose fermentation pathway of *Theranaerobium brockii*.** *Journal of Bacteriology* **141**(3): 1251–1257

Laub M T, McAdams H H, Feldblyum T, Fraser C H, Shapiro L. 2000. **Global analysis of the genetic network controlling a bacterial cell cycle.** *Science* **290**: 2144–2148

Loos H, Kramer R, Sahm H, Sprenger G A. 1994. **Sorbitol promotes growth of *Zymomonas mobilis* in environments with high concentrations of sugar: evidence for a physiological function of glucose-fructose oxidoreductase in osmoprotection.** *Journal of Bacteriology* **176**(24): 7688–7693

Martinez A, York S W, Yomano L P, Pineda V L, Davis F C, Shelton J C, Ingram L O. 1999. **Biosynthetic burden and plasmid burden limit expression of chromosomally integrated heterologous genes (pdc, adhB) in *Escherichia coli*.** *Biotechnology Progress* **15**: 891–897

Mauguin Ph, Bonfils C, and Nacfaire H. 1994. **Biofuels in Europe: developments, applications and perspectives 1994–2004.** In *Proceedings of the First European Forum on Motor Biofuels*, 9–11 May 1994, Tours (France)

McGowan F. 1991. **Controlling the greenhouse effect: the role of renewables.** *Energy Policy:* 111–118

Moniruzzaman M, Dien B S, Ferrer B, Hespell R B, Dale B E, Ingram L O, Bothast R J. 1996. **Ethanol production from AFEX pretreated corn fiber by recombinant bacteria.** *Biotechnology Letters* **18**: 985–990

Neale A D, Scopes P K, Wettenhall R E H, Hoogenraad N J. 1986. **Pyruvate decarboxylase of *Zymomonas mobilis*: isolation, properties and genetic expression in *Escherichia coli*.** *Journal of Bacteriolology* **169**(3): 1024–1028

Nomuro M, Bin T, and Nakao S I. 2002. **Selective ethanol extraction from fermentation broth using a silicate membrane.** *Separation Purifying Technology* **27**: 59–66

Ohta K, Beall D S, Mejia J P, Shanmugam K T, Ingram L O. 1991a. **Genetic improvement of *Escherichia coli* for ethanol production: chromosomal integration of *Zymomonas mobilis* genes encoding pyruvate decarboxylase and alcohol dehydrogenase II.** *Applied and Environmental Microbiology* **57**: 893–900

Ohta K, Beall D S, Mejia J P, Shanmugam K T, Ingram L O. 1991b. **Metabolic engineering of *Klebsiella oxytoca* M5A1 for ethanol production from xylose and glucose.** *Environmental Microbiology* **57**: 2810–2815

Osman Y A, Conway T, Bonetti S J, Ingram L O. 1987. **Glycolytic flux in *Zymomonas mobilis*: enzyme and metabolite levels during batch fermentation.** *Journal of Bacteriology* **169**(8): 3726–3736

Parker C, Barnell W O, Snoep J L, Ingram L O, Conway T. 1995. **Characterization of the *Zymomonas mobilis* glucose facilitator gene-product (Glf) in recombinant *Escherichia coli*: examination of transport mechanism, kinetics and the role of glucokinase in glucose transport.** *Molecular Microbiology* **15**: 795–802

Puls J. 1993. **Substrate analysis of forest and agricultural wastes.** In *Bioconversion of Forest and Agricultural Plant Residues*, pp. 183–209, edited by J N Saddler. Wallingford, UK: CAB International

Rogers P L, Lee K J, Skotnicki M L, Tribe D E. 1982. **Ethanol production by *Zymomonas mobilis*.** *Advances in Biochemical Engineering* **23**: 37–84

Scopes R K. 1983. **An iron-activated alcohol dehydrogenase.** *FEBS Letters* **68**(6): 2869–2876

Scopes R K, Testolin V, Stoter A, Smith K G, Algar E M. 1985. **Simultaneous purification and characterization of glucokinase, fructokinase, and glucose-6-phosphate dehydrogenase from *Zymomonas mobilis*.** *Biochemistry Journal* **228**: 627–634

Sharma S K, Kalra K L, and Grewal H S. 2002. **Fermentation of enzymatically saccharified sunflower stalks for ethanol production and its scale-up.** *Bioresources Technology* **85**: 31–33

Skotnicki M L, Lee K J, Tribe D E, Rogers P L. 1981. **Comparison of ethanol production by different *Zymomonas* strains.** *Applied and Environmental Microbiology* **41**(4): 889–893

Soccol C R, de Souza Vandenberghe L P, Medeiros A B P, Karp S G, Buckeridge M, Ramos L P, Pitarelo A P, Ferreira-Leitão V, Gottschalk L M F, Ferrara M A, da Silva Bon E P, de Moraes L M P, de Amorim Araújo J, Torres F A G. 2010. **Bioethanol from lignocelluloses: Status and perspectives in Brazil.** *Bioresource Technology* **101**(13): 4820–4825

Sprenger G A. 1996. **Carbohydrate metabolism in *Zymomonas mobilis*: a catabolic highway with some scenic routes.** *FEMS Microbiology Letters* **145**: 301–307

Sree N K, Sridhar M, Rao L V, Pandey A. 1998. **Ethanol production in solid fermentation using thermotolerant yeast.** *Process Biochemistry* **34**: 115–119

Tamarit J, Cabiscol E, Aguilar J, Ros J. 1997. **Differential inactivation of alcohol dehydrogenase isoenzyme in *Zymomonas mobilis* by oxygen.** *Journal of Bacteriology* **179**: 1102–1104

Tanaka K, Hillary Z D, and Ishizaki A. 1999. **Investigation of the utility of pineapple juice and pineapple waste material as low-cost substrate for ethanol fermentation by *Zymomonas mobilis*.** *Journal of Bioscience and Bioengineering* **87**(5): 642–646

Tao H, Gonzalez R, Martinez A, Rodriguez M, Ingram L O, Preston J F, Shanmugam K T. 2001. **Engineering a homo-ethanol pathway in *Escherichia coli*: increased glycolytic flux and levels of expression of glycolytic genes during xylose fermentation.** *Journal of Bacteriology* **183**: 2979–2988

Toma M M, Kalnenieks U, Berzins A, Vigants A, Rikmains M, Viesturs U. 2002. **The effect of mixing on glucose fermentation by *Z. mobilis* continuous culture.** *Process Biochemistry*: 1–4

Tripetchkul S, Hillary Z D, and Ishizaki A. 1998. **Strategies for improving ethanol production using *Z. mobilis*.** *Recent Research Developments in Agricultural & Biological Chemistry* **2**: 41–55

Uchiyama I, Higuchi T, and Kawai M. 2010. **MBGD update 2010: toward a comprehensive resource for exploring microbial genome diversity.** *Nucleic Acids Research* **38**: D361–D365

Underwood S A, Buszko M L, Shanmugam K T, Ingram L O. 2002. **Flux through citrate synthase limits the growth of ethanologenic *Escherichia coli* KO11 during xylose fermentation.** *Applied and Environmental Microbiology* **68**: 1071–1081

Underwood S A, Zhou S, Causey T B, Yomano L P, Shanmugam K T, Ingram L O. 2002. **Genetic changes to optimize carbon partitioning between ethanol and biosynthesis in ethanologenic *Escherichia coli*.** *Applied and Environmental Microbiology* **68**(12): 6263–6272

Warnick T A, Methe B A, and Leschine S B. 2002. ***Clostridium phytofermentans* sp. nov., a cellulolytic mesophile from forest soil.** *International Journal of Systematic and Evolutionary Microbiology* **52**: 1155–1160

Wiselogel A, Tyson S, and Johnson D. 1996. **Biomass feedstock resources and composition.** In *Handbook on Bioethanol: Production and Utilization*, edited by C E Wyman [Applied Energy Technology Series], pp. 105–118. Washington, DC: Taylor and Francis

Wood B E and Ingram L O. 1992. **Ethanol production from cellobiose, amorphous cellulose, and crystalline cellulose by recombinant *Klebsiella oxytoca* containing chromosomally integrated *Zymomonas mobilis* genes for ethanol production and plasmids expressing thermostable cellulase genes from *Clostridium thermocellum*.** *Applied and Environmental Microbiology* **58**: 2103–2110

Yamano L P, York S W, and Ingram L O. 1998. **Isolation and characterization of ethanol-tolerant mutants of *Escherichia coli* KO11 for fuel ethanol production.** *Journal of Industrial Microbiology and Biotechnology* **20**: 132–138

Yu Z and Zhang H. 2002. **Pretreatment of cellulose pyrolysate for ethanol production by *Saccharomyces cereviesiae, Pichia* sp.YZ-1 and *Zymomonas mobilis*.** *Biomass and Bioenergy* **24**: 257–262

Yu B, Zhang F, Zheng Y, Wang P. 1994. **Alcohol fermentation from the mash dried sweet potato with its dregs using immobilized yeast.** *Process Biochemistry* **31** (1): 1–6

Yuh C H, Bolouri H, and Davidson E. 1998. **Genomic cis-regulatory logic: experimental and computational analysis of a sea urchin gene.** *Science* **279**(5358): 1896–1902

Zaldivar J and Ingram L O. 1999. **Effect of organic acids on the growth and fermentation of ethanologenic *Escherichia coli* LY01.** *Biotechnology and Bioengineering* **66**: 203–210

Zaldivar J and Olsson N J. 2001. **Fuel ethanol production from lignocellulose: a challenge for metabolic engineering and process integration.** *Applied Microbiology and Biotechnology* **56**:17–34

Zaldivar J, Martinez A, and Ingram L O. 2000. **Effect of alcohol compounds found in hemicellulose hydrolysate on the growth and fermentation of ethanologenic *Escherichia coli*.** *Biotechnology and Bioengineering* **68**: 524–530

Zaldivar J, Nielsen J, and Olsson L. 2001. **Fuel ethanol production from lignocellulose: a challenge for metabolic engineering and process integration.** *Applied Microbial Biotechnology* **56**: 17–34

Zhang M, Eddy C, Deanda K, Finkelstein M, Picataggio S. 1995. **Metabolic engineering of a pentose metabolism pathway in ethanologenic *Zymomonas mobilis*.** *Science* **267**: 240–243

Zhou S, Davis F C, and Ingram L O. 2001. **Gene integration and expression and extracellular secretion of *Erwinia chrysanthemi* endoglucanase CelY (celY) and CelZ (celZ) in ethanologenic *Klebsiella oxytoca* P2.** *Applied Environmental Microbiology* **67**: 6–14

Bibliography

Cho J Y and Jeffries T W. 1999. **Transcriptional control of ADH genes in the xylose-fermenting yeast *Pichia stipitis*.** *Applied and Environmental Microbiology* **65**(6): 2363–2368

Douka E, Koukkou A I, Vartholomatos G, Frillingos S, Papamichael E M, Drainas C. 1999. **A *Zymomonas mobilis* mutant with delayed growth on high glucose concentrations.** *Journal of Bacteriology* **181**(15): 4598–4604

Gordon J L, Edge M, Fort T, Holden M. 2000. ***Applications and Innovations in Expert Systems III ES95***, pp. 205–219, edited by A Macintosh and C Cooper Lancashire, UK: Coopers & Lybrand

Graaf A A, Striegel K, Wittig R M, Laufer B, Schmitz G, Wiechert W, Sprenger G A, Sahm H. 1999. **Metabolic state of *Zymomonas mobilis* in glucose-, fructose-, and xylose-fed continuous cultures as analysed by 13C- and 31P-NMR spectroscopy.** *Archives in Microbiology* **171**: 371–385

Gunning R. 2000. **International Energy Agency, Photovoltaic Power Systems TASK 1.** UK expert. Harwel, UK: Energy Technology Support Unit. [Final report ETSU S/P2/000357/00/00]

Kalnenieks U, Galinina N, Irbe I, Toma M. 1995. **Energy coupling sites in the electron transport chain of *Zymomonas mobilis*.** *FEMS Microbiology Letters* **133**: 99–104

Kalnenieks U, Galinina N, Toma M, Skards I. 1996. **Electron transport chain in aerobically cultivated *Zymomonas mobilis*.** *FEMS Microbiology Letters* **143**: 185–189

Kim I S, Barrow K D, and Rogers P L. 2000. **Nuclear magnetic resonance studies of acetic acid inhibition of rec *Zymomonas mobilis* ZM4 (pZB5).** *Applied Biochemistry and Biotechnology* **84**: 357–370

Lawford H G and Rousseau J D. 1998. **Improving of fermentation performance of *Zymomonas mobilis* in acetic acid containing media.** *Applied Biochemistry and Biotechnology* **70–72**: 161–172

Lawford H G and Rousseau J D. 2002. **Performance testing of *Zymomonas mobilis* metabolically engineered for co-fermentation of glucose, xylose, and arabinose.** *Applied Biochemistry and Biotechnology* **98–100**: 429–448

Lee W C and Huang C T. 1999. **Modeling of ethanol fermentation using *Zymomonas mobilis* ATCC 10988 grown on the media containing glucose and fructose.** *Biochemistry Engineering Journal* **4**: 217–227

McMillan J D, Newman M M, Templeton D W, Mohagheghi A. 1999. **Simultaneous saccharification and cofermentation of dilute-acid pretreated yellow poplar hardwood to ethanol using xylose-fermenting *Zymomonas mobilis*.** *Applied Biochemistry and Biotechnology* **77**: 649–665

Minhas D and Grover A. 1999. **Transcriptional levels of genes encoding various glycolytic and fermentation enzymes change in response to abiotic stress.** *Plant Science* **146**: 41–51

Ohta K, Beall D S, Mejia J P, Shanmugam K T, Ingram L O. 1991. **Genetic improvement of *Escherichia coli* for ethanol production: chromosomal integration of *Zymomonas mobilis* genes encoding pyruvate decarboxylase and alcohol dehydrogenase II.** *Applied and Environmental Microbiology* **57**: 893–900

Raj K C, Talarico L A, Ingram L O, Maupin Furlow J A. 2002. **Cloning and characterization of *Zymobactor palmae* pyruvate decarboxylase gene (*pdc*) and comparison to bacterial homologues**

Zaldivar J, Martinez A, and Ingram L O. 1999. **Effect of selected aldehydes on the growth and fermentation of ethanologenic *Escherichia coli*.** *Biotechnology and Bioengineering* **65**: 24–33

Bioethanol production technologies: economics, environmental impact, and policy issues

K L Kadam
16696 W. Bayaud Dr.
Golden, Colorado, USA

Introduction

Ethanol production from corn and sugar cane is a well-established technology, with several plants located in the mid-western USA and Brazil using these crops. There are upper limits, however, to using corn as a feedstock, as the economics is tied to the by-products such as distiller's dried grain (DDG), the effective marketing of which is problematic due to potential market saturation. Corn-based ethanol can meet only about 12.8–17.8 billion gallons/year demand in the US on the basis of the estimates of the National Corn Growers Association, Chesterfield, Missouri, USA (NCGA 2005). Using lignocellulosic biomass as a substrate to make ethanol is also a promising approach, which does not suffer from these limitations. Although currently, a more expensive approach, it also offers concomitant environmental benefits, and India is focusing on cellulosic ethanol as a long-term strategy. Many sources of lignocellulosic biomass, such as agricultural residues, forestry residues, pulp and paper waste streams, and municipal solid waste, are abundant and underutilized. These can be converted to ethanol. Woody and herbaceous energy crops such as hybrid poplar and switch grass can also be used as renewable resources for ethanol production. Researchers at the Oak Ridge National Laboratory, USA, studied the logistics and economics of energy crops (Perlack, Walsh, Wright *et al.* 1996; Perlack and Wright 1995; Turhollow 1994; Wright and Hohenstein 1994); however, it appears that the first materials used for ethanol production will most probably be waste materials and residues. In this context, the availability of corn stover, rice straw, and bagasse has been discussed by Kadam (1999), Kadam, Forrest, and Jacobson (2000), and Kadam and McMillan (2003).

Compositions for some of the waste feedstock are shown in Table 1. The conversion technology is sensitive to the feedstock type. For example, softwoods in the forest residue are more difficult to pretreat compared to hardwoods or agricultural residues. Also, the total sugar content is important, as ethanol yield, a key economic parameter, depends on sugar content; the three types of feedstock are comparable in total sugar content. Agricultural residues such as rice straw, wheat straw, bagasse, and so on are relevant for India, although forest and mill residues may also be of interest. Excess bagasse is a potentially attractive feedstock, given that it is produced at sugar mills; some of which also produce ethanol from molasses.

Sugar cane and bagasse production

As shown in Table 2, Brazil and India are the world's two largest sugar cane (*Saccharum officinarum*) growers, with production of 514 million tonnes (MT) and 356 MT/year, respectively (FAOSTAT 2008). India's leading sugar cane-producing states are Uttar Pradesh, Maharashtra, and Tamil Nadu (Table 3, Figure 1). In the USA, sugar cane is grown commercially in Florida, Hawaii, Louisiana, and Texas.

Bagasse is the fibrous residue left after the extraction of sugar from the cane, and it can be a good substrate for bioethanol (that is, biomass-derived ethanol) production. Bagasse is preferably used by the sugar mills for steam and power generation to meet their internal needs; however, about 15%–25% of the bagasse is left after meeting the mills' energy requirements, and this excess is not burned in the mill boilers. Steam consumption in Indian sugar mills is as high as 50%–55% on cane compared to 40% in Hawaii. With improvements in or replacement of existing boilers, the excess bagasse figure could be higher. Verenium Corp. has a demonstration plant in Jennings, Louisiana, which uses bagasse. Similarly, bagasse can be a major feedstock in India.

Table 1 Feedstock composition of forest residue, rice straw, and bagasse (dry wt%)

Feedstock component	Forest residue[a]	Rice straw	Bagasse[b]
Glucan	40.5	38.9	40.6
Galactan	10.5	0.0	0.8
Mannan	4.0	0.5	0.2
Xylan	6.0	20.4	20.0
Arabinan	3.0	3.4	1.7
Lignin	26.5	13.5	25.5
Extractives and other	7.5	5.3	7.5
Ash	2.0	18.0	3.7
Total	100.0	100.0	100.0

[a] 70/30 mix of white fir and ponderosa pine.

[b] Hawaiian bagasse.

Table 2 Worldwide sugar cane production in 2008

Country	Production in million tonnes (MT)
Brazil	514.1
India	355.5
China	106.3
Thailand	64.4
Pakistan	54.8
Mexico	50.7
Colombia	40.0[a]
Australia	36.0
USA	27.8
The Philippines	25.3[a]
World	1557.7

[a] Food and Agriculture Organization of the United Nations (FAO) estimate. Aggregates are sum of available data.

Table 3 Sugar cane production in India, by state

	Production (MT)	
State	2006 (actual)	2008 (estimate)[a]
Uttar Pradesh	121.5	169.9
Maharashtra	34.7	48.5
Tamil Nadu	33.3	46.6
Andhra Pradesh	17.9	25.1
Karnataka	15.2	21.3
Tamil Nadu	13.3	18.6
Haryana	6.8	9.6
Uttarakhand	6.1	8.6
Punjab	5.3	7.4
Other	12.7	17.7
Total	254.2	355.5

[a]Using the same percentage as in 2006 and a total of 355.5 MT.

Technology options

The production of ethanol from biomass through the biochemical route requires the following basic steps: pretreatment to hydrolyse the hemicellulose; hydrolysis of cellulose to produce glucose; fermentation of sugars to ethanol; and ethanol recovery. Alternatively, biomass gasification followed by syngas fermentation (Klasson, Ackerson, Clausen, *et al.* 1992a, b) is a route towards producing fuel ethanol. Thermo-chemical conversion is yet another approach, and thermal, chemical and biological processings have been reviewed by Katzen and Hinman (1993). Table 4

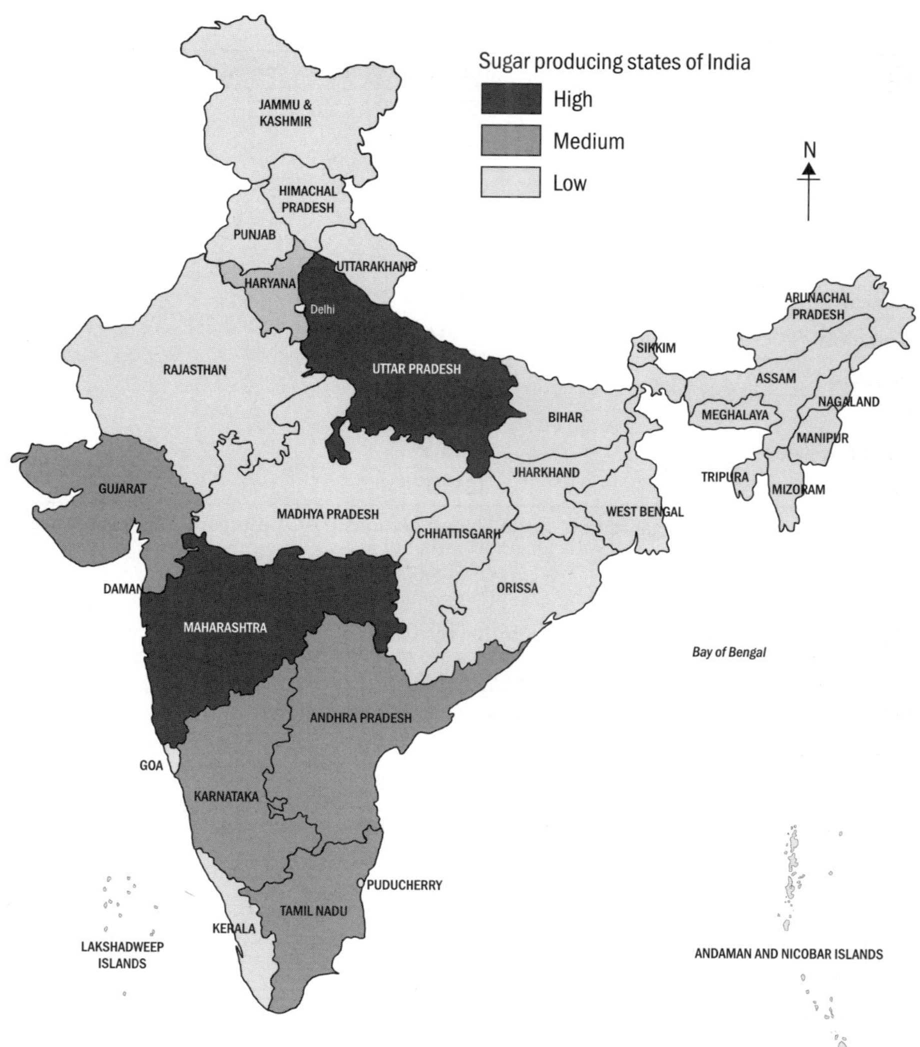

Figure 1 Sugar production in India

lists the awards announced in 2010 by the US Department of Energy (US DoE) for integrated biorefinery projects set up with an amount of $564 million for pilot-, demonstration- and commercial-scale facilities that involve ethanol and other biofuels (BiofuelsDigest 2009). There is a plethora of technologies, including (1) algae to fuels, (2) green gasoline via multi-step gasification, (3) green diesel using advanced pyrolysis and steam reforming, (4) fuel-grade ethanol using hybrid technology (through acetic acid fermentation), (5) ethanol production through gasification and catalytic processes, (6) ethanol production by combining biomass gasification and fermentation, (7) ethanol fuel through strong acid hydrolysis, (8) biochemical conversion processes, and so on. Furthermore, consolidated bioprocessing (CBP), formerly called direct microbial conversion, or DMC, in which

Table 4 Awards by the US Department of Energy for integrated biorefinery projects

Company	Project and technology
Algenol Biofuels, Inc.	Make ethanol directly from carbon dioxide and sea water using algae. Facility capacity: 100 000 gallons/year of fuel-grade ethanol
Amyris Biotechnologies, Inc.	Produce a diesel substitute through the fermentation of sweet sorghum. Pilot plant capacity to coproduce lubricants, polymers, and other petrochemical substitutes
Archer Daniels Midland	Use acid to break down biomass, which can be converted to liquid fuels or energy. Facility to produce ethanol and ethyl acrylate, a compound used to make a variety of materials
Clearfuels Technology, Inc.	Produce renewable diesel and jet fuel from woody biomass by integrating ClearFuels' and Rentech's conversion technologies
Haldor Topsoe, Inc.	Convert wood to green gasoline by fully integrating and optimizing a multi-step gasification process. Pilot plant capacity to process 21 metric tonnes of feedstock per day
ICM, Inc.	Modify existing corn ethanol facility to produce cellulosic ethanol from switch grass and energy sorghum using biochemical conversion processes
Logos Technologies	Convert switch grass and woody biomass into ethanol using a biochemical conversion process
Renewable Energy Institute International	Produce high-quality green diesel from agriculture and forest residues using advanced pyrolysis and steam reforming. Pilot plant capacity to process 25 dry tonnes of feedstock per day
Solazyme, Inc.	Validate projected economics of a commercial-scale biorefinery producing multiple advanced biofuels. Produce algae oil that can be converted to oil-based fuels
UOP, LLC	Integrate existing technology from Ensyn and UOP to produce green gasoline, diesel, and jet fuel from agricultural residue, woody biomass, dedicated energy crops, and algae
ZeaChem, Inc.	Use purpose-grown hybrid poplar trees to produce fuel-grade ethanol using hybrid technology. Additional feedstock such as agricultural residues and energy crops also to be evaluated in the pilot plant
Enerkem Corporation	Site at existing landfill and use feedstock such as woody biomass and biomass removed from municipal solid waste to produce ethanol and other green chemicals through gasification and catalytic processes.
INEOS New Planet BioEnergy, LLC	Produce ethanol and electricity from wood and vegetative residues and construction and demolition materials by combining biomass gasification and fermentation. Facility capacity to produce 8 million gallons/year of ethanol and 2 MW of electricity by the end of 2011
Sapphire Energy, Inc.	Cultivate algae in ponds to be ultimately converted into green fuels, such as jet fuel and diesel, using the Dynamic Fuels refining process
Bluefire, LLC	Construct facility to produce ethanol fuel from woody biomass, mill residue and sorted municipal solid waste. Facility capacity to produce 19 million gallons/year of ethanol.

cellulase production, hydrolysis, and ethanol fermentation are combined in one operation, is another possible mode (Hogsett, Ahn, Bernardez, *et al.* 1992); however, CBP is not considered here as it is not much advanced. CBP involves consolidating into a single process step four biologically mediated events: cellulase production, cellulose hydrolysis, hexose fermentation, and pentose fermentation. This requires development of micro-organisms that both utilize cellulose and other biomass components while also producing ethanol at sufficiently high yields and concentrations. It is not trivial to optimize all four steps in a single organism.

Despite the diversity of technologies for near-term, strong-acid hydrolysis, dilute-acid hydrolysis, and enzymatic hydrolysis are considered relevant for India. Bluefire, LLC, is constructing a demonstration plant using strong-acid hydrolysis. The Verenium Corp. demonstration plant uses dilute-acid hydrolysis of hemicellulose and separates C5 and C6 fermentations using *Escherichia coli* and *Klebsiella oxytoca*, respectively. The enzymatic conversion process selected by Abengoa Bioenergy New Technologies (ABNT) utilizes dilute-acid pretreatment followed by enzymatic hydrolysis (using Dyadic International's *Chrysosporium lucknowense* cellulase) and finally fermentation (using Cargill, Inc.'s yeast, which can ferment hexose and xylose sugars). ABNT has a biomass-to-ethanol pilot plant in York, Nebraska, adjacent to its existing 56 million gallons/year corn-based production plant, and a demonstration facility in Salamanca, Spain (Wooley 2009). Thus, these technologies are more advanced than the most mentioned in Table 4.

Hence, the current discussion is limited to the above mode of attack involving pretreatment, cellulose hydrolysis, and fermentation, for which there are different process configurations, both enzyme-based and non-enzyme-based, which can be used to achieve the overall goal. In the non-enzyme-based approach, acid is used for both hemicellulose and cellulose hydrolysis, and the mode is separate hydrolysis and fermentation (SHF). In the enzymatic approach, dilute-acid pretreatment is used to hydrolyse the hemicellulose portion. The saccharification (hydrolysis) of cellulose to cellobiose and eventually to glucose is catalysed by the synergistic action of cellulase and β-glucosidase enzymes. The mode of operation used is simultaneous saccharification and fermentation (SSF) or simultaneous saccharification and cofermentation (SSCF); cofermentation refers to the fermentation of both six-carbon (hexoses, that is, glucose, mannose, and galactose) and five-carbon (pentoses, that is, xylose and arabinose) sugars to ethanol. By analogy, separate hydrolysis and cofermentation (SHCF) is a possibility as well with the acid-based approach.

In an industrial process, the process configuration would depend on the enzyme and ethanologen characteristics. Cellulases with reduced end-product inhibition would favour SHF-type configuration, whereas current mesophilic cellulases with susceptibility to end-product inhibition call for SSF- or SSCF-type configuration (Figures 2 and 3). Higher thermotolerance of ethanologenic organisms (for example, *Bacillus stearothermophilus*) would

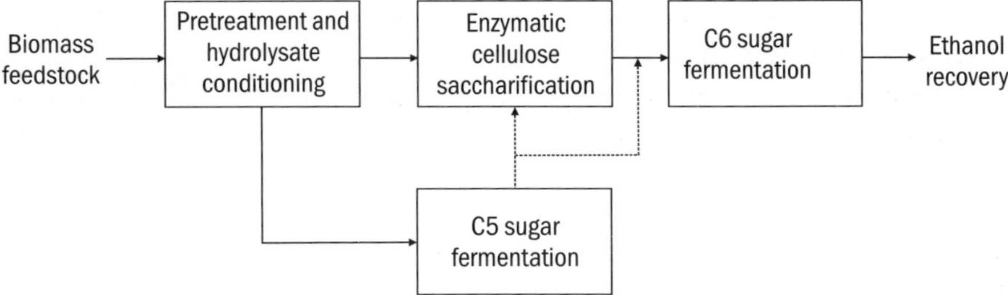

SHF with separate C5 and C6 sugar fermentation

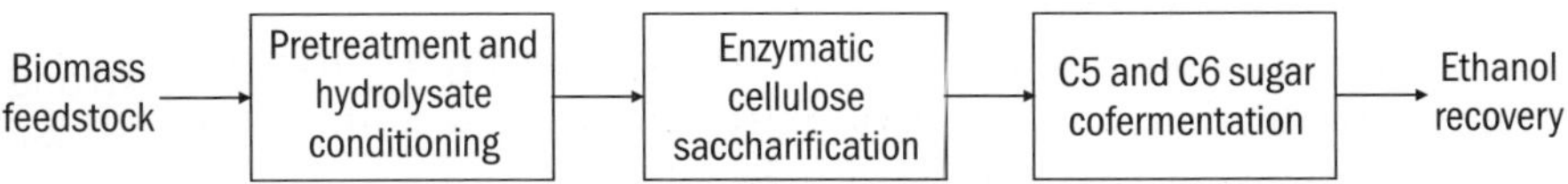

SHF with combined C5 and C6 sugar fermentation

Figure 2 Process configurations based on separate hydrolysis and fermentation

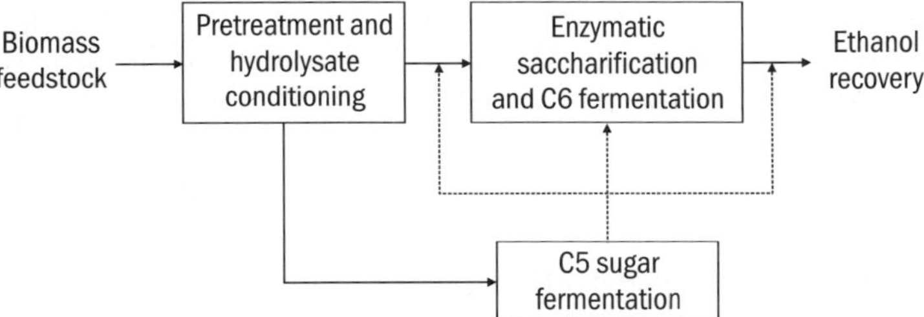

SSF with separate C5 and C6 sugar fermentation (SSF)

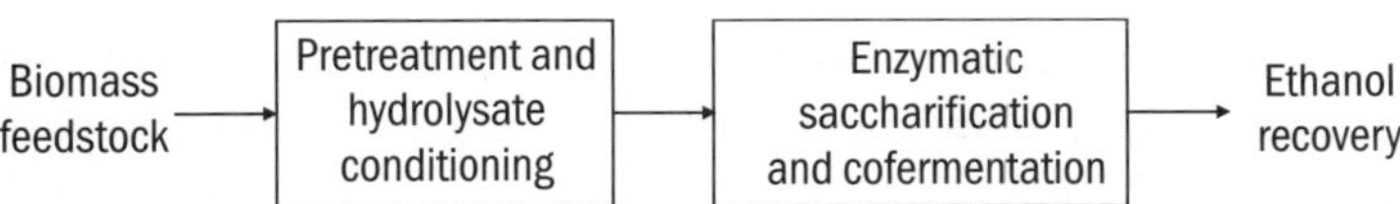

SSF with combined C5 and C6 sugar cofermentation (SSCF)

Figure 3 Process configurations based on simultaneous saccharification and cofermentation

allow high-temperature SSF. Each configuration has its advantages as discussed subsequently.

The benefits of SHF are as follows.

- It enables each step to be carried out at its respective optimal conditions (temperature, pH, and so on).
- It offers a huge benefit if significantly different optima exist.
- It produces intermediate sugar product.
- Sugars released by cellulose hydrolysis are decoupled from fermentation (unlike in SSF), and it is hence not captive to any specific fermentation/ product.

- These sugars can be sold or used to produce coproducts.
- It reduces fermentation mixing requirements.
- It produces less insoluble solids during fermentation, which reduces fermentation mixing requirements (or absence of solids if solid–liquid separation is used prior to fermentation).

The benefits of SSCF processing are as follows.

- Its proven "one-step" approach reduces process complexity.
- It minimizes enzyme inhibition by biomass sugars.
- There is no need for enzymes resistant to sugar inhibition.
- It provides an advantageous fermentation environment.
- It increases pentose utilization and strain robustness through sustained glucose production and utilization (strains with the ability to ferment pentoses at low glucose levels need to be selected).
- It improves process robustness by maintaining sugar concentrations at low levels, thus minimizing contamination.

Hybrid hydrolysis and fermentation (HHF) is another process configuration that begins like SHF and finishes like SSF/SSCF and combines the benefits of both (Figure 4). Higher enzyme thermostability would favour HHF type configuration.

In this chapter, the following different biomass-to-ethanol conversion technologies will be discussed.

- Concentrated sulphuric acid process (SHCF process configuration)
- Two-stage dilute-acid process (SHCF process configuration)
- Enzyme-based process (SSCF process configuration)

A recombinant *Zymomonas mobilis* (such as that from E.I. du Pont de Nemours & Co.) can be used to ferment both six- and five-carbon sugars. Other recombinant ethanologens such as a xylose-fermenting yeast (such as that from Cargill, Inc. or Royal Nedalco), *E. coli*, and *Klebsiella oxytoca* are also possible choices. For the following process descriptions, one of these ethanologens is used.

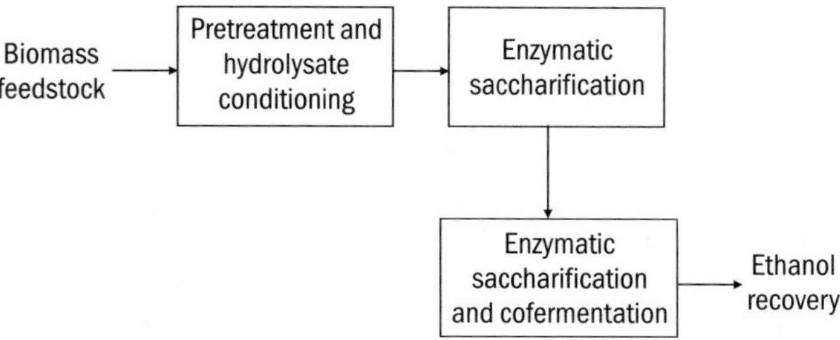

Figure 4 Hybrid hydrolysis and fermentation

Concentrated acid process

The Tennessee Valley Authority, Muscle Shoals, Alabama (TVA), was the original developer of the concentrated-acid hydrolysis technology, which has been further developed by Arkenol, Inc. of Mission Viejo, California (recently renamed Bluefire LLC), and Masada Resource Group of Birmingham, Alabama (Farone and Cuzens 1996a, b). These and other recent advances in the chromatographic separation of acid and sugars from the hydrolysate have improved the economics of the process (Hester 1997; Russo 1998). The acid-based technology presented here is a generic process that consists of four basic unit operations: hydrolysis, separation of the acid and sugars, ethanol fermentation, and product purification (Figure 5).

Hydrolysis

Prior to acid hydrolysis, the biomass is dried to a moisture content of approximately 10% and milled to an average size of 15 mm. The feedstock is mixed with concentrated sulphuric acid at a concentration of 70%–77%. The sulphuric acid to biomass ratio is 1.25 to 1. This step results in the disruption of the bonds between the crystalline cellulose chains, making the long-chain cellulose accessible for hydrolysis. The resulting acid concentration is 30%.

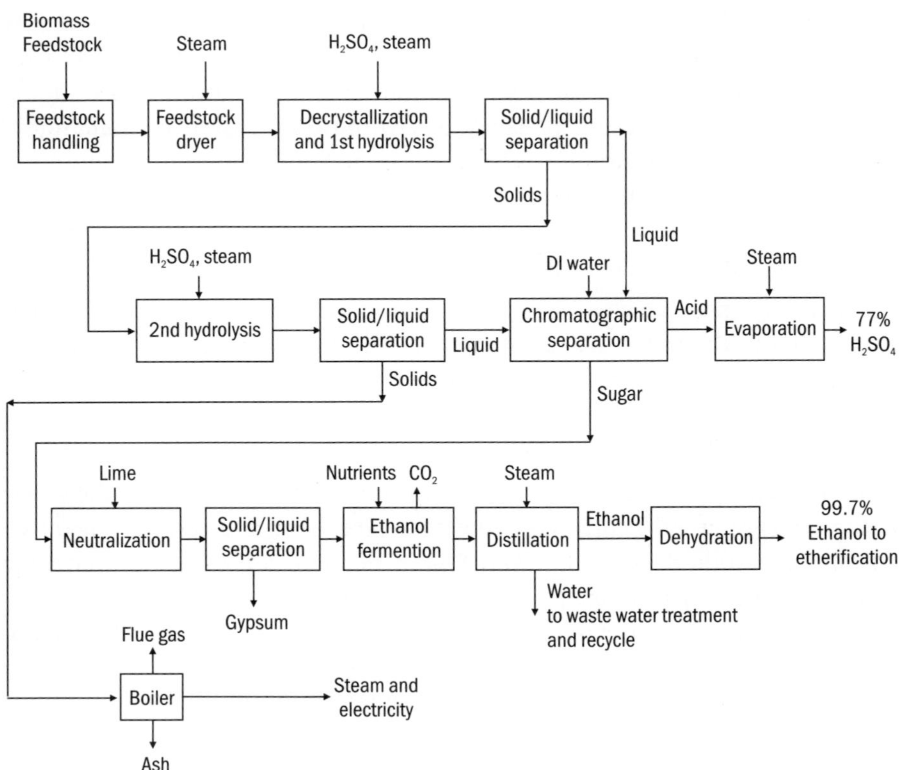

Figure 5 Concentrated acid process flow diagram

Decrystallization is performed at temperatures in the range of 30–50°C. The addition of acid to the biomass results in the formation of a thick gel. In the first hydrolysis, the mixture of acid and biomass is heated to 100°C for 60 minutes to hydrolyse the cellulose. The resulting gel is pressed to obtain an acid–sugar stream (approximately 17% sugar and 35% acid, depending on the feedstock composition).

The solids remaining after the first hydrolysis and solid/liquid separation are mixed with concentrated sulphuric acid until a concentration of 30% acid is again reached. The second hydrolysis step is very similar to the first one. The mixture is heated for 50 minutes at 100°C to effect further cellulose hydrolysis. The resulting gel is pressed to obtain a second acid–sugar stream (approximately 18% sugar and 30% acid, depending on the feedstock composition), and the streams from the two hydrolysis steps are combined. The remaining lignin-rich solids are collected and optionally pelletized for fuel.

Chromatographic separation of acid and sugars

The acid–sugar stream is further processed through a chromatographic separation column packed with a strong-acid polystyrene–divinylbenzene resin. Water is used as eluant. As a result of this chromatographic separation process, two streams are collected: a 25% concentrated-acid stream and a 12%–15% concentrated-sugar stream. The sugar recovery is 95%, whereas the acid recovery is 98%. The acid stream is concentrated and recycled. The sugar stream, which contains no more than 1% acid, can then be fermented after the pH is adjusted. Residual acid in the sugar stream is neutralized by adding lime, which forms a gypsum precipitate. Gypsum is removed by solid/liquid separation. This neutralization has the added benefit of precipitating unwanted metal hydroxides and other fermentation inhibitors.

Concentration and recycling of acid

The acid solution recovered from the separation unit can be concentrated and recycled to the earlier stages of the process. Concentration of the acid to 70%–77% is achieved through the use of a triple effect evaporator.

Ethanol fermentation and distillation

An appropriate ethanologen is mixed with nutrients and added to the sugar solution, where it efficiently converts both six- and five-carbon sugars to ethanol and carbon dioxide. The SHCF is carried out in continuous anaerobic fermenters. The flow of fermentation broth between fermenters is facilitated by gravity. Fermentation off gases, containing mostly carbon dioxide and ethanol vapour, are sent to the vent scrubber for ethanol recovery. The fermentation broth is sent to the distillation section for ethanol recovery.

Ethanol is separated from the fermentation beer by standard distillation technology and dehydrated with conventional molecular sieve technology. The 99.7% ethanol is sent to storage. The stillage from the distillation column is sent for waste water treatment and recycling. The dewatered ligneous residue, containing mostly lignin and cellulose, is pH-adjusted and burned on-site to cogenerate steam and electricity, which can be used by the process. The net electricity produced for this process is zero, and a supplemental fuel is needed to generate steam needed by the process.

Two-stage dilute-acid process

Figure 6 shows a simplified process flow diagram illustrating the major processing steps and flow paths in the plant. A process description is provided in the subsequent section, based on the work of Kadam, Wooley, Aden, *et al.* (2000).

First-stage hydrolysis

The feedstock, of about 15 mm size, is conveyed to the first-stage acid impregnator, where it is heated to approximately 50°C and soaked in water containing 0.7% sulphuric acid. From the impregnator, the biomass enters the first-stage hydrolysis reactor, where the temperature is raised to 190°C (167 psia, 1150 kPa). At the liquid-phase acid concentration of 0.7%, a residence time of 3 minutes is required to achieve the efficient hydrolysis of

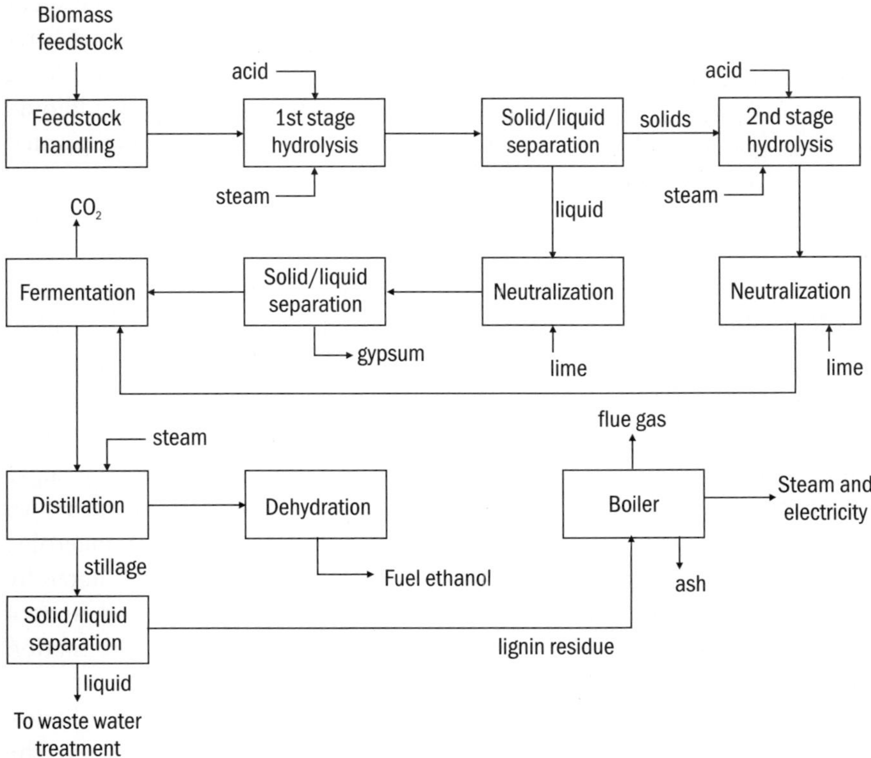

Figure 6 Two-stage dilute-acid process flow diagram

hemicellulose. Approximately, 80% of the hemicellulose and 20% of the cellulose are hydrolysed in the first reactor. The stream leaving the first hydrolysis reactor (hydrolysate) contains about 30% total solids (suspended and dissolved solids).

The hydrolysate pressure is reduced in the flash tank, lowering the temperature of the hydrolysate to about 130°C. Following a residence time of 2 hours in the flash tank to hydrolyse most of the oligosaccharides to monosaccharides, the hydrolysate enters the countercurrent slurry washer, where the solids and liquids are separated, and the solids are washed to remove the sugars and other soluble compounds. Efficient washing is critical to maximize sugar recovery and to minimize the dilution of the sugar stream. Too much washwater will increase the size and the cost of the fermenters and the distillation system. A washwater flow rate of three to four times the dry weight of solids being washed is used. The combined liquid hydrolysate and washwater are sent to the first-stage pH adjustment tank. The solid stream (30% solids) from the washer is further dewatered by a screw press. The pressate is returned to the slurry washer, and the solids, now at about 45% solids, are sent to the second-stage acid impregnator.

Second-stage hydrolysis

The acid-impregnated material then enters the second-stage hydrolyser. In the second hydrolysis reactor, the temperature is raised to 220°C (322 psia, 2220 kPa), and the prevailing acid concentration in the liquid phase is 1.6%. A residence time of 3 minutes is required at these conditions to achieve the necessary reactions. About 60% of the remaining cellulose is hydrolysed, with about 70% going to glucose and 30% degraded to hydroxymethyl furfural (HMF) and other degradation products. Following the second hydrolysis, the hydrolysate slurry is flashed but—unlike in the first hydrolysis step—not washed.

Ethanol fermentation and distillation

Lime is added to the first-stage pH adjustment tank to neutralize the sulphuric acid and raise the pH of the hydrolysate to about 5.5. This causes most of the calcium sulphate (gypsum) to precipitate out of the solution. The calcium sulphate is removed by a filter and sent for disposal. The relatively solid-free sugar stream is cooled to 35°C and sent to the fermenters. Ammonium sulphate and nutrients like corn steep liquor are added to the two fementers.

The hydrolysate slurry from the second-stage pH adjustment tank is cooled to 35°C before it enters the fermentation section. The fermentation section has a total residence time of 32 hours, with an overall ethanol yield of 90% of the six-carbon sugars and 75% of the xylose entering the fermenters. A recombinant, xylose-fermenting yeast can be used in this case.

The fermentation broth is sent to the distillation/molecular sieve dehydration systems, where the ethanol is separated from water. The 99.7%

ethanol is denatured with 5% gasoline and sent to storage. The stillage from the distillation column is processed using centrifuges and evaporators. Lignin separation is accomplished in solid-bowl decanting centrifuges. The lignin stream, containing about 45% solids, is sent to the boiler/power generation section. The liquid stream is sent to a waste water treatment system, and reclaimed water is recycled to the process. See Kadam, Wooley, Ferraro *et al.* (1999) for a description of the waste water treatment system.

Enzymatic process

A generalized process is described based on *Trichoderma reesei* derived cellulases for cellulose hydrolysis and an appropriate recombinant ethanologen for the cofermentation of six- and five-carbon sugars to ethanol. The enzyme-based process consists of four basic unit operations: pretreatment, cellulase production, ethanol production, and product purification (Figure 7).

Feedstock preparation and pretreatment

The biomass is milled to an average size of 15 mm. A screw feeder conveys the biomass from the storage bunker to the acid impregnator. Dilute sulphuric acid and low-pressure steam are also fed to the acid impregnator. The acidic slurry is discharged from the acid impregnator into the pretreatment reactor. High-pressure steam and additional dilute sulphuric acid are fed to the reactor, where hemicellulosic sugars are hydrolysed to their respective monomers/

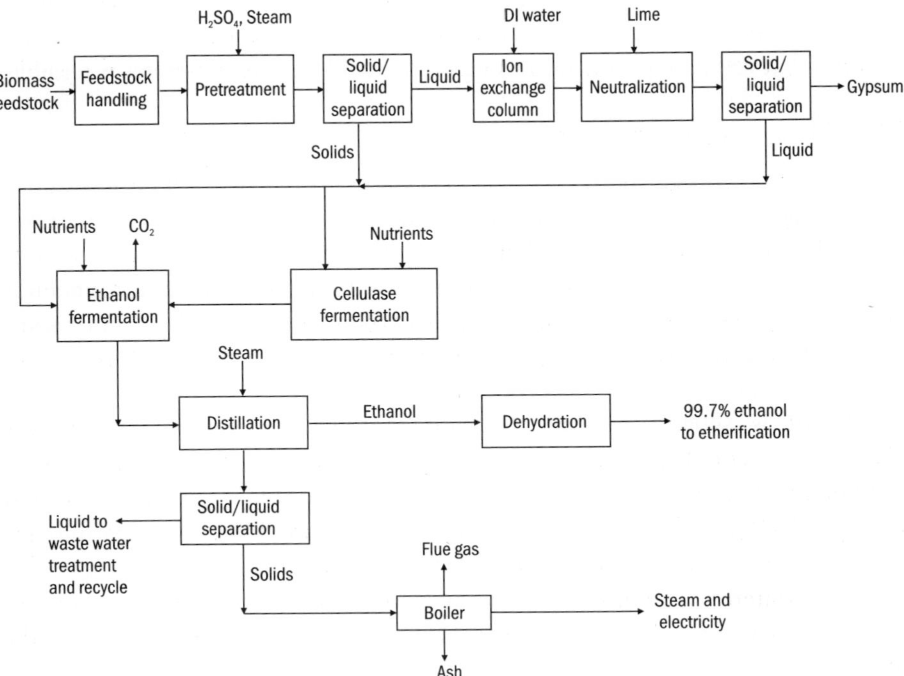

Figure 7 Enzymatic process flow diagram

oligomers (temperature = 160–180°C, liquid phase acid concentration = 0.7%–1.0% wt). The hydrolysed mash is discharged from the acid hydrolysis reactor into the lower pressure flash drum, where cooling quenches the reactions. The hydrolysate is separated from the solids in a solid/liquid separation step. The hydrolysate is then pumped to the neutralization and detoxification section. This involves continuous ion exchange using a weak-base anion resin, followed by overliming. The process primarily removes acetic acid and other chemicals that could be toxic to the micro-organisms used.

Lime is used to neutralize the detoxified hydrolysate; the neutralization reaction produces calcium sulphate, which is removed by solid/liquid separation. The neutralized hydrolysate is pumped through a heat exchanger, where it is cooled, using cooling tower water, to fermentation temperature. The hydrolysate and solids from the solid/liquid separation step are then pumped to the ethanol fermentation section.

Cellulase production

Cellulase production is by *T. reesei* using a slipstream of pretreated biomass as a carbon source. The fermentation is conducted in a fed-batch mode at 28°C and pH 5. Corn steep liquor is used as a source of nitrogen and micronutrients. For a low-cost product such as ethanol, the enzyme need not be processed to a great extent for it to be useful. Whole broth from cellulase fermentation is actually more effective for the SSCF process. In this process, the whole fermentation broth is used as a source of cellulase enzyme. Because enzyme production is via the fed-batch mode, and the SSCF is a continuous process, a surge storage tank is necessary. This is a generic process, which assumes that cellulase production using pretreated biomass substrates is feasible.

Ethanol fermentation and distillation

Using an appropriate ethanologen, the SSCF process converts cellulose and five-carbon sugars to ethanol and carbon dioxide. Cellulase catalyses the hydrolysis of cellulose to glucose. The SSCF operation takes place in continuous anaerobic fermenters. Gravity drives the flow of fermentation broth between fermenters. Fermentation exhaust gases consisting of carbon dioxide and ethanol vapour are sent to the vent scrubber for ethanol recovery. The SSCF broth is pumped to the distillation section for the recovery of ethanol.

Ethanol is separated from the fermentation beer by well established distillation technology and dehydrated using conventional molecular sieve technology. The still bottoms are collected, and the 99.7% ethanol is sent to storage. The lignin residue is further dewatered by solid/liquid separation. The liquid stream is sent for waste water treatment and recycling.

The dewatered ligneous residue is burned on-site to cogenerate steam and electricity, which can be used by the process. The excess electricity generated is sold to the grid.

Comparison of ethanol production technologies

Wilke, Yang, Sciamanna, *et al.* (1981) carried out one of the first process development and cost analyses for the conversion of biomass to sugars and ethanol (based on the SHF mode). Hinman, Schell, Riley, *et al.* (1992) and Wright (1988) later reported on the economics of SSF. The projected cost of ethanol production from cellulosic biomass from the earlier estimates has been reduced by almost a factor of four over the last 20 years (Wyman 1999, 2001). Wingren, Galbe, and Zacchi (2003) discussed SHF and SSF economics using enzymes in both configurations, and SSF was found to be less expensive by about 10%; their estimate of \$0.56–0.67/l ethanol is high compared to others. Despite such analyses, a single study for evaluating all of the above three technologies does not exist. A qualitative comparison of the three technologies is provided in Table 5. A study by Sivers and Zacchi (1995), however, can be used as a surrogate, as they evaluated technology options that are similar, using a pine feedstock: concentrated-acid process (CHAP), SO_2/dilute-acid process (CASH), and enzymatic hydrolysis process. This is a useful study in that it evaluates these options (see Table 6 for process details) for the conversion of pine to ethanol using a uniform platform. Conversion of Swedish kronors (SEK) to American currency is as follows: 8 SEK = 1 US\$, for the 1992/93 timeframe of the study.

Table 5 Comparison of ethanol production technologies

	Concentrated sulphuric acid	Two-stage dilute acid	Enzyme-based
Technological maturity	High	High	Low
Potential for cost reduction	Low	Low	High
Energy consumption	Very high	Low	Low
Environmental impact	High	Low	Low
Approximate cost, \$/gal[a]	~1.20	~1.20 (Kadam, Wooley, Aden, *et al.* 2000)	1.16–1.44 (Wooley, Ruth, Glassner, *et al.* 1999)
Strengths	Proven technology with commercial plant built in Germany in 1937. Robust process and can handle heterogeneous waste streams such as municipal solid waste (MSW), which have tipping fees	Proven technology with extensive application to wood; first commercial process dating back to 1898	Potential for yields higher because of less severe processing conditions. Beneficial lower-cost cellulases under development

[a]These are only approximate estimates; basis for estimation differs.

Table 6 Summary of main process details for CHAP, CASH, and enzymatic processes

	CHAP	CASH	Enzymatic
Process steps			
First hydrolysis	Concentrated HCl (41%, 30°C)	Steam (188°C) + SO_2 (2.5% of DM)	Steam (210°C) + SO_2 (2.6% of DM)
Second hydrolysis	n/a	Dilute HCl (0.2%, 230°C)	Enzymes
Fermentation	Biostil	Cascade system	Cascade system
Overall yields			
Wood to hexoses (%)[a]	85	61	81
Wood to pentoses (%)[a]	88	54	69
Wood to ethanol (%)[b]	80	57	76

[a]Percentage of theoretical, including first hydrolysis, first wash, second hydrolysis, and second wash.

[b]Percentage of theoretical.

Increased sugar yield, energy integration or larger plant capacity are improvements that reduce the ethanol production cost for all the three processes. Integration of the ethanol plant with already existing facilities, such as a pulp and paper plant or a heating plant, could further reduce the ethanol production cost. The major conclusion drawn from this study is that none of the three processes can be eliminated on the basis of one being less economical than the other two. The ethanol production price varies between 4.0 SEK/l and 4.3 SEK/l ($1.89–2.04/gallon), and the difference is within the margin of uncertainty in the various assumptions underlying the technical and economic calculations. However, this does not take into account the potential for cost reduction. Although acid-based technologies are technologically more mature, enzymatic technologies are roughly equivalent in costs and are forecast to achieve significant cost reduction as these develop further (Lynd, Cushman, Nichols, *et al.* 1991). Wooley, Ruth, Glassner, *et al.* (1999) and Wooley, Ruth, Scheehan, *et al.* (1999) state that bioethanol could cost in the range $1.16–1.44/gallon, depending on the specific technology and the availability of low-cost feedstocks. They identify research strategies and goals to improve the yield and performance in the conversion process, which can translate into cost savings of 40 ¢/gallon over the next 10 years. Estimates of potential cost reduction for all three technologies are presented in Figure 8.

Cardona and Sánchez (2006) used a process simulator and literature data to assess the energy consumption for several process configurations. SSCF configuration with water recycling showed the best results with energy consumption of 42.0 MJ/l ethanol. If pervaporation replaces azeotropic distillation as the dehydration method, further energy savings can be realized. They calculated the net energy value of 17.7–18.9 MJ/l ethanol,

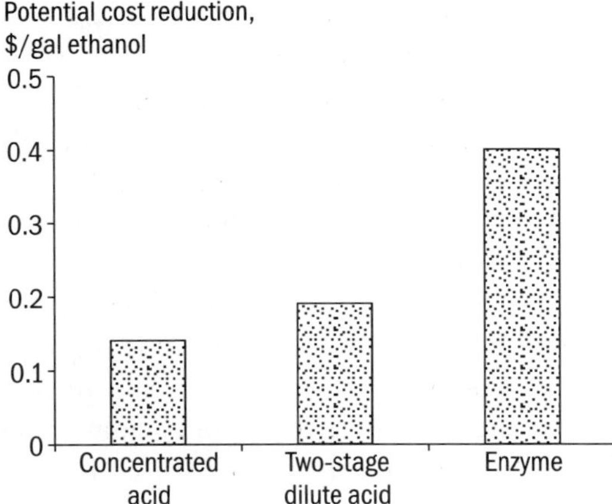

Figure 8 Potential for cost reduction of the three technologies

Table 7 Comparison of different process configurations

Process	Ethanol yield, litre/dry wood tonne	Distillation energy, MJ/l ethanol	Evaporation energy, MJ/l ethanol
SHF	270.2	6.4	7.3
SSF	283.7	5.9	6.7
SSCF	310.9	4.4	3.9

SHF – separate hydrolysis and fermentation; SSF – simultaneous saccharification and fermentation; SSCF – simultaneous saccharification and cofermentation

indicating energy efficiency of lignocellulosic ethanol. Piccolo and Bezzo's (2007) comparison of yield and energy consumption for different process configurations is summarized in Table 7. The SSCF configuration appears to be the best solution with respect to both ethanol yield and energy costs. Their simulation results showed that burning residual lignin and biogas obtained during waste water treatment generates all the steam required by the process and produces an electric energy surplus: a plant processing 1000 kilotonnes/year of biomass can deliver about 16.9 MW of electric energy to the grid.

Environmental impact of ethanol production technologies

The previous section attempted to elucidate the relative operational and economic advantages of the three processes. Although this is somewhat problematic due to the paucity of data on a constant basis, it is possible to compare the environmental impact of these technologies using life cycle assessment (LCA), and such data is readily available. LCA is a technique for assessing the environmental aspects and potential impacts associated with a product. Basic aspects of LCA methodology are discussed in the literature (Heijungs, Guinée, Huppes, *et al.* 1992; SETAC-Europe 1992).

Based on the LCA studies, we can attempt a comparison of the three processes. A study by Kadam (2002) compared the introduction of an E10 fuel blend (containing 10% bioethanol by volume) for use in motor vehicles in Mumbai (alternative scenario) with the gasoline being consumed (current scenario, that is, status quo). All environmental flows were examined over the product life cycle, from production and extraction of raw materials through intermediate conversion processes, transportation, distribution, and use. As this study evaluated both dilute-acid and enzyme processes for ethanol production, we can extract the relevant data to compare them. The environmental flows were used to arrive at impact indicators, which give a broader insight into the environmental impacts of the competing options by examining the potential impacts beyond the initial release. The indices calculated were greenhouse potential, natural resources depletion, acidification potential, eutrophication potential, and human toxicity potential (see Kadam [1999] for definition). The environmental impact indicators of ethanol production using the two processes are summarized in Table 8. It is evident from the table that for all the indicators, dilute-acid process fares better than the enzyme process. This is because the former sends a much higher proportion of the biomass to the boiler for electricity production, which in turn offsets large amounts of emissions. Table 8 also normalizes the data to the enzyme process and also lists process energy, which is not a standard impact indicator but gives us an insight into the process.

Another LCA for California compared biomass disposal through burning or diverting it for ethanol production (Kadam, Camobreco, Glazebrook, *et al.* 1999). This study evaluated both concentrated-acid and enzyme processes for ethanol production, and again we can extract relevant data to compare these two processes. Similar to the indicators presented in Table 8, there are impact indicators for concentrated-acid and enzyme processes, with forest residue as a feedstock, which are similar in total sugars and lignin to bagasse and were normalized to the enzyme process. These normalized impact indicators along with those in Table 8 were combined and are shown in Figure 9.

Table 8 Environmental impact assessment of dilute-acid and enzyme technologies

Impact parameter	Unit	Enzyme process	Dilute-acid process	Dilute acid/ enzyme ratio
IPCC – greenhouse effect	kg equivalent CO_2	–73.2	–151.9	0.48[a]
Depletion of non-renewable resources	Fraction of reserve	1.5×10^{-13}	9.5×10^{-14}	0.62
Air acidification	g equivalent H^+	154.6	120.0	0.78
Eutrophication	g equivalent PO_4	596	441	0.74
Human toxicity	kg	5.7	4.3	0.76
Process energy	MJ	12 625	13 270	1.05

IPCC – Intergovernmental Panel on Climate Change

[a]As the dilute-acid process offsets more CO_2, that is, has a higher negative number, its greenhouse gas impact is higher.

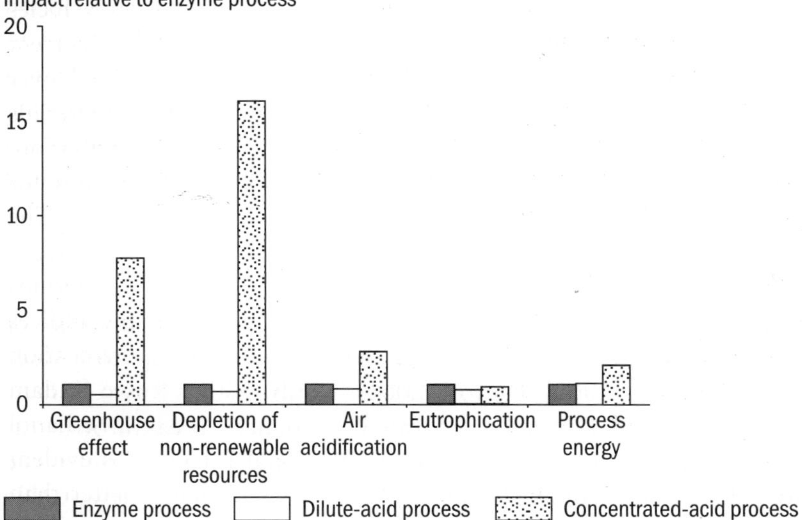

Figure 9 Relative environmental impact of ethanol production options using the enzyme process as a baseline

As mentioned above, the dilute-acid process results in the generation of excess electricity, which offsets emissions. Conversely, the concentrated-acid process is a net consumer of energy, which results in very high process energy input (the deficit made up by fossil fuels), leading to higher values for greenhouse effect and other impact parameters. It should be pointed out that the normalized impact indicator for the depletion of non-renewable resources involves a ratio of two very small fractions, which renders the normalized indicator very high for the concentrated-acid process. Reduced greenhouse potential for the dilute-acid and enzyme processes provides an additional driver for the deployment of the bioethanol technology in India through the Clean Development Mechanism, as discussed further in the next section.

Bioethanol production in India

India is the fourth largest ethanol producer after Brazil, the USA, and China, the major ethanol substrate being molasses, which is a byproduct of sugar manufacture. India's average annual ethanol output amounts to 1900 million litres, with a distillation capacity of 2900 million litres/year. For a 5% ethanol blend in petrol nationally, the ethanol required was 640 million litres of ethanol in 2006/07 and would be 810 million litres in 2011/12 (Gonsalves 2006).

The state of Maharashtra leads in both agriculture and industrial growth in India, and it is the second largest sugar cane producing state in the country. Based on the data available for the Vasantdada Shetkari Sakhar Karkhana Ltd's (VSSK) sugar mill in Sangli, Maharashtra, a typical large sugar mill with 5000 tonnes/day capacity produces 284 422 tonnes of bagasse per year (Smouse, Staats, Rao, *et al.* 1998). This mill is considered as a typical candidate mill as far as ethanol production is concerned. An ethanol plant could be centrally located to utilize bagasse from several sugar mills within a reasonable distance, for example, a 50-km radius. Similar ethanol-producing facilities are possible in Uttar Pradesh and Tamil Nadu, and these could provide ethanol for an E10 blend, respectively, for Delhi and Chennai. Table 9 lists the largest sugar mills in Uttar Pradesh.

Ethanol production

Under contracts with the US DoE, Genencor International, a division of Danisco A/S, and Novozymes Biotech Inc. have been working to reduce enzyme costs (Canning 2002; Dutton 2001; Lamb 2001; Russo 2001). Recently, Novozymes and Genencor, separately announced breakthroughs at the Renewable Fuels Association's 15th Annual National Ethanol Conference in Orlando, Florida (SustainableBusiness 2010). Novozymes' new Cellic CTec2 enzymes would enable the production of cellulosic ethanol at a price below $2.00/gallon for the initial commercial-scale plants and are slated for operation in 2011. Novozymes claimed that the new enzyme can be used with various types of feedstock, including corn cobs and stalks, wheat straw, sugar cane bagasse, and woodchips. Advances in enzyme development have reduced the enzyme cost for cellulosic ethanol by 80% over the past two years, and enzyme cost is now down to approximately 50 ¢/gallon of cellulosic ethanol, according to Novozyme's figures. Genencor introduced its own version of an enzyme designed to do the same thing. Genencor did not give any specific cost figures, but stated that its Accellerase DUET product line will achieve higher sugar and biofuel yields, often at three-fold lower dosing, and it is feedstock-

Table 9 Largest sugar mills in Uttar Pradesh

Sugar mill	Crushing capacity, tonne/day
Balrampur Chini Mills Ltd (Babhnan, Mankapur, Gonda)	3500–5500
DSM Sugar Kashipur Ltd (Kashipur, Nainital)	2500–5000
Dwarikesh Sugar Industries Ltd (Village Bundki, Nagina, Bijnor)	2500–5000
Gangeshwar Ltd (Ramkola, Hata, Deoria)	2500–5000
Kanoria Sugar Gen. Mfg. Co. Ltd. (Captainganj, Hata, Padranuna)	3000–8000
Kesar Enterprises Ltd (Baheri, Barreilly)	3500–6000
The Bazpur Co-op. Sugar Factory Ltd (Bazpur, Nainital)	4000–5000
The Dhampur Sugar Mills Ltd (Dhampur, Bijnor)	9000–10 000
The Kisan Co-op. Sugar Factory Ltd (Sarsawa, Chilkana, Saharanpur)	2500–5000
The Mahalakshmi Sugar Mills Co. Ltd (Iqbalpur, Roorkee, Haridwar)	3000–10 000
The Oudh Sugar Mills Ltd (Hargaon, Sitapur)	5000–10 000
The Simbholi Sugar Mills Ltd (Simbholi, Garh Mukeshwar, Ghaziabad)	5000–10 000

and pretreatment-flexible. Higher performance at lower dose would lower enzyme costs for producers. Earlier, Genencor had announced reduction of enzyme costs by an order of magnitude (Genencor 2003). Although this represents a big drop, the costs are in the range of 20–25 ¢/gallon of ethanol, while the Wooley, Ruth, Glassner, *et al.* (1999) benchmark for enzyme cost is about half of this. Efficiency of enzymatic hydrolysis constitutes a major cost for the overall bioconversion process (Gregg, Boussaid, Saddler, *et al.* 1998), and further cost reductions are necessary to render the process economical. Hence, the dilute-acid process, which has the lowest environmental burdens of the three, may be better suited for India, which lacks a full-fledged enzyme industry, in the near term.

Ethanol capacity

Assuming that the VSSK sugar mill is a typical large sugar mill, bagasse produced by such a mill is about 285 000 tonnes/year. If 20% of the bagasse is excess, then each mill has an extra 57 000 tonnes/year. If three mills combine their excess bagasse, 171 000 tonnes/year of bagasse is available for ethanol (at 50% moisture, 85 500 tonnes/year of dry bagasse). This corresponds to a 23 million litres/year (ca. 6 million gallon/year) ethanol facility, which is fairly large by Indian standards. The annual petrol (gasoline) consumption in Mumbai is estimated to be 500 million litres. Thus, the plant could meet about half of the ethanol demand, assuming that all gasoline is sold as E10 fuel—a blend of 90% gasoline and 10% ethanol by volume. If the other three large mills in the area also contribute 20% of their bagasse, all the ethanol needed in Mumbai could be supplied by building a second plant or a larger single plant.

Bagasse availability could be increased beyond the current levels by introducing energy conservation measures in the sugar factories to reduce steam consumption, thereby reducing in-house bagasse usage. Moreover, an alternative feedstock that is available is cane trash, also known as cane residue or farm waste. For every 100 tonnes of clean sugar cane, about 28 tonnes of cane trash is left in the field and simply burned (Rao 1997). Based on this ratio, cane trash availability in Maharashtra alone can be estimated at approximately 16 MT/year (wet basis). However, the evaluation of cane trash for ethanol production has not been attempted.

Policy issues

Using biomass-derived fuel ethanol is a strategy that is relevant to India, not only in terms of greenhouse gas (GHG) emissions and improved air quality in major urban centres such as Mumbai, but also as an instrument to drive rural economic development and improve the country's energy security by reducing its exposure to risks associated with foreign oil. Petroleum consumption is growing at the rate of 8%–10% per annum. The Indian Ministry of Petroleum and Natural Gas (MoPNG) had approved the use of ethanol in gasoline blends, removing a potential regulatory hurdle (Uppal 2003; Winrock-International

2000). India launched 5% ethanol-blended petrol from 1 January 2003, in some states. To revive such mandatory blending, the central government is reworking the formula for pricing ethanol in order to facilitate ethanol supply from sugar factories to oil marketing companies, and it has recently exempted 5% ethanol-blended gasoline from additional excise duty of Rs 0.75/litre (Jagannathan 2009). For the long term, higher blending ratios of biofuels, similar to those in Brazil, are being sought. Taking lessons from the problems encountered in its previous attempt, this time, the government has specified the roles of the prime minister and each ministry, cabinet officer, and committee, along with the policies and measures to be instituted, which are as follows (BiofuelsDigest 2010).

- National Biofuel Coordination Committee under the prime minister for broader policy perspective.
- Biofuel Steering Committee under the cabinet secretary to oversee the implementation of policy.
- Target of 20% blending of biofuels in diesel and ethanol by 2017.

The approach adopted towards the goal of 20% blending of biofuels in diesel and ethanol by 2017 is more aggressive than that used by the US Environmental Protection Agency (USEPA). USEPA, under the Energy Independence and Security Act (EISA) of 2007, is responsible for the regulations regarding renewable fuels used as transportation fuel in the USA. The Renewable Fuel Standard Program will increase the required volumes of renewable fuel to 36 billion gallons by 2022 (USEPA 2010). This is about 20% of the *current* gasoline consumption. As mentioned earlier, corn-based ethanol can supply only, on average, about 15 billion gallons/year (NCGA 2005). The rest needs to come from ligoncellulosics-based fuels.

Clean Development Mechanism and Emissions Trading

The Kyoto Protocol provides two mechanisms, GHG Emissions Trading (ET) and Clean Development Mechanism (CDM); the latter accords an opportunity to developing countries that did not accept binding emissions reductions at Kyoto to participate in GHG mitigation. The CDM was created as a successor to "Joint Implementation" (JI), which constitutes a bilateral agreement between two entities towards a GHG mitigation venture. However, collaborative projects to reduce emissions or sequester carbon in developing countries are now to be implemented through the CDM route (Toman and Cazorla 2000).

The CDM poses some special coordination issues by virtue of its unique structure, which differs from that of the other two trading mechanisms under the Kyoto Protocol. For example, the CDM, like JI but unlike ET, is a project-based form of credit trading, and unlike both JI and ET, trading under this mechanism is to take place between a party with a firm national emission cap (Annex-I party) and the one without such a cap (Non-Annex-I party). Article 12 of the Kyoto Protocol subjects the projects under the CDM to a number of eligibility criteria, foremost being that the projects have to meet the overall objective of additional financing for sustainable development (for Non-Annex-I countries) and cost-effective emission reductions (for Annex-I countries) (Aslam 2000).

In terms of ET, the World Bank's Prototype Carbon Fund (PCF), launched in January 2000, is the first global GHG emissions trading fund (World Bank 2003). The fund is designed to facilitate investment in cleaner technologies for developing nations and simultaneously to provide industrialized nations emission trading credits that can be used in any future trading programme under the Kyoto Protocol. India's favourable approach to the CDM has been discussed by Hausker and McGinty (2000). The deployment of the bioethanol technology in India would then be a sound project for CDM or ET.

Copenhagen Accord

The United Nations Climate Change Conference (Conference of the Parties [COP] 15) took place in Copenhagen, Denmark, in December 2009. One of the aims of the conference was to negotiate a legally binding agreement that would be in place when the first commitment period of the Kyoto Protocol expires in 2012. Mitigation actions for Annex-I and Non-Annex-I countries are discussed subsequently.

Annex-I mitigation

Annex-I (developed) Parties commit to implement individually or jointly quantified economy-wide emissions targets for 2020. This provides a useful start by way of setting up a non-binding "depository" for the US emission reduction target of the US. However, targets in this Appendix are not legally binding and are not subject to an international compliance mechanism like a target under the Kyoto Protocol (Hobley 2010; Rose and Hobley 2010).

Non-Annex-I mitigation

Non-Annex-I (developing) Parties will implement mitigation actions, and nationally appropriate mitigation actions (NAMAs) seeking international support will be recorded in a registry along with relevant technology, finance, and capacity-building support. Supported NAMAs will be subject to international measurement, reporting and verification (MRV) in accordance with the guidelines adopted by the COP. Domestic MRV will be possible in developing countries; except where a NAMA is supported, the MRV guidelines will be "international". This helps in assuaging concerns about national sovereignty and ensuring that when developed countries provide support to developing countries, they will be able to verify that such support leads to genuine emissions reductions.

For India, the following applies.

Emissions reductions by 2020:	20%–25%
Conditions:	Will endeavour to reduce emissions intensity of gross domestic product by 20%–25% of 2005 levels by 2020, not including emissions from the agricultural sector
Base year:	2005
Reduction type:	Carbon intensity
% of global GHGs:	4.32

Deploying bioethanol projects, for example, ethanol from bagasse, will help India meet its goals.

Conclusion

India has an abundance of biomass that can be exploited to make fuel ethanol. Three conversion technologies are discussed from the viewpoint of maturity, currents costs, as well as long-term potential for cost reductions and environmental impact. The technologies discussed—concentrated-acid process, two-stage dilute-acid process, and enzyme-based process—are all applicable to the Indian scene. The dilute-acid and enzyme processes are more suitable in terms of potential economics and environmental impacts. There is a fundamental difference in net energy consumption between the concentrated-acid and the other processes, and these translate into concomitant environmental impacts. For the near term, the dilute-acid process can be deployed, as it exhibits lower burdens for key environmental criteria, both regulated and unregulated. In deploying the bioethanol option, for example, ethanol from bagasse, the lower greenhouse potential can be used to advantage. Specifically, the CDM or GHG trading can be used to help deploy the bioethanol option in India. Additional drivers for the ethanol scenario are the lower values observed for the following impact assessment categories: eutrophication potential, depletion of natural resources, air acidification potential, and human toxicity potential. This information would be useful to public policy-makers.

References

Aslam M A. 2000. **CDM: Coordination with the other Kyoto instruments. Key issues and options.** [Proceedings of 5th international conference on greenhouse gas control technologies, Cairns, August 13–19]

BiofuelsDigest. 2009. **19 integrated biorefinery projects to receive up to 564 million for pilot, demonstration and commercial-scale facilities.** 7 December 2009. Details available at <http://www.biofuelsdigest.com>, last accessed in March 2010

BiofuelsDigest. 2010. **India's 20% biofuels mandate.** 4 January 2010. Details available at <www.biofuelsdigest.com>, last accessed in March 2010

Canning K. 2002. **Transforming chemical production through bioprocessing – choosing the renewable route.** *Chemical Processing* **65**(4): 24–27

Cardona C A and Sánchez O J. 2006. **Energy consumption analysis of integrated flowsheets for production of fuel ethanol from lignocellulosic biomass.** *Energy* **31**(13): 2447–2459

Dutton G. 2001. **Bioindustry takes another look at biomass – new policies encourage growth of technology.** *Genetic Engineering News* **21**(5): 17

FAOSTAT. 2008. ***Sugarcane Production.*** Food and Agricultural Organization of United Nations: Economic and Social Department, the Statistical Division. Details available at <http://faostat.fao.org/>

Farone W A and Cuzens J E. 1996a. **Method of producing sugars using strong acid hydrolysis of cellulosic and hemicellulosic materials.** US patent 5,562,777

Farone W A and Cuzens J E. 1996b. **Method of separating acids and sugars resulting from strong acid hydrolysis.** US patent 5,580,389

Genencor. 2003. ***Genencor Exceeds Research Goal in Effort to Convert Biomass to Ethanol.*** Palo Alto: Genencor International

Gonsalves J B. 2006. **An assessment of the biofuels industry in India.** United Nations Conference on Trade and Development, Geneva

Gregg D J, Boussaid A, and Saddler J N. 1998. **Techno-economic evaluations of a generic wood-to-ethanol process: effect of increased cellulose yields and enzyme recycle.** *Bioresource Technology* **63**(1): 7–12

Hausker K and McGinty K. 2000. ***India's Reappraisal of the Clean Development Mechanism.*** Washington, DC: Resources for the Future

Heijungs R, Guinée J B, Huppes G, Lankreijer R M, Udo de Haes H A, Wegener A, Sleeswijk, Ansems A M M, Eggels P G, van Duin R, de Goede H P. 1992. ***Environmental Life Cycle Assessment of Products.*** Center of Environmental Science, University of Leiden, the Netherlands

Hester R D. 1997. **Process for separating acid-sugar mixtures using ion exclusion chromatography.** US patent 5,628,907

Hinman N D, Schell D J, Riley C J, Bergeron P W, Walter P J. 1992. **Preliminary estimate of the cost of ethanol-production for SSF technology.** *Applied Biochemistry and Biotechnology* **34**(5): 639–649

Hobley A. 2010. **The Copenhagen Accord: an update.** Details available at <http://www.nortonrose.com/knowledge/publications/2010>, last accessed in March 2010

Hogsett D A, Ahn H J, Bernardez T D, South C R, Lynd L R. 1992. **Direct microbial conversion – prospects, progress, and obstacles.** *Applied Biochemistry and Biotechnology* **34**(5): 527–541

Jagannathan P. 2009. **Fuel doping: government to rework pricing formula for ethanol.** *Economic Times,* December 23 2009. Details available at <http://economictimes.indiatimes.com/news/news-by-industry/energy/oil-gas/Fuel-doping-Govt-to-rework-pricing-formula-for-ethanol/articleshow/5367793.cms>, last accessed in March 2010

Kadam K L. 1999. **Environmental life cycle implications of using bagasse-derived ethanol as a gasoline oxygenate in Mumbai (Bombay).** National Renewable Energy Laboratory (NREL) Report No. NREL/TP-580- 28705. Golden: NREL

Kadam K L. 2002. **Environmental benefits on a life cycle basis of using bagasse-derived ethanol as a gasoline oxygenate in India.** *Energy Policy* **30**(5): 371–384

Kadam K L and McMillan J D. 2003. **Availability of corn stover as a sustainable feedstock for bioethanol production.** *Bioresource Technology* **88**(1): 17–25

Kadam K L, Forrest L H, and Jacobson W A. 2000. **Rice straw as a lignocellulosic resource: collection, processing, transportation, and environmental aspects.** *Biomass and Bioenergy* **18**(5): 369–389

Kadam K L, Camobreco V J, Glazebrook B E, Forrest L H, Jacobson W A, Simeroth D C, Blackburn W, Nehoda K C . 1999. ***Environmental life cycle implications of fuel oxygenate production from California biomass.*** National Renewable Energy Laboratory (NREL) Report No. NREL/TP-580-25688. Golden: NREL

Kadam K L, Wooley R J, Ferraro F M, Voiles R E, Ruocco J J, Varani F T, Putsche V L. 1999. **Wastewater treatment for a biomass-to-ethanol process: system design and cost estimates.** In *Proceedings of the Fourth Biomass Conference of the Americas: Biomass—a growth opportunity in green energy and value-added products,* pp. 699–705, edited by R O Overend and E Chornet. Oxford: Pergamon Press

Kadam K L, Wooley R J, Aden A, Nguyen Q A, Yancey M A, Ferraro F M. 2000. **Softwood forest thinnings as a biomass source for ethanol production: a feasibility study for California.** *Biotechnology Progress* **16**(6): 947–957

Katzen R and Hinman N D. 1993. **Thermal, chemical, and biological processing – introduction.** *Applied Biochemistry and Biotechnology* **39**: 3–4

Klasson K T, Ackerson C M D, Clausen E C, Gaddy J L. 1992a. **Biological conversion of synthesis gas into fuels.** *International Journal of Hydrogen Energy* **17**(4): 281–288

Klasson K T, Ackerson M D, Clausen E C, Gaddy J L. 1992b. **Bioconversion of synthesis gas into liquid or gaseous fuels.** *Enzyme and Microbial Technology* **14**(8): 602–608

Lamb C. 2001. **Davis biotech firm lands $14.8M ethanol grant.** *Sacramento Business Journal,* 12 January 2001. Details available at <http://www.accessmylibrary.com/coms2/summary_0286-10567489_ITM>, last accessed in March 2010

Lynd L R, Cushman J H, Nichols R J, Wyman C E. 1991. **Fuel ethanol from cellulosic biomass.** *Science* **251**(4999): 1318–1323

NCGA (National Corn Growers Association). 2005. ***Corn-to-ethanol capacity estimates for 10% fuel ethanol inclusion nationwide by 2015.*** Chesterfield: National Corn Growers Association. Details available at <www.ncga.com>

Perlack R D and Wright L L. 1995. **Technical and economic-status of wood energy feedstock production.** *Energy* **20**(4): 279–284

Perlack R D, Walsh M E, Wright L L, Ostlie L D. 1996. **The economic potential of whole-tree feedstock production.** *Bioresource Technology* **55**(3): 223–229

Piccolo C and Bezzo F. 2007. **Ethanol from lignocellulosic biomass: a comparison between conversion technologies.** In *17th European Symposium on Computer Aided Process Engineering – ESCAPE17,* edited by V Plesu and P S Agachi. Amsterdam: Elsevier Science B.V.

Rao P J M. 1997. ***Industrial Utilization of Sugarcane and its Co-products.*** Delhi: ISPCK Publishers and Distributors

Rose N and Hobley A. 2010. **Summary of Copenhagen outcomes.** Details available at <http://www.lexology.com/library>, last accessed in March 2010

Russo L J. 1998. **Method for the separation of acid from sugars.** US patent 5,968,362

Russo E. 2001. **Turning trash into treasure – can organic waste become the nation's next big power source?** *Scientist* **15**(14): 1–4

SETAC (Society of Environmental Toxicology and Chemistry)-Europe. 1992. ***Life-cycle Assessment.*** Brussels: SETAC-Europe

Sivers M V and Zacchi G. 1995. **A techno-economical comparison of three processes for the production of ethanol from pine.** *Bioresource Technology* **51**: 43–52

Smouse S M, Staats G E, Rao S N, Goldman R, Hess D. 1998. **Promotion of biomass cogeneration with power export in the Indian sugar industry.** *Fuel Processing Technology* **54**(1-3): 227–247

SustainableBusiness. 2010. **Novozymes, genencor unveil new enzymes for cellulosic ethanol.** 17 February 2010. Details available at <sustainablebusiness.com>, last accessed in March 2010

Toman M and Cazorla M. 2000. ***The Clean Development Mechanism: a primer.*** Washington, DC: Resources for the Future

Turhollow A. 1994. **The economics of energy crop production.** *Biomass and Bioenergy* **6**(3): 229–241

Uppal J. 2003. **Editorial in biofuels India: sustainable fuel for the transport sector,** vol. 1. In *Biofuels India: sustainable fuel for the transport sector*

USEPA (United States Environment Protection Agency). 2010. **Renewable fuel standard 2.** Details available at <http://www.epa.gov/otaq/renewablefuels/index.htm. Washington, DC>, last accessed in March 2010

Wilke C R, Yang R D, Sciamanna A F, Freitas R P. 1981. **Raw-materials evaluation and process-development studies for conversion of biomass to sugars and ethanol.** *Biotechnology and Bioengineering* **23**(1): 163–183

Wingren A, Galbe M, and Zacchi G. 2003. **Techno-economic evaluation of producing ethanol from softwood: comparison of SSF and SHF and identification of bottlenecks.** *Biotechnology Progress* **19**(4): 1109–1117

Winrock-International. 2000. ***Indian Government Approves the Use of Ethanol.*** New Delhi, India: Winrock International

Wooley R. 2009. ***Abengoa Bioenergy New Technologies Development of Enzymatic Cellulosic Ethanol.*** Nashville: AIChE

Wooley R, Ruth M, Glassner D, Sheehan J. 1999. **Process design and costing of bioethanol technology: a tool for determining the status and direction of research and development.** *Biotechnology Progress* **15**(5): 794–803

Wooley R, Ruth M, Sheehan J, Majdeski H, Galvez A. 1999. **Lignocellulosic biomass to ethanol process design and economics, utilizing co-current dilute acid prehydrolysis and enzymatic hydrolysis: current and futuristic scenarios.** National Renewable Energy Laboratory (NREL) Report TP-580-26157. Golden: NREL

World Bank. 2010. ***Prototype Carbon Fund.*** Washington, DC. Details available at <http://wbcarbonfinance.org/Router.cfm?Page=PCF>, last accessed in March 2010

Wright J D. 1988. **Ethanol from biomass by enzymatic-hydrolysis.** *Chemical Engineering Progress* **84**(8): 62–74

Wright L L and Hohenstein W G. 1994. **Dedicated feedstock supply-systems.** *Biomass and Bioenergy* **6**(3): 159–241

Wyman C E. 1999. **Biomass ethanol: technical progress, opportunities, and commercial challenges.** *Annual Review of Energy and the Environment* **24**: 189–226

Wyman C E. 2001. **Twenty years of trials, tribulations, and research progress in bioethanol technology – selected key events along the way.** *Applied Biochemistry and Biotechnology* **91**(3): 5–21

Bio-hydrogen: applications and future prospects

P M Sarma

Fellow, Microbial Biotechnology, The Energy and Resources Institute, New Delhi, India

and

B Lal

Director, Environmental and Industrial Biotechnology Division
The Energy and Resources Institute, New Delhi, India

Introduction

Energy is vital to global prosperity, yet the use of fossil fuels as the primary energy source contributes to global climate change, environmental degradation, and other associated health problems. Growing environmental concerns and energy insecurity have led to an increasing interest in research for alternative sources of energy. In this context, hydrogen (H_2) holds a promising option as an energy carrier of the future. H_2 has the highest gravimetric energy density among known fuels and is compatible with electrochemical and combustion processes for energy conversion, without giving rise to carbon-based emissions.

Presently, H_2 is mainly used as a chemical feedstock in petrochemicals to upgrade low-quality coal and heavy oils. According to the Atomic Energy Authority's (AEA 2002) technology report, ammonia production facilities, oil refineries, and methanol production plants consume up to 62.4%, 24.3%, and 8.7%, respectively, of the total world production of H_2 (Akkerman, Janssen, Rocha, *et al.* 2002). However, being a clean, environmentally benign energy carrier, it has range of applications in the transport, building, industry, and utility sectors, as well as in space travel. It is also used in fuel cells to generate electricity. In addition, the increasing threat of global warming and the expected shortfall in all kinds of fossil fuels warrant more vigorous uses of H_2 as a clean energy carrier for a sustainable global economy. For example, a 5% H_2 addition in gasoline/air mixture can reduce the oxides of nitrogen (NO_x) emissions by 30%–40%, while an engine running on pure H_2 would produce only water and negligible NO_x, including drastic cuts in carbon dioxide (CO_2)

emissions. In addition, the high energy content per unit weight (122 kJ/g) in H_2 offers security, diversity, and flexibility as a fuel, especially by reducing dependency on less stable oil-exporting countries.

At present, the production of H_2 for these applications is primarily made possible through high-energy-intensive and expensive non-biological routes such as the processing of fossil fuels and electrolysis. But to qualify as an alternative renewable energy source, the processes that can produce H_2 naturally through biological routes are gaining importance. H_2 produced by photobiological or biomimetic means is important because water can be used as the source of electrons, and sunlight as the driving force. This process of H_2 production traditionally targeted the cyanobacterial cultures. Review of the more recently published works bring to the fore the importance of alternative approach like dark fermentation for bio-hydrogen production, due to the dual benefits of energy production as well as waste reduction. This chapter discusses the current trends, including the concepts in modern bioreactor engineering, recombinant deoxyribonucleic acid (DNA) technologies, and protein engineering techniques, in bridging the technological gaps in biological H_2 production. In addition, the various non-biological H_2 production technologies, along with merits, limitations, and advances made in dark fermentation for the production of H_2 from different substrates, including different types of waste, are also discussed.

Non-biological hydrogen production

H_2 is contained in several natural resources such as water, natural gas, coal, and methanol. Presently, over 95% of the H_2 production is based on processing of fossil fuels, which make significant contribution to global warming and carbon emissions. Therefore, several alternative processes are in various stages of research and development (R&D) so as to enable the use of more cleaner non-fossil fuel sources such as biomass, solar, wind, geothermal, hydropower, tidal, and nuclear. All these non-conventional sources require some kind of conversion from their original form, but no technology is currently available on a commercial scale for H_2 production from biomass.

Steam reforming of natural gas

Steam reforming is the most common, fully developed, and well-established commercial process for the merchant H_2 production from natural gas. It is an exothermic catalytic conversion of light hydrocarbons with the aid of water vapour. In the first step, H_2, CO_2, and CO (carbon monoxide) are produced when natural gas is exposed to high-temperature steam. In the second step, the CO is converted to additional H_2 using steam. The process is carried out at 850°C and 2.5 MPa, and results in approximately 70%–90% H_2 yield. It is the preferred choice for commercial H_2 production because of its low cost. Nuclear-heat-assisted steam reforming also produces H_2 from natural gas.

The technology has limited applications for H_2 production on a large scale because of the following demerits.

- Conversion of carbon into CO_2 leads to greenhouse effects.
- Sequestering of CO_2 increases the production costs of H_2 by at least 25%–50% based on the CO_2 capture efficiencies of 70%–97%.
- Only large centralized units can afford to undertake sequestering due to the additional costs and infrastructure requirements.

Electrolysis

During electrolytic splitting, when electric current is passed through water, H_2 and O_2 accumulate at the cathode and anode, respectively. According to an estimate, the cost of H_2 production through electrolysis is about \$30/million Btu (British thermal unit), which includes handling costs and production inefficiencies. The comparative costs for natural gas and gasoline are \$3/million Btu and \$9/million Btu, respectively. Similarly, the conversion of electricity to 1 million Btu of H_2 is equal to \$15 of electric energy. Therefore, electrolytic H_2 production can be ensured only under the scenario of lower electricity costs, and excess generation of electricity in the H_2-starved areas can result by utilizing local resources such as solar, wind, hydro, and other renewables. In addition, affluent societies willing to pay a premium cost for a clean environment can be served through such environment-friendly and sustainable technological options. In order to realize the true benefits of electrolysis through efficiency improvement, several modifications in the conventional electrolysis and advanced electrolytic processes are being done. For example, instead of electricity, if raw heat is used to decompose water into H_2 and O_2, the process could be made more efficient. Similarly, an advanced electrolysis of water can be performed as high-pressure water undergoes electrolysis at 5 MPa through special materials. Alternatively, high-temperature water electrolysis (900–1000°C) can be advantageous because part of the energy can be supplied through high-temperature heat, resulting in reduced electricity consumption. Heat from solar concentrators or waste heat from power stations can also be utilized with an efficiency of about 88%–95%. Similarly, both heat and electricity can be supplied by the nuclear sources to further improve the economics.

Gasification

Partial oxidation of heavy hydrocarbons (residual oil from the treatment of crude oil) with O_2 and steam results in the H_2 production, the process being called gasification. It is a well-established technology with 75%–86% and 45%–50% of operating efficiencies for residual oil and coal, respectively. The coal-rich countries, namely, South Africa and China, largely depend on this process due to high prices of natural gas and oil. Low-value agricultural residues and industrial solid waste rich in carbon can also be subjected to partial oxidation for the generation of H_2.

Pyrolytic cracking

High temperature converts hydrocarbons, biomass, methane, and municipal solid waste (MSW) into H_2 and solid carbon. The technology is attractive because of the economic benefits and flexibility in the variety of feedstock utilization. In addition, solid carbon black that is produced can be easily used in battery, plastic, and tyre-manufacturing industries.

Biological hydrogen production

In spite of the extensive experience in and demonstrated low costs of non-biological H_2 technologies, there is a growing realization and consensus that the cost of development and benefits accrued due to large-scale application of alternative H_2 technologies such as bio-hydrogen can be negligible, and even be offset, if the global warming phenomenon manifests due to the continued CO_2 emissions through steam reforming of natural gas and partial oxidation of heavy hydrocarbons and coal. The main routes of biological H_2 production include direct and indirect photo-biolysis, photo-fermentation, and dark fermentation, which involve various cyanobacteria, and photosynthetic and anaerobic bacteria. H_2 production during decomposition of formate by bacteria was reported by Strickland (1929). Gaffron and Rubin (1942) subsequently reported the photochemical H_2 production with micro algae. The cyanobacteria split water into H_2 and O_2 during photosynthesis. Anaerobic bacteria use organic substrates as the sole source of e^- (electrons) and energy, and convert these into H_2. The reaction is rapid and does not require solar radiation, thus is useful for the treatment of large quantities of waste. The properties of photosynthetic bacteria lie in between the properties of cyanobacteria and anaerobic bacteria, as they convert the organic substances into H_2 in the presence of light energy (McKinlay and Harwood 2009). Figure 1 shows an integrated scheme for biological H_2 production and utilization of the various process products. Details on biochemical mechanisms, and merits and demerits of individual biological H_2 processes are discussed in the following paragraphs.

Biological H_2 production through bio-photolysis, photo-fermentation, and dark fermentation is mediated by three types of complex metallocluster-containing enzymes harboured by different types of algae and bacteria. The principal enzymes of these pathways are nitrogenase, Fe-hydrogenase, and Ni–Fe hydrogenase. Nitrogenase reduces a wide variety of substrates using e^- derived from reduced ferredoxin or flavodoxin. The N_2-fixing activity of nitrogenase is expressed in the presence of N_2, while under illumination, it reduces protons to H_2 in the absence of N_2 (Lambert and Smith 1981). Nitrogenases require elaborate enzyme machinery and an enormous energy input for biosynthesis. At least 20 gene products are necessary for the co-factor synthesis, insertion, transport, and activation. This elaborate genetic mechanism leads to extremely slow turnover rates (6.4 seconds). The relatively low turnover numbers, adenosine triphosphate (ATP)-dependent catalysis, and

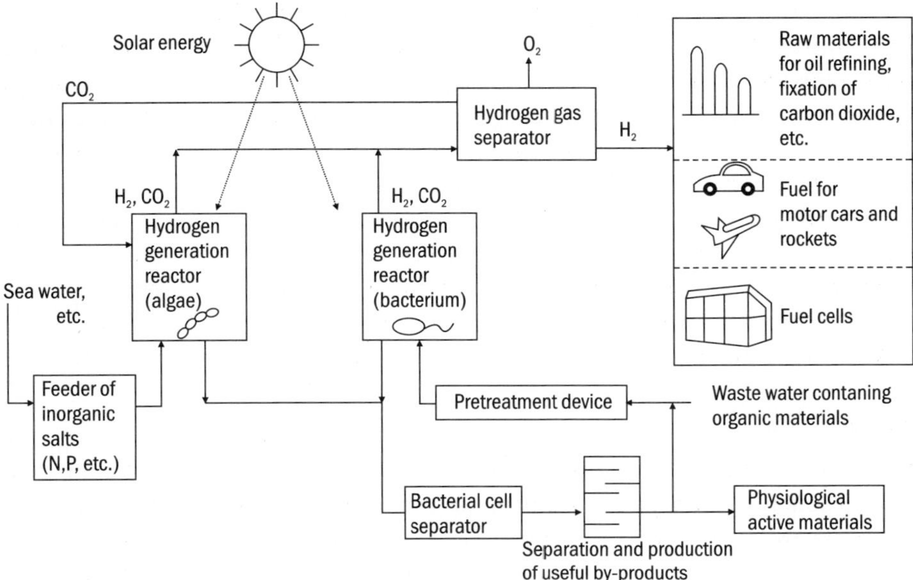

Figure 1 Biological production of hydrogen by environmentally acceptable technologies

instability render nitrogenases metabolically unattractive for large-scale H_2 production (Das and Veziroglu 2008).

The next important enzyme called Fe-hydrogenase was isolated from *Clostridium pasteurianum* about three decades ago. It contains an iron–sulphur (Fe–S) centre with one of the Fe atoms complexed with the CO and CN (Peter, Lanzilotta, Lemon, *et al.* 1998; Pierik, Hulstein, Hagen, *et al.* 1998; Posewitz, Mulder, Peters 2008). The highly reactive nature of the Fe–S cluster makes the existence of O_2-insensitive hydrogenases questionable. The Fe-hydrogenases are the best among the three types of enzymes for H_2 production, with very high turnover numbers (6000/second and 9000/second), as estimated for *C. pasteurianum* and *Desulfovibrio* sp., respectively (Hallenbeck 2009; Hallenbeck and Benemann 2002).

An additional category of enzymes called nickel–iron hydrogenases comprises heteromeric proteins containing a small and a large subunit. The small subunit contains three Fe–S clusters, two 4Fe–4S clusters, and one 3Fe–4S cluster. The larger subunit has a unique metallo-centre consisting of Ni–Fe, which coordinates with the two CN and one CO (Pierik, Roseboom, Happe, *et al.* 1999). The cytoplasmic location of the Ni–Fe hydrogenases helps in removing the fermentatively generated excess reducing power in the strict anaerobic bacteria at the rate of 300–400 mol/min/mg (Heinekey 2009; Rousset, Montet, Guigliarelli, *et al.* 1998). The periplasmic location is involved in the oxidation of H_2. Several micro-organisms also possess Ni–Fe or Ni–FeSe hydrogenases called uptake hydrogenases. The primary function is to derive the reducing power from H_2. The e^- is then used to reduce NAD (P) through the quinone system. The synthesis of Ni–Fe hydrogenase requires several gene products encoding for Ni–Fe clustering, synthesis of CN and CO, cluster insertion, and protein maturation. Several auxiliary enzymes and post-translational modifications are required for the synthesis of enzyme in the active form. The purified hydrogenase has an H_2 evolution rate of 65 mol/min/mg, with a turnover rate of 98/second (Rousset, Montet, Guigliarelli, *et al.* 1998). Therefore, this enzyme is a better candidate for H_2 production than nitrogenases. The H_2 evolution process is not limited by either the quantity or the activity of these enzymes. In different microbial systems, differences in the catalytic activities of these enzymes and higher catalytic activity compared to H_2 production have been recorded. Therefore, some unknown metabolic barriers are responsible for the low H_2 evolution rates.

Bio-photolysis

During the splitting of water by direct bio-photolysis, solar energy is converted into H_2 energy. Both photo-autotrophic (micro algae and cyanobacteria) and photo-heterotrophic bacteria produce H_2 (Benemann, Berenson, Kaplan, *et al.* 1973; Benemann and San Pietro 2001). In a few initial reports, Spruit (1958) mentioned simultaneous H_2 and O_2 production during photosynthesis in *Chlorella*. During bio-photolysis, the light energy of the photons is stored as chemical energy in the basic unit of the photosynthetic apparatus called the photosystem (PS). The PS consists of two parts: (1) an antenna complex made up of carotenoid and chlorophyll *a* and *b* molecules that absorb the photons and (2) a reaction centre made of specialized P680 molecules to convert photon energy into chemical energy. The direct bio-photolysis (oxygenic photosynthesis) in green algae involves two PS that operate in a series: PS I and PS II. The reaction centres of PS I and PS II are called P700 and P680, respectively. P700 absorbs light most strongly at 700 nm, whereas P680 absorbs the excitation energy of each photon with a wavelength smaller than 680 nm. The PS II, thus having greater energy, generates a strong oxidant capable of liberating e^- from water. Under anaerobic conditions with light, the e^- is used to reduce protons to H_2 by hydrogenase. In this process, no extra ATP energy is needed (Das and Veziroglu 2008).

Photo-autotrophic hydrogen production

The normal growth of photo-autotrophic micro algae and cyanobacteria takes place through the fixation of CO_2 into organic compounds in the presence of sunlight. The green algae studied for the H_2 production include *Scenedesmus obliquus*, *Chlamydomonas reinhardtii*, and *Chlamydomonas moewusii*. The photo-autotrophic production of H_2 is attractive from a practical viewpoint because both water and light energy are unlimited. This zero CO_2 process also helps in mitigating the greenhouse effect. In spite of these advantages, photo-autotrophic H_2 production is faced with several challenges. For example, the cost involved in large areas capturing sufficient light energy is prohibitively high. In addition, the extreme sensitivity of hydrogenase towards O_2 results in very low conversion efficiencies (less than 1.5% of the solar spectrum) for the conversion into H_2. Higher conversion efficiencies are achievable by the reduction of the partial pressure of O_2 to less than 0.1% (equal to less than 1 mM O_2 in liquid phase). However, due to the high cost, it is not possible to maintain the O_2 pressure less than near normal atmospheric condition by using inert gas as diluent. Several alternate remedies have been investigated to bridge the vast differences in the O_2 concentration tolerable by hydrogenase and the achievable levels of O_2. Akkerman, Janssen, Rocha, *et al.* (2002) reported conversion efficiency as high as 3%–10% by efficient and immediate removal of O_2. Under low light intensities and very low partial pressure of O_2, *C. reinhardtii* has a solar energy conversion efficiency of 10% (Greenbaum 1988). Use of O_2 absorbers that can reversibly (haemoglobin) or irreversibly (glucose, glucose oxidase, or dithionite) bind to O_2 and thus stimulate H_2 production in algae has been suggested by Rosenkranz and Krasna (1984). Similarly, several green algae involved in direct bio-photolysis are known to possess inducible and reversible Fe-hydrogenase. Mutation and protein engineering to achieve the O_2-tolerant hydrogenases have also been attempted in the recent years (Ghirardi, Togasaki, and Seibert 1997; McTavish, Sayavedra-Soto, and Arp 1995; Nagy, Meuser, Plummer, *et al.* 2007; Seibert, Flynn, and Ghirardi 2001). These mutated hydrogenases actually help in reducing the O_2 levels during H_2 production by an increased respiratory activity rather than due to actual increased O_2 tolerance. Therefore, H_2 yield from a per unit of substrate is reduced to half because an equal quantity of substrate is lost during respiration. This leads to an additional drain of substrate and renders the overall process economically non-viable. In order to look for more feasible options, elucidation of an intricate genetic regulatory networking of the H_2 metabolism is, therefore, essential. Alternately, the direct bio-photolysis carried out by the cell-free photosynthetic components offers novel, reliable, and more robust systems than the whole-cell systems under oxic conditions.

Photo-heterotrophic H_2 production

Hydrogen production in the N_2-fixing, photo-heterotrophic bacteria and heterocystous cyanobacteria is mediated by oxygen- and ammonium-sensitive nitrogenase in the absence of nitrogen (Benemann, Miyamoto, and Hallenbeck 1980; Benemann 1997, 1998; Li and Fang 2009). Organic substrate and light energy are essential for this metabolism to take place. These cyanobacteria have two types of cells. H_2 evolution takes place in the heterocysts. This natural spatial separation of H_2 protects the enzyme from the exposure to O_2 (Das and Veziroglu 2001). The important heterocystous cyanobacteria studied for H_2 production include *Anabaena azollae*, *Anabaena variabilis*, *Anabaena cylindrica*, *Nostoc muscorum*, *Nostoc spongiaeforme*, and *Westiellopsis prolifica*. Non-heterocystous, N_2-fixing bacteria of interest are *Plectonema boryanum*,

Oscillatoria Miami BG7, *Oscillatoria limnetica*, *Synechococcus* sp., *Aphanothece halophytico*, *Mastidocladus laminosus*, and *Phormidium valderianum*. The main advantage of photo-heterotrophic H_2 production is the lack of exposure of nitrogenase to O_2 because the photosynthesis and H_2 production stages are separated in time. In addition, the photo-heterotrophic purple bacteria can use a much wider part of the solar spectrum (400–950 nm) compared with micro algae (400–700 nm). In spite of these advantages, the inability to utilize sunlight fully is the main bottleneck in the photo-heterotrophic H_2 production. In the initial reports, the photochemical efficiency of 0.2% has been reported for *A. cylindrica* (Miyamoto, Benemann, and Hallenbeck 1979; Miyamoto, Hallenbeck, and Benemann 1979). In the recent past, the solar conversion efficiency in the outdoor micro algal cultivation is computed to be about 3%. Conversion efficiencies of 3.5% and 1.5% were estimated for the marine cyanobacterium, *Synechococcus* sp. *Miami* BG043511, based on the photosynthetic active radiation (PAR) and total solar energy, respectively. The low conversion efficiency under full sunlight in both micro algae and photosynthetic bacteria (pigment antennae system) is due to a lower rate of electron transfer between the PS II and PS I (dark reaction) during oxygenic photosynthesis. The 10-times higher light capturing ability of chlorophyll results in over 90% photons being captured by chlorophyll, and the other pigments being wasted as heat. The slower rates of dark reaction during intense sunlight also result in a low algal biomass. Therefore, 10% conversion efficiency is feasible only at low light intensities and lower H_2 production rates (Uyara, Eroglub, Yücelc, *et al.* 2007).

Maximization of photon utilization efficiency by flashing light effect was initially demonstrated by Kok (1953). To achieve this clever flashing light effect, the commercial algal bioreactors require extensive mixing power. From a practical viewpoint, mixing velocities greater than 25–35 cm/second are not economical for the production of H_2 for fuel applications (Hallenbeck 2001). In yet another approach, algal mutants with less light-harvesting chlorophyll, phycobiliproteins (cyanobacteria), and antennae were created, leading to lesser wastage of photons (Nakajima and Ueda 1999, 2000; Nakajima, Tsuzuki, and Ueda 2001; Polle, Kanakagiri, Benemann, *et al.* 2001). A 50% increase in productivities at higher light intensities when compared with the wild-type *Chlamydomonas perigranulata* has been reported with the mutated algal antennae.

In contrast to direct bio-photolysis, the concept of indirect bio-photolysis is controversial and still in the early stages of understanding (Ghirardi, Kosourov, Tsygankov, *et al.* 2000; Manisha and Banerjee 2008). The process temporally and/or spatially separates O_2 and H_2 evolution coupled to the CO_2 fixation/evolution. H_2 evolution in *C. reinhardtii* was reported by the elimination of the dark fermentative stage and replacement of N_2 by sulphur as the limiting nutrient. Under these conditions, no induction in carbohydrate biosynthesis was found. Reduced O_2 evolution due to the uptake during respiration was also reported (Melis, Zhang, Forestier, *et al.* 2000).

In the new concepts, a biomass is selected to yield high carbohydrate storage capacity at 10% solar efficiency in open production ponds and is concentrated in the settling ponds. Initial dark fermentation is carried out to yield two acetates and four H_2/glucose stored in the algal cell. Later, a photo-bioreactor is used to convert two acetate molecules into 8 mol of H_2. Additional units for the treatment of excess biomass through methane production, inoculum preparation, recycling of CO_2 back to the biomass production ponds, gas handling, and the gas separation may be integrated (Redwood, Paterson-Beedle, and Macaskie 2009). The cost estimate for such a conceptual design based on the requirements for at least two photons for the evolution of each H_2 from acetate has been provided by Benemann (1998). It is shown that an increase in the solar conversion efficiencies and increased carbohydrate storage under N_2-limiting conditions are important for realizing the benefits from this process. The indirect bio-photolysis process can be simplified by growing selected high-carbohydrate-accumulating algae in open ponds. Similarly, the ability of the green algae and the cyanobacteria to endogenously convert stored starch and glycogen, respectively, into H_2 under obligate anaerobic conditions can eliminate the need for photo-bioreactors.

Photo-fermentation

Several photosynthetic bacteria possess nitrogenases capable of H_2 evolution. Examples of such bacteria are, *Rhodobacter sphaeroides*, *Rhodobacter capsulatus*, *Rhodobacter sulfidophilus*, *Rhodopseudomonas sphaeroides*, *Rhodopseudomonas palustris*, *Rhodopseudomonas capsulata*, *Rhodospirillum rubrum*, *Chromatium* sp. *Miami* PSB 1071, *Chlorobium limicola*, *Chloroflexu aurantiacus*, *Thiocapsa roseopersicina*, and *Halobacterium halobium*. The photo-fermentative purple bacteria have only one photosystem and use e^- generated from organic carbon or hydrogen sulphide (H_2S). In the absence of N_2, nitrogenase uses the electron and ATP to reduce protons to H_2. The ATP is used to transport the e^- to ferredoxin (Fd). The major component of organic acids can be transformed into H_2 and CO_2 without any evolution of O_2 (anoxygenic photosynthesis). Therefore, the sensitivity of nitrogenase towards O_2 is not a matter of concern.

The near-stoichiometric conversion of agricultural waste, food processing industrial waste, and organic acids into H_2 has been achieved using the photosynthetic antennae centres of these bacteria. The pigment apparatus of these bacteria is optimum for low light intensities, and conversion efficiencies of 100% of solar energy into H_2 have been achieved. However, while making these estimates, energy content of the organic matter and achievable conversion efficiencies at full solar irradiations have largely been overlooked. *R. sphaeroides* RV, a powerful H_2 producer, showed 7% energy conversion efficiency in the presence of lactate and glutamate (Miyake 1990) and was found to be suitable for photo-bioreactors (Nakada, Nishikata, Asada, *et al.* 1996; Tsygankov, Hirata, Miyake, *et al.* 1994). An H_2 production rate of 3.9 mmol H_2/g-VSS-day from multiple substrates with purple non-sulphur bacteria has been recorded in a continuous stirred tank reactor (CSTR) (Cheng, Lin, Tseng, *et al.* 2003). For the photo-biological production of H_2 (direct bio-photolysis, indirect bio-photolysis, and photo-fermentation), several prototype bioreactors have been evaluated (Kapdan, Kargia, Oztekina, *et al.* 2009; Lehra and Postena 2009). The biomass production of photo-autotrophs is governed by the depth and volume of the photic zone. The photic zone, in turn, depends on the algal concentration, dimensions, and operations of the reactor, specific absorption coefficient of the algae, and wavelength of the incoming light. The photosynthetic efficiency (PE) and productivity of micro algal cultivations in outdoor enclosed reactors are also affected by the light regime inside the reactors.

Performances of photo-bioreactors and the photo-heterotrophic bacteria differ due to the differences in photochemical efficiency, absorption coefficient, and size, in comparison to the micro algal photo-bioreactors. Therefore, for photo-biological H_2 production, ideal photo-bioreactors are designed to target maximum conversion efficiencies with the following essential features.

- An enclosed photo-bioreactor to ensure proper collection of gases.
- A sterilizable reactor to maintain a pure culture.
- A bioreactor with high surface-to-volume ratio for efficient utilization of solar energy for increased productivity.
- A reactor with the provision for either dilution of light and/or its distribution over the largest reactor volume, or exposing the culture to light only for a very short duration by providing mixing devices.
- A single unit bioreactor as scale-up for practical ease of sterilization and operation with one control mechanism.

At present, even at a maximum PE at all light intensities, a very large photo-bioreactor surface is required for a reasonable H_2 production. According to an estimate, in the Netherlands and in southern Spain (an average of 2.96 m^2/day and 16.88 m^2/day), 420 h and 250 h, respectively, would be needed for the production of 1 GJ H_2/year. In general, there are very low achievable conversion efficiencies and poor process economics for all photosynthetically derived H_2 production systems (direct bio-photolysis, indirect bio-photolysis, and photo-fermentation). Therefore, H_2 as a fuel on a global scale ultimately has to be derived from organic substrates, such as residues and industrial waste through dark fermentation.

Dark fermentation

Owing to the dual benefits, that is, energy production as well as waste reduction, a review of the recently published works highlighted the dark fermentative approach of bio-hydrogen production. The process involves catabolism of pyruvate under anaerobic conditions. The sources of pyruvate during glycolysis could be carbohydrate formed by photosynthesis or the organic waste generated by anthropogenic activities. The important challenges faced in this approach are low hydrogen conversion efficiency and unstable hydrogen production. Another limiting factor of bio-hydrogen production pertains to the enzymes (hydrogenases) as they are sensitive to oxygen. Further, for a net energy gain, the process should be operated at maximum possible total solid concentration by carefully curtailing the end-product inhibition.

Substrates

One of the most easily biodegradable carbon sources is glucose, which can be obtained abundantly from different sources. Theoretically, bioconversion of 1 mol of glucose yields 12 mol of hydrogen gas, but the highest hydrogen yield obtained from glucose is about 2.0–2.4 mol/mol (Das and Veziroglu 2008; Morimoto, Atsuko, Atif, *et al.* 2004). The carbohydrate-rich, low cost, and easily fermentable biomass without needing much pretreatment is ideally suited for this purpose. Other factors such as the sustainable supply of the high-density biomass with high fermentation and conversion efficiencies and least generation of waste streams are equally important. Industrial and domestic waste streams; agricultural and forest residues; dedicated sugar-containing crops (sweet sorghum and sugar beet); waste cereal crops (wheat, rice, and corn); lignocellulosic fodder crops; and grasses can serve as feedstock for a sustainable H_2 production. Although not significant from a practical viewpoint, defined carbohydrate substrates (glucose, sucrose, starch, and celluloses); malate; and lactate have also received much attention for bio-hydrogen production under laboratory studies.

As mentioned earlier, glucose fermentation for H_2 production was also initially reported by various authors, and conversion of glucose to H_2 was attempted with *Clostridium beijerinckii*, *R. sphaeroides*, anaerobic bacteria, and *Bacillus megaterium* by Cha and Jeong (2003). Chen and Lin (2003) obtained a yield of 4.52 mol H_2/mol sucrose in a CSTR with 8 h hydraulic residence time. This yield is higher than that reported in other studies such as 3.47 mol H_2/mol sucrose in CSTR (Chen, Lin, and Chang 2001) and 1.5 mol H_2/mol sucrose (Taguchi, Yamada, Hasegawa, *et al.* 1996) at the same retention time. With optimization of parameters for efficient conversion of sucrose, the yield of 4.8 mol H_2/mol sucrose was achieved by Lin and Lay (2004). Among the Indian strains, *Enterobacter cloacae* IIT-BY 08 produced 6 mol H_2/mol sucrose, which is the highest yield among the other tested carbon sources (Kumar and Das 1999). Taguchi, Chang, Takiguchi, *et al.* (1992) used glucose or starch to produce H_2 using *C. beijerinckii.* Clostridial strains are also reported to utilize simple sugars to produce hydrogen. A yield of 2.14 mol H_2/mol of hexose from a feedstock containing 0.75% soluble starch and mixed consortium was reported by Lay (2000).

Among the other sources, use of sweet potato starch residues has been reported by Yokoi, Saitsu, Uchida, *et al.* (2001) and Zhu, Wakayama, Asada, *et al.* (2001). A sugar factory waste water fed laboratory continuous reactor was reported to yield 2.52 mol H_2/mol of glucose when a mixed consortium was used (Ueno, Otsuka, and Morimoto 1996). Kotaya and Das (2007) reported H_2 production from biowaste with *Bacillus* strains. Similar study was also reported by Porwala, Kumara, Lala, *et al.* (2008). Food industry waste constitutes a major fraction of the MSW and has high carbohydrate content in the form of simple sugars, starch, and cellulose, which together make the solid food waste a potential feedstock for biological hydrogen production. Kapdan and Kargi (2006) had reviewed these aspects in detail. Lignocellulose materials were used as matrices with *E. cloacae* strain IIT-BT 08 by Kumar and Das (2001). Among other substrates, in a recent investigation, glycerol was considered as one of the candidates because it forms 10% by-product accumulation in a bio-diesel production process (Maru, Medinaa, Sueirasa, *et al.* 2009).

Effect of environmental factors

A variety of factors affect the bio-hydrogen production. In a recent communication, Wang and Wana (2009) summarized several factors influencing fermentative hydrogen production. The reviewed factors included inoculum, substrate, reactor type, nitrogen, phosphate, metal ion, temperature, and pH. The detailed biochemical, physiological, and reducing power generating mechanisms of most fermentative H_2 producers are already known (Hallenbeck 2001). As described by many investigators, H_2-consuming processes leading to the formation of alcohol, lactic acid, and propionate should be discouraged (Lin and Cheng 2006). There might be a requirement for the elimination of *Clostridium propionicum,* which can be achieved by heat treatment. Substrate limitation of H_2 for *C. propionicum* or glucose for *Clostridium butyricum* can create harmony between these two types of bacteria for H_2 production. On the contrary, at a very low hydraulic retention time (HRT), a slight change in temperature, even for a short time, can stimulate sporulation in *Clostridia*, leading to a washout of vegetative cells, instability, and a complete breakdown of the process. The metabolic pathways can also be controlled by maintaining the delicate balance between an appropriate growth phase, pH, temperature, and HRT. For example, *Clostridia* produce H_2 and volatile fatty acids (VFA) during the log phase, and alcohol is produced in the late log phase (Lay, Lee, and Noike 1999; Lay 2001). In *Clostridium acetobutyricum*, at pH >5, sufficient iron and phosphate in the medium with glucose concentrations of less than 2.5% (w/v) result in high H_2 yields, while exactly the reverse conditions cause the solvent production. Van Ginkel, Sung, and Lay (2001) and van Niel, Budde, De Haas, *et al.* (2002) have provided useful information on the effects of pH and substrate concentrations. For all heat-treated sludge inocula containing spore-forming H_2 producers, addition of amino acids was essential for the germination of spore into vegetative cells. Similarly, addition of an appropriate N_2 source (urea, ammonium salts, peptone, and polypeptone) was reported to stimulate the H_2 production from carbohydrates. An absolute need for 0.1% polypeptone for both *C. butyricum* and a co-culture of *C. butyricum* and *Enterobacter aerogenes* for spore germination under fed-batch condition has also been demonstrated. To prevent sporulation in *Clostridia*, a constant supply of glucose and casein hydrolysate during start-up is also essential (Cohen, Distel, van Deursen, *et al.* 1985). A shift to non-spore-forming propionate producers can occur if a butyrate and H_2-producing, non-pasteurized activated sludge were to be subjected to an interrupted feed. The H_2 production is also influenced by phosphate concentration (Pana, Fana, Xinga, *et al.* 2008). Under the phosphate limitation, *C. pasteurianum* preferentially produced butanol, ethanol, and 1, 3-propandiol from glucose. Similarly, the medium supplemented with $FeCl_3$ at 800 mg/l is shown to give maximum H_2 yield, while growth inhibition of *C. pasteurianum* and increase in ethanol and butanol were reported to occur at an iron content of <0.56 mg/l (Lee, Miyahara, and Noike 2001). Cohen (1982) demonstrated that maximum H_2 yields could be achieved if the redox potential is maintained at –300 mV to favour the butyrate production. Lower dissolved H_2 concentrations can also be achieved by purging of CO-free N_2, argon, or even fuel cell exhaust gas (Mizuno, Dinsdale, Hawkes, *et al.* 2000). Alternately, under favourable cost-economics, operation of the process at thermophilic temperatures may lead to reduced dissolved H_2 levels (Cheng, Lin, Tseng, *et al.* 2003).

A doubling in the daily H_2 production rate in a continuous laboratory-scale reactor using starch has been observed with increasing agitation from 100 revolutions per minute (RPM) to 700 RPM (Lay 2000). The operating temperature in a complex interaction with pH and HRT is reported to cause a shift in the product formation. For example, in a two-stage digestion of glucose with activated sludge inoculum at 6 h HRT, optimum butyrate–H_2 production occurred at 30°C, while acetate and propionate formation was favoured at a temperature higher than 30°C (Chen and Lin 2003). Similarly, for a continuous thermophilic H_2 production from glucose, 52°C was reported as the optimum temperature (Zoetemeyer, Vandenheuvel, Cohen 1982). H_2 production at 60°C in the presence of cellulose, yeast extract, and in a peptone medium under continuous operation has also been reported (Ueno, Haruta, Ishii, *et al.* 2001). Recently, Jayasinghearachchi, Sarma, Singh, *et al.* (2009) found that *Enterobacter* strains can produce H_2 in thermophilic conditions. Analysis of microbes in the two-phase anaerobic H_2 reactor by the denaturing gradient gel electrophoresis (DGGE) directly showed the shift of the dominant butyrate producers (*C. butyricum* and *Butyrivibrio fibrisolvens*) to *Megasphaera elsdenii* (Hwang, Jang, and Kim 2003). This study, for the first time, suggested that *M. elsdenii* and *Linmingia china* played a major role in H_2

production than the much-known *Clostridia*. The efficiency of the H_2 production process also depends on the biomass concentration in the reaction vessel. Cell biomass concentrations ranging between 1.5 g/l and 2.5 g/l have been reported with pure carbohydrate substrates (Chen, Lin, and Chang 2001). Several efforts are being made to optimize efficient mass transfer of reactants in an immobilized packed-bed reactor. Similarly, the size and density of the immobilizing beads, reduction in the reactor volumes by maximization of cell concentrations or cell loading, and increase in productivity are receiving due attention. Additional studies, however, are essential to prevent the leakage of immobilized cells due to ionic influences, initial cell concentrations, and cell distribution.

Scope for H_2 production by dark fermentation

In the reported literature, bacterial strains in the dark fermentation process are known to produce the best mol per mol yield of H_2 (Fang, Zhu, and Zhang 2003). *Clostridium* sp. is reported to yield between 1.61 mol and 2.36 mol of H_2/mol glucose (Lin and Chang 1999; Taguchi, Mikukami, Saito-Taki, *et al.* 1995). Hyper-thermophiles such as *Caldicellulosiruptor saccharolyticus* and *Thermotoga elfii* gave 3.3 mol of H_2 with sucrose and glucose, respectively, accounting for 83% of the theoretical maximum (van Niel, Budde, De Haas, *et al.* 2002). An H_2 production rate of 0.136 ml/g VSS/h was recorded with *C. beijerinckii* in batch reactors, with 99% degradation efficiency of glucose at 5.7 pH and 12-h HRT (Cha and Jeong 2003). The ideal goal for future research in the field of dark fermentation is to achieve 12 mol of H_2/mol of glucose. Near-stoichiometric yields of H_2 from glucose can be achieved by the micro algal oxidative pentose phosphate pathway. However, in practice, no such high H_2-yielding dark fermentation pathways are known to exist. The thermodynamics also favour the conversion of 1 mol of glucose to 2 mol of acetate and 4 mol of H_2 as a more feasible option. In addition, stoichiometric yields are possible only near the equilibrium at very slow rates and at a low partial pressure of H_2 (Woodward, Orr, Cordray, *et al.* 2000). In future, reverse e^- flow, a respiration-driven H_2 evolution, may replace the photosynthetically driven H_2 production. In the absence of light, endogenous fermentation in the high carbohydrate storage micro algae is supposed to yield 10 mol H_2/mol of glucose under limited N_2 and O_2. In addition, genetically engineered metabolic pathways can offer possibilities for the coupling of the respiratory process to low redox e^- transport pathways. With optimization of parameters for efficient conversion of sucrose, the yield of 4.8 mol H_2/mol sucrose was achieved by Lin and Lay (2004). Among the Indian strains, *E. cloacae* IIT-BY 08 produced 6 mol H_2/mol sucrose, which is the highest yield among the other tested carbon sources (Kumar and Das 2000).

In yet another approach, a hybrid system comprising non-photosynthetic and photosynthetic bacteria is proposed for an enhanced H_2 production. In this system, the anaerobic bacteria decompose a variety of carbohydrates to obtain both energy and e^-. As the organic acids formed by anaerobic digestion cannot be decomposed to obtain H_2 any more due to the positive free energy, the complete degradation of glucose to H_2 and CO_2 is impossible by anaerobic digestion alone. Under these conditions, incorporation of photosynthetic bacteria leads to the utilization of light energy to overcome the positive free-energy reaction. This mechanism not only reduces the light energy demand of the photosynthetic bacteria but also increases H_2 production.

Potential of organic waste and biomass for hydrogen production

Though waste biomass could be made available for almost a third of the price of conventional fuels, it is a documented fact that the use of agricultural and forest residues for direct burning results in health hazards, greenhouse effect, and acid rain Therefore, conventional approaches are presently undergoing a thorough reassessment for the utilization of biomass. The cleaner technologies for biomass use as fuel include gasification, co-firing, biomethanation, bioethanol, and bio-hydrogen. Energy-producing biomass could be in the form of solid, liquid, gas, electricity, and even chemicals. Biomass was expected to contribute between 10% and 20% of the total energy in co-firing with coal (NREL 2000). Several types of dedicated energy crops, such as switch grass, willow crops, hybrid poplars, alfalfa, and corn stovers have been identified for H_2 production. A comparison of energy costs for different biological processes and the non-biological H_2 processes has been made by several investigators. The direct production of H_2 from biomass eliminates the need for electrolysis, resulting in higher system efficiencies. The cost at ~\$40/million Btu of fermentative H_2 production at 10% conversion efficiency is considered to be unattractive. However, by adopting immobilized systems, an increase in the conversion efficiencies up to 28.34% has been proposed (Kumar and Das 1999, 2000). In contrast, photo-biological H_2 production at 10% conversion efficiency is estimated to cost ~\$10/million Btu. It is attractive when compared with the fermentative ethanol production that costs ~\$31.5/million Btu at 1%–30% conversion efficiencies. Similarly, H_2 production by dark fermentation of the broad-spectrum fermentable waste biomass is also more cost-competitive when compared with ethanol production by yeast. The latter process can use only selected feedstock. In addition, lower costs due to fewer requirements for pretreatment in comparison to ethanol production from biomass offer greater net energy gains. The costs involved in carbon sequestration are also eliminated due to the carbon-neutral H_2 fuel production from biomass. In spite of these advantages, the diversion of biomass for H_2 production by dark fermentation and/or electricity generation would depend on the technological maturity and land availability. In addition, short duration, high density, fast growing, and easily biodegradable dedicated energy crops would be needed for the successful implementation of biomass as a future H_2 provider. Similarly, long-term R&D efforts would be essential for the assessment of local and global sustainable resources. The development of integrated energy crop systems is also necessary due to the uncertainty in the availability and heterogeneous nature of biomass. Identification of secondary uses of energy crops and by-products arising from the H_2 production units is also important.

Technological bottlenecks

The fossil origin and its associated CO_2 emissions, along with the shortfall of conventional fossil fuels, are the major limiting factors for sustainable non-biological H_2 production from the current practices. Low conversion efficiencies of electrolytic processes and high cost of electricity in comparison to fossil fuels are the major barriers to the widespread usage of electrolytic production of H_2. Despite these limitations, over 35% improvement in efficiencies is expected through advanced electrolytic techniques. Production of H_2 with the electricity generated from solar energy, nuclear power, and renewables is among the few feasible choices. The cost obstacles and engineering challenges, however, remain to be resolved.

In the field of biological production of H_2, the main technological hindrances in the bio-photolytic routes include several natural factors, for example, uneven distribution of solar energy results in very high and low energy densities, and diffused nature of the solar energy is also a constraint. Apart from these, other major hindrances are as follows.

- Typically less than 1% solar conversion efficiencies.
- Oxygen sensitivity of enzymes hydrogenases and nitrogenases.
- Requirement of extra ATP.
- Low turnover rates of the enzymes.
- Light saturation effect.
- High conversion efficiencies achievable only at lower light intensities.
- Unfavourable bioreactor economics due to increased bioreactor volumes, and construction and operational (mixing, gas exchange, and compression, cooling) costs.
- High costs of production per unit of H_2 output.

In case of bacterial dark fermentation route, the major technical bottlenecks and the possible research initiatives are listed below.

- The H_2 yield of 2–3 mol H_2/mol of glucose that needs to be increased for a successful commercial production of H_2 from biomass.
- The physiological, physico-chemical, and reactor configuration details for the maximum obtainable yield of H_2 from microbial metabolism are still not clear.
- Qualitative control of fermentation is not easily achievable. Fermentation to acetate and butyrate should be favoured over alcohol, lactic or propionic acid fermentation to improve the overall yield of H_2 from the unit substrate.
- Bacteria involved in the H_2 consumption (methanogenesis) need to be eliminated for better H_2 yields.
- The processes for continuous and sustainable H_2 production with a wide variety of biomass using either the pure cultures or the mixed consortia are still in nascent stages.

- Biomass feedstock may require pretreatment.
- Ability to utilize non-sterile feedstock is necessary for handling and for the overall economics of the process.
- Stability of H_2-producing mixed consortia is of paramount importance for predictable process performances.
- Detailed information is not available on the feasibility of scale-up of the laboratory studies to a pilot scale and/or full commercial scale.

Conclusion

Energy is vital to global prosperity, yet increased use of fossil fuels as the primary source leads to global climate change and environmental degradation. Increasing environmental concerns and energy insecurity have led to an increasing interest in research for alternative sources of energy. In this context, H_2 holds a promising option as an energy carrier of the future. Apart from the issues pertaining to its storage and usage, one of the major hindrances in transition to H_2-based economy lies in the economics of cheap H_2 production. It is predicted that for a successful transition, extraction of H_2 by natural gas reforming may be expected to serve as a bridging technology because of the lowest cost. However, the long-term future of the gas-reforming process may face challenges due to the increase in the cost of the feedstock, interruption in supplies, and CO_2 emissions. Under such a scenario, together with the increasing realization of the impacts of global climate change and growth in demand for transport vehicles, alternatives such as advanced electrolysis and bio-hydrogen production could receive the much-needed boost.

Biological systems provide a wide range of processes to generate H_2, and include direct bio-photolysis, indirect bio-photolysis, photo-fermentation, and dark fermentation. However, each process comes with its own limitations. For example, conversion efficiencies under direct bio-photolysis are less than 1% on a solar spectral basis. The photo-bioreactor economics do not meet the present economic needs. Similarly, the photo-bioreactor design and optimization of the photo-autotrophic process are practical only when the O_2 inhibition of hydrogenase can be controlled. To produce and capture H_2 and O_2, a completely enclosed system is required, and separation of these gases is also a practical necessity. Similarly, renewable biomass or waste water based dark fermentative H_2 production, or a hybrid system consisting of anaerobic bacteria and photosynthetic bacteria could provide a sustainable global supply of H_2, once the low yield problems and engineering obstacles are overcome.

For this purpose, modern molecular engineering, protein engineering, and simple alterations in the physiological conditions are being explored so as to improve the overall economics of the process. Although advances in bioreactor designs are expected to make renewable H_2 available in the next two decades, bio-hydrogen alone may not be sufficient to fuel the future global economy. Integration of dark and light fermentation and integration with more recent techniques of microbial fuel cells are a few practical options. Therefore, an uninterrupted supply of H_2 will depend on several alternative H_2 production technologies. Thus, a polyphasic approach can help in addressing the current issues related to H_2 and renewable energy, and it will be a combination of processes that would give us economical H_2 for utilization as an energy source.

References

AEA (Atomic Energy Authority) Technology Environment 2002. **The feasibility, costs, and markets for hydrogen production.** In *A study for British Energy undertaken by AEA Technology Environment.* [Final Report prepared by Paul Witkiss and Nikolas Hill, September 2002]

Akkerman I, Janssen M, Rocha J, Wijffels R H. 2002. **Photo-biological hydrogen production: photochemical efficiency and bioreactor design.** *International Journal of Hydrogen Energy* **27**: 1195–1208

Benemann J R. 1997. **Feasibility analysis of photobiological hydrogen production.** *International Journal of Hydrogen Energy* **22**: 979–987

Benemann J R. 1998. **Processes analysis and economics of bio photolysis of water: a preliminary assessment.** In *Photo Production of Hydrogen* [IEA/H2/10/TR2-98] [Report on the International Energy Agency Hydrogen Programme, Subtask B, Annex 10]

Benemann J R and San Pietro A. 2001. **Workshop on biohydrogen R & D** [28–29 September 2000, Bethesda, MD, USA] Final Report submitted to the US Department of Energy, Hydrogen R&D Program, May 2001

Benemann J R, Miyamoto K, and Hallenbeck P C. 1980. **Bioengineering aspects of biophotolysis.** *Enzyme and Microbial Technology* **2**: 103–111

Benemann J R, Berenson J A, Kaplan N O, Kamen M D. 1973. **Hydrogen evolution by a chloroplast–ferredoxin–hydrogenase system** *Proceedings of the National Academy of Sciences of the United States of America* **70**: 2317–2320

Cha G C and Jeong T Y. 2003. **Kinetics of hydrogen production from glucose by hydrogen producing bacteria** In *Proceedings of the First NRL International Workshop on Innovative Anaerobic Technology*, pp. 11–18 [Organized by the Advanced Institute of Science and Technology, Korea, 12 December 2003]

Chen C and Lin C Y. 2003. **Using sucrose as a substrate in an anaerobic hydrogen producing reactor.** *Advances in Environmental Research* **7**: 695–699

Chen C C, Lin C Y, and Chang J S. 2001. **Kinetics of hydrogen production with continuous anaerobic cultures utilizing sucrose as limiting substrate.** *Applied Microbiology and Biotechnology* **57**: 56–64

Cheng S S, Lin C Y, Tseng I C, Lee C M, Lin H I, Lin M R, Chen S T, Chen S D, Liu P W. 2003. ***Biohydrogen Production Mechanisms and Processes Application on Multiple Substrates.*** Korea: Advanced Institute of Science and Technology, pp. 33–40. [Proceedings of the First NRL International Workshop on Innovative Anaerobic Technology, 12 December 2003]

Cohen A. 1982 **Optimization of anaerobic digestion of soluble carbohydrate-containing waste waters by phase separation.** Universiteit van Amsterdam, Laboratorium voor Microbiologie [PhD thesis]

Cohen A, Distel B, van Deursen A, van Andel J G. 1985. **Role of anaerobic spore-forming bacteria in the acidogenesis of glucose. Changes induced by discontinuous or low-rate feed supply** *Antonie van Leeuwenhoek Journal of Microbiology* **51**: 179–192

Das D and Veziroglu T N. 2001. **Hydrogen production by biological processes: a survey of literature** *International Journal of Hydrogen Energy* **26**: 29–37

Das D and Veziroglu T N. 2008. **Advances in biological hydrogen production processes.** *International Journal of Hydrogen Energy* **33**: 6046–6057

Fang H H P, Zhu H, and Zhang T. 2003. **Effect of pH on hydrogen production from glucose by a mixed culture.** *Bioresource Technology* **82**: 87–93

Gaffron H and Rubin J. 1942. **Fermentative and photochemical production of hydrogen in algae** *Journal of General Physiology* **26**: 219–240

Ghirardi M L, Kosourov S, Tsygankov A, Seibert M. 2000. **Two-phase photo biological algal H2-production system** In *Proceedings of the 2000 DOE Hydrogen Programme Review* [NREL/CP-570-28890] Golden, Colorado: National Renewable Energy Laboratory

Ghirardi M L, Togasaki R K, and Seibert M. 1997. **Oxygen sensitivity of algal H2 production.** *Applied Biochemistry and Biotechnology* **63**: 141–151

Greenbaum E. 1988. **Energetic efficiency of hydrogen photo evolution by algal water-splitting.** *Biophysical Journal* **54**: 365–368

Hallenbeck P C. 2001. **Integration of hydrogen evolving systems with cellular metabolism: the molecular biology and biochemistry of electron transport factors and associated reductases.** In *Biohydrogen II,* pp. 171–184, edited by J Miyake, T Matsunaga, and A San Pietro. Amsterdam: Pergamon Elsevier Sciences

Hallenbeck P C. 2009. **Fermentative hydrogen production: principles, progress, and prognosis.** *International Journal of Hydrogen Energy* **34**: 7379–7389

Hallenbeck P C and Benemann J R. 2002. **Biological hydrogen production; fundamentals and limiting processes** *International Journal of Hydrogen Energy* **27**: 1185–1193

Heinekey M D. 2009. **Hydrogenase enzymes: recent structural studies and active site models.** *Journal of Organometallic Chemistry* **694**: 2671–2680

Hwang H M, Jang N J, and Kim I S. 2003. **Monitoring of kinetics and bacterial community structure in two-phase anaerobic biohydrogen process,** pp. 19–30 In *Proceedings of the First NRL International Workshop on Innovative Anaerobic Technology.* Korea: Advanced Institute of Science and Technology

Jayasinghearachchi H S, Sarma P M, Singh S, Aginihotri A, Mandal A K, Lal B. 2009. **Fermentative hydrogen production by two novel strains of *Enterobacter aerogenes* HGN-2 and HT 34 isolated from sea buried crude oil pipelines.** *International Journal of Hydrogen Energy* **34**: 7197–7207

Kapdan I K and Kargi F. 2006. **Bio-hydrogen production from waste materials.** *Enzyme and Microbial Technology* 38: 569–582

Kapdan I K, Kargia F, Oztekina R, Arguna H. 2009. **Bio-hydrogen production from acid hydrolyzed wheat starch by photo-fermentation using different *Rhodobacter* sp.** *International Journal of Hydrogen Energy* **34**: 2201–2207

Kok B. 1953. **Experiments on photosynthesis by *Chlorella* in flashing light.** In *Algal culture: from laboratory to pilot plant*, edited by J S Burlew. Washington, DC: Carnegie Institute of Washington

Kotaya S and Das D. 2007. **Microbial hydrogen production with *Bacillus coagulans* IIT-BT S1 isolated from anaerobic sewage sludge.** *Bioresource Technology* **98**: 1183–1190

Kumar N and Das D. 1999 **Enhancement of hydrogen production by *Enterobacter cloacae* IIT-BT 08.** *Process Biochemistry* **35**: 589–594

Kumar N and Das D. 2000. **Enhancement of hydrogen production by *Enterobacter cloacae* IIT-BT 08.** *Process Biochemistry* **35**: 589–594

Kumar N and Das D. 2001 **Continuous hydrogen production by immobilized *Enterobacter cloacae* IIT-BT 08 using lignocellulosic material as solid matrices.** *Enzyme and Microbial Technology* **29**: 280–287

Lambert G R and Smith G D. 1981. **The hydrogen metabolism of cyanobacteria (blue-green algae).** *Biological Reviews* **56**: 589–660

Lay J J. 2000. **Modelling and optimization of anaerobic digested sludge converting starch to hydrogen.** *Biotechnology and Bioengineering* **68**: 269–278

Lay J J. 2001. **Biohydrogen generation by mesophilic anaerobic fermentation of microcrystalline cellulose.** *Biotechnology and Bioengineering* **74**: 280–287

Lay J J, Lee Y J, and Noike T. 1999. **Feasibility of biological hydrogen production from organic fraction of municipal solid waste.** *Water Research* **33**: 2579–2586

Lee Y J, Miyahara T, and Noike T. 2001. **Effect of iron concentration on hydrogen fermentation.** *Bioresource Technology* **80**: 227–231

Lehra F and Postena C. 2009. **Closed photo-bioreactors as tools for biofuel production.** *Current Opinion in Biotechnology* **20**: 280–285

Li R Y and Fang H P. 2009. **Heterotrophic photo fermentative hydrogen production.** *Critical Reviews in Environmental Science and Technology* **39**: 1064–3389

Lin C Y and Chang R C. 1999. **Hydrogen production during the anaerobic acidogenic conversion of glucose.** *Journal of Chemical Technology and Biotechnology* **74**: 498–500

Lin C-Y and Cheng C-H. 2006. **Fermentative hydrogen production from xylose using anaerobic mixed microflora.** *International Journal of Hydrogen Energy* **31**: 832–840

Lin C Y and Lay C H. 2004. **Carbon/nitrogen ratio effect on fermentative hydrogen production by mixed microflora.** *International Journal of Hydrogen Energy* **29**: 41–45

Manisha S and Banerjee R. 2008. **Comparison of biohydrogen production processes.** *International Journal of Hydrogen Energy* **33**: 279–286

Maru B T, Medinaa F, Sueirasa J E, Stchigel A M. 2009. **Biohydrogen production from different biodegradable substrates through dark fermentation.** *New Biotechnology* **25**: 216–217

McKinlay J B and Harwood C S. 2009. **Photobiological production of hydrogen gas as a biofuel.** *Current Opinion in Biotechnology* **21**: 244–251

McTavish H, Sayavedra-Soto L A, and Arp D J. 1995. **Substitution of *Azotobacter vinelandii* hydrogenase small sub-unit cysteines by serines can create insensitivity to inhibition by O_2 and preferentially damage H_2 oxidation over H_2 evolution.** *Journal of Bacteriology* **177**: 3960–3964

Melis A, Zhang L, Forestier M, Ghirardi M L, Seibert M. 2000. **Sustained photobiological hydrogen gas production upon reversible inactivation of oxygen evolution in the green alga *Chlamydomonas reinhardtii*.** *Plant Physiology* **122**: 127–135

Miyake J. 1990. **Application of photosynthetic systems for energy conversion.** In *Hydrogen Energy Progress VIII*, pp. 755–764, edited by T N Veziroglu and P K Takashashi [Proceedings of *Eighth WHEC*] New York: Pergamon Press

Miyamoto K, Benemann J R, and Hallenbeck P C. 1979. **Solar energy conversion by nitrogen-limited cultures of *Anabaena cylindrica*.** *Journal of Fermentation Technology* **57**: 287–293

Miyamoto K, Hallenbeck P C, and Benemann J R. 1979. **Effects of nitrogen supply on hydrogen production by cultures of *Anabaena cylindrica*.** *Biotechnology and Bioengineering* **21**: 1855–1860

Mizuno O, Dinsdale R, Hawkes F R, Hawkes D L, Noike T. 2000. **Enhancement of hydrogen production from glucose by nitrogen gas sparging.** *Bioresource Technology* **73**: 59–65

Morimoto M, Atsuko M, Atif A A Y, Ngan M A, Fakhru'l-Razi A, Iyuke S E. 2004. **Biological production of hydrogen from glucose by natural anaerobic microflora.** *International Journal of Hydrogen Energy* **29**: 709–713

Nagy L E, Meuser J E, Plummer S, Seibert M, Ghirardi M L, King P W, Ahmann D, Posewitz M C. 2007. **Application of gene-shuffling for the rapid generation of novel [FeFe]-hydrogenase libraries.** *Biotechnology Letters* **29**: 421–430

Nakada E, Nishikata S, Asada Y, Miyake J. 1996. **Hydrogen production by gel-immobilized cells of *Rhodobacter sphaeroides*—distribution of cells, pigments, and hydrogen evolution.** *Journal of Marine Biotechnology* **4**: 38–42

Nakajima Y and Ueda R. 1999. **Improvement of micro algal photosynthetic productivity by reducing the content of light-harvesting pigment.** *Journal of Applied Psychology* **11**: 195–201

Nakajima Y and Ueda R. 2000. **The effect of reducing light-harvesting pigment on marine madrigal productivity.** *Journal of Applied Psychology* **12**: 285–290

Nakajima Y, Tsuzuki M, and Ueda R. 2001. **Improved productivity by reduction of the content of light-harvesting pigment in *Chlamydomonas perigranulata*.** *Journal of Applied Phycology* **13**: 95–101

NREL (National Renewable Energy Laboratory). 2000. **Biopower fact sheet: pioneering energy crops in the Northeast** [DOE/GO-102000- 1053, June 2000] Colorado: Project update, Salix consortium; document prepared for US Department of Energy by the NREL

Pana C M, Fana Y T, Xinga Y, Houa H W, Zhanga M L. 2008. **Statistical optimization of process parameters on biohydrogen production from glucose by *Clostridium* sp.** *Bioresource Technology* **99**: 3146–3154

Peter J W, Lanzilotta W N, Lemon B J, Seefeldt L C. 1998. **X-ray crystal structure of the Fe-only hydrogenase (CpI) from *Clostridium pasteurianum* to 1.8 angstrom resolution.** *Science* **282**: 1853–1858

Pierik A J, Hulstein M, Hagen W R, Albracht S P J. 1998. **A low-spin iron with CN and CO as intrinsic ligands forms the core of the active site in [Fe]-hydrogenases.** *European Journal of Biochemistry* **258**: 572–578

Pierik A J, Roseboom W, Happe R P, Bagley K A, Albracht S P J. 1999. **Carbon monoxide and cyanide as intrinsic ligands to iron in the active site of [NiFe]-hydrogenases-NiFe$(CN)_2$CO, biology's way to activate H_2.** *Journal of Biological Chemistry* **274**: 3331–3337

Polle J, Kanakagiri S, Benemann J R, Melis A. 2001. **Maximizing photosynthetic efficiencies and hydrogen production in micro algal cultures.** In *Biohydrogen II: An Approach to Environmentally Acceptable Technology*, pp. 111–130, edited by J Miyake, T Matsunaga, and A San Pietro. New York: Pergamon Elsevier Science

Porwala S, Kumara T, Lala S, Rania A, Kumara S, Cheema S, Purohit H J, Sharma R, Singh S K, Kalia V C. 2008. **Hydrogen and polyhydroxybutyrate producing abilities of microbes from diverse habitats by dark fermentative process.** *Bioresource Technology* **99**: 5444–5451

Posewitz M C, Mulder D W, and Peters W. 2008. **New frontiers in hydrogenase structure and biosynthesis.** *Current Chemical Biology* **2**: 178–199

Redwood M, Paterson-Beedle M, and Macaskie L. 2009. **Integrating dark and light bio-hydrogen production strategies: towards the hydrogen economy.** *Reviews in Environmental Science and Biotechnology* **8**: 149–185

Rosenkranz A and Krasna A J. 1984. **Stimulation of hydrogen photo production in algae by removal of oxygen by reagents that combine reversibly with oxygen.** *Biotechnology and Bioengineering* **26**: 1334–1342

Rousset M, Montet Y, Guigliarelli B, Forget N, Asso M, Bertrand P, Fonteccilla-Camps J C, Hatchikian E C. 1998. **[3Fe-4S] to [4Fe-4S] cluster conversion in *Desulfovibrio fructosovorans* [NiFe] hydrogenase by site-directed mutagenesis.** *Proceedings of the National Academy of Sciences of the United States of America* **95**: 11625–11630

Seibert M, Flynn T, and Ghirardi M L. 2001. Strategies for improving oxygen tolerance of algal hydrogen production. In *Biohydrogen II*, pp. 67–77, edited by J Miyake, T Matsunaga, and A San Pietro Amsterdam: Pergamon Elsevier Science

Spruit C P. 1958. **Simultaneous photo production of hydrogen and oxygen by *Chlorella*.** *Meded Landbouwhogeschool Wageningen* 58: 1–17

Strickland L H. 1929. **Efficient hydrogen production from starch by a bacterium isolated from termites.** *Journal of Fermentation and Bioengineering* 73: 244–245

Taguchi F, Chang J D, Takiguchi S, Morimoto M. 1992. **Efficient hydrogen production from starch by a bacterium isolated from termites.** *Journal of Fermentation and Bioengineering* 73: 244–245

Taguchi F, Mikukami N, Saito-Taki T, Hasegawa K. 1995. **Hydrogen production from continuous fermentation of xylose during growth of *Clostridium* sp. Strain No. 2.** *Canadian Journal of Microbiology* 41: 536–540

Tsygankov A A, Hirata Y, Miyake M, Asada Y, Miyake J. 1994. **Photo bioreactor with photosynthetic bacteria immobilized on porous glass for hydrogen photo production.** *Journal of Fermentation and Bioengineering* 77: 575–578

Ueno Y, Otsuka S, and Morimoto M. 1996. **Hydrogen production from industrial waste water by anaerobic microflora in chemostat culture.** *Journal of Fermentation and Bioengineering* 82: 194–197

Ueno Y, Haruta S, Ishii M, Igarashi Y. 2001. **Microbial community in anaerobic hydrogen-producing microflora enriched from sludge compost.** *Applied Microbiology and Biotechnology* 57: 555–562

Uyara B, Eroglub I, Yücelc M, Gündüzc U, Türkerd L. 2007. **Effect of light intensity, wavelength and illumination protocol on hydrogen production in photobioreactors.** *International Journal of Hydrogen Energy* 32: 4670–4677

Van Ginkel S, Sung S, and Lay J J. 2001. **Biohydrogen production as a function of pH and substrate concentration.** *Environmental Science and Technology* 35: 4726–4730

van Niel E W J, Budde M A W, De Haas G G, van der Wal F J, Claassen P A M, Stams A J M. 2002. **Distinctive properties of high hydrogen producing extreme thermophiles, *Caldicellulosiruptor saccharolyticus* and *Thermotoga elfii*.** *International Journal of Hydrogen Energy* 27: 1391–1398

Wang J and Wana W. 2009. **Factors influencing fermentative hydrogen production: a review.** *International Journal of Hydrogen Energy* 34: 799–811

Woodward J, Orr M, Cordray K, Greenbaum E. 2000. **Enzymatic production of biohydrogen.** *Nature* 405: 1014–1015

Yokoi H, Saitsu A, Uchida H, Hirose J, Hayashi S, Takasaki Y. 2001. **H2 production by immobilized cells of *Clostridium butyricum* on porous glass beads.** *Biotechnology Techniques* 11: 431–433

Zhu H G, Wakayama T, Asada Y, Miyake J. 2001. Hydrogen production by four cultures with participation by anoxygenic phototrophic bacterium and anaerobic bacterium in the presence of NH4 +. *International Journal of Hydrogen Energy* 26: 1149–1154

Zoetemeyer R J, Vandenheuvel J C, and Cohen A. 1982. **Influence of temperature on the anaerobic acidification of glucose in a mixed culture forming part of a two-stage digestion process.** *Water Research* 16: 313–321

Bibliography

Noike T, Ko I B, Lee D Y, Yokoyama S. 2003. **Hydrogen production from organic waste.** *International Journal of Hydrogen Energy* 26: 547–550

Methanogenesis from agro-industrial residues: potential and prospects

D Somayaji

EID Parry (India) Ltd, Cuddalore, Tamil Nadu, India

Agro-industrial processes produce the largest quantities of residues, varying from 5% to 50% of the raw material processed. For many years, energy and environmental issues of worldwide concern have been underscoring the need for attending to these residues immediately. All over the world, methanogenic anaerobic digestion occurring in natural and man-made lagoons, landfills, and waste dumps releases biologically generated methane (CH_4) to the tune of over 250 million tonnes (MT) annually. This anthropogenic CH_4 was equivalent to 16% of the energy consumed in the US in 1990. The CH_4 gas traps 30 times more heat than the carbon dioxide (CO_2) and contributes to 18% of the global warming. A 20% reduction in the global warming may be achieved by converting organic wastes and residues to biofuels and chemicals. This clearly establishes the resource value of these residues, which if untapped can cause havoc, in terms of global warming and soil and water pollution. For over five decades, concerted efforts have been made to use anaerobic digestion to produce CH_4 and manure from various waste residues. The biologically amenable physico-chemical properties of these residues underscore their value as resources rather than pollutants (Ghosh 1997).

This approach to waste management is being encouraged worldwide to overcome the environmental problems, mitigate global warming, and promote energy independence. Anaerobic digestion can play an important role in this regard due to its ability to stabilize heterogeneous organic wastes and residues to biomethane—a clean renewable energy source for captive use. The applicability of conventional anaerobic digestion to high solids content biomass and waste feeds, and the status of development of high-solids anaerobic digestion processes are discussed in this article. A simple and cost-effective anaerobic digestion technology, exhibiting high organic reduction and net energy recovery efficiencies, needs to be commercialized (Ghosh 1997).

An annual production of 5 MT of sugar cane pressmud from 425 sugar mills in the country can generate over 1700 million cubic metres of biogas annually, in addition to the stabilized organic manure to the tune of 1 MT, through anaerobic methanogenesis. Methanogenesis is a complex biological process that utilizes a consortium of anaerobic micro-organisms to act in concert to hydrolyse complex organics to simple monomers and thereafter to volatile fatty acids (VFAs). The VFAs are then converted to CH_4 and CO_2. This process has been applied for treating various types of sludge in unmixed or mixed, hermetically sealed tanks called anaerobic digesters, which harbour the complex microbial population required for sequential conversion of substrates to CH_4. In order to overcome the complexity of the biological process, several designs and configurations for the process of anaerobic digestion such as continuous stirred tank reactors (CSTRs), phase separation, and combinations thereof have been demonstrated to have a profound impact on the performance of the process. This article enumerates the process, potential, and prospects of anaerobic methanogenesis from agro-industrial residues to realize value from waste.

Biomethanation

Biomethanation is a biological process of generating CH_4 from organic matter, mediated by micro-organisms. In fact, no known microbe can produce CH_4 from substances other than acetate, CO_2, hydrogen, methanol, formate, carbon monoxide (CO), and some methylated amines. As the available substrates are much more complex, it is imperative to have a population of different groups of organisms, which can produce simpler end-products from these complex substrates. In practice, it is also necessary to have a group of bacteria capable of producing acetate, CO_2, and H from the end-products of the complex substrates. There are three distinct trophic groups of micro-organisms involved in the process of biomethanation: (1) hydrolytic fermentative micro-organisms, (2) acetogenic fermentative bacteria, and (3) methanogenic bacteria (Ziekus 1980).

Hydrolytic fermentative micro-organisms

The first fermentative or hydrolytic groups of organisms are able to attack polymeric molecules despite appearing as solid material (Somayaji, Reddy, Bardiya, *et al.* 1996). These micro-organisms have elaborate hydrolytic enzymes, which hydrolyse polymeric substrates into low molecular weight material or even monomers, for example, proteins to amino acids, polysaccharides to oligo and monosaccharides, and lipids to free fatty acids. Though a diverse substrate specificity exists, many of these bacteria have a wide substrate range and short generation time. They, in competition with major fast-growing fermentative bacteria, are unable to utilize polymeric material but yet are able to take up smaller soluble molecules for their metabolism and growth (Figure 1). As a result, the short-chain fatty acids with 2–5 (or more) carbon (C) atoms are produced. The primary acids produced are acetic acid, followed by propionic and butyric acid, and small quantities of formic, valeric, isovaleric, and caproic acids. Ammonia, sulphate, iso-acids, and certain aromatic compounds are also produced from amino acids (Figure 1b). Because many organic acids are produced during this step, the process is often called acidogenesis (Nyns 1986). The acidogenic population is by far the largest of the trophic groups of bacteria in anaerobic digestion. A population of 10^8–10^9 hydrolytic bacteria per millilitre of mesophilic sewage sludge has been documented by several investigators (Table 1) (Somayaji, Reddy, Bardiya, *et al.* 1996). Enumeration studies of hydrolytic bacteria that utilize specific C substrates demonstrated 10^7 proteolytic and 10^5 cellulolytic bacteria in sewage sludge and 10^4–10^5 hemicellulose-degrading bacteria in rumen digestion (Somayaji, Reddy, Bardiya, *et al.* 1996).

Another important aspect of hydrolytic fermentation is that many of these bacteria, in the presence of autotrophic methanogens and sulphate-reducing bacteria, produce H for the disposal of excess electrons generated during energy-yielding oxidation of organic material

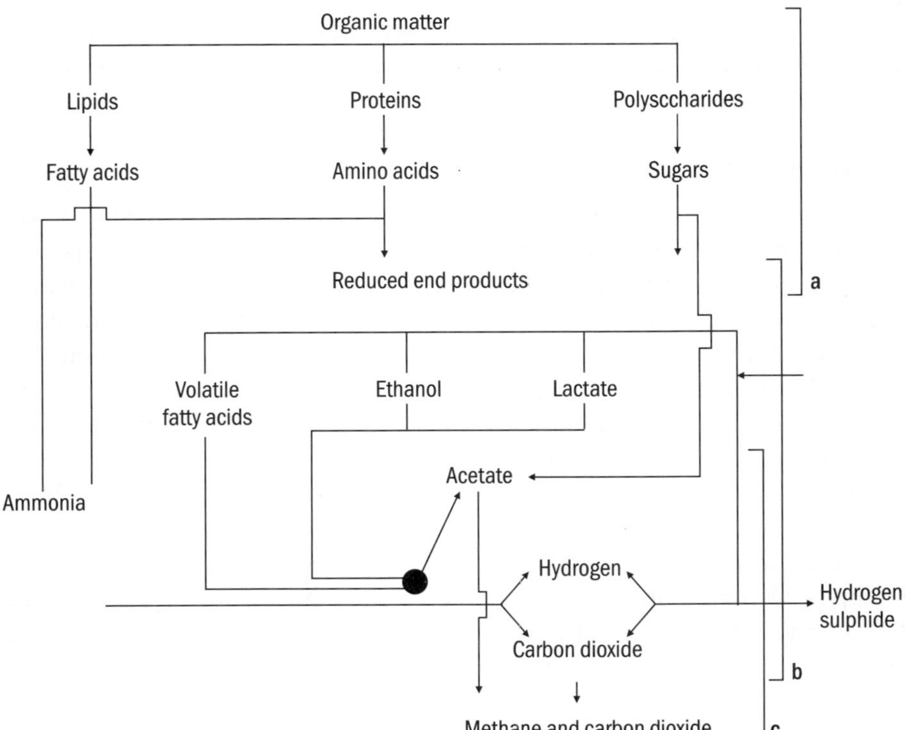

Figure 1 Actualized flow-sheet of methanogenesis
Source Nyns (1986)

Table 1 Bacterial population from anaerobic digesters

Group	Cells/ml
Total hydrolytic bacteria	10^8–10^9
Proteolytic	10^7
Cellulolytic	10^5
Hemicellulolytic	10^6–10^7
Hydrogen-producing acetogenic bacteria	10^6
Homoacetogenic bacteria	10^5–10^6
Methanogens	10^6–10^8
Sulphate reducers	10^4

Source Somayaji, Reddy, Bardiya, *et al.* (1996)

(Figure 1). However, accumulation of the H due to the inhibition of autotrophic methanogens results in thermodynamically unfavourable conditions for the disposal of electrons as H molecules. This results in the formation of products such as propionate, butyrate, succinate, and ethanol, which act as the H sink. Unionized organic acids, H ions, and H are inhibitory to the other groups of bacteria, while the acidogenic population is much more tolerant to these intermediates (Somayaji, Reddy, Bardiya, *et al.* 1996). Certain anaerobic bacteria that utilize polymeric molecules are *Fibrobacter succinogens, Ruminococcus albus, Ruminococcus flavefaciens, Butyvivibrio fibrisolvens, Prerotella ruminicola,* and *Selenomonas ruminatium* (Somayaji, Reddy, Bardiya, *et al.* 1996).

Acetogenic fermentative bacteria

Acetogenic bacteria obtain energy from the oxidation of organic acids, alcohols, and the VFAs with more than two C atoms, for example, caproate, butyrate, and propionate. Individual species of acetogenic bacteria are usually very specialized and ferment only a limited number of simple compounds to acetate, formate, bicarbonate, and H (Figure 1b). These conversions are extremely sensitive to H and growth of acetogenic bacteria, and can occur only at a less than 10^{-5} atmosphere partial pressure of H. Some acetogens also ferment xenobiotic compounds such as polyethylene glycol (PEG) and phenolic compounds (Somayaji, Reddy, Bardiya, *et al.* 1996). Some acetogenic bacteria produce acetate from compounds such as trimethoxy benzoate, trimethoxy cinnamate, and some methoxylated aromatics. Examples of some important acetogenic bacteria are *Acetobacterium woodii* (Bache and Pfenning 1981), *Clostridium aceticum* (Braun, Mayer, and Gottschalk 1981), *Clostridium formicoaceticum* (Andeersen, Gottschalk, and Schlegel 1970), *Eubacterium limosum, Sporomusa termitida* (Breznak, Switzer, and Seitz 1988), *Acetogenium kivui, Acetoanaerobium noterae, Butyribacterium methylotrophicum* (Zeikus, Lynd, Thomson, *et al.* 1980), and *Peptostreptococus productus*. Some acetogens are found to dechlorinate tetrachloromethane through the reductive process. A lot of potential exists for the efficient exploitation of this capability of acetogens in the degradation of such chemicals in industrial as well as municipal wastes (Somayaji, Reddy, Bardiya, *et al.* 1996).

The chemical-free energy change is positive for most of these acetogenic reactions under standard conditions. The acetogenic bacteria suffer from a thermodynamic product inhibition by the H (Wolin 1976). Hence, the growth rate of the slowest growing trophic groups' anaerobic digestion is totally dependent on the simultaneous removal of the reduced end-products such as acetate and H. For this reason, acetogens are always found in co-cultures with those bacteria (usually methanogens) that provide them with favourable growth conditions, while their metabolic partners are provided with C and energy sources (Somayaji, Reddy, Bardiya, *et al.* 1996).

Methanogenic fermentative bacteria

Methanogenic bacteria (methanogens) are the ultimate trophic group in the process of anaerobic digestion. They produce the most reduced form of C—CH_4. As stated earlier, methanogens cannot use the fermentation end-products of the first trophic group of the anaerobic digestion process, that is, hydrolytic fermentation, either for their growth or for CH_4 production. They use terminal end-products of acetogenic fermentation, namely, acetate, formate, methanol, methylamine, H, and CO_2 as a C and energy source (Ziekus 1980). A few methanogens derive energy from the CO_2 by reduction using the H (Figure 1c). Autotrophic methanogens assimilate CO_2 as a C source. The other microbial species are aceticlastic, and chemo-organotrophic archaebacteria split acetate into CH_4 and CO_2 and obtain their energy (Somayaji, Reddy, Bardiya, *et al.* 1996; Nyns 1986). Acetate is an anion and when split into CH_4 and CO_2, requires a pairing cation. When the pairing cation is either $NH^+{}_4$ or H_3O^+, bicarbonate and carbonic acid are produced.

During the carbonic acid formation, CO_2 escapes in the evolved biogas, ultimately resulting in alkalinity or reinforced buffering capacity to combat the acid produced by the previous groups (Nyns 1986). In anaerobic digesters, more than 60% of the biological CH_4 originates from acetate; thus, aceticlastic methanogens perform the important ecophysiological function of the C and electron removal while maintaining a neutral pH (Ziekus 1980).

Barker (1956) postulated that the pathway of CH_4 formation was initiated by the attachment of CO_2 to a carrier molecule and that all the subsequent reduction steps involved one C intermediate attached to a carrier molecule. The most thoroughly studied system is the reduction of CO_2 with H by *Methanobacterium thermoautotrophicum*. Its metabolism encompasses nearly the entire spectrum of reactions involved in methanogenesis. In the elucidation of this central pathway, along with conventional co-enzymes, several surprising molecules were discovered. The enzyme-mediated reactions, which involve the six new co-enzymes, are only partially understood (Somayaji, Reddy, Bardiya, *et al.* 1996; Whitman 1985). Major taxonomic groups of methanogenic bacteria are isolated from various sources, and these include *Methanosaeta soehngenii* (Somayaji, Reddy, Bardiya, *et al.* 1996), *Methanothrix* sp. (Ahring and Westermann 1984), *Methanosarcina* sp., and *Methanosarcina mazei*.

Factors affecting biomethanation

Biomethanation as an industrial process is gaining importance, and its success depends on the control over a diverse microbial population and its biological activity. Besides the microbial population, environmental factors that govern the process of methanogenesis are anaerobiosis, temperature, pH, substrate composition, C:N (carbon:nitrogen) ratio, micronutrients, presence of toxic materials in the substrate biomass, and accumulation of inhibitory by-products or intermediates of the process.

Anaerobiosis

Methanogenesis is a strict anaerobic microbial process. The major trophic groups and methanogens in this process die in the presence of O_2. Though the major microbial activity in methanogenesis is anaerobic, many workers have reported the presence of aerobic and facultative anaerobic bacteria (Hobson, Bousfield, and Summers 1981; Somayaji, Reddy, Bardiya, *et al.* 1996). The number of facultative anaerobic bacteria in the piggery waste digesters was found to be the same as that of the anaerobes (Hobson, Bousfield, and Summers 1981). These bacteria do not have a role in the main degradable reactions of digestion but probably play a role in the anaerobic digestion as scavengers of O_2, rendering the system suitable for the growth of the methanogens (Hobson, Bousfield, and Summers 1981). Therefore, the requirement for anaerobiosis in the digesters can be formulated in a dynamic way by minimizing the diffusion of air into the digestion liquor. Most digesters are designed in such a way that their ecosystem is closed to the atmospheric air so that the microflora in the digester is not exposed to the O_2 tension (Nyns 1986).

Temperature

Temperature in all microbial systems enhances the microbial activity to a certain optimal level. Methanogenesis occurs practically in an optimal way in two ranges of temperatures—mesophilic at 30–37°C and thermophilic at 50–65°C. Adaptation of a mesophilic digester to thermophilic conditions is possible (in about 10–20 days) as it always contains about 10% of the thermophilic micro-organisms (Nyns 1986). Thermophilic digestion offers a number of advantages. The rapid metabolic activity helps in reducing the retention time and increasing the loading rates, hence leading to small digester volume for plant design. In addition to the increased rates of CH_4 production, which is reported to be 1.5 times faster than the mesophilic digestion, thermophilic digestion improves dewaterability and enhances the removal of pathogenic organisms (Buhr and Andrews 1977). However, changes in the overall process efficiency due to increased metabolic activity are balanced by a corresponding increase in the rates of microbial inactivation (Somayaji, Reddy, Bardiya, *et al.* 1996). Hence, the optimal process temperature represents the best compromise between these two factors. The required optimal digestion temperature may also depend on the waste being treated. For example, a higher conversion of proteins was observed at 30–50°C, and conversely, lower temperatures were found suitable for lipid digestion (Somayaji, Reddy, Bardiya, *et al.* 1996). Low-temperature waste digestion is slower than mesophilic and thermophilic digestion.

pH

A pH value of 6.5–7.7 has been found to be optimum for the H-utilizing bacterium *Methanobacterium ruminantium* seen in digesters and rumen. A pH of 7.2 for the optimum running of the methanogenic stage of a two-stage laboratory digester was used; Cohen, Zoetemeyer, Deursen, *et al.* (1979) used the value 7.8 for the same. Maintenance of a near-neutral pH is due to the conversion of acid intermediates to CH_4 by the combined effect of acetogenic and methanogenic bacteria. It should be recalled that the methanogenic process is alkalizing. The major controlling buffer in this pH range is the carbonate–bicarbonate system. The VFAs and ammonia have also been known to regulate the buffering capacity of the digester (Capri and Marais 1975).

Sensitivity of anaerobic digestion to pH levels is due to the critically pH-sensitive methanogenic population. However, acid-tolerant methanogens do exist in a peat bog environment (Clarke and Speece 1974). Methanogenesis beyond the range of 6.5–8.0 pH has been found to be less yielding (Nyns 1986).

Substrate composition

The biomass substrate used in methanogenesis comprises various C sources such as cellulose, hemicellulose, lipid, lignin, pectin, protein, starch, and other organic material at varied concentrations. For a pure substrate, it is possible to calculate the amount of gas that may be produced on the basis of a simple C balance. Theoretical yield of biogas (m^3/kg volatile solids [VS] destroyed) from various components of organic matter is 0.886 (carbohydrates), 1.535 (fat), and 0.587 (proteins), with a CH_4 content of 50%, 70%, and 84%, respectively (Burford and Varani 1976).

Substrates for methanogenesis

Recently, due to the widespread use of anaerobic digestion for waste treatment as well as biogas production, a wide variety of substrates has been tried and commercially exploited. The range of biomass material that can be broken down by anaerobic digestion is very broad. More than 30 types of waste with the potential for CH_4 generation have been listed in Table 2 (Kubler and Wild 1992; NAS 1977; Mallik, Singh, and Tauro 1989; Somayaji, Reddy, Bardiya, *et al.* 1996). Domestic waste such as cattle manure, sewage sludge, domestic garbage, and night soil; agricultural waste such as piggery, poultry, and crop residues; and industrial waste like distillery waste have been successfully used for biomethanation in the past (Hobson, Bousfield, and Summers 1981).

Till recently, biogas substrates were those that were either natural habitat for the bacteria involved in anaerobic digestion (namely, cattle manure, sewage sludge, and food waste) or those that supported anaerobic digestion on their own as they contained all the necessary nutrient components. In recent years, fruit and vegetable processing waste such as mango peel, tomato processing waste, orange peel, pineapple processing waste, and banana peel, and agricultural crop residues such as rice straw and wheat straw have been successfully demonstrated for their potential biomethanation (Somayaji 1992; Somayaji and Khanna 1994). Somayaji, Reddy, Bardiya, *et al.* 1996 established a high biomethanation potential for canteen waste. By far, the maximum biomethanation potential recorded for a solid substrate biomethanation is 0.75 m^3/kg of organic matter in food market waste (Somayaji, Reddy, Bardiya, *et al.* 1996). Table 3 summarizes the application of many such substrates for CH_4 generation at different operating conditions.

Carbon:nitrogen ratio

Extra attention has to be paid to the N content of the substrate biomass for two reasons. First, the N is required for cell growth—a phenomenon necessarily linked to CH_4 production—and second, the nitrogenous compounds contribute to the buffering of the digesting liquor by releasing ammonium cation. Investigations show that a C:N ratio of about 20

Table 2 Potential organic matter for methane (CH_4) generation

Broad type	Example
By-products and wastes from agriculture-based industries	Oil cakes, bagasse, rice bran, tobacco wastes and seeds, wastes from fruit and vegetable processing, press-mud from sugar factories, tea wastes, and cotton dust from textile factories
Crop wastes	Sugar cane trash, weeds, corn and related crop stubble, straw, and spoiled fodder
Wastes of animal origin	Cattle-shed wastes (dung, urine, and litter); poultry litter; sheep and goat droppings; slaughter-house wastes (blood and meat); fishery; leather; and wool wastes
Wastes of human origin	Solid wastes from municipal operation and night soil
Forest litter	Twigs, bark, branches, and leaves
Wastes from aquatic plants	Marine algae, sea weeds, and water hyacinths

Source Somayaji, Reddy, Bardiya, *et al.* (1996)

Table 3 Potential biogas production from various substrates

Substrate	Operation			Biogas	
	HRT[a] (days)	Solid concentration (%)	Temperature (°C)	Yield added (m^3/kg VS)	Methane (%)
Sewage sludge	25	6.0	35	0.52	68
Domestic garbage	30	5.0	35	0.47	—
Piggery waste	20	6.5	35	0.43	69
Poultry waste	15	6.0	35	0.50	69
Cattle waste	30	10.0	35	0.30	58
Canteen waste	20	10.0	30	0.60	50
Food-market waste	20	4.0	35	0.75	62
Mango-processing waste	20	10.0	35	0.45	52
Tomato-processing waste	24	4.5	35	0.63	65
Lemon waste	30	4.0	37	0.72	53
Citrus waste	32	4.0	37	0.63	62
Banana peel	25	10.0	37	0.20	55
Pineapple waste	30	4.0	37	0.37	60
Mixed feed of fruit waste	20	4.0	37	0.62	50

[a] Hydraulic retention time

Source Somayaji (1992)

is required for maximum methanogenic performance. Glucose and urea were used to adjust the C:N ratio of swine manure to 20–25. Low CH_4 yield at a low C:N ratio was reported because of ammonia inhibition. A concentration above 3000 mg/litre of ammonia–nitrogen is supposed to be toxic at all the pH levels. The maximum CH_4 yields have been reported at C:N ratios between 16 and 19 (Somayaji, Reddy, Bardiya, *et al.* 1996). Decreasing the CH_4 yields has also been reported when the C:N ratio increased above 20, probably because of the low N availability. The findings of Somayaji, Reddy, Bardiya, *et al.* (1996); and Somayaji and Nand (1988) endorse the above result in the studies related to the anaerobic digestion of cattle manure and molasses, and mango processing waste, respectively.

Micronutrients

Bacteria in the anaerobic digestion process also require phosphorus (P) and metal ions for optimum growth. Somayaji, Reddy, Bardiya, *et al.* (1996) have reported that nickel and cobalt ions are necessary at concentrations of 100 nM and 50 nM, respectively. Molybdenum ions at a concentration of 50 nM may enhance the joint effect of nickel and cobalt. Iron cations up to 2 nM and copper ions at a concentration of 4 nM are necessary to enhance the performance of aceticlastic methanogenic archaebacteria. An increase of up to three-fold can be achieved by incorporating the above minerals into a digester deficient in these components (Nyns 1986; Somayaji, Reddy, Bardiya, *et al.* 1996). The requirement of P has also been studied and a C:P ratio of 100:200 is said to be optimum. The required optimum C:N:P ratio for the enhanced yield of CH_4 was found to be 100:2.5:0.5 (Somayaji 1992).

Toxins and inhibitors

It has been observed that the metal ions exert a toxic effect when they exceed the required concentration. Researchers proposed the critical value of pS > 14.0 as a threshold value, indicating the inhibitory presence of heavy metal ions such as Zn^{2+}, Cd^{2+}, Fe^{2+}, or Cu^{+} and Cu^{2+}. Somayaji, Reddy, Bardiya, *et al.* (1996) described the concentrations inhibitory to anaerobic digestion; accumulation of NH_4^+, SO_2, HSO_3^- , and SO_3^{-2} was reported to inhibit a methanogenic bacterial community at a concentration beyond 1 mg/litre of digesting liquor. The presence of organic compounds (amino derivatives, higher alcohols [5–12 carbon], and ketones [5–8 carbon]) is inhibitory only when their concentration exceeds their solubility in water. Petrochemicals are toxic when they are present as derivatives of Cl, |NH_2, |C = O, di|COOH, though accumulation is possible with the severely increased time requirements. However, chloroform ($CHCl_3$), an analogue of CH_4, is a very potent inhibitor of the methanogenic archaebacteria (Nyns 1986).

Intermediates

Isobutyric acid, in particular, is inhibitory at concentrations as low as 50 parts per million (ppm) (Somayaji, Reddy, Bardiya, *et al.* 1996). Toxic effect of the VFAs at high concentrations is attributed either to the toxicity of the VFAs themselves or to the decrease in pH brought about by them. However, inhibition due to the VFAs in the anaerobic digestion is varied.

Hydraulic retention time

The hydraulic retention time (HRT) expresses the volume of fluids in the reactor per volume of fluid passing into and out of the reactor on a daily basis. It indicates the contact time allowed between the substrate and the micro-organisms in the system. Maintaining an optimum retention time is an important factor, resulting in efficient conversion of organic matter into CH_4. A lengthy HRT results in inefficient use of the digester capacity as it allows the bacterial population to proceed beyond the phase of exponential multiplication. If the HRT is too short, the rate of bacterial multiplication may not be sufficient to compensate for the bacteria discharged into the effluent from the digester, resulting in an unstable bacterial population (Hawkes 1979; Somayaji, Reddy, Bardiya, *et al.* 1996). The HRT maintained in conventional anaerobic digesters utilizing raw sewage solids and organic refuse is for 15–30 days (Fair, Geyer, and Okum 1971; Pfeffer and Liebman 1975; Vesilind 1977; Eckenfelder and Shathanam 1981; Garrison, Nagel, and Easley 1982; Golueke, Diaz, Savage, *et al.* 1982). Cellulosic material requires a minimum HRT of 10 days, and the lignaceous material was reported to be non-degradable even at an HRT of 30 days (Golueke, Diaz, Savage, *et al.* 1982). However, an optimum retention time for a given substrate in an industrial application has to be designated not only on the basis of fermentability of the substrate but also on the basis of the quantum of material that needs to be treated and the possible economy of biomethanation. It is known that with an increase in retention time, there will be an increase in the yield of biogas (m^3/kg of solids added) with a high conversion efficiency. The specific rate of gas production (volume of gas/volume of reactor/day) decreases with an increase in the retention time, and thus, the specific rate of gas production is inversely proportional to the retention time. For example, in the case of mango peel biomethanation, it was found that at a 30-day retention time, gas yield of 0.6 m^3/kg of solids added with a specific rate of 2.8 (volume/volume of reactor [v/v]/day) was achieved, while at a 15-day retention time, a gas yield of 0.33 m^3/kg of solids added with a specific rate of 2.3 (v/v/day) was achieved. Hence, to achieve maximum gas production with higher conversion efficiency, it is essential to arrive at a particular retention time, which serves both the objectives (Somayaji and Nand 1989; Somayaji 1992).

Loading rate

The loading rate is the amount of raw material introduced into the digester per unit of time. It concerns the interaction between the substrate and the micro-organisms, and directly influences the equilibrium between different physiological groups of bacteria involved in the process. With certain substances, the adverse effects of overloading are immediate and drastic (for example, sewage sludge), and the effects are less drastic when the substrate is heterogeneous in nature (for example, organic refuse) (Somayaji, Reddy, Bardiya, *et al.* 1996). A high loading rate leads to an unfavourable pH, resulting in a low CH_4 production. An appropriate loading rate is required to maintain the balanced activity of different physiological groups, leading to a stable digestion process. The loading rate of organic matter is expressed in terms of kilograms of VS per cubic metre of digester or percentage of the VS. The reported loading rate for various substrates is 0.6–1.6 kg VS/m^3 for sewage sludge (Vesilind 1977), 1.27–4.80 kg VS/m^3 for organic refuse (Pfeffer and Liebman 1975; Vesilind 1977; Eckenfelder and Shathanam 1981; Garrison, Nagal, and Easley 1982; Golueke, Diaz, Savage, *et al.* 1982), 2.56–3.50 kg VS/m^3 for piggery waste (Maynell 1976), 1.12–27.24 kg VS/m^3 for cattle manure (Price and Cheremisinoff 1981), and 3.50–4.25 kg VS/m^3 for fruit and vegetable waste (Somayaji, Reddy, Bardiya, *et al.* 1996). In an industrial application, it is essential to note that an optimum loading rate be decided on the basis of the moisture (water) content of the given substrate. As in the case of retention time, to achieve maximum gas production with higher conversion efficiency, it is essential to arrive at a particular loading rate that meets both the objectives. A higher loading rate is inversely proportional to the yield of biogas (m^3/kg of solids added) and the conversion efficiency, while it is directly proportional to the rate of gas (v/v/d) production (Somayaji and Nand 1989; Somayaji 1992).

Indicators for monitoring the digestion process

Recently, efforts have been made to identify suitable indicators for the process evaluation. An ideal indicator should be easy to measure, should enable detection of imbalance at an early stage, and should be in a direct relation to reflect the metabolic status of the system. Further, the relative change of parameters following a perturbation is significant as compared to the background fluctuations and analysis accuracy.

Several parameters have been suggested as stress indicators. The process indicators can either be indicators indirectly depicting the metabolic activity of the bacteria in the reactor (Table 4) or indicators of the numbers and metabolic state of the bacteria in the reactor (Table 5).

Two-phase anaerobic digestion

The diverse nature of the requirement of trophic microbial population in the anaerobic digestion, inter-species electron flow, and apparent limitations of the methanogenic process have led to the proposal of phase separation. In phase 1, the organic matter is hydrolysed and acidified, and in phase 2, it is subjected to

Table 4 Traditional indicators

Indicator	Principle
Biogas production	Specific gas production
Biogas composition	Concentrations of methane (CH_4) and carbon dioxide (CO_2)
pH	Decrease in the pH due to the accumulation of volatile fatty acids (VFAs)
Alkalinity	Changes in buffer capacity
Total VFA	Total concentration of the VFA (usually determined by titration)
Individual VFA	Accumulation of the different VFAs (usually determined by gas chromatography)
Chemical oxygen demand or volatile solid reduction	Degradation efficiency
Hydrogen concentration	Accumulation of hydrogen, a key intermediate
Carbon monoxide (CO) concentration	Accumulation of CO, a precursor of acetate/formate

Source Ahring and Angelidaki (1997)

Table 5 Alternative indicators based on direct measurement of bacterial numbers and/or activity

Indicator	Principle
Bacterial numbers	
Bacteria counting	Numbers of different groups of bacteria
Molecular methods	
Genetic probing	Specific probes against 16S ribosomal (r) RNA for specific groups or individual bacteria
Immuno-techniques	Specific antibody probes for identification and qualification of specific bacteria
Biochemical-based indicators	
Bacterial membrane lipids	Special lipids can be used to identify different bacterial groups and their numbers
Enzyme activity	Activity of specific enzymes can indicate substrate turn in the bacterial ecosystem
Adenosine triphosphate (ATP) F_{420}	Concentration of the ATP indicates the general microbiological activity. F_{420} is a characteristic co-enzyme found in methanogens and can quantify active methanogens
Nicotinamide adenine diphosphate (NADH)	NADH activity is correlated with bacterial activity
Bacterial activity	
Methanogenic activity	CH_4 production rate during degradation of different substrates can estimate the activity of the corresponding bacterial group

Source Ahring and Angelidaki (1997)

biomethanation. It would be advantageous, as methanogenesis is a step-wise process (Ziekus 1980; Somayaji, Reddy, Bardiya, *et al.* 1996; Thiel and Ziekus 1988 a, b). The environmental and operational parameters may be optimized separately to improve the overall kinetics of the process. This would not only protect the limiting microbial population from the inhibitory effects of intermediates but also enhance overall performance of the anaerobic digestion. A comparison of the two-phase and conventional anaerobic digestion of soft drink waste water indicates that in the two-phase system, the loading rate could be increased up to 4.8 kg VS/m^3/day, while the high-rate system could sustain only 0.64 kg/m^3/day. The rate of gas production increases to 2.74 volumes as compared to 0.4 volume per liquid volume per day obtained in conventional biomethanation. The two-phase system has also recorded 96% of the chemical oxygen demand (COD) removal, whereas it was 84% in the conventional system (Somayaji, Reddy, Bardiya, *et al.* 1996).

Many studies on biological systems have been conducted in the CSTRs. These studies have demonstrated that subtle differences in process configuration such as phasing, staging, biofilms, granules, gas phase management, and combinations thereof can profoundly impact the anaerobic process performance.

It has been demonstrated in a number of cases that careful phasing/staging of the process profoundly improves performance of the microbial process. It is possible that either the liquid or the biomass moves together as is the case with a dispersed growth CSTR, or uncoupling may occur. The consortia of micro-organisms can be coupled or uncoupled by phase separation, and the influent BOD/day mixed liquor biomass in the tank and environmental conditions can be coupled or uncoupled for the liquid and/or biomass. In some cases, without staging, it has been impossible to achieve the desired microbial transformation.

Following are some examples of improved performance due to phasing/staging.

- *Two-phase anaerobic digestion* Anaerobic digestion involves an acid formation phase with complex substrates and a terminal CH_4 formation phase for all substrates. Massey and Pohland (1978), and Cohen, Zoetemeyer, van Deursen, *et al.* (1982) have demonstrated improved process performance through separate phasing. Separate phasing optimizes environmental conditions for each phase/stage because in the single phase/stage processes, both classes of organisms are forced to operate under a common environment.
- *Staging of gas phase* Harper and Pohland (1987) demonstrated that staging of the gas phase in anaerobic treatment effectively alters the H_2 concentration and enables improved utilization of the propionate.
- *Low substrate affinity of microbiota* Some microbial processes, especially anaerobic processes, manifest high K_s values (>2000 mg/litre), which may be further increased at reduced temperature. Thus, excessively large single-stage reactors are required to achieve low-effluent substrate concentrations. Staging or plug-flow configurations can reduce the reactor volume requirements by an order of magnitude when compared to one-stage CSTRs while achieving the same effluent concentrations.
- *Methanogen predominance* Two broad classes of methanogens are responsible for the conversion of acetate to CH_4 under anaerobic conditions. These classes are *Methanosaeta (Methanothrix)* with K_s= –30 kg/litre and k=–2/day and *Methanosarcina* with K_s=–300 mg/litre and k=–6/day. Methanosaeta tends to predominate at low substrate concentrations, while *Methanosarcina* predominates at higher substrate concentrations. Staging enables *Methanosarcina* to predominate in the first stage and *Methanosaeta* to predominate in the second stage, thus producing high quality, low effluent concentrations with a minimum of total reactor volumes. This has also been observed in the structure of granules.
- *Aerobic activated sludge and anaerobic suspended growth reactors selector stage* Provision of a relatively short HRT stage (selector) at the head of an activated sludge process maintains the return sludge in contact with a relatively high substrate concentration. Consequently, a shift in predominance of microbial population from filamentous (poor settling) to flocculent (good settling) can be realized, whereas a one-stage CSTR favours

the predominance of filamentous biomass. Similar improvements in the ability of the biomass to settle were noted when a selector stage was applied to a CSTR anaerobic suspended growth process.

Two-phase digestion

Massey and Pohland (1978) and other researchers reported the advantages of two-phase digestion. In order to examine the importance of reactor configuration in microbial selection, Anderson, Kasapgil, Ince, *et al.* (1994) studied the changes in microbial populations in two-phase digestion. They demonstrated that two-phased systems have several advantages over single-stage processes: (1) selection and enrichment of different bacteria in each reactor; (2) increased process stability; and (3) the methanogenic phase buffered by the prior acid phase.

Speece, Metin, Goksel, *et al.* (1997) operated single- and two-phase CSTRs under the conditions of shock loading and found that while both systems accumulated volatile acids, the two-phase system returned to typical operating conditions in one-fifth of the time required by the single-phase system.

Wiegant, Henning, and Lettinga (1986) studied the staging of propionate degradation in thermophilic anaerobic treatment and compared the single- and two-stage upflow anaerobic sludge blanket (UASB) reactors. It was found that significantly better results were obtained in the two-stage process, that is, there was 10%–13% better treatment efficiency. They concluded that the improvement was due to the removal of the biogas evolved in the first stage. Propionate COD accounted for 80% of the VFA–COD in the effluent of the first-stage reactor in their study and is often considered to be the limiting step in the methanogenesis of acidified waste waters. Acetate is known to be inhibitory to propionate degradation (Kus and Wisemann 1995). Mawson, Earl, and Larson (1991) found that 1000–2000 mg/litre of acetate significantly inhibited the degradation of 500 mg/litre propionic acid. Wiegant, Henning, and Lettinga (1986) hypothesized that staging would not improve the performance if toxicity was a problem.

Van Lier, Boersma, Debets, *et al.* (1994) studied thermophilic anaerobic treatment in compartmentalized upflow reactors. They reported superior performance of the plug-flow configuration as evidenced by the very low effluent volatile acid concentrations, high degree of biomass retention, and stable reactor performance. The group used either sucrose or a mixture of the VFAs as feed with the result that sucrose fostered granule development, whereas the VFA substrate resulted in almost no granule formation. Sucrose was converted in the first compartment, with butyrate and acetate conversion occurring in subsequent compartments. Propionate was the most difficult intermediate to degrade, but in the last compartment, even propionate degraded almost completely. Staging of the thermophilic process was able to reduce the high steady-state VFA concentrations typically found in the thermophilic CSTR. It has been established that propionate conversion limits the thermophilic CSTR processes. Propionate conversion is thermodynamically the most difficult reaction in the entire digestion process, but Van Lier, Boersma, Debets, *et al.* (1994) showed that unfavourable conditions for propionate conversion characterized only the first stage. In the staged reactor, H was effectively removed from each compartment by venting, thus precluding its passage to the subsequent stages. As both acetate and H have a detrimental effect on propionate conversion, the CSTR configurations cannot accommodate high organic loading rates (OLRs) for prolonged periods without an increase in the VFA concentration in the effluent. However, when using staged processes, the low effluent VFA concentrations can be maintained at a very high OLR.

Gas phase management

Harper and Pohland (1987) proposed gas phase management in a staged process configuration. Venting of the gas phase from each stage resulted in a more efficient and stable performance as compared to a system in which gas from each stage was passed through the subsequent stage.

High solids anaerobic digestion

Most natural high-solids fermentation processes responsible for anthropogenic CH_4 emissions are uncontrolled, slow to start, and require a very long time to achieve completion. Impeded fermentation results from mass-transfer and moisture limitations as well as from the VFA accumulations resulting from an imbalance between acid production and acid conversion to CH_4. A number of process configurations have been proposed and tested during the past 20 years to conduct high-solids anaerobic digestion. Notable among these are the leach-bed, two-phase or LanFilgas process, BIOCEL process (Braber 1995), National Renewable Energy Laboratory (NREL) high-solids system (Rivard, Himmel, Vinzant, *et al.* 1990), Valorga process (Begouen, Thiebaut, Pavia, *et al.* 1988), Dranco process (De Baere and Verstraete 1984), and the BTA (Braber 1995).

As mentioned above, one of the significant problems with high-solids anaerobic digestion is the VFA accumulation (sour digestion), which leads to severe slowing down of the methanogenesis process. This problem has been observed in our study for the past 15 years. For example, volatile acid accumulation up to 9000 mg/litre was observed during the digestion of a municipal solid waste (MSW) bed with the solids concentration of about 21% total solids (TS). A recent laboratory work with the anaerobic digestion of a 15% solids content bed of simulated Indian solid waste showed the VFA accumulation of up to 17 mg/litre with a concomitant pH reduction to 4. The low pH and highly acidic environment led to a cessation of hydrolytic and acidification reactions, presumably due to the end-product inhibition of fermentative acidogens (Ghosh 1997). Syntrophic and acetoclastic methanogenic activities were also severely inhibited under this sour digestion condition. (The reason for a higher VFA accumulation with nutrient-rich Indian MSW was that it had less paper products and more food waste components than the American MSW.) An innovative biogasification process that could successfully deal with the VFA accumulation problem is the anaerobic digestion process (Ghosh 1997).

A process design approach that can effectively manage the aforementioned problems of high-solids digestion includes the following steps.

- Initiate hydrolytic bioleaching of feed solids by applying an appropriate anaerobic inoculum atop the bed of solid waste; the inoculum volume is adjusted according to the field capacity of the solid bed.
- Induce accelerated bioleaching and acidification of bed solids by recirculation of the percolated culture through the solid bed at an appropriate frequency of liquid application.
- Develop an enrichment culture of hydrolytic acidifying organisms in the solid-bed reactor.
- Recycle a part of the acidic bioleachate from time to time through a separate packed-bed CH_4 fermenter (anaerobic filter) and back to the top of the solid-bed reactor in order to alleviate sour digestion conditions.

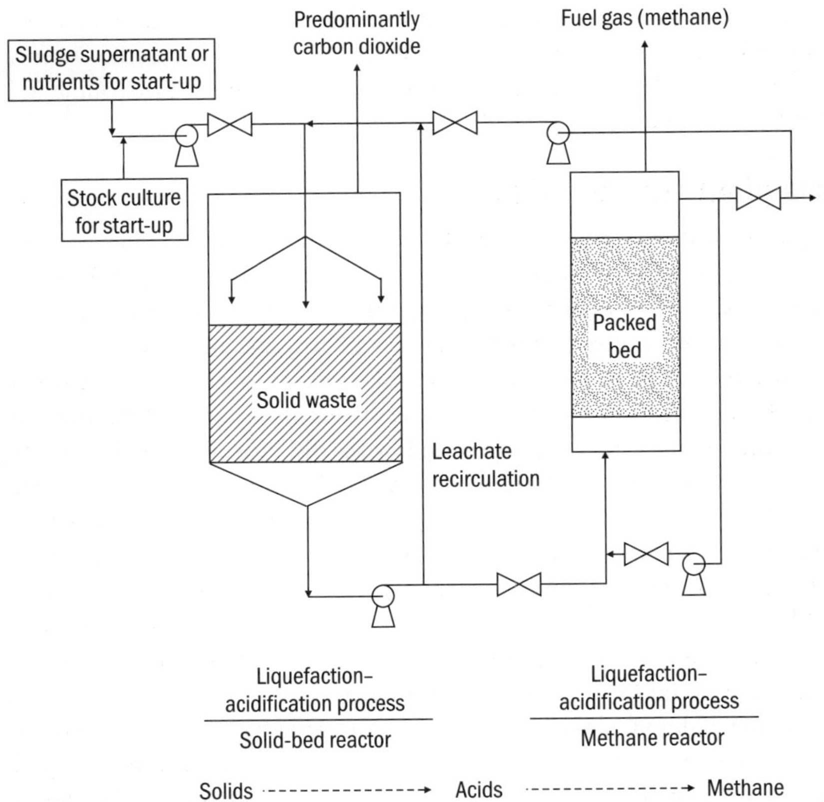

Figure 2 Schematic representation of leach bed two-phase anaerobic digestion process

The overall digestion process consists of an acid-phase solid-bed reactor operated in tandem with a separate packed-bed CH_4 fermenter (anaerobic filter) and back to the top of the solid-bed reactor to alleviate sour digestion conditions (Figure 2).

Depending on the nature of the solid feed, pH of the solid bed may drop to very low levels (4–5) owing to the VFA accumulations of 9000–17000 mg/litre. Acidogenic fermentation of the solid bed, characterized by the evolution of the CO_2, H, and N concentrations of 95 mol%, 30 mol%, and 25 mol%, respectively, was detected in the head gases of the solid-bed acid-phase reactor. The CH_4 fermentation virtually stops under the extreme sour-digestion condition that develops in the high-solids, acid-phase reactor.

The CH_4-phase, packed-bed reactor was operated at a low HRT of eight hours to exhibit the VFA and COD reductions of about 80%. The CH_4 content of the CH_4 digester gases ranged from 50 mol% to 75 mol%, depending on the stage of operation. Biomethanation of the solid feed was completed within a period of one to three months, depending on the physico-chemical characteristics of the feed and the mode of reactor operation. The overall system gas and CH_4 yields of 0.38 m^3/kg VS and 0.21 m^3/kg VS, respectively, were obtained with the American MSW, indicating an 81% conversion of the MSW biodegradables. Other high-solids digestion processes that have been modelled after the two-phase bioleaching approach include the sequential batch anaerobic composting (SEBAC) process (Chynoweth, Owens, O' Keefe, *et al.* 1992), and systems proposed by the Snell Environmental Group (1988) and Hofenk, Lipps, Rijkens, *et al.* (1984).

Full-scale plants in operation

The Valorga full-scale plant in Tilburg, the Netherlands, is designed to process 52 000 tonnes per year of organic MSW collected separately. The Valorga digestion process is a semi-continuous, dry, one-step, plug-flow type process. Its main characteristic is the complete absence of any mechanical equipment inside the reactors. The waste to be treated consists of food and garden waste. The characterization of the waste stream reveals a seasonal fluctuation in quantity and quality, that is, correlated with the seasonal garden waste production. The CH_4 yield is 210–290 m^3 standard temperature and pressure (STP)/mg of VS, as it is related to waste composition. During the dull winter period, the waste contains proportionally more food waste, which is more biodegradable than lignocellulosic garden waste. The biological process was stable and based on volatile acidity, alkalinity, and ammonia measurements in the effluent. The organic residue, after de-watering and storage under aerobic conditions, had been taken as the soil conditioner.

Results of the Valorga plant in Tilburg clearly demonstrate that anaerobic digestion is a reliable industrial process for the treatment of organic solid waste. Despite the annual variation in waste quantity and quality, the biological process remains well-balanced. The overall treatment exhibits a largely positive energy balance. A key factor towards the reliability of the Valorga process is the complete absence of any mechanical device inside the reactors, which prevents the wearing off and other maintenance problems (de Laclos, Des bois, and Saint-Joly 1997). Table 6 lists some commercial full-scale plants under operation.

Sugar cane pressmud as a case study

Sugar cane pressmud is one of the three major by-products in the cane sugar manufacture (such as bagasse, molasses, and pressmud). Pressmud is composed of precipitated impurities contained in the cane juice after removal by filtration. In this review, the present status of pressmud utilization is examined. This by-product is obtained in the form of cakes with varying levels of moisture. The cake generally contains colloidal organic matter anions that precipitate during clarification, and certain other non-sugars present in the juice are occluded in these precipitates (Paturau 1982). The cake thus obtained has a high moisture content of 65%–80%. The quantity of pressmud generated from a given amount of cane crushing and its composition vary greatly with the locality, variety of cane, milling efficiency, method of clarification, and so on. Pressmud generation from sugar cane is considered normal when it is about 3.0%. According to Mittal (1996), there is an estimated generation of 5 MT of pressmud from 425 sugar mills in the country. The composition analysed for pressmud obtained from Nellikuppam sugar factory for two seasons is given as TS 25%–26%, moisture 74%–75%, ash 20%–22%, cellulose 13%–18%, hemicellulose 5%–15%, total sugars 1.5%–6%, fats and waxes 12%–23%, and crude protein 8%–10%.

Table 6 Full-scale plants under commercial operation across the world

Plant	Phase	Type	Hydraulic retention time (days)	Annual turnover (tonnes)	Operating since
Valorga (Amiens, France)	Single	Mesophilic dry digestion	15–20	55 000	1988
Dranco (Brecht, Belgium)	Single	Thermophilic dry digestion	18–21	15 000	1992
BTA (Helsingor, Denmark)	Two	Mesophilic wet digestion	4–6	20 000	1991
BIOCEL (Flevoland, The Netherlands)	Single	Mesophilic dry digestion	14–21	25 000	1993
Avecon (Vassa, Finland)	Single	Mesophilic wet digestion	15–20	21 000	1990
Kompogas (Rumlang, Switzerland)	Single	Thermophilic dry digestion	13–25	10 000	1992
EID Parry (India) Ltd, Nellikuppam, India	Single	Mesophilic wet digestion	15–20	9 000	1996

Biomethanation of sugar cane pressmud

Sugar cane pressmud with its rich organic matter content along with phosphate, potash, and N has a high potential for commercial biomethanation. Although little effort has gone into biomethanation, it is evident from these results that utilization of pressmud through the biomethanation route is beneficial (Somayaji, Reddy, Bardiya, *et al.* 1996; Mittal 1996). Laboratory trials carried out on pressmud biomethanation recorded 48 litres/kg feed and 160 litres/kg TS added at a retention time of 30 days, with a loading rate of 2.1 kg VS/m^3/day (2.6 kg TS). Pressmud as an additive to cow dung slurry for biomethanation was studied. Based on Burford and Varani (1976), the estimate of theoretical biomethanation potential for a given tonne of pressmud containing a representative 38% carbohydrates, 14% fats and waxes, and 15% crude protein is about 640 m^3 when converted completely, and it is equivalent to 67% conversion of the TS added, while other components like lignin and ash do not contribute to biomethanation (Table 7).

EID Parry (India) Ltd patented (Indian Patent No. 186 848, dated 2 August 1995) its in-house designed anaerobic digestion plant for pressmud and related raw materials for biomethanation and compost production. A full-scale plant with a capacity of handling 30–50 tonnes of pressmud per day is under operation since 1996. The plant produces about 4000 m^3 of biogas daily at the rate of 530 m^3/tonne of dry pressmud. In 2002/03, the plant produced about 0.82 million m^3 of biogas with over 55% CH_4 and 8600 tonnes of stabilized organic compost under two brand names—Garden Bloom and Farm Boon—for various agricultural and gardening applications.

Conclusion

Ever since the beginning of the human civilization, waste has been a necessary evil of human life. The concentration of waste now threatens to go beyond the earth's regeneration capacity. Within the regeneration capability, the waste from one process serves as raw material for another process. Biomethanation through anaerobic digestion of these residues is one such means to convert hazardous waste into a useful resource. Anaerobic digestion not only prevents pollution but also transforms a disposal problem into CH_4, besides providing nutrient-rich manure. Biogas produced by methanogenic microbes through the

Table 7 Theoretical potential of pressmud biomethanation

Convertible component	m^3 equivalent /tonne	Per cent content in pressmud	Tonnes/ tonne of pressmud	Biogas equivalent (m^3)
Carbohydrates	886	38	0.38	336.68
Fats and waxes	1535	14	0.14	214.90
Crude proteins	587	15	0.15	88.05
Total	—	67	—	639.63

anaerobic digestion process is quite similar to the natural gas extracted from the wellhead, and is suitable for use in boilers/engines/generators to produce steam/electricity. When biogas is used to produce electricity, there is the added potential for harvesting thermal energy from the engine's exhaust and cooling system. Biogas can also be burned in boilers to produce hot water and steam for sanitary washing or in gas-fired absorption chillers used for refrigeration.

Promising future waste-to-profit activities may enhance the economic performance of the overall waste management system. One new end-use application that can provide added value to co-products is the fuel cell for generating electricity and process heat. Fuel cells appear to offer a particularly clean and effective method for converting biogas into electricity and process heat. Instead of the 20%–25% electrical conversion efficiency from today's engine/generator technologies, fuel cells can produce electricity from biogas with 40%–50% conversion efficiency, which is a two-fold increase in productivity. When combined with the process heat recovery, the overall fuel cell efficiency is greater than 80%. Because of high efficiency, fuel cells emit less CO_2 per kilowatt-hour of electricity than the current conversion technologies. The process also produces a minimal amount of CO and NO_x, an issue of extreme importance to areas with substandard classifications under the Clean Air Act Amendments of 1990. With no moving parts, fuel cells also operate very quietly.

Use of solid and liquid residues from the anaerobic digestion process has beneficial environmental impacts. Anaerobic digestion enriches alternative agricultural practices when the digestion residues are applied to crops or used as dedicated feedstock for alternative fuels such as bio-diesel or ethanol.

Anaerobic digestion is a technically feasible option as it provides benefits to the environment through energy and nutrient recycling while also mitigating odour, pathogens, and atmospheric CH_4.

The future

In the past, anaerobic technologies emphasized energy production from animal manure and organic waste in combination with animal manures. Today, energy production and recovery are still important, but recognition of anaerobic digestion as an inexpensive technology to stabilize organic waste reduces the BOD and suspended solids with minimal sludge production, and reduces odour, which is almost as important as energy production. In future, the driving forces for the use of anaerobic digestion are expected to continue drifting away from energy production. Inexpensive organic stabilization, pathogen reduction, and odour control will be important reasons to use anaerobic digestion, particularly in the developing world. Energy savings in anaerobic treatment versus aerobic treatment will become more important in energy-deficient countries. Use of more complex anaerobic digestion processes for industrial waste treatment will increase. The anaerobic digestion can decompose some organic toxics and hazardous materials in co-digestion schemes, and this potential will be realized. Odour control and recovery of nutrients for export to farms away from livestock concentrations will be the other important functions of animal manure. Energy production and use will continue to be a component of the benefit analysis for anaerobic digestion, but perhaps not result in the most important benefit. A future application in many new developments is the digestion of the MSW. Biological treatment processes such as anaerobic digestion and composting offer the only route for recycling organic matter and nutrients from the putrescible fraction of the MSW. However, the MSW composting technologies require energy inputs of 30–35 kWh/tonne of the MSW input, whereas anaerobic digestion is a net energy-producing process that can generate 100–150 kWh/tonne of the MSW input.

The MSW digestion poses many technical problems. A number of anaerobic digestion systems have been developed for low- or high-solids content waste, operated at medium or high temperature. The high solids allow a higher loading rate of organic materials, while high temperatures allow shorter retention times. The higher temperatures also increase the destruction of pathogens present in the MSW. Co-products produced by the anaerobic digestion of the MSW include compost and liquid leachate that have value as soil-conditioner and fertilizer. The amount and quality of these products depend entirely on the quality of the MSW feedstock, the method of digestion, and the extent of any post-treatment refining processes. Many countries have developed standards for the use of MSW co-products, and any use of these co-products must strictly comply with consumer quality standards.

References

Ahring B K and Angelidaki I. 1997. **Monitoring and controlling the biogas process.** *Proceedings of the Eighth International Conference on Anaerobic Digestion* (Vol. 1), Sendai, 25–29 May

Ahring B K and Westermann P. 1984. **Isolation and characterization of a thermophilic, acetate-utilizing methanogenic bacterium.** *FEMS Microbiology Letters* **25**: 47–52

Andreesen J R, Gottschalk G, and Schlegel H G. 1970. ***Clostridium formicoaceticum*** **nov. spec. isolation, description, and distinction from *C. aceticum* and *C. thermoaceticum*.** *Archives of Microbiology* **72**: 154–174

Anderson G K, Kasapgil B, and Ince O. 1994. **Microbiological studies of two phase anaerobic digestion start-up.** *Water Research* **28**: 2383–2392

Bache R and Pfenning N. 1981. **Selective isolation of *Acetobacterium woodii* on methoxylated aromatic acids and determination of growth yields.** *Archives of Microbiology* **130**: 255–261

Barker H A. 1956. ***Bacterial Fermentations***. New York: John Wiley and Sons

Begouen O, Thiebaut E, Pavia A, Peillex J P. 1998. **Thermophilic anaerobic digestion of solid wastes by the Valorga process,** pp. 789–792. In *Fifth International Symposium on Anaerobic Digestion: poster papers,* edited by A Tilche and A Rozzi. [Bologna, Italy]

Braber K. 1995. **Anaerobic digestion of municipal solid waste: a modern waste disposal option on the verge of breakthrough.** *Biomass & Bioenergy* **9**(1–5): 365–376

Braun M, Mayer F, and Gottschalk G. 1981. ***Clostridium aceticum*** **(Wieringa), a microorganism producing acetic acid from molecular hydrogen and carbon dioxide.** *Archives of Microbiology* **128**: 288–293

Breznak J A, J M Switzer, and Seitz H-J. 1988. ***Sporomusa termitida* sp. nov., an H_2/CO_2-utilizing acetogen isolated from termites.** *Archives of Microbiology* **150**: 282–288

Buhr H O and Andrews J F. 1977. *Water Research* **11**: 129

Burford J L and Varani F T. 1976. **Energy potential through bioconversion of agricultural wastes.** Final reports to the Four Corner Regional Commission by Biogas of Colarado Inc. and Colarado Energy Research Institute

Capri M G and Marais G von R. 1975. **pH adjustment in anaerobic digestion.** *Water Research* **9**: 307

Chynoweth D P, Owens J, O' Keefe D, Earl J F K, Bosch G, Legrand R. 1992. **Sequential batch anaerobic composting of organic fraction of municipal solid waste.** *Water Science Technology* **25**: 327–339

Clarke R H and Speece C E. 1974. ***Proceedings of the Fifth International Conference: International Association Water Pollution Research,*** edited by S H Jenkins. Oxford: Pergamon Press

Cohen A, Zoetemeyer R J, van Deursen A, van Andel J G. 1979. **Anaerobic digestion of glucose with separated acid and methane formation.** *Water Research* **13**: 571–580

De Baere L and Verstraete W. 1984. **High-rate anaerobic composting with biogas recovery.** *Biocycle* (March): 30–31

de Laclos H F, Serge Desbois, and Claude Saint-Joly. 1997. **Anaerobic digestion of municipal solid organic waste: Valorga full-scale plant in Tilburg, the Netherlands.** In *Proceedings of Eighth International Conference on Anaerobic Digestion* (Vol. 2). 232 pp.

Eckenfelder W W Jr and Shathanam C J. 1981. ***Sludge Treatment***, edited by M l Dekker. New York. 591 pp.

Fair G M, Geyer J C, and Okun D A. 1971. ***Elements of Water Supply and Waste Water Disposal*** (2nd edn). New York: John Wiley and Sons. 752 pp.

Garrison W E, Nagel C A, and Easley S. 1982. **Design and operation of anaerobic sludge digestion system.** *Water Engineering and Management Reference Handbook*, p. 82 R/68

Ghosh S. 1997. **Anaerobic digestion for renewable energy and environmental restoration.** *Proceeding in Eight International Conference on Anaerobic Digestion* (Vol. 1)

Golueke C G, Diaz L F, Savage G M, Lofrinz D J, Jones D S. 1982. ***Laboratory and Pilot Studies of Refuse Digestion.*** Richmond, California: California Recovery Systems Inc.

Harper S R and Pohland F G. 1987. **Enhancement of anaerobic treatment efficiency through process modification.** *Journal of Water Pollution Control Federation* **59**: 152–161

Hawkes D L. 1979. ***Anaerobic Digestion,*** edited by D A Stafford, B I Wheatley, and D E Hughes. London: Elsevier Applied Science Publishers. 131 pp.

Hobson P N, Bousfield S, and Summers R. 1981. ***Methane Production from Agricultural and Domestic Wastes.*** London: Elsevier Applied Science Publishers

Hofenk G, Lipps S J J, Rijkens J, Coetberg J W. 1984. **Two-phase anaerobic digestion of solid organic wastes yielding biogas and compost** [EC Contract Final Report ESE-E-R-040-NL]

Kubler H and Wild M. 1992. **Anaerobic digestion of solid waste.** *Proceedings of the International Symposium on Anaerobic Digestion of Solid Waste,* Venice, Italy, 14–17 April, pp. 535–538, edited by F Ceechi, J Mata-Alavarez, and F G Pohland

Kus F and Weismann U. 1995. **Degradation of kinetics of acetate and propionate by immobilized anaerobic mixed cultures.** *Ater Research* **29**: 1437–1443

Mallik R K, Singh R, and Tauro P. 1989. **Fuel gases from biogas digesters.** In *Biological Waste Treatment*, pp. 1–34, edited by A Mizrahi. New York: Alan R Liss, Inc.

Massey M L and Pohland F G. 1978. **The phase separation of anaerobic stabilization by kinetic controls.** *Journal of Water Pollution Control Federation* **50**: 2204–2222

Mawson A J, Earl R L, and Larson V F. 1991. **Degradation of acetic and propionic acids in the methane fermentation.** *Water Research* **25**: 1549–1554

Maynell P J. 1976. ***Methane: planning a digester*** (CTT Series No. 2). Chalmington, Dorchester: Prism Press

Mittal K M. 1996. ***Biogas Systems, Principles and Applications.*** New Delhi: New Age International (P) Ltd

NAS (National Academy of Science). 1977. ***Methane Generation from Human, Animal and Agricultural Wastes.*** Washington, DC: NAS

Nyns E J. 1986. **Biomethanation processes.** *Biotechnology* 8, p. 207, edited by J Rehn and G Road. Weinheim, Germany: VCH Publishers Inc.

Paturau J M. 1982. ***By-products of the cane sugar industry.*** Amsterdam: Elsevier

Pfeffer J T and Liebman J C. 1975. **Biological conversion of organic refuse to methane** (NSF/RANN/Se/G1. 3919/75/2). *Urban Report,* UILU-Eng. 75–2019 September. University of Illinois, Department of Civil Engineering

Price E C and Cheremisinoff P N. 1981. ***Biogas Production and Utilization.*** Michigan: Ann Arbor Science Publishers. 146 pp.

Rivard C J, Himmel, M E, Vinzant T B, Adney W S, Wyman W S, Grohmann C E. 1990. **Anaerobic digestion of processed municipal solid wastes, using a novel high solids reactor: maximum solids level and mixing requirements.** *Biotechnology Letters* **12**(3): 234–240

Snell Environmental Group. 1988. **Rapid methane generation from solid waste.** [US Department of Energy Technical Report No. DE-FG02-81R10326]

Somayaji D. 1992. **Microbiological studies on biomethanation of mango peel with special reference to methanogens.** University of Gulbarga [PhD Thesis]

Somayaji D and Khanna S. 1994. **Biomethanation of rice and wheat straw.** *World Journal of Microbiology and Biotechnology* **10**: 521–523

Somayaji D and Nand K. 1988. **Anaerobic digestion of mango peel: influence of nitrogenous sources on methane production.** *Proceedings: Annual Conference of Society of Biological Chemists*, New Delhi

Somayaji D and Nand K. 1989. **Biogas production from mango processing wastes: optimization of nutritional parameters.** *30th Annual Conference of Association of Microbiologists of India*, Annamalai Nagar

Somayaji D, Reddy M R V P, Bardiya N, Mohan K, Khanna S. 1996. **Methanogenesis.** In *Wealth from Waste*, edited by S Khanna and K Mohan. New Delhi: Tata Energy Research Institute

Speece R E, Metin D, Goksel D, Henry Z, DiStefano T. 1997. **The role of process configuration in the performance of anaerobic systems.** In *Proceedings of Eighth International Conference on Anaerobic Digestion* (Vol. 1), 25–29 May, Sendai, Japan

Thiel J H and Ziekus J G. 1988a. ***Hand Book of Anaerobic Fermentations***, edited by L E Erickson and D Y Fung. New York: Marcel Dekker, Inc. 537 pp.

Van Lier J B, Boersma F, Debets M M, Lettinga G. 1994. **High rate thermophilic anaerobic waste water treatment in compartmentalized upflow reactors.** *Water Science Technology* **30**: 251–261

Vesilind P A. 1977. ***Treatment and Disposal of Waste Water Sludge*** (2nd edn). Michigan: Ann Arbor Science Publishers. 322 pp.

Whitman W B. 1985. **Methanogenic bacteria.** In *The Bacteria* **8**: pp. 3–84, edited by C R Woese and R S Wolfe. New York: Academic Press, Inc.

Wiegant W M, Henning M, and Lettinga G. 1986. **Separation of propionate degradation to improve the efficiency of thermophilic anaerobic treatment of acidified waste waters.** *Water Research* **20**: 517–524

Wolin M J. 1976. ***Microbial Production and Utilization of Gases***, edited by H G Schlegel, G Gottschalk, and N Pfenning, p. 141. Germany: Goltz Virlag E

Ziekus J G. 1980. ***Anaerobic Digestion***, edited by D A Stafford, B L Wheatley, and D E Hughes. London: Elsevier Applied Science Publishers. 61 pp.

Ziekus J G, Lynd L G, Thomson T E, Krzycki J A, Weimer P J, Hegge P W. 1980. *Current Microbiology* 3: 381–386.

Bibliography

Burges A. 1968. **The role of the microflora in the decomposition and synthesis of soil organic matter.** *Transactions of Ninth International Congress of Soil Science* 2: 29–35

Burton J C. 1967. **Rhizobium culture and use.** In *Microbial Technology,* edited by J Peppler. New York: Reinhold Publishing Co.

Cooper P F and Atkinson B. 1981. **Biological fluidized bed treatment of waste water.** Chichester: Ellis Horwood

Jewell W J, Switzenbaum M S, and Morris J W. 1979. *Proceedings of the 52nd Water Pollution Control Federation Conference*, 11 October, Houston

Grin P, Roersma R, and Lettinga G. 1981. *Paper presented at the Second International Symposium on Anaerobic Digestion.* Germany: Travemunde

Lettinga G, van Velson L, De Zeeuw W, Hoboma S W. 1980. ***Anaerobic Digestion***, edited by D A Stafford, B L Wheatley, and D E Hughes. London: Elsevier Applied Science Publishers. 61 pp.

Morgan H. 1981. ***Food Industry Wastes: disposal and recovery.*** London: Elsevier Applied Science Publishers

Polpresert C. 1996. ***Organic Waste Recycling Technology Management.*** Chichester: John Wiley & Sons Ltd

van der Meer R R. 1979. ***Anaerobic Treatment of Waste Water Containing Fatty Acids in Up-flow Reactor.*** Delft, the Netherlands: University Press

van der Berg L and Lenz C P. 1980. ***Proceedings of the Seminar/Workshop. Anaerobic Filters: an energy plus for waste water treatment.*** Florida: Department of Energy

Warth W and Schnell R. 1989. **BTA process for the utilization of organic wastes suited for practical use.** *Recycling International* 1: 256–271

Wheatley B L and Hughes D E. 1980. ***Anaerobic Digestion***, edited by D A Stafford, B L Wheatley, and D E Hughes. London: Elsevier Applied Science Publishers. 61 pp.

Biomass-gasifier-based decentralized energy solutions: technology, potential, problems, and research

S Dhingra

Fellow, BETA, The Energy and Resources Institute, New Delhi, India

Worldwide, photosynthetic activity is estimated to lead to the storage of 17 times as much energy as is consumed annually by all the countries in the world. Even though the energy required for the collection, processing, and conversion of biomass into other useful forms is taken into account, biomass still holds the promise to meet the complete energy needs of the world, if managed and used effectively and sustainably.

The term "biomass" generally refers to the renewable organic matter generated by plants through photosynthesis, wherein solar energy combines with CO_2 and moisture to form carbohydrates and oxygen. Materials having combustible organic matter are referred to as biomass. Coal is the end product of a sequence of biological and geological processes that biomass undergoes. Biomass contains C, H, and O, which are oxygenated hydrocarbons. It generally contains a high level of moisture and volatile matter but has a low bulk density and calorific value.

Biomass sources

Biomass can be classified into two types: woody and non-woody. Non-woody biomass comprises agro-crop and agro-industrial processing residue. Municipal solid waste, and animal and poultry waste are also referred to as biomass, as they are biodegradable in nature. The main biomass sources are as listed below.

- Wood and sawdust
- *Agricultural residues* Rice husk, bagasse, groundnut shells, coffee husk, straws, coconut shells, coconut husk, arhar stalks, jute sticks, and so on.

- *Aquatic and marine biomass* Algae, water hyacinth, aquatic weeds and plants, sea grass beds, kelp, coral reef, and so on.
- *Waste* Municipal solid waste, municipal sewage sludge, animal waste, industrial waste, and so on.

Due to the diversity of biomass, sufficient data and documentation as regards its availability and consumption/utilization pattern are not easily available. Although biomass meets a major part of the total energy requirements, it does not find an appropriate place in the overall energy balance of India, probably due to reasons cited above.

Potential availability of agro-residues

Agricultural residues can be divided into two groups: crop residues and agro-industrial residues. Crop residues are the plant materials left behind in the farm after the removal of the main crop produce. The remaining materials could be of different sizes, shapes, forms, and densities, examples being straw, stalks, sticks, leaves, haulms, fibrous materials, roots, branches, and twigs. The agro-industrial residues are the by-products of the post-harvest processes of crops such as cleaning, threshing, linting, sieving, and crushing. These could be in the form of husk, dust, straws, and so on. The major crop residues produced in India are straws of paddy, wheat, millet, sorghum, pulses, and oilseed crops; maize stalks and cobs; cotton stalks; jute sticks; sugar cane trash; mustard stalks; and so on. The agro-industrial residues are groundnut shells, rice husk, bagasse, cotton waste, coconut shell, and coir pith (Dhingra, Mande, Kishore, *et al.* 1996).

The quantity of agricultural residues produced differs from crop to crop and is affected by seasons, soil types, and irrigation conditions. Production of agricultural residues is directly related to the corresponding crop production and ratio between the main crop produce and the residues, which vary from crop to crop and, at times, with the variety of the seeds in one crop itself.

Thus, for known amounts of crop production, it may be possible to estimate the quantity of agricultural residues produced using the residue to crop ratio (Vimal and Tyagi 1984). It may be noted that with improved agricultural farming techniques, the production of crops has increased consistently during the past three decades. Correspondingly, the availability of agricultural residues is also changing with time. The estimated data for various agro-residues (year 2008/09) at the national level is given in Table 1.

Most agricultural residues are not found throughout the year but are available only at the time of harvest. This makes collection easy, but creates storage problems if the residues have to be saved for use during other months of the year, especially due to their low bulk density. The amount available depends on the harvesting time, storage-related characteristics, storage facility, and so on. The broad periods of availability of some important agricultural residues are presented in Table 2.

Table 1 Agricultural crop residue production in India

Crop	Type of residue	Production (in MT) 2008/09
Rice	Straw + husk	178.5
Wheat	Straw	124.16
Jowar	Stalk	14.48
Maize	Straw	49.32
Bajra	Stalk + cobs	17.8
Ragi	Stalk	4.08
Sugar cane	Bagasse + leaves	108.50
Soyabean	Stalk	17.25
Rapeseed and mustard	Oilseed	14.62
Jute	Waste	15.41

Source Indiastat and Ravindranath, Somashekar, Dasappa, *et al.* (2005)

Table 2 Seasonal availability of agricultural residues

	Availability period											
Residue	January	February	March	April	May	June	July	August	September	October	November	December
Arhar stalk	■	■	■							■	■	■
Maize stalk								■	■	■	■	■
Maize cobs								■	■	■	■	■
Cotton stalk	■	■	■							■	■	■
Mustard husk				■	■	■						
Jute sticks							■	■	■	■	■	■
Rice husk						■	■	■	■	■	■	■
Groundnut shell				■	■	■				■	■	■

Source Iyer, Rao, and Grover (2002)

In India, normally two crops—kharif and rabi—are taken into consideration. Therefore, availability of crop residues is expected to be spread evenly over the year. As a result, crop residues of one kind or the other are available throughout the year. Since crop production depends on the agro-climatic conditions, all agro-residues are not available in all parts of the country. An indication of the availability of various types of agro-residues in India, by state, is given in Table 3.

However, agro-residues do suffer from two major constraints: high moisture content and relatively low bulk density. These constraints inhibit their economical transportation over long distances, thereby necessitating

Table 3 Availability of various crop residues in India, by region

Region	Crop residues available
North (Punjab)	Sawdust rice husk, rice straw, and mustard stalk
North East (Assam)	Jute and mesta sticks, and rice husk
West (Gujarat)	Cotton stalks, arhar stalks, castor waste, groundnut shell, maize stalks, and maize cobs
Central (Maharashtra)	Sawdust, groundnut shell, arhar stalks, cotton stalks, sugar cane waste, and maize waste
South (Tamil Nadu)	Rice husk, sawdust, groundnut shell, tamarind shell, coir pith, sugar cane waste, and coffee husk
South (Karnataka)	Rice husk, sawdust, groundnut shell, coffee waste, coir pith, and sugar cane waste

their utilization near the sources of production. Unlike fossil fuels, strategy for agro-residues utilization has to be different. These are, therefore, most appropriate for decentralized technological applications in rural environments. The processing of the agricultural produce and utilization of agro-residues, therefore, can contribute their maximum share to rural development.

With worldwide shortage of wood, especially in the developing countries, and the need for reforestation to maintain the global ecological balance, there is an increasing demand for the proper utilization of agro- and forestry residues to replace wood. However, at present, these residues are either grossly underutilized or completely unutilized due to in situ burning in fields as a means of disposal. Development of technologies to utilize this major resource and its management need to be emphasized to meet the growing energy demands at the domestic level as well as those of rural small-scale/cottage industrial sectors.

Biomass electricity technologies

The technologies used for generating power from biomass range from conventional to the cutting edge research. The choice of technology can drastically affect the economics of biomass electricity and limit the size of the plant generating it. Table 4 summarizes various technological options available for biomass-based electricity generation.

Decentralized biomass-based power generation

Conventional biomass-based thermal power plants are normally suitable for large megawatt-level capacity. However, as biomass generally has low energy density, it favours small, dispersed plants for its conversion into other energy carriers. Thus, a modern megawatt/gigawatt-capacity conventional steam-cycle electricity plant is unlikely to be fuelled by biomass since it would draw its fuel from an enormous area, incurring prohibitive transportation costs. It is also clear from Table 4 that for decentralized small-capacity power generation

Table 4 Comparison of biomass electricity generation systems

Alternative technology	Rankine cycle		Otto/Diesel cycle		Sterling cycle	
	Steam	Organic	Wood-based	Briquetted biomass	Air	H_2/He
Acceptability of different biomass	Yes	Yes	No	Yes	Yes	Yes
Constraints related to installed capacity	Heavy	Moderate	Moderate	Moderate	Heavy	Moderate
Cost constraints	Moderate	Heavy	Moderate	Moderate	Moderate	Heavy
Constraints related to collection of biomass	Heavy	Not severe	Not severe	Not severe	Not severe	Not severe
Level of development within the country	Very little	Very little	High	Fair	Fair	—
Indigenous development	Good	Poor	Good	Good	Good	Very poor
Amenability for decentralized use	Poor	Good	Good	Good	Good	Good
Cost of maintenance	High	High	Moderate	Moderate	Moderate	High
Energy requirement for pre-processing	Low	Low	Moderate	High	Low	Low
Space requirement	High	High	Low	Moderate	Low	Low
Water requirement	High	High	Low	Low	Low	Low

using biomass, Otto or Diesel cycle internal combustion (IC) engines and Sterling cycle systems are more suitable. Biomass-gasifier-based system converts solid biomass into more user-friendly gaseous form, which can be used directly in IC engines to generate power.

The gasification process

Gasification of biomass takes place in four distinct stages, that is, drying, pyrolysis, oxidation/combustion, and reduction, in that order. Biomass is fed at the top of the hopper. As the dried biomass moves down, it is gradually heated up. Biomass starts releasing volatiles at above 200°C and continues to do so till it reaches the oxidation zone where the pyrolyzed biomass is completely consumed, releasing ash. Once the temperature reaches 400°C, the structure of wood breaks down, releasing combustible gases, along with water vapour, methanol, acetic acid, and tars as part of the pyrolysis process. These products of pyrolysis are drawn towards the oxidation zone, where a predetermined quantity of air is supplied to carry out combustion in order to raise the temperature. A portion of the pyrolysis gases and char burns as the temperature rises from about 900°C to 1200°C in the oxidation zone. Partial oxidation of biomass by gasifying agents (air or O_2) takes place in the oxidation zone, resulting in the release of producer gas, cracked and un-cracked pyrolysis products, and water vapour (steam), which pass through the reduction zone containing mostly charcoal. The principal chemical reactions taking place in a gasifier are given in Table 5.

In the reactions in Table 5, devolatilization takes place in the pyrolysis zone, oxidation in the oxidation zone, and the rest is distributed in the partial oxidation and reduction zones. The low heating value of the producer gas (about 10%–15% of natural gas) is mainly due to the diluting effect of nitrogen present in the partial combustion/oxidation air. An efficient gasifier produces a clean burning producer gas over a range of selected operating conditions. If all the above-mentioned processes take place efficiently, the energy content of the producer gas should lie between 70% and 78% of the energy content of the biomass consumed in the gasifier.

With the help of the above reaction scheme, and with the knowledge of the equilibrium constants, it is possible to predict the equilibrium composition of the gaseous products. The equilibrium composition of a given solid fuel depends on the air supply per unit weight of the biomass. A dimensionless parameter, known as the equivalence ratio (ER), is applied to characterize the air supply conditions, and is usually defined as follows.

$$\text{Equivalence ratio} = \frac{\text{Weight of oxygen / weight of dry fuel}}{\text{(Weight of oxygen / weight of dry fuel) stoichiometric}}$$

Table 5 Principal reactions in a gasifier

Reaction type	Reactions	Enthalpy (kJ/mol)
Devolatilization	C + heat $\rightarrow$ CH_4+condensable hydrocarbons + char	
Steam-carbon	C + H_2O+ heat $\rightarrow$ CO+H_2	131.4
Reverse Boudouard	C + CO_2 + heat $\rightarrow$ 2CO	172.6
Oxidation	C + O_2 $\rightarrow$ CO_2 + heat	–393.8
Hydrogasification	C+ $2H_2$ $\rightarrow$ CH_4 + heat	–74.9
Water gas shift	H_2O + CO $\rightarrow$ H_2 + CO_2 + heat	–41.2
Methanation	$3H_2$ + CO $\rightarrow$ CH_4 + H_2O + heat	–206.3
	$4H_2$ + CO_2 $\rightarrow$ CH_4 + $2H_2O$ + heat	–165.1

The denominator in the equation is the oxygen required for the complete combustion of the fuel, and it varies from fuel to fuel. The mole fraction (or volume fraction) of the various components of the producer gas as a function of the ER is shown in Figure 1. It is generally observed that for effective gasification, the ER should be in the range of 0.2–0.4. If the ER value is below 0.2, pyrolysis predominates the process and if the ER value is above 0.4, combustion predominates (Kaupp and Goss 1984).

In a real gasifier, however, the composition of the gas is far from the equilibrium value and is generally dependent on the gasifier design.

The gasification process is mainly influenced by two parameters: properties of the biomass and the gasifier design. Biomass properties such as energy content, density, moisture content, volatile matter, fixed carbon and ash content, and the particle size and geometry of biomass affect the gasification process. While designing a gasifier, the design of the oxidation zone is most important, as the quality and quantity of the producer gas depend on it. Complete conversion of the biomass depends on the residence time of biomass in the oxidation and reduction zones. Size, geometry, location, and number of air tuyers are the critical parameters for the design of gasifiers.

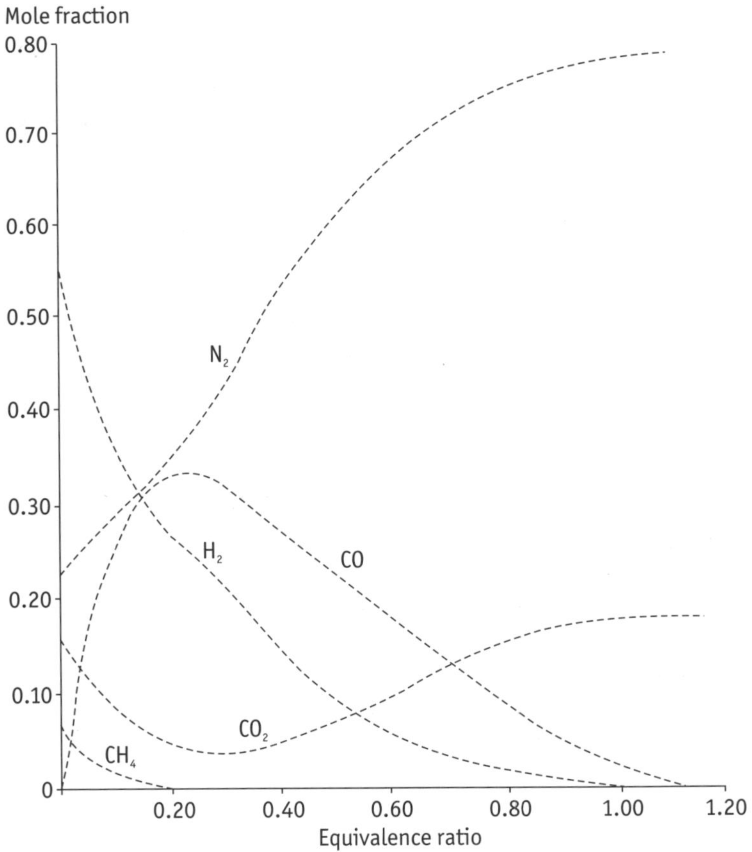

Figure 1 Effect of air–fuel ratio on gas composition

Types of gasifier

According to the flow of gas/air introduced into the gasifier, gasifiers can be classified as follows.

- Updraft gasifier
- Downdraft gasifier
- Cross-draft gasifier

Updraft gasifier

In the counter-current moving bed reactor, also called the updraft gasifier, the air flows counter to the downward fuel flow and enters into the gasifier from below the grate and flows in the upward direction within the gasifier (Figure 2). An updraft gasifier has distinctly defined zones for partial combustion, reduction, and distillation/devolatilization. The gas produced in the reduction zone leaves the gasifier reactor together with the pyrolysis products and the steam from the drying zone. The resulting combustible producer gas is rich in hydrocarbons (tars) and, therefore, has a relatively higher calorific value.

Downdraft gasifier

The most common gasifier is the downdraft gasifier, or co-current type. In this gasifier, the pyrolysis zone is above the combustion zone, and the reduction zone is below the combustion zone. Fuel is fed at the top. The flow of air is

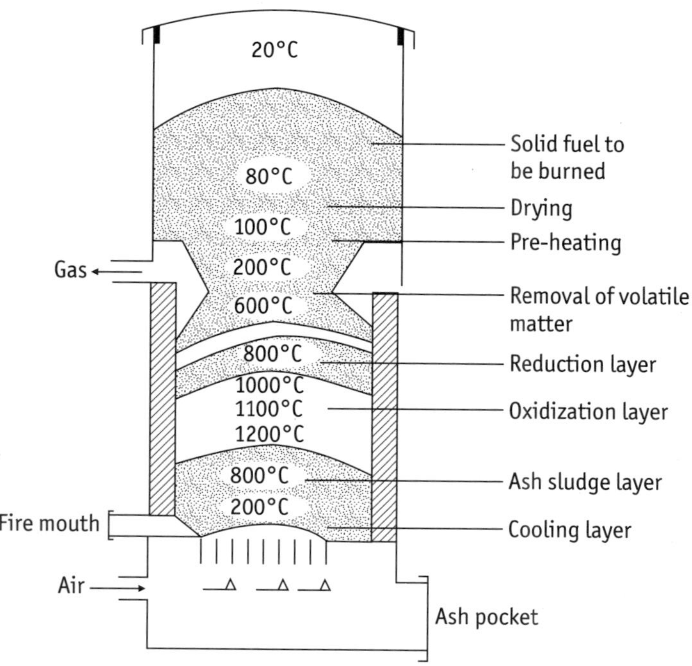

Figure 2 An updraft gasifier

downward through the combustion and reduction zones (Figure 3). The term co-current refers to the fact that the movement of air is in the same direction as that of the fuel.

The essential characteristics of the downdraft gasifier is that it is designed so that the tars given off in the pyrolysis zone travel through the combustion zone, where they will be broken down or burned (provided the gasifier is working properly). As a result, energy is released, and the mixture of gases in the exit stream is relatively clean. The arrangement of combustion zone is thus a critical element in the downdraft gasifier.

Cross-draft gasifier

In a cross-draft gasifier, air enters from one side of the gasifier reactor and leaves from the other side (Figure 4). Cross-draft gasifiers have very few applications and have a distinct advantage other than good permeability of the bed.

Biomass-gasifier-based thermal systems

A biomass gasifier for heat application is simple in construction. It consists of a biomass gasifier, a blower, and a burner with a furnace. For heat applications, any of the aforementioned three types of biomass gasifiers can be used. The producer gas released from the gasifier can be efficiently used with a good degree of control to meet heat demands in ovens/burners, boilers or kilns for thermal applications.

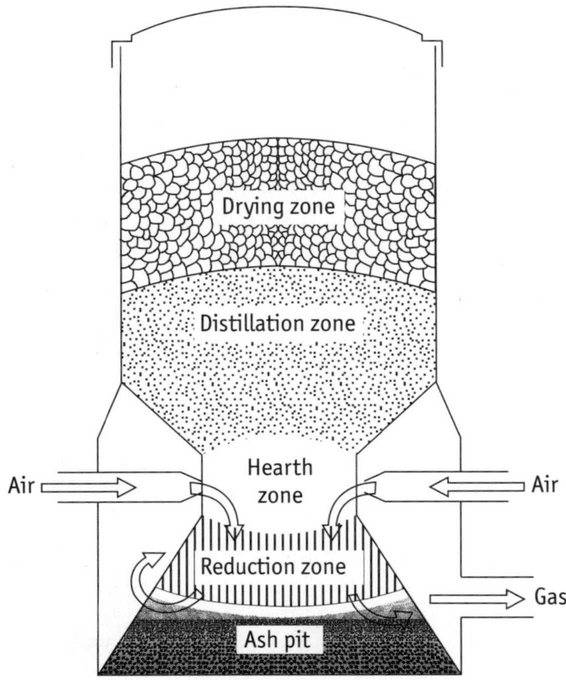

Figure 3 A downdraft gasifier

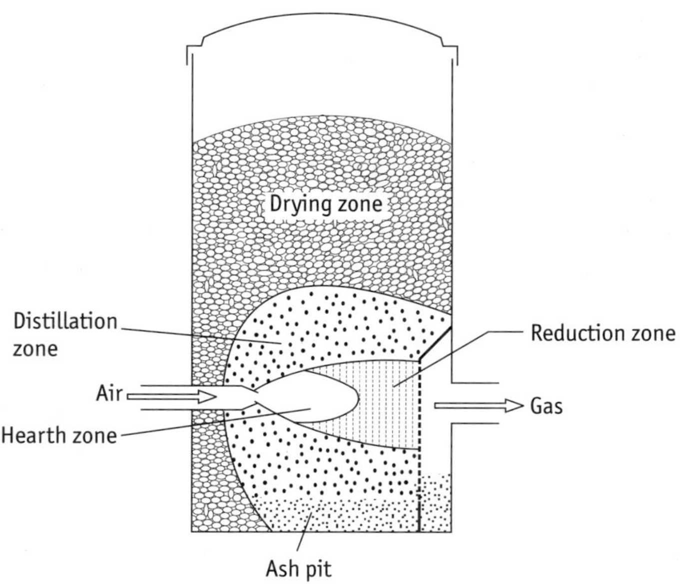

Figure 4 A cross-draft gasifier

Biomass-gasifier-based power generation systems

A general scheme of a biomass-gasifier-based power system explaining the working arrangement is shown in Figure 5. A gasifier-based power generation system consists of a biomass gasifier that converts the solid fuel into combustible gas. The hot combustible gas coming out of the gasifier carries impurities like dust particles and tar vapours, which need to be cleaned in the cleaning train before supplying to the IC engine for power generation. In general, biomass gasifier is coupled to a series of filters such as gravity filters, wet scrubbers, cyclone separators, and bag house filters. A paper filter is used as a safety filter to ensure that the clean gas with permissible levels of tar and dust particulate is supplied to the engine. The gas and air mixture is introduced at the air intake manifold of the engine. Valves are provided both for the air and the gas in order to vary the quantity of gas and air to obtain the optimum mixture.

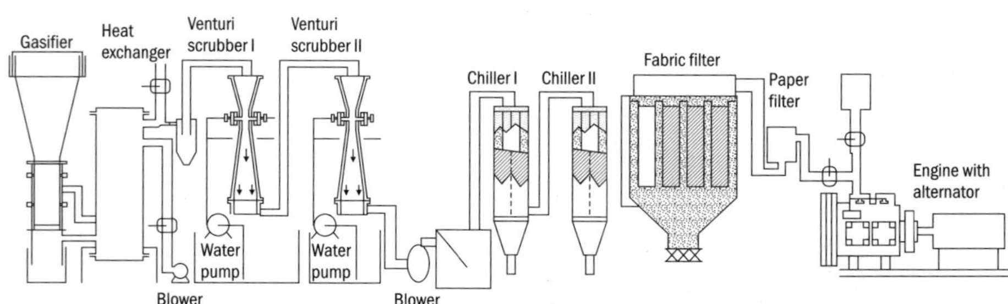

Figure 5 A general scheme for biomass-gasifier-based power system

The gas–air mixture is in turn supplied to a natural gas engine, which converts the chemical energy in the gas to mechanical energy by rotating a shaft. The engine shaft is in turn coupled to the shaft of an alternator that converts the mechanical energy into electrical energy. The electrical energy so produced by the alternator is distributed through electrical conductors to the connected load to power the bulbs, motors, and other electrical appliances.

Review of biomass gasifier development in India

In India, the development of the woody gasifier system was initiated first at the Jyoti Solar Research Institute, Vadodara (presently Sardar Patel Renewable Energy Research Institute [SPRERI], Vallabh Vidyanagar, Gujarat) in the early 1980s for a 5 horsepower (hp) irrigation pumping system. From then, R&D work was initiated by many R&D institutions throughout India. The major institutes that are pursuing active research in this field are IISc, Bengaluru; TERI, New Delhi, IIT, Delhi; Ankur Scientific, Vadodara; and others.

In 1987, the first pilot projects were started. Biomass gasifiers were used in a dual-fuel system for pumping water or for power generation. In dual-fuel systems, the producer gas is combusted together with diesel in a slightly modified diesel engine (Talib, Goss, Flanigan, *et al.* 1989).

In the same year, the Department of Non-conventional Energy Sources (DNES, now the Ministry of New and Renewable Energy [MNRE]) initiated the National Biomass Gasifier Programme. According to the DNES (MNRE), 1370 gasifiers were installed between 1987 and 1993, with a total capacity of 8472 kW, mainly for irrigation purpose. There were different categories: small-scale gasifiers for direct driving pumps or for power generation (3.7–15 kW) and larger gasifiers for power generation (40 kW or 100 kW). The majority of the systems implemented were small-scale gasifier systems that adopted the design of IISc and IIT Delhi. The focus of the programme was on providing small farmers in remote rural areas with an independent power system. The main interest of the government was to achieve a reduction in diesel consumption and an expansion of the centralized power system. IISc researchers evaluated the field performance of the gasifiers based on their technology. Results were disappointing. The gasifiers did not last very long: their average life span was 300 hours.

In 1988, IISc installed a 20-kW gasifier system in Hosahalli village. The project was funded, organized, and monitored by ASTRA, the Application of Science and Technology to Rural Areas centre of IISc. ASTRA provided the technical infrastructure and equipment, instructed the villagers how to cultivate the biomass, and trained the operators. The performance was satisfactory, with an average overall diesel replacement of more than 65% and specific fuelwood consumption of 1.5–2.0 kg/kWh.

Another experiment was the installation of an 80-kW_e dual-fuel gasifier system in Orchoa in 1996. The system provided captive power for a hand-made paper factory. The feedstock for the paper pulp, Ipomea weed, was also used in the gasifier. Technology is being updated continuously and IISc developed open top, twin air entry, re-burn gasifier (2005). There are more than 40 plants successfully operating in India and overseas for heat and power applications.

The recent development in the gas cooling and cleaning system produces dry producer gas, with the tar and particulate level in the range of ppb levels. This has been made possible by Cn patented technology. In this process, hot high efficiency cyclones are used to remove dry particulates from the gas, and ejector scrubbers are used to cool and clean the gas. The gas is de-humidified or dried using the principle of condensate nucleation so as to reduce moisture and the fine contaminants. Also, IISc is working in close association with Cummins India, which is offering its gas engine generator sets for producer gas operation.

A newcomer in the gasifier domain in 1990s was TERI. It focused on thermal applications and identified several promising industrial sectors. TERI implemented thermal gasifiers of various capacities in various industries. These industries included the silk reeling (production of silk yarn) and silk dyeing, cardamom drying, and salt extraction industries. A prototype of a gasifier system was developed and tested in the silk reeling industry in cooperation with practitioners in the mid-1990s. Although the field experiments were successful, the manufacturers did not succeed in attracting customers for the gasifier system.

Another design was developed for the drying of cardamom in Sikkim in 1996. The gasifier replaced the traditional wood ovens with very low thermal efficiencies (5%–15%). The design was improved and tested in the field. Results were promising, and a large-scale diffusion programme was started. Other experiments were more successful. TERI installed an updraft 500 kW_{th} gasifier in the salt extraction industry in 1997. This gasifier system performed so well that in 2003, a second gasifier system was installed. Several other gasifiers are also installed in industries like wax melting, bakery, foundry, khoya making industry, namkeen making industry, and so on. These gasifier systems replaced the traditional wood-fired oven and furnaces that were fuelled by light diesel oil.

TERI also worked for power gasifier, and the R&D efforts at TERI resulted in the development of a downdraft throatless gasifier design with a multi-fuel capability (1996). The absence of throat helps in operating the gasifier with a relatively large fuel size. TERI had also developed a small 7-kVA combined briquetting gasification system using biomass briquettes, made up of sawdust, coconut pith, mustard stalk, pearl millet stalk, and groundnut shell. The system was aimed at operating the irrigation system. After gaining experience from small-scale system, TERI put developmental efforts to evolve a small (10–40 kW_e) gasifier-based power plant for rural and decentralized applications. The evolution of the gas clean-up system went through several stages of design modifications and testing. The cooling–cleaning train includes cyclone, venturi scrubbers, and filters. Capacity extension activities result in the design of

100-kW$_e$ gasifier system. The system is performing satisfactorily and is presently under the process of automization.

TERI is presently working with Danish Technical University (DTU) for the development of two-staged biomass gasification system. The technological efforts should primarily be directed at developing low-tar producing gasifiers, so that maintenance and operational problems are drastically reduced. The two-stage process is characterized by the pyrolysis and gasification processes taking place in separate reactors with an intermediate high-temperature tar-cracking zone. This allows for a very fine control of the process temperatures, resulting in extremely low tar concentrations in the producer gas.

Other institutes such as IISc, IIT Delhi, and SPRERI are also focusing towards the reduction of tar content in the producer gas. Producer gas cooling and cleaning systems designed by CIAE and SPRERI (2008) were tested for long duration for operating a 100% producer gas generator set so as to evaluate their performance.

Thermo-chemical characterization studies of a variety of biomass materials relevant to combustion/gasification were carried out at IIT Delhi, and it was found that arhar stalk, groundnut shell, mustard husk, bagasse, cotton stalk, jute stick, coconut coir, corn cob, bamboo dust, acacia, and pine needle are suitable for gasification (Iyer, Rao, and Grover 2002).

Gasifying rice husk in biomass gasifier is a promising application, and it could become a relatively large source of energy in rural areas. Ankur Scientific Energy Technology (ASCENT) started to experiment with the gasification of rice husk in 1993. ASCENT manufactures two types of gasifiers—WBG (for wood fuels) with 3–500 kW capacity, and FBG (for fine biomass material like rice husk) with 60–300 kW capacity. Recently, they have started developing a 100% gas-engine-based system using Kirloskar gas engines in which liquefied petroleum gas (LPG) is used as an alternative fuel for starting the system to ignite the gasifier (Jain 2001).

Gasifier market is always attractive to the entrepreneurs. Many new entrepreneurs, district rural agencies, and village panchayats have come forward for the development of biomass gasifiers. Grain Processing Works, Kolkata, began with rice husk gasification and now manufactures a wide range of gasifiers, especially rice-husk-based systems, for both thermal as well as power applications. Other major power gasifier manufacturers include Associated Engineering Works, Tanuku; Cosmo Pvt. Ltd, Raipur; and NETPRO Ltd, Bengaluru. Entrepreneurs like DESIpower, Devaria Brothers, Pointec, and so on are actively participating in the installation of gasifier with the help of Bihar Renewable Energy Development Agency (BREDA), West Bengal Renewable Energy Development Agency (WBREDA), Gujarat Energy Development Agency (GEDA), and other state nodal agencies. Apart from these, there are a large number of gasifier manufacturers in India, marketing gasifier systems of different designs and capacities for thermal applications.

Gasifier system for decentralized power generation

It is a well-established fact that access to quality, reliable, and affordable energy is critical for promoting economic and social development in rural areas. The energy scenario in rural India is characterized by low quality of fuels, low efficiency of use, and low reliability of electricity supply and access, leading to lower productivity from the use of land, water, and human effort, resulting in low quality of life and environmental degradation.

As per the Ministry of Power (MoP), Government of India, estimates, about 8% power deficit is prevalent in the country. This has strained the availability of power in rural areas. On the rural electrification front, the Ministry estimates that 74% of villages are electrified, with still 154 230 villages remaining unelectrified. Of these, about 18 000 villages have difficulty in linking with the grid and hence getting electrified by a decentralized source of energy. Even in the so-called grid-linked villages, the quality and availability of power fare poorly. In order to mitigate the power crises, the MoP has launched an ambitious programme called "Power for all" by 2012. The underlying approach of this programme is to strengthen power generation in the country by utilizing locally available resources in a sustainable manner. Another in-built strategy for decentralized power generation is the reduction of transmission and distribution losses. This, therefore, opens up a huge potential for small-scale distributed power generation systems.

Biomass gasification technology holds promise as a decentralized power generation system in terms of improving the quality of life, which includes supply of quality electricity on demand for illumination, providing hygienic drinking water and irrigation water, and possibility for rural micro enterprises.

The supply of generated power to agro-enterprises is likely to enhance the rural energy security. Assured power supply using locally available biomass would ensure smooth operation of agro-industries and increased farm income and better livelihood for the rural people. Primary and secondary processing of agricultural produce at the production catchments would also reduce energy demand and cost of transportation of the produce to the market. Utilization of gas obtained from biomass is environment-friendly as no extra CO_2 is introduced into the atmosphere in contrast to fossil-fuelled combustion systems.

Biomass for power generation has been recognized as an important component of the renewable energy programme in India, and this is reflected in the priority attached to it by the MNRE. There are niches with substantial potential for the use of biomass for power generation: for example, bagasse cogeneration in sugar mills, decentralized gasifier-based diesel cogeneration systems, and so on.

Biomass gasifier programme in India

In India, the biomass gasification programme is governed by the MNRE (previously MNES). Biomass gasifier systems in the capacity range of 1 kW to

1 MW have been successfully implemented in the country. The biomass gasifier programme has twin objective of attaining better quality and favourable economics. Biomass power generation systems, linked to plantation on wastelands, simultaneously address the vital issues of wasteland development, environmental restoration, rural employment generation, and power generation.

The MNRE launched an ambitious research, development, and demonstration programme on biomass gasifiers in the early 1986. The earliest wood-based model of a 3.7-kW gasifier for irrigation purpose was primarily based on the social determinant of providing insurance against reduced diesel availability for millions of diesel pump sets in the country, so that the crops and the farmers' financial inputs would not be lost. Initially, the costs of the gasifier system (including pump) were completely covered by the government. Later, this subsidy was reduced to 40% of the costs of the pump (Mukunda, Dasappa, Paul, *et al.* 1994).

The MNRE is setting up biomass gasifiers in the country under different schemes. These are the Village Energy Security Programme (VESP), Remote Village Electrification (RVE) Programme, and so on. In 2003/04, the RVE Programme was launched for the electrification of those unelectrified remote census villages/hamlets, where grid connectivity was not likely to reach by 2012. During 2005/06, the MoP released the guidelines with respect to village electrification via the decentralized distribution generation (DDG) mode under the ambit of Rajiv Gandhi Grameen Vidyutikaran Yojana (RGGVY), in the Eleventh Plan, that is, 2007–12. The objective is the electrification of about 115 000 unelectrified villages and providing electricity connections to a whopping 23.4 million below the poverty line households. The capital subsidy for the project under the scheme would be routed through Rural Electrification Corporation (REC). As for the ownership of these DDG projects, these would be owned directly by the state government implementation agencies. The RVE Programme was thus suitably modified to align it with the RGGVY and would now be extended only to those villages/hamlets that are not likely to receive grid connectivity under the said RGGVY scheme. The objective of VESP is not only limited to rural electrification but would go beyond electrification by addressing the total energy requirements for cooking electricity and motive power. VESP is currently being taken up through test projects under the RVE Programme.

Under the MNRE's RVE Programme, 2071 census villages and 272 hamlets have been electrified during the Tenth Plan. About 3330 unelectrified census villages and 830 unelectrified hamlets of electrified villages have so far been provided with basic electricity facilities. System installation work is currently going on in 1670 remote villages and 630 remote hamlets. Of the villages taken up for electrification through renewable energy, 296 have been covered through distributed generation technologies of biomass gasification and small hydro. Rest 6431 are being/have been covered under solar photovoltaic technologies.

The various programmes provide financial support to manufacturers and users of biomass gasifiers in India. A package of incentives, which includes fiscal concessions such as accelerated depreciation, concessional custom duty, excise duty exemption, sales tax exemption, income tax exemption on projects for power generation for 10 years, and so on, is available to biomass power projects. The state electricity regulatory commissions (SERCs) of Andhra Pradesh, Karnataka, Maharashtra, and Uttar Pradesh have announced preferential tariff. Higher subsidies are given in high thrust areas like hilly regions and the north-eastern states. Table 6 lists the biomass power/cogeneration projects commissioned, by state.

Problems and research needs

In this era, gasification has evolved as a start-of-the-art technology in India, and it can compete with some of the best technologies in the world. Serious R&D efforts were made only during the last 10 years, achieving some breakthroughs, but a few of the problems faced by gasification power plants still require research.

Biomass availability and pretreatment

One of the major obstacles for industrial applications of biomass gasification is related to scattered distribution of biomass resources and the inferior properties of biomass as compared to coal (for example, lower bulk density and lower ash melting point, potentially causing slagging/corrosions problem in gasification reactor). Biomass production, harvesting, and transportation are labour intensive, as biomass is widely dispersed. There is no rigid fuel

Table 6 List of biomass power/cogeneration projects commissioned, by state (as on 30 November 2008), in MW

State	Total (MW)
Andhra Pradesh	334.25
Chhattisgarh	156.10
Gujarat	0.50
Haryana	6.00
Karnataka	274.28
Madhya Pradesh	1.00
Maharashtra	155.50
Punjab	28.00
Rajasthan	31.30
Tamil Nadu	308.70
Uttar Pradesh	372.50
Total	1677.13

Source MNRE

supply system existing in the sites near to gasification plant or even in the country itself. Besides, the availability, pretreatment (biomass preprocessing and processing), and processing of biomass are essential. These processes need additional facilities and investment. Biomass material usually requires extensive processing to make it compatible with the conversion process. However, the small-scale biomass processing technology with an output of 100 kg/hour or so, suitable for rural areas, is still not well established and is still under R&D.

Gasifier design

The other problem of utilizing biomass for gasification is concerned with operations. For a continuous operation, a continuous flow of gases and biomass is essential. But due to the various unique characteristics of biomass, for example, low ash melting point, the flow gets hampered. For example, during its movement through the various zones of the gasifier, at high temperature, ash tends to melt and forms a molten slag, which subsequently agglomerates and forms huge lumps (clinker) above the grate. Restoring the efficient ash removal systems can solve the problem of mash clinkering.

Bridging is another problem encountered in gasifier design. Bridging or arch formation occurs typically just above the combustion zone. The fuel in the immediate vicinity of the air tuyeres is consumed rapidly, and a hollow space is formed above the air-entry zone. The arch so formed is sufficiently strong to stop the biomass above from falling into the combustion zone. With no fuel supply and only air supply, the gas quality deteriorates rapidly until the production of gas is stopped completely. The factors causing arch formation are many, some of them being (1) shape and size of the fuel, (2) void fraction, (3) geometry of the gasifier, (4) design of the air inlet, and (5) nature of the fuel itself. A systematic study of the flow of biomass inside the gasifier under various operating conditions and for different types of biomass fuels has not been carried out so far, and this should be one of the subjects of future R&D efforts.

Tar and other contaminants

In addition to producer gas, biomass gasification systems also produce some contaminants in the product gases, such as tar, NH_3, H_2S, HCl, and SO_2. Tar is one of the major problems to deal with during gasification. Tar condenses at reduced temperature, thus causing frequent equipment shutdown for maintenance and repair. Physical tar separation involves extraction units, typically bag filters, wet scrubbers, and so on. They are efficient and require little maintenance but entail important filter disposal and effluent treatment expenses.

Reactor design is very crucial for gasification in terms of efficiency, heating value of product gas, and tar formation. The design of low tar gasifier is another major area of research, which needs to be explored. The two-stage gasification has been reported to be most effective in eliminating the tar content. A typically two-stage biomass gasifier for biomass has been studied in the Asian Institute of Technology (AIT). The gasifier at AIT includes two air intakes (primary and secondary air). A different kind of two-stage gasifier system has been designed by the DTU. In this gasifier, pyrolysis product and char from the pyrolysis unit enter from the top of the gasifier unit (the char bed), where the pyro gases are mixed with preheated steam and air, passing through the char bed. The tar content in clean gas is reported to be 15 mg/m^3n. At TERI, efforts to develop a two-stage system are under way. TERI, along with DTU, has developed a two-stage system for low tar content. The performance evaluation of the system is under progress. In the IISc design, the part of the air required for gasification is supplied from the open top gasifier.

Gas cleaning–cooling system

For almost all the gasifier processes, the gas clean-up and conditioning are the largest cost components because of the stringent requirements of the end application units for producer gas. However, as the permissible level of contaminants is not spelt out, the design of cooling and cleaning equipment would remain highly arbitrary at present. Most of the gasification plants adopt wet scrubbing method for the removal of contaminants from the producer gas. Waste water contains acetic acid, phenols, and many other oxygenated compounds that are soluble or insoluble in water. As a result, a large quantity of waste water is generated, which is hazardous in nature. The waste water treating facilities would have occupied a large area and entailed additional cost because of which many users adopted no measures to save and cut down the operational cost. Consequently, many environmental problems have resulted. But as the pollution norms are becoming stringent, there is an urgent need to minimize the water requirement for gas cooling and cleaning. Development of appropriate technology for the treatment of contaminant-laden waste water is one research area, which needs to be focused for the safety of the environment and humans.

Engine operation with 100% producer gas

The most common problem with the application of producer gas in combustion engines is the lifetime of the engine. Engine wear is a result of unwanted components such as tar-dust and corrosive components in the gas. Tar as such does not create abrasive wear but tends to settle down on valves, piston rings, and so on, thereby obstructing the operation. Sometimes tar becomes bituminous, which can cause problems, particularly during the start-up of a cold engine. Dust (soot, carbon, and ash particles) leads to the wearing out of the cylinder wall and piston rings. Sometimes these particles can also come in contact with lubricating oil and damage the bearings. The corrosive components of gas (CO_2, organic acids, and ammonia) do not create problems when the engine is at its working temperature, but during cooling down, condensate can form, resulting in fast corrosive wear of the cylinder and piston rings. There is a need to carry out a systematic study of engine-wear problems and analyse the effect of gas impurities (quality and quantity) on the life of the engine.

A gasifier, when used in conjunction with a diesel engine, is essentially a device for saving diesel, and the dependence on diesel is substantial. Petrol engines, however, can be made to run completely on producer gas, albeit with loss of power. However, engines cannot be started with gas; diesel or petrol must be used for starting. The use of petrol engines is not common in rural areas. Further, petrol is more expensive than diesel. Due to these considerations, it seems highly desirable to develop engines starting on diesel and running on 100% producer gas. This means that engines with the combined features of compression ignition and spark ignition would have to be developed. Efforts in this direction are continuing, but no commercial prototype has been developed so far.

Conclusion

Biomass gasification is one of the most attractive renewable energy technologies in India, and its contribution to overall energy supply is likely to increase rapidly in future. Biomass gasifiers are ideal for decentralized applications of power generation in rural areas, where it is either too expensive to extend the grid or the power demand is low. The MNRE is implementing many programmes such as DDG, RVE, and VESP to promote the rural electrification, and several biomass gasifiers have been implemented in rural areas in recent years.

In spite of all the above facts, the gasification technology is considered as a cumbersome technology. The major reason for the unacceptability of the biomass gasifier as a decentralized power plant is its cumbersome long duration operation, and the inefficient cooling–cleaning system of gasifier results in lower efficiency and the presence of high tar and particulate content in producer gas. The gas cleaning equipment require frequent shutdown for maintenance and capital investment, and maintenance and operation are noticeably costly. Therefore, there is a need to develop low-tar gasifier designs and gas cleaning–cooling systems with minimal waste water generation.

Moreover, there is no organized market for biomass trading, and this is a major barrier in developing biomass-based energy systems on a large scale. A key factor that can influence the future of biomass energy is the development of a market for biomass energy resources and services. Long-term penetration of biomass energy in the industrial and power sectors depends on the cost of delivered energy as well as reliability of technologies. The future of biomass energy lies in its use with modern technologies.

References

Dhingra S, Mande S, Kishore V V N, Joshi V. 1996. ***Briquetting of biomass: status and potential.*** [Regional Wood Energy Development Programme in Asia Report No. 23]. In *Proceedings of the International Workshop on Biomass Briquetting*, pp. 24–30. Bangkok: Food and Agriculture Organization

Iyer P V R, Rao T R, and Grover P D (eds). 2002 ***Biomass: thermo-chemical characterization***, revised 3rd edn [The Gasifier Action Research Project, sponsored by the Ministry of Non-conventional Energy Sources]. New Delhi: Indian Institute of Technology

Jain B C. 2001. **Product development in the context of biomass gasification.** In *Renewables Product and Markets*, pp. 121–124, edited by M Vipradas. New Delhi: The Energy and Resources Institute

Kaupp A and Goss J R. 1984. ***Small-scale Gas Producer Engine System.*** A GATE/ GTZ, Gmbh Publication

Mukunda H S, Dasappa S, Paul P J, Rajan N K S, Shrinivasa U. 1994. **Gasifiers and combustors for biomass: technology and field studies.** *Energy for Sustainable Development* 1 (September 3).

Ravindranath N H, Somashekar H I, Dasappa S, Reddy C N J. 2004. **Sustainable biomass power for rural India: case study of biomass gasifier for village**

Talib A, Goss J R, Flanigan V J, Grover P D, Mathur H B, Durga Prasad M B. 1989. **Development and field testing of small biomass gasifier–engine systems in India: a joint project by an American and Indian team.** *Biomass* 19, 99–122

Vimal O P and Tyagi P D. 1984. ***Energy from Biomass.*** New Delhi: Agricole Publishing Academy

Industrial solid waste

S Pandey

Fellow, Centre for Environmental Studies, The Energy and Resources Institute, New Delhi, India

Introduction

Solid waste generally refers to any garbage, refuse, sludge or other discarded or salvageable materials, including solid, liquid, and semi-solid materials, resulting from industrial, commercial, mining, and agricultural operations as well as from community activities. It does not include solids or dissolved materials in domestic sewage, dissolved or suspended solids in industrial waste water effluent or other common water pollutants. Solid waste management refers to the scientific management of waste after generation, reuse, and recycling, up to treatment and disposal. In this chapter, management of industrial solid waste is discussed with reference to the hazardous waste generated from industrial sources in a manner so as to avoid its adverse impact on human health and the environment.

The World Health Organization (WHO), United Nations Environment Programme (UNEP), and the World Bank have formulated a working definition of hazardous waste as "any waste, excluding domestic and radioactive waste, which, because of its quantity and physical, chemical, and infectious characteristics, can cause a significant hazard to the human health and/or the environment when improperly treated, stored, transported, or disposed."

This definition is now also adopted by the Hazardous Wastes (Management, Handling and Transboundary) Rules, 2008, notified by the Ministry of Environment and Forests, Government of India The definition of hazardous waste as per this rule is as follows.

Hazardous waste means any waste that, by reason of any of its physical, chemical, reactive, toxic, flammable, explosive or corrosive characteristics, causes danger or is likely to cause danger to health or the environment, whether by itself or in combination with other wastes or substances. Such waste includes the following.

1 Wastes listed in Column 3 of Schedule 1 of the Hazardous Waste (Management, Handling and Transboundary) Rules, 2008.
2 Wastes having constituents listed in Schedule 2 of the rules if their concentration is equal to or more than the limit indicated in the said schedule, and wastes listed in Lists "A" and "B" of Schedule 3 (Part A) applicable only in case(s) of import or export of hazardous wastes in accordance with Rules 12, 13 and 14 if they possess any of the hazardous characteristics listed in Part B of Schedule 3.

Rapid urbanization and industrialization in India have resulted in an increased need for proper disposal of industrial waste. There has been a significant increase in the units in industrial sectors, namely, pesticides, drugs and pharmaceuticals, petroleum refining, textiles, dyes, fertilizer, tanneries, paint, chloralkali, and so on, which have a major potential for the generation of hazardous waste.

Hazardous waste from these sectors contains heavy metals, cyanides, pesticides, complex aromatic compounds, and other chemicals that are toxic to humans, plants or animals, and are flammable, corrosive or explosive, or have a high chemical reactivity. Contribution of the industrial sector to hazardous waste is also critical due to the wide geographical spread of industrial units in the country, thereby leading to region-wide impacts. Hazardous waste, when not adequately handled and disposed, can cause immediate short-term public health problems as well as long-term environmental contamination and degradation of natural resources. Proper treatment and disposal of such hazardous waste are much more expensive and complex than dealing with the conventional or non-hazardous waste/residues.

Experience from a number of developed countries suggests that cleaning up of hazardous waste is much more expensive in the long term as compared to its prevention at the source. For instance, in the US, cleaning of improperly managed waste has been estimated to cost 10–100 times as much as proper early management (HWA 2002). It is, therefore, essential for the industrialized states in India to improve the monitoring and enforcement measures, as well as the physical infrastructure needed for their effective implementation while promoting recycling, recovery of materials, and avoidance of generation of hazardous waste.

Industrial growth and solid waste generation

Enhancing the economic growth is imperative for India's progress, and industrial activity is the key. Growth of the industrial sector has quadrupled in the past five decades since Independence. It is important to recognize that while India's economic growth over the past 20 years has been quantified as

163%, pollution has increased by approximately 475%, with the industrial pollution alone accounting for 247% of this pollution.

It is noteworthy that as higher levels of economic growth are achieved, generation of hazardous waste is also increased. As per the Confederation of Indian Industry (CII) estimates, in France, every 1% growth in the gross domestic product (GDP) in the past has resulted in a 3% growth in hazardous waste. Therefore, it becomes important for a country like India to devise an effective strategy for hazardous waste management in the early stages itself so as to tackle the problems of rapid economic growth.

Industrial solid waste generation in India

Industrial solid waste can be divided into two types based on its chemical nature—non-hazardous and hazardous. Major generators of non-hazardous industrial solid waste in India are thermal power stations producing coal ash; mineral exploration activities producing mine tailings; steel mills producing blast furnace slag and steel melting slag; non-ferrous industries such as aluminium, zinc, and copper producing red mud and tailings; sugar industries generating pressmud; pulp and paper industries producing lime sludge; and fertilizer and allied industries producing gypsum. As these waste products are generated in huge quantities—about 200 million tonnes (MT) per annum (Pappu, Saxena, and Asolekar 2007)—their recycle/reuse potential should be explored; otherwise, a huge land area would be needed for their disposal. The quantum of industrial waste produced per annum, along with its origin, is given in Table 1.

The exercise to make an inventory of the hazardous waste generation in different states of India was initiated by the Central Pollution Control Board (CPCB) in 1993. The current information on the total hazardous waste generated from industries and the facilities available for its disposal in the

Table 1 Industrial waste produced in million tonnes per annum (MTPA) in India and its origin

Waste	Quantity (MTPA)	Source/origin
Mine tailings	80	Iron, copper, and zinc mines
Coal combustion residues	112	Thermal power plants
Construction waste	14.5	Construction industries
Stone dust	6	Marble cutting, polishing, and processing
Red mud	5.5	Mining and extraction of alumina from bauxite
Phospho-gypsum	6	Phosphoric acid plant, ammonium phosphate
Steel and blast furnace slag	11	Conversion of pig iron to steel and manufacture of iron

Source Pappu, Saxena, and Asolekar (2007)

respective states in India has been collected by the Ministry of Environment and Forests (MoEF), Government of India, through the respective state pollution control boards (SPCBs). As per the current MoEF (2008) estimates, about 6.23 MT of hazardous waste is generated in the country from 36165 hazardous waste generating units (CPCB 2009). The present generation of hazardous waste (in MTPA) in selected states in India is shown in Figure 1.

The waste stored/stocked in industrial units in the states that still do not have common hazardous waste treatment and disposal facilities is also not part of this waste inventory. In addition, there are about 3 million small-scale units in various sectors all over the country. Also, there are a number of units in the small-scale and unorganized sector, which are handling hazardous waste without pollution control safeguards. They generate hazardous waste in quantities that are not regulated by the rules; however, their cumulative impact cannot be ignored.

Management of industrial waste

An efficient waste management system comprises the management of solid waste from the time it is generated till the final treatment and disposal. The interim stages involve its collection with and without segregation, storage, and transportation. The usual treatment and disposal practices adopted for industrial solid waste comprise the following options (Figure 2).

- Source reduction
- Recycling and reuse
- Thermal treatment
- Landfilling

These options, in the order of preference for implementation, are shown in Figure 3.

Source reduction

Rather than managing waste after it is generated, source reduction is designed to change the way products are made and used in order to minimize waste

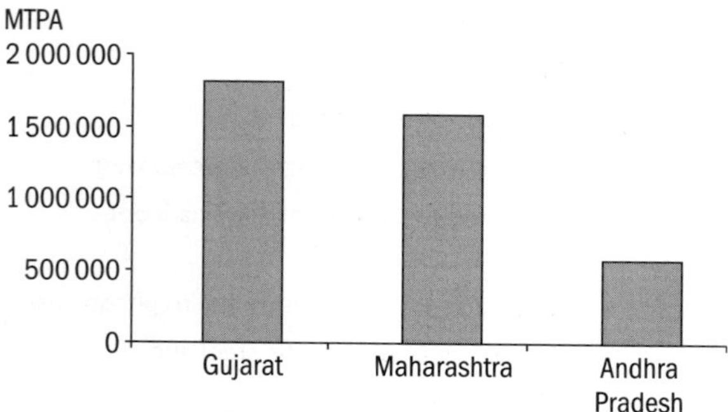

Figure 1 Hazardous waste generated in million tonnes per annum (MTPA) in selected states in India

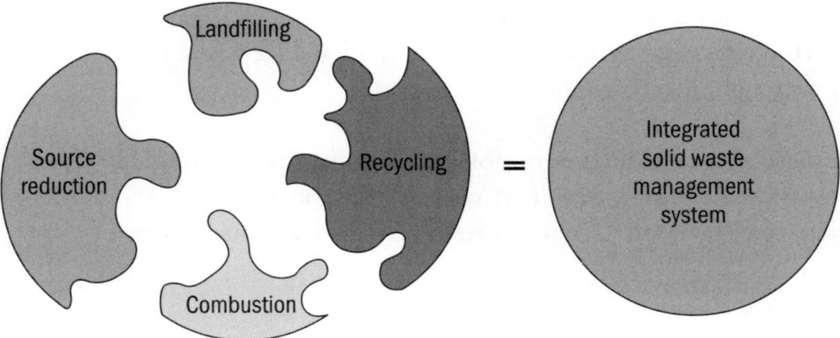

Source reduction, landfilling, recycling, and combustion are all pieces of the solid waste management puzzle. Source reduction and recycling are preferred elements of the system.

Figure 2 Solid waste management hierarchy

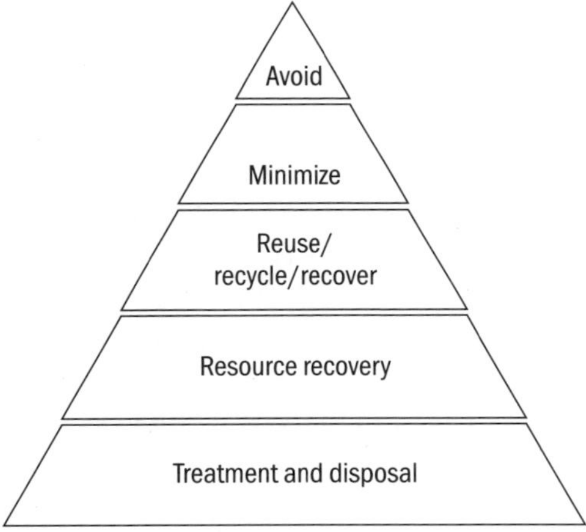

Figure 3 Preferred option for waste management

generation. Source reduction, also called prevention, is defined as the design, manufacture, and use of products in a way so as to reduce the quantity and toxicity of waste produced when products reach the end of their useful lives. The ultimate goal of source reduction is to decrease the amount and toxicity of the waste generated. The steps needed for the successful achievement of the goals of waste minimization begin with an audit. The waste minimization audit consists of planning the entire programme and fully characterizing all types of waste. From there, waste minimization options are identified, evaluated, selected, implemented, monitored, and communicated. Waste minimization can be practised by implementing any of the following options.

- Change in the methods and practices of production
- Input substitution

- Process technology and equipment change
- On-site/off-site recovery, reuse, and recycling, including waste exchange
- Product redesign and reformulation

Though these source reduction options do not directly generate any revenue from waste, they eventually result in cost savings, in terms of avoidance of waste, which, once generated, requires resources for its management.

Reuse and recycle

Packaging reduction is a common manufacturing and retail business opportunity. While many forms of packaging create solid waste, reducing packaging costs can lead to savings in several ways. As packaging does not add to the product value, reduced packaging leads to savings for the manufacturer, in terms of product cost, and also for the consumer, in terms of money. Examples of retail packaging reduction include eliminating plastic bags for undergarments on display, reducing material used in containers of canned goods, eliminating multiple layers of packaging materials, and providing products in bulk or in concentrated forms.

The benefits and opportunities for manufacturers are basically the same as those for retailers. For example, a manufacturer of small auto parts was shipping parts in Gaylord boxes. The boxes were closed and mounted on wooden pallets with steel banding. This created three types of solid waste for customers, and also the company had to continually buy the packaging material. It switched to returnable plastic pallets with folding sides. After an initial investment, the company eliminated the constant cost of buying disposable packaging. In addition, customers were more satisfied when waste was eliminated.

Many materials that are disposed of either as hazardous or non-hazardous waste have the potential to be reclaimed for another application. For instance, contaminated material may be of an adequate quality to serve as a solvent or cleaning material for less sensitive applications. In the printing industry, toluene is used both as a cleaning agent for the presses and as a thinning agent for inks prior to being placed on the presses. If this solvent is used for cleaning a specific ink, the same stock can be used for thinning the same colour ink.

The acid recovery process used by the steel industry provides an example of the reuse of the liquid waste stream. Iron scales are removed by an acid-cleaning process, referred to as pickling. The pickling liquor, which is a hazardous waste, can be used in the neutralization and treatment of distillery waste water.

Office paper constitutes a continuous stream of waste, which can be reduced by reusing the back of the paper to draft material or create notepaper. Boxes can be sent to corrugated manufacturers for recycling, but can often be reused before being recycled. A carton company was created for collecting cardboard boxes, cleaning them, and selling them for reuse. When waste cannot be reused, the possibility of recycling it should be explored. Solvents are the most common among the industrial waste that is recycled. Azeotropic distillation is commonly used to enhance the solvent recovery. It comprises adding a substance (usually steam) to the mixed or contaminated solvent to form an azeotrop with the solvent to be recycled. The azeotrop will have a lower boiling point than the original mixture, and the recovery of the solvent is thus facilitated. Packaged solvent distillation units are available in the US and European markets, and usually consist of simple distillation units applicable to low-boiling point solvents.

When we consider industrial waste, there are essentially two categories that can be profitably reused or recycled. One category is the metal-bearing waste from which precious metals can be recovered, provided the concentration is high enough. The other category is organic waste, which, because of its higher calorific value, if not contaminated with polychlorinated biphenyl (PCB) and chlorine, can be used as a fuel for energy recovery.

Waste oil recycling

Another industrial waste with reportedly high levels of recycling is waste oil. Oil that has become contaminated with hazardous materials may require disposal as a hazardous waste. This oil is currently reclaimed in India either by acid bleaching by the informal sector or by vacuum distillation by the more organized sector. The Hazardous Wastes (Management, Handling and Tranboundary) Rules, 2008, make it mandatory for the re-refining and recycling used/waste oil industry to apply for authorization from the CPCB for waste/used oil recycling/refining as per the approved environmentally sound technologies. As on date, there are 191 such recyclers registered under this programme[1].

As of now, three different techniques are used in India for used/waste oil recycling.

- Acid/clay technique
- Vacuum distillation/clay technique
- Vacuum distillation/hydro-treatment technique

The evaluation of these three techniques for their suitability is given in Table 2.

Waste utilization in cement kilns

In the past, the most traditional methods for managing flammable hazardous waste comprised disposing it in landfills or neutralizing its harmful constituents and properties by burning it in incinerators. Both these waste management techniques have the inherent disadvantages of environmental pollution and economic unviability. A waste management alternative to these

---

[1] <www.cpcb.nic.in>

Table 2 Comparison of three techniques for used/waste oil recycling

Item	Acid/clay	Vacuum distillation/ clay	Vacuum distillation/ hydro
Lube yield[a]	Low	Medium	Medium
Bright stocks[b]	Recovered	Lost	Lost
Utilities[c]	Low	Low	High
Overall energy[d]	High	Low	Medium
Hazardous chemicals[e]	Sulphuric acid	Caustic	Caustic
Waste stream			
Acid sludge	Most	None	None
Oily clay	Most	Some	None
Caustic sludge or spent caustic	None	Some	Some
Process water	Low	Medium	High

[a]The oil yield in the acid/clay process is low because of the losses to the acid sludge. The two distillation processes do not recover the bright stock, and this is reflected in their moderate lube oil recovery.
[b]They are recovered only in the acid/ clay process. This process would be favoured in unusual situation where used oil contains extremely high proportions of bright stock.
[c]These refer to total external energy requirements (power plus fuel).
[d]This is the total external energy (utilities) plus potential energy lost in non-recovered oil.
[e]In the acid/clay process, the operators are exposed to the risk of handling acid and the resulting acid sludge.

conventional methods should include a comprehensive approach in which the quantity of waste originally present is reduced, waste products are recovered and used beneficially, and harmful characteristics of the remaining waste are neutralized. Faced with these stringent requirements, many hazardous waste generators have begun to rely on a waste management technology called co-processing. Co-processing, for example in the cement industry, is defined as "using the cement-manufacturing process to recycle, reuse, or treat waste while simultaneously manufacturing cement in a single combined operation." By employing co-processing, the following two goals are accomplished in one process.

1 An efficient and cost-effective method of waste recycling, treatment, and destruction is realized.
2 Cement, a vital product in the construction and maintenance of the society's infrastructure, is produced.

The use of waste-derived fuels, originally developed in Canada and Europe, in the cement kilns was perfected in the US during the 1970s by Systech Environmental Corporation, Xenia, Ohio. This firm's involvement with co-processing was the subject of a US Environmental Protection Agency (USEPA)-funded study on waste-derived fuels. Conclusions reached at in this study indicated that waste could be effectively eliminated when burned as fuel and that waste fuels burned cleaner than the traditional fossil fuels. In 1986, when Systech became a subsidiary of LaFarge Corporation, a cement and construction materials manufacturer, they were the innovators of the on-site co-processing facility concept. Basically, a co-processing facility combines a treatment storage and disposal facility used to convert flammable hazardous waste into fuel, with a conventional cement production facility. As a consequence, effective generation and disposal of waste-derived fuels are achieved, while safe handling of flammable hazardous waste is not compromised.

Before the actual co-processing is initialized, the flammable hazardous waste that can be beneficially utilized in the cement kilns must be identified. The next step would be to evaluate the waste required to be used as fuel at the cement kiln for its physical and chemical properties. By evaluating the physical, chemical, and thermodynamic properties of waste, specific requirements regarding the energy value and compatibility with the cement-making process can be met. The objective is to attain both an optimum blending of waste with other waste and the most favourable matching of the resulting secondary fuel with its desired application. Once this analysis and acceptance stage is complete, waste is classified and separated. Solid waste is crushed into fine particles with a grinder and then suspended into organic liquid waste. This mixture is then blended with other suitable liquid waste to produce secondary solvent. When production operations are ready to use the fuel, it is polish-filtered and screened to remove any solid material that might adversely affect the valves and nozzles or settle at the bottom of the tanks. The cement manufacturing process is not only the second stage of co-processing but is also an ideal method for recycling waste by recovering its energy value. Much of the waste burned as fuel and recycled in cement-making has high energy content. Such waste is commonly generated during the manufacture or use of everyday goods and services, some of which include the following.

- By-products of pharmaceutical, cosmetic, and electronics manufacture
- Solvents and inks used to print newspaper and other publications
- Solvents used to recycle paper
- Dry-cleaning solvents
- Paint thinners, paint sludge, and paint residues
- Sludge from petroleum industry
- Used motor oil
- Agricultural waste
- Scrap tyres

Generally, these waste products are organic compounds, the constituent elements of which are primarily carbon and hydrogen.

The prerequisites for hazardous wastes to be utilized as supplemental fuels include a certain degree of combustibility and energy content. Recent regulations mandate that hazardous waste fuels possess a minimum energy content of 5000 Btu (British thermal unit) per pound. Hazardous waste generally has an energy content exceeding 10 000 Btu/lb. Acting as a supplemental fuel, this waste replaces significant amounts of traditional fossil fuels such as coal, natural gas, oil, and petroleum coke. Other restrictions govern the chlorine and metal content of the hazardous waste. When chlorinated waste, such as that containing carbon tetrachloride or trichlorobenzene, is burned in a cement kiln, hydrogen chloride (HCl) is produced. The HCl gas reacts with potassium and sodium oxide in the kiln to form alkali salts. These salts then volatilize in the burning zone and condense in the cooler regions of the kiln. Blockages can form in the kiln system, which ultimately hinder the cement-making process if large amounts of alkali salts are formed as a consequence of the waste's high chlorine content. In addition, setting of cement can be negatively impacted if the concentration of certain metals exceeds 0.1%. For the manufactured cement to comply with the performance standards set by the American Society for Testing and Materials (ASTM), control of the metal concentrations is imperative. Metals that are generally present in raw materials or fuel and must be monitored include arsenic, cadmium, chromium, lead, nickel, thallium, and zinc. Furthermore, cement plants avoid burning the PCB waste and do not accept dioxins, pesticides or radioactive waste.

Organic waste blending system

The organic waste blending (OWB) system is designed to handle most types of organic waste generated in industries. The purpose of the system is to process the waste to generate a low-grade fuel for use in cement kilns.

The waste is blended to give a suitable specification controlling the following parameters.

- Calorific value
- Viscosity
- Solids content
- Ash content
- Particle size
- Halogen content
- Metals content

Maceration and blending

The maceration system consists of a small tank equipped with a heavy-duty macerator (sometimes referred to as a "hydropulper"), which has a re-circulation system. The system is equipped with a drum hoist to enable the drums to be decanted into the maceration tank. The maceration tank is made of 316 stainless steel and has a conical bottom. The tank should have a maceration pump underneath the re-circulating tank contents. Blending tanks are made of stainless steel and equipped with pumped (centrifugal) re-circulation systems. A typical layout of the waste-blending system is shown in Figure 4.

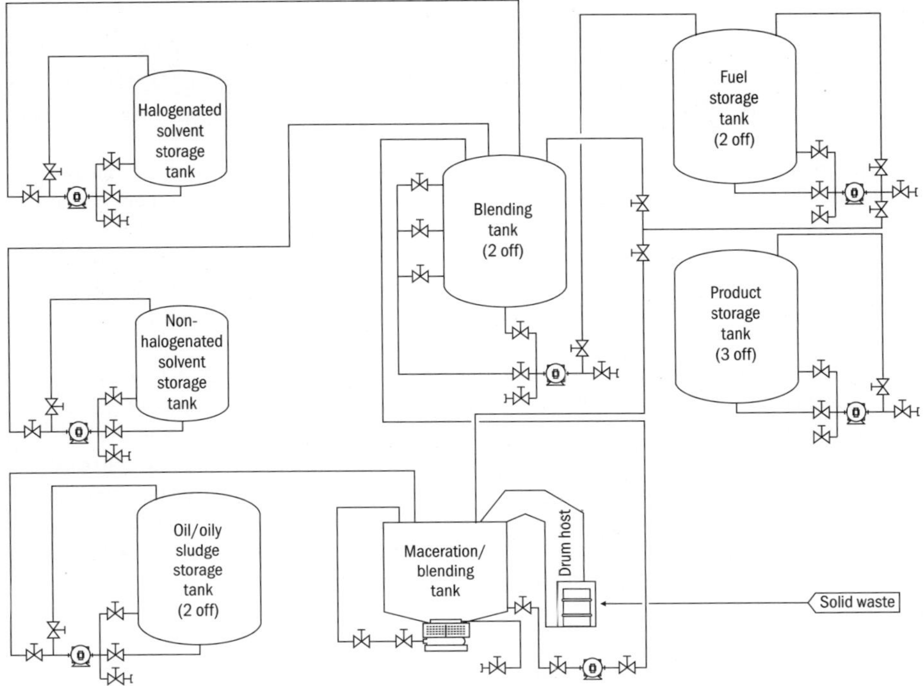

Figure 4 Layout of the waste blender

As of now, the CPCB has carried out trial runs for the co-incineration of high calorific value hazardous waste streams such as effluent treatment plant (ETP) sludge, TDI tar waste, paint sludge, refinery sludge, and tyre chips in cement kilns.

It is reported that a Grasim Cement unit based at Tamil Nadu is using waste stocks such as a mix of de-oiled rice bran, used tyres, and municipal solid waste as alternative fuel, along with coal, for cement manufacturing, thus reducing the emission of CO_2 into the environment (Box 1).

Heavy metal utilization/recovery in industry

Sources of heavy metals

Heavy-metal-containing waste is generated from industries and due to its environmental significance, it is necessary to handle and manage it in a safe and proper manner. The important sources of metal-bearing waste are as follows.

- Metal-finishing industry involving electroplating, polishing, spray painting, vacuum-metalizing, and so on.
- Electronic industry manufacturing printed circuits boards.
- Pickling of copper and its alloys.
- Galvanizing units (hot dip).

Box 1 Co-processing of waste in the manufacture of Portland cement at Grasim Industries Ltd-Cement Division South (GIL-CDS), Tamil Nadu, India

Grasim Industries Ltd-Cement Division South (GIL-CDS), in their cement manufacturing facility, located in the central part of Tamil Nadu in Reddipilayam village, Perambular district, have replaced some portion of the fossil fuels, namely, coal, lignite, and pet coke, used in the kiln for clinker formation with alternative fuels derived from agriculture by-products, mainly de-oiled rice bran (DORB), tyres, and municipal solid waste (MSW). This intervention has resulted in significant saving on non-renewable fossil fuel and net reduction of CO_2 emissions from clinker formation.

GIL-CDS has contacted Trichy Municipal Corporation, which estimated daily capacity of 350 MT for treating MSW. DORB is available in abundance in the vicinity of the plant, and its utilization has helped preventing uncontrolled burning of the MSW in the area.

Tyre is shredded and the tyre chips and agricultural waste are commonly collected in the existing belt conveyor by cross belt conveyor and fired in the calciner. MSW is processed and made into pellets in the processing unit, which is installed in the landfill site. All the alternative fuels are stored in a concrete floor with properly covered shed. Figure A shows the flowchart and the project boundary.

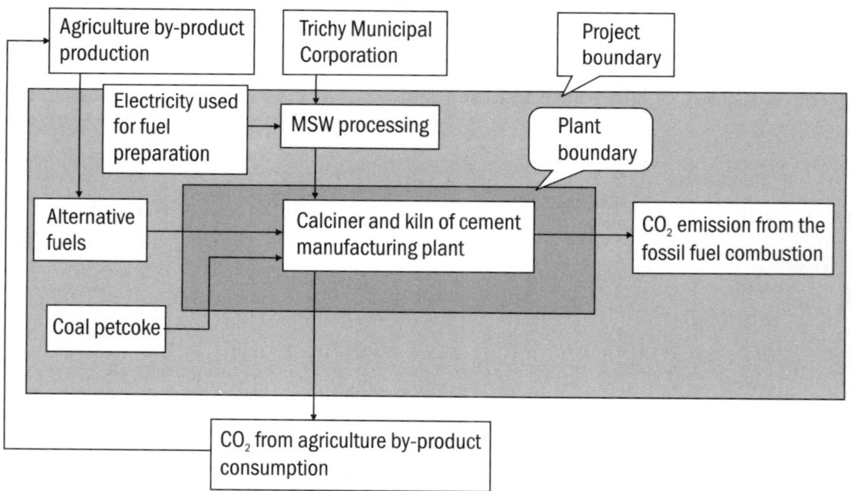

Figure A Process flowchart and the project boundary

Before the implementation of the project, the plant was using 58% coke, 32% petcoke, and 10% lignite. Project inception has helped to significantly reduce the use of these non-renewable energy sources. In 2010/11, the plant proposed to use 5% coal, 61.1% petcoke, 0% lignite, 13.8% DORB, 12.2% tyres, and 6.9% MSW. The project has been registered by the Clean Development Management (CDM) Executive Board in 2006 and promises to save 519319 tonnes of CO_2 in the atmosphere during the crediting period of 10 years.

Source http://cdm.unfccc.int/Projects/DB/DNV-CUK1142971280.51/view; planningcommission.nic.in/aboutus/committee/wrkgrp11/wg11_cement.pdf, last accessed on 18 February 2010

- Anodizing units.
- Units with plating on plastic operations.
- Electrochemical machining units.
- Grinding and polishing operations.
- Units producing and processing metals such as zinc, iron, aluminium, and so on (melting, smelting, and refining).
- Inorganic chemical manufacturing.
- Printing industry.
- Electric arc furnaces.
- Industries using metal salts, catalysts, and so on.
- Tanneries.

Resource recovery options

For metal-bearing waste, the aim should be to reduce the quantity of waste arising from industrial operations and then to examine the opportunities to recover or reuse it.

Recovery of metal from effluents

Two types of processes are involved in the recovery of metals from effluents.

1 Effluent treatment process in which metals are recovered in a more concentrated form.
2 Metal reclamation process in which metals are recovered in a reusable or saleable form from the concentrate produced in the above process or through direct electrolytic recovery.

Following are the various processes available for the recovery of metals from effluents.

- *Ion exchange* It is an extensively used process, the main applications being recovery of precious metals from the effluent, and from the processes wherein conservation of high-quality water is required. Continuous ion exchange processes, eliminating the need for alternate loading, and elution cycles in fixed beds have proved advantageous in the hydrometallurgical processing. Ion exchange works on the principle that the resin gives back an ion in exchange for the ion to be removed and is accomplished by developing either a macro-porous resin or a metal-specific resin. The resin has to be regenerated once it reaches its capacity, and the ions removed come along with the regenerant stream. The regenerant can be treated or reused after concentration, if needed.
- *Electrolytic recovery* Tin, gold, silver, and zinc are recovered in integrated treatment systems by electrolytic methods, and electrolysis is widely used for regenerating copper-pickling solutions. The concentration of metals in rinse waters and process effluents is generally too low for efficient and economic metal recovery, but electrode systems with increasing efficiency are available, allowing wider applications. It is unlikely that the metal-

removal efficiency is enhanced sufficiently so as to permit direct discharge of the effluent, but the method may be useful when water is reused within the plant, for example, in static drag-out tanks. The metal is normally recovered in a reusable form.

- *Evaporation* This process is useful for recovering values by concentration or crystallization from a relatively concentrated solution, particularly where waste heat is available. It has been employed for recycling chrome drag-out solutions, generally using climbing-film evaporators under the reduced pressure.
- *Cementation* Copper, silver, and other noble metals are recovered by reduction with iron or aluminium scrap. It is unlikely that effluents fit for discharge would be produced without further treatment, and water recycling within the plant would not be possible. It is a simple process and might prove attractive to the small operators.
- *Solvent extraction* Although metals can be brought down to a low level by multi-stage solvent-extraction processes, and a concentrated metal strip solution can be obtained, the small aqueous solubility of the extractants leads to organic contamination of effluents and results in high costs due to solvent loss. The solubility of the extractant is reduced markedly in the presence of concentrations of neutral salts, and effluents might be reused within the process where presence of such salts can be tolerated.
- *Adsorption* The term "adsorption" covers the use of active carbon or inorganic materials on which metal ions may be adsorbed in the presence of organic materials, possibly as organo-metallic complexes. Recent developments have included the adsorption of organic complexing or chelating agents on clay-like substrates. The products can be used as highly selective ion exchangers, which can remove heavy metals in the presence of alkali metal or alkaline-earth ions, thus increasing the effective capacity for effluent treatment.
- *Reverse osmosis* In reverse osmosis, high pressure is applied (200 psi, pounds per square inch) on dilute waste to allow water to pass through a membrane but to retain the metals and other solutes. While passing through the membrane, two streams, namely, a concentrated metal solution and a dilute stream, are generated.

Reverse osmosis is especially suited for handling nickel solutions and is a preferred option because of the favourable economics associated with the recovery and reuse of expensive nickel chemicals. The use of ion exchange membranes with the electrolytic driving forces is potentially capable of solution regeneration and metal recovery. The process would be applicable to metal-finishing effluents from chromating, etching, and electroplating. A comparison of the chemical process technologies for the recovery of heavy metal ions is given in Table 3.

Recovery of metals from sludge

The chemically precipitated sludge from effluent plants presents treatment and disposal problems due to its low solid content and mixed composition, but in certain cases, it can provide opportunities for metal recovery. Reclamation of valuable metals from sludge is made difficult and expensive due to the mixed composition and physical form of these wastes. Separation of process streams at the source, together with reliable and efficient de-watering equipment at a moderate cost, would be needed. At present, the intrinsic metal values in sludge are easily exceeded due to the costs of transport and recovery processing, except for a few metals such as gold, silver and, to a lesser degree, copper and nickel.

Table 3 Comparison of chemical process technologies for the recovery of heavy metal ion

Process	Chemical / energy input	Metal reclamation	Major advantages	Major disadvantages
Chemical precipitation	Precipitant; flocculent; acid; base mixing; fluid handling	Metal sludge	Well-established; low effluent concentrations	High chemical requirements; several process elements
Electrolytic recovery	Electric power	Solid metal scrap	Well-established; direct recovery of solid metal; no chemical consumption	Energy intensive; high capital costs; reduced efficiency at dilute concentrations
Ion exchange/ regenerable adsorbents	Regenerant solutions; fluid handling	Concentrated soluble metal stream	Highly selective; effective <100 parts per million (ppm)	Chemical regeneration requirements, adsorbent expense, prone to fouling in mixed waste streams
Disposable adsorbents	Replacement adsorbent;` fluid handling	Metal immobilized on solid adsorbent	Simple metal removal process; low adsorbent costs; effective <100 ppm	Selectivity; recurring cost of new adsorbent; disposal cost of spent adsorbent
Liquid membranes	Extractant for liquid-supported membrane; fluid handling	Concentrated soluble metal stream	Selective; continuous concentrated metal solution recycle	Membrane durability; fouling
Liquid–liquid extraction	Stripping solvent; makeup extraction solvent; fluid handling	Concentrated soluble metal stream	Selective, continuous concentrated metal solution recycle	Capital costs; solvent emissions to air / water, solvent disposal problem

For recovery of metals, the following important factors have to be considered.

- Economic feasibility for adopting recovery rather than disposal.
- Technical feasibility.
- Availability and reliability of suitable techniques to concentrate the low concentrations of valuable metals present in effluents.
- Increasing cost of water and effluent treatment. Sludge generated by chemical precipitation during effluent treatment can be made more suitable for metal recovery in two respects.
  1 By reducing the number of metals present in the sludge.
  2 By increasing the solid content of the sludge.

Except for the sludge produced in the integrated treatment plants, which is reused within the process, there is little evidence of large-scale metal recovery from chemically precipitated effluent sludge. Nickel and copper are, however, recovered by a few metal-finishing companies in the developed countries, and a recovery process based on the sludge and concentrated solutions containing nickel and copper is now being followed. The processing costs are high, and the process can be carried out economically only with relatively high-grade feedstock. Copper sludge produced from pickle effluents and containing as little as 8%–10% copper has been treated in copper blast furnaces, but difficulties have been experienced because of dust formation. Transport and process costs make it necessary to raise the solid content of the sludge to at least 50% for economic recovery, and the availability of cheap or waste heat for sludge drying at the point of origin is desirable. The presence of calcium carbonate in sludge is advantageous. In the absence of calcium carbonate, solid content greater than 70% is necessary for economic operation.

Recycling of lead

Mostly lead is mined as a co-product of zinc; so the price as well as supply and demand are highly affected by the zinc market. The annual demand for lead is nearly 160 000 tonnes. Approximately 3 MT of lead is recycled annually, of which about 75% scrap comes from the waste batteries. As per the CPCB estimates, the present secondary lead refining capacity in the country is about 0.9 MT per annum.

The current secondary lead refining is a batch process based on traditional pyrometallurgical methods. Batch refining is carried out in a hemispherical vessel, usually stirred to mix the reactants. The metal is present in molten form, while the reaction products float out and are recovered from the surface. These systems typically have no pollution control system. However, lead with a purity of 99.99% is recovered in these units. Traditionally, the processing of lead-acid batteries is carried out using pyrometallurgy. A flow diagram of a typical plant is shown in Figure 5 (Boralkar and Biswas 2000). Pollution is controlled by installing cyclone and bag filters. There are about

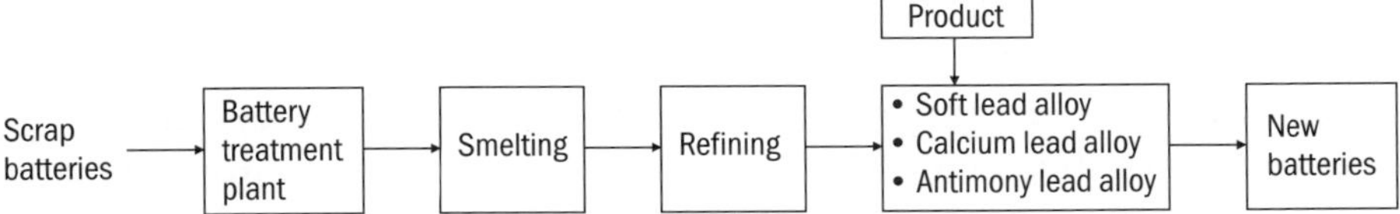

Figure 5 Process flow for secondary lead smelter

270 lead recycling units registered with the CPCB/MoEF as re-processors with environmentally sound technology/processes in the country.

Recycling of zinc

About 15%–20% of the demand for zinc in India is met through secondary production. The current zinc production capacity in India is 0.2 MT per annum in the primary sector and about 0.4 MT per annum in the secondary sector, as per the data on authorized units registered with the CPCB. Most of the secondary zinc units use both the mechanical and electrolytic methods, while some units recover zinc from zinc ash by mechanical methods and sell metal fines (mainly 50%–60% zinc oxide) to zinc chemical manufacturers. As of now, there are about 175 zinc metal and 117 mixed metal (nickel, zinc, copper, and brass) recyclers registered with the CPCB/MoEF.

Electrolytic extraction process for zinc ash

The process of extraction of zinc ash, a raw material, consists of mainly the following stages.

- *Material preparation* Zinc ash is generally available in the form of lumps and chips. It is, therefore, first crushed and pulverized to separate out metallic zinc from the fine ash. On melting and casting, metallic zinc is obtained. Fine ash is subjected to calcination in an oil-fired rotary kiln at 900°C. The rotary kiln is equipped with cyclones, a settling chamber, and a scrubber. Calcined ash lumps are pulverized to get a particle size of 100 mesh.
- *Leaching* The calcined fine ash is then treated with sulphuric acid (H_2SO_4) and/or spent electrolyte generated during electrolysis for the leaching operation. During this operation, compressed air and pyrolusite (MnO_2) are added to oxidize ferrous ions to ferric ions. Leaching of zinc is continued till a pH of 4.5 to 5 is achieved, at which the oxidized impurity of iron is hydrolysed to ferric hydroxide precipitate. The clear solution of zinc sulphate is then sent for purification.
- *Purification* In the first stage, copper is cemented out with the addition of zinc dust/powder. The resultant pulp is filtered to remove copper as copper cement, and the filtrate is taken to second-stage purification. In the second stage, the solution is treated with dimethyl glycol (DMG) to remove the impurity of nickel. In the third stage, activated charcoal is added to remove organic impurities.

- *Electrolysis and melting* The purified solution mixed with spent electrolyte is electrolysed using lead anodes and aluminium cathodes. Zinc metal deposits on the cathode, and oxygen is given off at the anode, and H_2SO_4 is regenerated in the process. The zinc deposit is stripped off manually after every 24 hours and melted in an oil-fired crucible furnace. Molten zinc metal is cast as zinc ingots, and the spent acid is reused in the process.
- *Bleed off* To lower the impurities in the system, some zinc sulphate ($ZnSO_4$) is bled off from time to time. This is used to manufacture $ZnSO_4$ crystals.

Thermal treatment of waste

Exposing waste, especially the organic fraction, to a high temperature in the presence or absence of oxygen results in its partial or complete destruction. An incinerator, which is an enclosed device using controlled flame combustion, is commonly used to thermally break down the waste material. Heat recovery during the incineration process is the major economic consideration if any energy is to be recovered from the waste. The heat recovery system presently available is costly, and there must be a practical use for the recovered heat, preferably at the site of incineration itself. Important factors for the incineration system to work effectively are the three "Ts"—temperature, turbulence, and time (Figure 6). A typical layout of the incineration system is shown in Figure 7. As an alternative, waste-derived fuel, which can be made from waste oils, solvents, and distillation bottoms, by either blending different wastes that have high calorific values or by mixing with coal or fuel oil, can be used in firing boilers, cement kilns or furnaces. The waste thus becomes the supplementary fuel for the kiln or furnace (Figure 8).

Land disposal of waste

Despite the effectiveness of source reduction, reuse, recycling, and waste-treatment options, there will always be some waste that cannot be diverted from the landfills. An engineered landfill is the recommended process for the disposal of hazardous waste (Figure 9). A landfill involves permanent

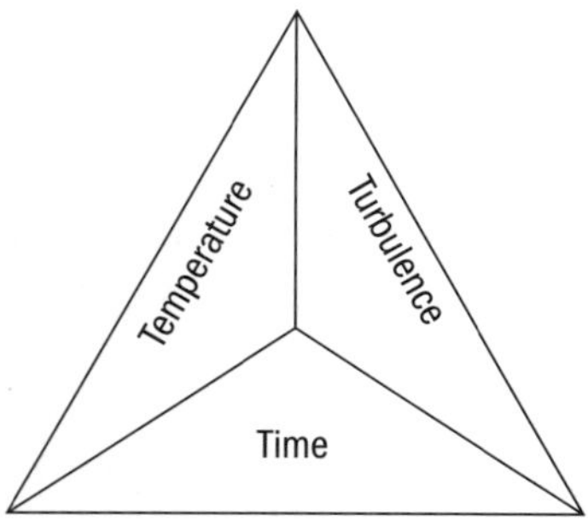

Figure 6 Elements for effective combustion of waste

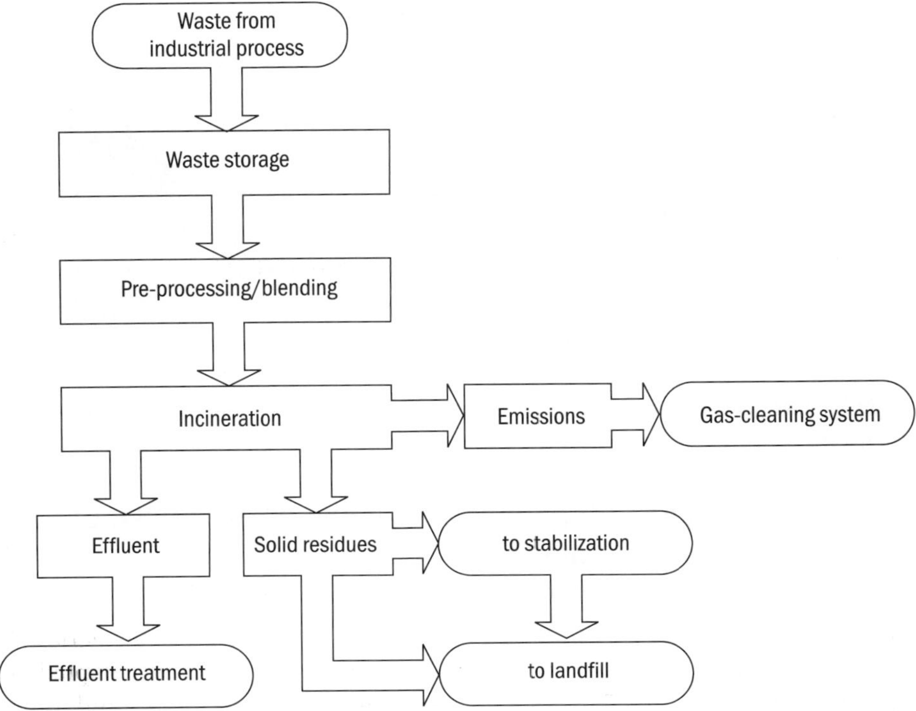

Figure 7 Layout of the incineration system

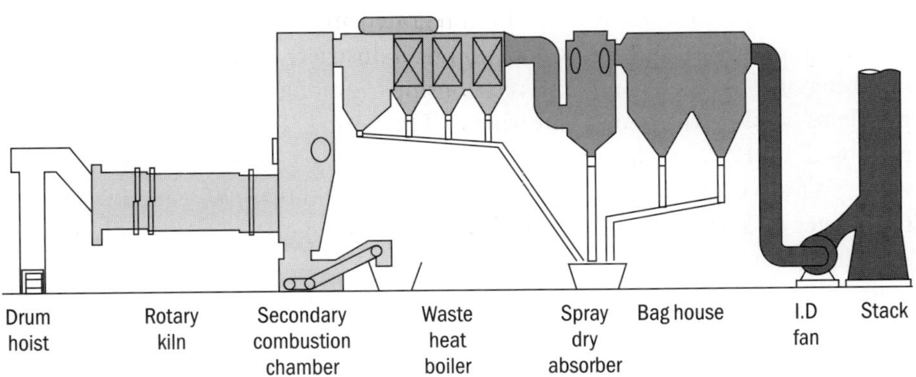

Figure 8 Layout of the kiln

emplacement of hazardous waste. The waste material, with or without stabilization, will have to be disposed of in relatively shallow trenches. A landfill facility has several protective elements: liners to protect groundwater and subsurface soil contamination; leachate collection system to treat the leachate generated during the land filling process; gas collection system to collect and to either flare or use any gas emitted from the landfill; and finally a top cover to provide protection against rain. Landfill disposal provides an attractive alternative in terms of gas collection and utilization when the

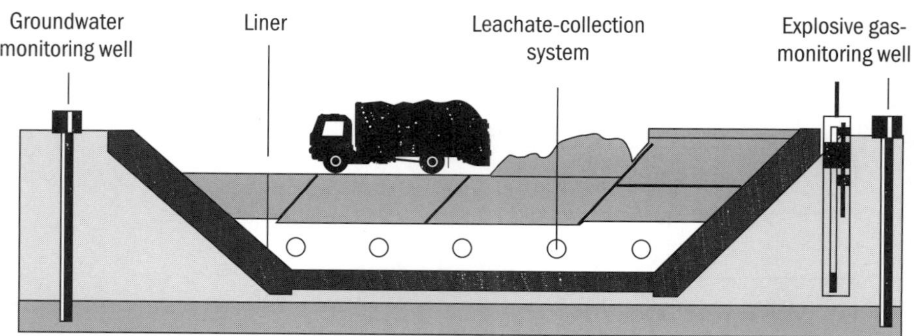

Figure 9 Cross-section of a landfill

organic fraction of the waste is also disposed of, and the energy recovered can be used economically, preferably near the landfill itself.

Industrial ecology and industrial symbiosis: extension of waste-producer responsibility

Industrial ecology is one of the exciting concepts linked to pollution-prevention and waste-minimization initiatives undertaken by industries to improve efficiencies and reduce the waste generated. It is being adopted by industrial estates in many developed and developing economies and refers to a framework within which the industry can preserve and enhance the environment through wiser material use, waste reduction, and pollution prevention. This framework can be integrated in the manufacturing sectors, service industries, symbiotic co-location of industries, comparative experiences in different nations, and in relationships among global environmental problems and performance measures. The term "industrial ecology" was proposed in 1989 in *Scientific American* by Robert Frosch.[2] Frosch's vision was, "Why would not our industrial system behave like an *ecosystem* where the wastes of a species may be *resource* to another species? Why would not the outputs of an industry be the inputs of another, thus reducing the use of raw materials and pollution, and saving on waste treatment?"

Industrial ecology proposes not to see an industrial system (for example, a plant, an eco-region or national or global economy) as being separate from the biosphere, but considers it as a particular case of an ecosystem based on infrastructural capital rather than natural capital. The idea is that if natural systems do not have waste in them, we should model our systems after them to ensure sustainability of the systems. The term "industrial symbiosis" is used to distinguish that subset of industrial ecology, which is concerned with the creation and optimization of resource exchange networks in which "waste" from one industrial operation is utilized as an alternative input for other industrial operations or applications in the community. Industrial symbiosis

[2] <http://www.devensec.com/ ecoreport.html>

deals with the exchange of waste, by-products, and energy among closely situated firms. Because of the many links among the firms, an industrial area is transformed into an industrial ecosystem, or industrial symbiosis. Chertow (2000) provides the following definition.

> *Industrial symbiosis engages traditionally separate industries in a collective approach to competitive advantage involving physical exchange of materials, energy, water, and/or by-products. The keys to industrial symbiosis are collaboration and the synergistic possibilities offered by geographic proximity.*

A historic example of how an industrial ecosystem can develop is found at Kalundborg in Denmark. At the core of this industrial system is the Asnaes Power Station (coal-fired), which provides steam to the Statoil Refinery and Novo Nordisk Pharmaceutical Plant, and Statoil Refinery provides fuel gas and cooling and waste utility water to the Asnaes Power Station. The adjacent Gyproc Wallboard plant utilizes fuel gas from the refinery and scrubber sludge from the power station. Fly ash from the power station is further processed as cement and road aggregate. Waste heat, both from the refinery and the power station, was initially provided to greenhouses, but recently, it is being provided to fish farms and is used for district heating. Finally, sulphur from the refinery is shipped to the Kemira Acid Plant, while Novo Nordisk's pharmaceutical plant provides treated sewerage sludge as fertilizer to the neighbouring farms. The current system is the result of a dynamic evolution over about four decades. The "twinning" between the three key process industries (power station, refinery, and pharmaceutical plant) is generally regarded as the critical success factor at Kalundborg because each provides large volumes of waste streams with valuable by-products on a continuous basis (Ehrenfeld and Nicholas 1997). The exchange of residual products between the companies is presented in Figure 10. In the Indian context, the most notable example of eco-industrial networking is efforts by small-scale industrial units in the Naroda Industrial Estate (NIE) in Gujarat. In the NIE, approximately, 26% of the industries fall under the chemicals category, predominantly dyestuff and dye intermediates. Other types of industries are plastics (5%), pharmaceuticals (3%), pesticides (1%), engineering (24%), textiles (5%), and trading companies (9%). About two-thirds of these industrial units are members of the Naroda Industry Association (NIA). The NIA has set up the Naroda Enviro Projects Ltd (NEPL) as a separate company to operate a landfill site for hazardous solid waste from the estate. The landfill site became operational in 1997. The NEPL has also taken the responsibility for the construction and running of a common effluent treatment plant (CETP), which came online in September 1999.

In the eco-industrial networking exercise, carried out by the NIA and Martin Wilderer, University of Kaiserslautern, Germany, with the support of the CII, the following important waste streams were identified at the NIE, which had high potential use by other units in the area.

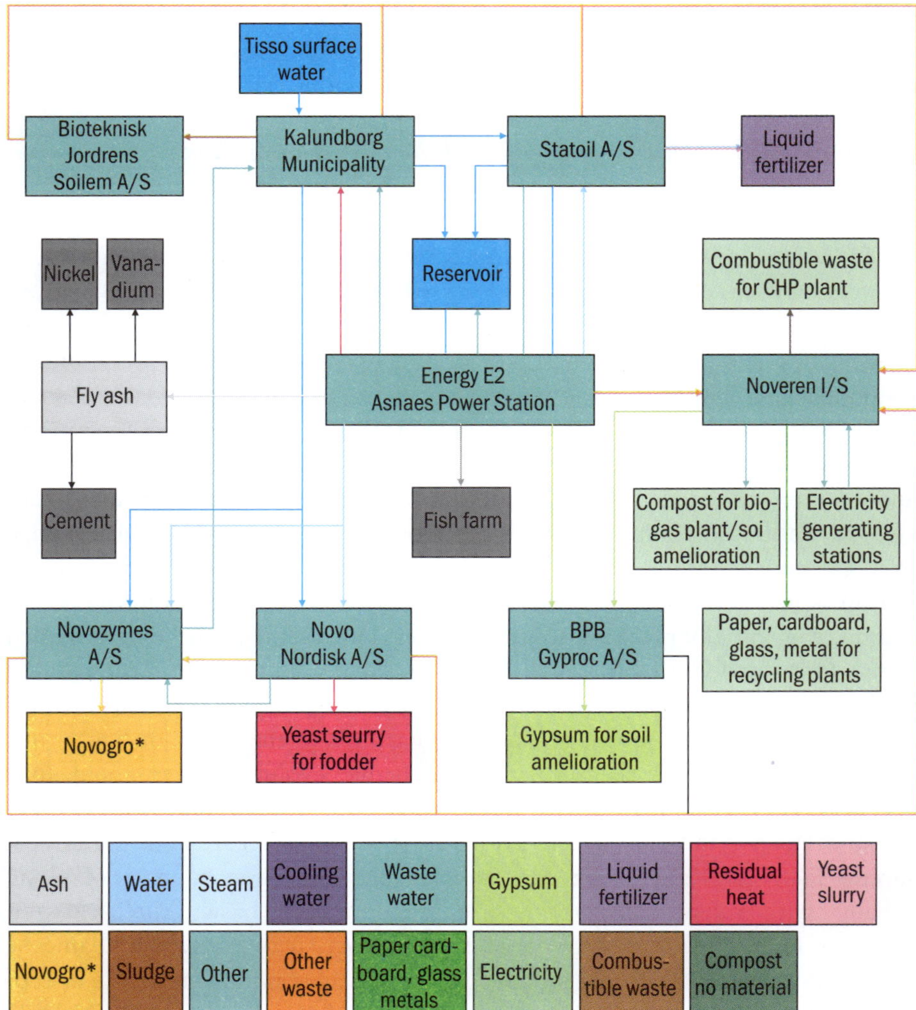

Figure 10 Exchange of residual products between three companies—Asnaes Power Station, Statoil Refinery, and Novo Nordisk Pharmaceutical Plant

- Chemical gypsum
- Biologically degradable waste
- Mild steel scrap
- Spent sulphuric acid
- Iron sludge

Efforts to utilize the above-listed waste material have been initiated by the members of the NIA under their cleaner production initiatives. Other examples of industrial networking from the developed and developing countries are presented in the Annexure.

ANNEXURE

Case studies: industrial symbiosis

Burnside Industrial Park, Halifax, Nova Scotia, Canada

Summary

Burnside Industrial Park is located in the Halifax Regional Municipality, Nova Scotia, Canada. It encompasses an area of 1200 ha, of which three-quarters has been currently developed. There are approximately 1300 companies and 17000 employees in the park, and it is one of the five largest parks in Canada.

Burnside Industrial Park is one of the seven parks operated by the municipality, of a total of 51 in Nova Scotia, and approximately 1000 in Canada. The United Nations Environment Programme (UNEP), Division of Technology, Industry, and Economics (DTIE) reports that there are more than 12000 industrial parks or estates around the world, and some estimates are as high as 20000.

Industrial development and environmental management

Municipalities in Canada, such as the Halifax Regional Municipality, have zoned areas for commercial and industrial development. In some cases, these are advertised as industrial parks. This approach is a means for optimizing infrastructural costs and facilitating industrial development through specific zoning standards. In Canada, the three levels of government, namely, local, provincial, and national, support industrial development through a variety of policies and incentives.

Ensuring protection of the environment (air, water, land) is the responsibility of the three levels of the government in Canada. The primary actors are the federal and provincial governments, which are assigned these responsibilities under the constitution. Some of these responsibilities are assigned to the municipal governments by the provinces. This is reflected in issues such as solid waste management, urban run-off waste water treatment, and nuisances such as noise. Recently, municipalities have begun addressing issues such as pesticide use and the greenhouse gas (GHG) emissions. These responsibilities extend to industrial parks or zones under the control of municipalities.

On the basis of a provincial solid waste/resource management framework, the Halifax Regional Municipality has implemented a solid waste management system, which is widely viewed as the most sophisticated in Canada. This system includes diversion of glass, some plastics, paper, cardboard, and aluminium; organic and compostable waste; and construction and demolition debris from the landfill. The municipality is also a member of the

Partners for Climate Protection Programme of the Federation of Canadian Municipalities, the goal of which is to reduce the GHG emissions by 20% of the 1990 levels. Finally, the municipality has adopted a sewer use by-law to limit the discharge of certain materials into the sewers and a pesticide use by-law, which will, within four years, for aesthetic purposes, ban the use of pesticides within the city. There are federal and provincial requirements, which also apply to industries located within the municipality, such as those in the Burnside Industrial Park. In particular, a relatively new legislation will require companies to adopt pollution-prevention plans and restrict the use of a number of toxic chemicals. Unfortunately, small- and medium-sized companies in Canada do not merit the attention of regulatory agencies unless they fall under certain sectors such as metal plating and printing. The fact that 1300 companies are located in a relatively compact area, with resulting cumulative impacts, should be a cause for greater concern.

Burnside Industrial Park

Burnside is designated primarily for light manufacturing, distribution, and commercial activities. One section of the park is designated as a business park and attracts computer, health, and technology companies. Companies range from local to multinational. The scale of businesses extends from a few employees to 250 people. A majority of the companies located in the park is small businesses. Although not specifically designated as such, another section has attracted many large trucking companies and their maintenance facilities. There is no "worker housing" at present, but there are two hotels in the park and a third is in the pipeline. Housing is an option that is under consideration for park expansion. Table I lists a few sectors represented in Burnside.

Development standards

From the inception of "Burnside as an industrial ecosystem" project, the municipality has strengthened the standards that apply to the park. The following objectives and covenants are given in the standards.

- Protect property values and enhance the investment of businesses located in the park by providing a well-planned and maintained development.
- Create an attractive and efficient business environment through sound land use, planning, and environmental management standards.
- Ensure harmonious relationships among users.

Two of the standards worth noting here are as follows.

...incorporating conservationist principles, preservation, and upgrading of existing topography, vegetation, and cover in conjunction with supplementary planting and treatment of a consistent and compatible nature. Interesting, distinctive, and unique site features and relationships in the existing environment should be maintained.

Table I Sectors represented in Burnside Industrial Park

Accommodation	Distribution
Adhesives	Door and window manufacturing
Air conditioning	Electrical equipment
Automotive repair	Environmental services
Beverage products	Furniture manufacturing
Building materials	Food equipment
Business Centres	Industrial equipment
Business firms	Steel fabrication
Carpeting and flooring	Machine shops
Chemicals processing	Medical equipment
Commercial cleaners	Paint recycling
Clothing manufacturers	Paper/cardboard products
Communications equipment	Printing
Computer assembly and repair	Metal plating
Construction	Refrigeration
Containers and packaging	Transportation
Dairy products	Warehousing

and

...the supplementary planting of native species.

The covenants are intended "to ensure that the park continues to be developed in a manner consistent with superior aesthetic and environmental protection standards and with the declared intention of creating a pleasant and harmonious environment for the park's residents." The requirements apply to architecture, landscaping, signage, protection of natural areas, particularly streams, lakes, and wetlands, and require buffer zones of undisturbed habitat or suitable green spaces around all watercourses.

Environmental management

Businesses in the park must satisfy the federal, provincial, and municipal regulations that apply to them. These include requirements to prevent the discharge of pollutants that may be hazardous to fish, restrictions of specified toxic chemicals such as the PCBs and ozone-depleting substances, and separation of solid waste that is compostable or recyclable for diversion from the landfill. These requirements are enforced through regulation or encouraged through fees such as tipping fees and sewer use fees.

Additionally, companies located in Burnside must abide by the particular covenants and standards intended for the industrial park. All these requirements, whether regulations or economic instruments, along with the increasing environmental awareness of people in Canada, have created an atmosphere conducive to enhanced networking and application of new environmental management strategies based on the industrial ecology.

Burnside as an industrial ecosystem

Largely because of its diversity, the park was selected as a working laboratory for assessing strategies that would create interrelationships between companies as well as between companies and their environment. In an initial study, a research team identified a number of strategies, guidelines, potential symbiotic relationships, and support systems that would facilitate the transformation of Burnside. Box I lists these support systems, while Box II lists the strategies that have been the focus of our continuing work.

Box I Support systems for designing, operating, and transforming parks, estates, or clusters into eco-industrial systems

- Information clearing house
- Materials exchange
- Environmental audit capability
- Educational and training programmes
- Applied research programme
- Enforcement of standards and regulations

Source Côté, Raymond, Ellison, *et al.* (1994)

Box II Strategies for designing, operating, and transforming industrial parks into eco-industrial systems

- Site selection
- Site design
- Building design and construction
- Conservation of energy resources
- Substitution of materials
- Using environment-friendly products
- Environmental reviews or audits
- New product development
- Packaging waste reuse and reduction
- Standardization
- Inventory control
- Co-location of businesses
- Material cycling and exchanges
- Information collection and management
- Encouraging "scavengers and decomposers"
- Cascading materials
- Feedback and communication

Source Côté, Raymond, Ellison, *et al.* (1994)

Support systems

All functions identified have been put in place, one way or the other. The Eco-efficiency Centre is primarily an information clearing house and a networking mechanism. Originally, creation of a computer-based information and decision-support system was envisaged, but it became clear that a "hands-on" approach was required, at least during the first few years. The Centre also conducts environmental reviews and encourages companies to join an eco-business programme, adopting an environmental code or policy, setting objectives and targets, and competing for reduction or conservation awards. Recently, the Centre has begun testing the efficient entrepreneur calendar and assistant, developed jointly by the Wuppertal Institute and UNEP–DTIE to encourage companies to track their performance. Both the Dalhousie University and the Eco-efficiency Centre act as educators in this endeavour. For the past seven years, they have written "The Burnside Ecosystem" column in the monthly park newspaper. The Centre prepares and publishes a series of fact sheets on various generic and sector-specific topics, which are distributed to businesses in the park. In addition, there is a just concluded agreement with the Metropolitan Chamber of Commerce, which will result in the publication of three or four articles per year in their monthly magazine. Under the supervision of professors, students undertake projects in collaboration with companies in the park. For example, during the past five years, papers on implications of environmental management systems with a gap analysis have been written and presented to 25 businesses in the park.

The Centre encourages material exchange and symbiotic relationships between companies and supports the Clean Nova Scotia project in the province-wide waste material exchange. There is a possibility that the Centre will be taking up this function during the next few months. Finally, the project and the Centre encourage professors and students to collaborate with companies or sectors on applied research projects. One such study involves the integration of an environmental management system as the natural step into furniture manufacturing owned by a multinational corporation. Another project is a demonstration of an engineered wetland for the treatment of landfill leachate and run-off.

Strategies

The Centre and the university have continued to promote and investigate the strategies identified at the beginning of the project. Companies are encouraged to substitute chemicals and use environmentally appropriate cleaning products. At least one company has established, and is now expanding, a manufacturing facility in the park. The Centre conducts environmental reviews of companies and encourages formal auditing, which can be undertaken by environmental consulting firms located in the park. Packaging of various types and configurations forms the largest quantity of waste materials in the park.

The Centre is encouraging the reuse of as much of this packaging as possible, along with the recovery and recycling of the remainder of this packaging.

Two key components of a mature natural ecosystem are scavengers and decomposers. Arguably, they are the most important groups of species in an ecosystem, as they recover and cycle nutrients within the system. The educational and informational efforts of the Centre, along with the regulatory framework and the economic instruments, have resulted in the proliferation of scavengers and decomposers, as might be seen in a maturing ecosystem (Box III).

The future

Burnside is composed of old areas, which are now being redeveloped (brownfield), sections which are relatively new, and areas yet to be developed (greenfield). Companies come and go, and in some years, as many as 10% of the businesses in the park will undergo some change: setting up in new premises, closing, opening, and expanding. Therefore, there are continuing opportunities to investigate the application of strategies in the transformation of the park into an industrial ecosystem.

In addition, the project and the Centre will continue to foster additional networking among companies within the sectors, within mall units, and with common issues or materials. One issue that can be explored for networking is transportation. Car-pooling and public transportation must be encouraged to improve efficiency and reduce the environmental impact. As listed in Box IV, there have been some successes in the transformation of the park, but the challenges to the evolution of the Burnside Industrial Park continue: limited

Box III Examples of "scavengers" and "decomposers" species, which have evolved or migrated into the park

Recover	**Rental**
Fine paper	Construction equipment
Cardboard	Tools
Glass bottles	Pallets
Metals	Communications equipment
Batteries	Photocopiers
Chemicals	Uniforms
Re-manufacturing	**Repair**
Toner cartridges	Computers
Furniture	Electronic equipment
Printer ribbons	Trucks and cars
Tyre re-treading	Furniture
Computers	Buildings
Automotive parts	
Reuse	**Recycling**
Building materials	Paint
Tools	Waste oils
Packaging	Solvent

Box IV Examples of the application of the concept in Burnside Industrial Park

Environmental design of the park
- Spectacle lake park and trails
- Use of wetlands

Scavengers and decomposers
- Happy Harry's used building materials
- Miller composting facility

Information and education
- Eco-efficiency centre in Burnside
- Burnside news

Pollution control charges
- Landfill bans
- Sewer use by-laws

Cycling of materials and cascading
- Scotia recycling
- Enviro-systems

Environmentally sensitive products
- Bebbington industries
- Swish maintenance

Economic instruments
- Tipping fees
- Water rates

enforcement of current regulations, getting the economic instruments right, expanding partnerships, and maintaining the momentum of a very complex initiative.

Eco-industrial development

Burnside Industrial Park has served as an eco-industrial laboratory for the past nine years. While investigating the application of eco-industrial development strategies, collaboration with the municipality in an attempt to transform the existing infrastructure and companies while influencing the nature of the new infrastructure and industrial operations was done. Collaboration and networking have occurred at different levels: (1) the university and the municipality in the study and application of industrial ecology to an industrial park; (2) the university, the private electric utility company, and governments in the creation of the Eco-efficiency Centre; (3) material exchanges between two or more companies; and (4) establishment of new companies to take advantage of opportunities in reuse, rental, repair, re-manufacturing, and recycling. Two aspects that we want to emphasize in future are supply chain management of materials within the park and partnerships or joint ventures in the recovery of wasted materials. We also wish to collaborate with industry groups and governments in environmental management training, which will result in an increasing awareness and testing of eco-efficiency analysis tools, leading to an increased understanding of benefits.

While an interest in eco-industrial development and a willingness to participate are important for launching projects, the main success factor is a continuing commitment by a group of partners from the government, industry, academia, and community organizations. Eco-industrial developments are not short-term initiatives; without such commitment, projects cannot be sustained.

Industrial networking in Laem Chabang Industrial Estate, Thailand

Summary

The case profile was prepared by the Environmental Enhancement Centre of the Industrial Estate Authority of Thailand (IEAT) for the workshop, and is reproduced here from the final report of the workshop (UNEP/IEAT 1997). During the late 1980s and early 1990s, Thailand enjoyed an annual economic growth rate of more than 8%, as exports changed from agricultural to industrial products. The basis for this shift to industrial manufacturing was a process of industrial decentralization, involving the development of 23 industrial estates spread throughout the country. Of these 23 estates, consisting of 1200 factories, nine are owned by the state, while the remaining are joint ventures.

These industrial estates are covered by the IEAT, which was set up in 1972 by the Government of Thailand. The IEAT is a state enterprise attached to the Ministry of Industry, with a mandate to promote industrial development in a systematic way through the use of zoning combined with environmental protection. One of the goals of the IEAT and the government has been to decentralize industrial activities away from the heavily congested Bangkok area. On the environment side, the IEAT has the responsibility for controlling industrial pollution and environmental management within all the industrial estates. In this respect, it has set up monitoring systems and enforcement mechanisms in all the industrial estates (Homchean 1996). One of the industrial estates covered by the IEAT is the Laem Chabang industrial estate situated about 130 km south-east of Bangkok. It is connected to Bangkok and the other provinces by the main regional highway and the railway. In addition, there is a deep seaport owned by the Port Authority of Thailand, which was constructed as part of the overall development plan. Laem Chabang industrial estate is surrounded by hills on the three sides, with the Gulf of Thailand bordering the fourth side (essentially south/south-west). While there are two oil refineries located in the north of the estate, the area is not merely an industrial centre. Agriculture remains an important part of the local economy, while tourism is extremely important further down the coast in Pattaya.

Laem Chabang Industrial Estate is divided into a general industrial zone (GIZ), with 45 tenant companies, and an export processing zone (EPZ) with 51 companies. The companies are, in general, concerned with light manufacturing—electronics, automotive parts, electrical appliances, and metals and plastics.

A central waste water treatment plant is provided (as is the case with all estates in Thailand), which is operated by an outside contractor under a mandate from the IEAT. A consultancy firm monitors the effluent and ambient environmental quality at the estate. The cost of treatment to the companies is based on the biochemical oxygen demand (BOD) and the volume of waste water. If companies generate waste water that contains chemicals such as heavy metals, they are obliged to have, in addition, their own in-house pretreatment facilities.

The Laem Chabang Industrial Estate has access to its own incinerator, although at the time of the study (1997), it was not functioning, and each company had outside contractors to remove the non-hazardous waste. Hazardous waste, on the other hand, is disposed of by a company that is partly owned by the government. An Occupational Health and Safety Centre has been established within the industrial estate as part of a programme sponsored by the government. The goal of this programme is four-fold.

1 Provide information and demonstrations on industrial safety.
2 Carry out training on health and safety issues for employees within the estates.
3 Provide facilities for health examinations for workers.
4 Carry out regular examination of the conditions in the workplace.

Laem Chabang Industrial Estate profile

The Laem Chabang Industrial Estate is situated on the eastern coast of Thailand in Sriracha district, Chonburi Province, about 130 km south-east of Bangkok. The estate was established as part of the Eastern Seaboard Development Programme, which was started in 1981 to promote industry-led economic development in the eastern coastal region.

Estate characteristics

The Laem Chabang Industrial Estate is divided into a GIZ of 370 ha and an EPZ of 176 ha. The site also contains commercial and residential areas. Almost 100 industries are tenants at the estate, of which 45 are located in the GIZ and 51 in the EPZ. Tenants at the estate are mostly light manufacturing industries, the major ones dealing with electronics, automotive parts, electrical appliances, metals, and plastics. The partial list of raw materials used by the enterprises at the estate includes plastic resins, food raw materials, textile cord, natural and synthetic rubber, aluminium, stainless steel, copper and zinc ingots, sodium silicate glass, silica sand, feldspar, limestone, gypsum, and more finished intermediate goods.

Infrastructure

The estate has 20 km of concrete roads servicing the industries and the facilities. Electricity is provided through double 115-kV transmission lines to a 115-kV/22-kV power substation constructed at the north-west corner of the estate, and an 85-MW substation. Water is supplied from the nearby Nong Kho Reservoir to an on-site water treatment plant that has a capacity of 27 000 m^3/day, then onto industrial users through a water transmission distribution system. Fire hydrants are provided near road crossings and along the roads.

Environmental situation

Waste water

A central waste water treatment plant with a capacity of 20 500 m³/day is located near the power substation. The plant has a maximum hourly flow rate of 2430 m³ and is designed for the influent/effluent characteristics, shown in Table II.

Table II Influent/effluent characteristics (mg/l)

Parameter	Influent	Effluent
Biochemical oxygen demand	500	20
Suspended solids	250	30
Total suspended solids	170	20
Total Kjeldahl nitrogen	40	<10
Phosphorus	8	<2

The waste water treatment system is operated by a contractor retained by the IEAT. Treatment charges are based on the BOD concentration and the volume of waste water. In the case of waste water contaminated with chemicals or heavy metals, tenant factories are required to have on-site pretreatment facilities. Effluent discharged to the receiving waters by the central waste water treatment plant must comply with the Ministry of Science, Technology and Environment (MoSTE) effluent standards.

Air pollution

Factories are required to install air pollution equipment, and the emissions must conform to MoSTE emission standards.

Solid waste disposal

The estate has an incinerator with a capacity of 8 tonnes/day for disposing of solid, non-hazardous waste.

Hazardous waste treatment centre

The Industrial Works Department has set up a joint venture company with General Environmental Conservation Co. Ltd (GENCO) to establish hazardous waste treatment and disposal centres in four regions to primarily service the industrial sector.

Thailand's existing hazardous waste treatment centre in Bangkhuntien district, Bangkok, and the first landfill in Rajburi Province (west of Bangkok) were operationalized by the Industrial Works Department before the operation of the GENCO began; they are now operated by the company. The first hazardous treatment centre and landfill set up by GENCO is in the IEAT's

Map Ta Phut Industrial Estate, located in Rayong Province, about 60 km from Laem Chabang. It services industries primarily located in the Eastern Seaboard.

Waste oil recovery plant in Eastern Seaboard

Waste oil has been illegally discharged from tankers and freighters polluting and damaging marine resources. The Harbor Department, the government agency responsible for Thailand's water territory, has issued a new regulation that all vessels must discharge their waste oil into the serviced barges before leaving the port. The department also contracts a company to provide such a service. The collected oil is transported to a recovery facility on shore, which might be in the Map Ta Phut Industrial Estate.

Other case studies

Other case studies relating to waste recycling/reuse are given in Table III.

Table III Case studies – waste utilization/recycle

Case study title	Copper recovery from graphite sludge
Community/company	Poco Graphite metal alloy company and copper market
Summary	During the machining of copper-impregnated graphite electrodes, water is used as a cutting fluid, to cool parts and to capture machining residue. The resultant waste water is sent through an evaporator, producing sludge containing about 50% copper. A metals recovery company is purchasing 37 500 pounds per year of the sludge and selling it to overseas customers.
Unique innovation	By-product synergy: pulling copper-laden waste water out of the waste stream and recovering the metal for reuse
Impact	Prior to this synergy, Poco Graphite had been centrifugally separating the waste stream and sending 412 500 gallons/day of waste water into the city's water treatment system. The sludge had been landfilled.
Benefits	• Energy saving through reuse of 18 750 pounds of copper/year. • Avoided 37 500 pounds of toxic sludge/year from being landfilled.
Barriers/challenges	Informational and awareness barriers
Cross reference links	BPS, metal reuse, separation
Replicability	*Potential for replicability is high where metal-laden sludges are produced.*
Case study title	Steel slag reused for production of Portland cement
Community/company	Chaparral Steel and Texas Industries
Summary	Steel slag replaced a portion of lime stream required in the production of high-quality Portland cement, resulting in increased productivity from the cement plant and higher value for slag; significant reductions of air emissions (carbon dioxide and nitrogen oxides).

Contd...

Table III Contd...

Unique innovation	By-product synergy: steel slag in the place of raw lime in cement production
Impact	Estimated value of this synergy operating at one cement plant and one steel mill is $10 million per year.
Benefits	• Reduced disposal costs for steel slag • Reduced raw material cost for cement plant • Reduced global warming emissions (carbon dioxide and nitrogen oxides) from cement plant
Barriers/challenges	Acceptance by the cement industry of such fundamental synergy.
Cross reference links	BPS, steel, cement, carbon dioxide
Replicability	*This process has been patented and is being implemented around the world.*
Case study title	Reuse of whitewater residuals
Community/company	The Dow Chemical Company and an electricity generation company.
Summary	The Dow Chemical Company currently produces approximately 10000 to 15000 gallons per week of ultra-filtration water containing recovered latex solids ("whitewater") from latex processing at a plant in New Jersey. The whitewater, which has high organic content, is stored in tanks and then disposed of at the Publicly Owned Treatment Works (POTW). The whitewater has exhibited film-forming properties that would be useful to a nearby electric utility to control dust associated with fly ash.
Unique innovation	By-product synergy: reuse of spent waste water in place of fresh water for dust control
Impact	Whitewater production at latex plants across country could be reused for dust control
Benefits	• Reduced energy and emissions associated with the reduction in waste water loads to POTW. Potential land use and other benefits associated with the reducing capacity of POTW (that is, potential capacity expansion no longer necessary). • Significant reduction in water consumption. • Reduction in energy and emissions associated with transporting fly ash because the whitewater spray enables each truck to carry more fly ash. • Increased energy and emissions associated with transporting the whitewater from the chemical company to electric utility.
Barriers/challenges	Before the whitewater can be used in this manner, the electric utility must obtain a beneficial use determination from the state.
Cross reference links	BPS, water, whitewater
Replicability	*The replicability of this synergy is high. US power plants produce millions of tonnes of coal fly ash annually. In addition to the electric utility industry, mining, mineral processing, agriculture, and steel and paper industries are faced with controlling or suppressing dust associated with bulk solids. Quantities of whitewater, similar to the quantity of whitewater in this case study, can be expected from companies that are involved in latex processing (for example, paint production); however, the specific whitewater composition may vary from plant to plant.*

Contd...

Table III Contd...

Case study title	Auto shredder residue mined for metal reclamation
Community/company	Texas Industries
Summary	A steel mill employed a large throughput float–sink separation system developed for the food processing industry to extract an additional 15% metals from auto shredder residue (ASR) landfilled nearby for 25 years. ASR is the unused remainder from auto shredding operations for steel reuse.
Unique innovation	By-product synergy: Technology developed to separate ripe from overripe vegetables was successfully used to extract additional metal content from landfilled ASR.
Impact	After adapting the process and building a large separation process adjacent to the steel mill in Midlotheon, Texas, the plant began recovering 15% additional metals from the ASR: aluminium, copper, magnesium, and tin. This stream amounted to 18 000 tonnes annually. These reclaimed metals streams are sold back into the commodity markets replacing virgin metals and saving all related energy costs required to produce them.
Benefits	• Reduced air emissions and energy use associated with metal recovery versus mining. • Reduced ASR disposal and related costs.
Barriers/challenges	Investment in separation technology.
Cross reference links	BPS, ASR, separation
Replicability	*Between 3 million tonnes and 5 million tonnes of ASR is produced in the US annually.*
Case study title	**Reuse of catalysts and slag containing vanadium**
Community/company	Refinery and metal supply companies.
Summary	Transfer of multiple materials containing vanadium at low levels from a refinery in Delaware to a metal alloy company in Ohio. The materials include spent process catalysts and a by-product of a petroleum process termed "gasifier slag", both of which are currently landfilled. The metal alloy company requires vanadium for use as a raw material in its alloy additives production process. The spent catalysts and gasifier slag from the oil refinery have the added benefit of containing carbon, which can be used as an energy source during the alloy additives production process.
Unique innovation	By-product synergy: vanadium by-product of the refining process used in metal alloy production
Impact	Approximately 70% of the US vanadium market is accounted by secondary sources of vanadium, and spent catalysts account for a very small portion of that vanadium supply
Benefits	• Savings for the refinery through the transfer of material for reuse rather than landfilling it. • Increased production at the metal alloy plant. • Reduced air emissions and energy associated with producing the vanadium the metal alloy company would have purchased. • Reduced amount of spent catalyst and gasifier slag landfilled and reduction in the various risks associated with the spent catalyst and the gasifier slag disposal

Contd...

Table III Contd...

	• Reduced amount of energy needed in the metal alloy smelting operation due to the use of catalysts and gasifier slag containing carbon and vanadium.
Barriers/challenges	Some vanadium-containing catalysts may be covered under Resource Conservation and Recovery Act (RCRA).
Cross reference links	BPS, vanadium, spent catalyst
Replicability	*The reuse of vanadium-bearing spent catalyst has the potential for wide-scale applicability. In 1999, 106 refineries produced approximately 34 445 tonnes of spent catalysts, of which 37% were landfilled, 8% were incinerated, and 55% were regenerated. Most of this material contains at least some vanadium.*

References

Boralkar D B and Biswas D. 2000. **Implications of the Basel Convention in Indian context.** Paper presented at the *International conference on hazardous waste management,* January 27, Chennai, organized by Anna University

Chertow M R. 2000. ***Industrial symbiosis: literature and taxonomy.*** *Annual Review of Energy and Environment* 25: 313–337.

Central Pollution Control Board (CPCB) 2009. ***National Inventory of Hazardous Wastes Generating Industries and Hazardous Waste Management in India.*** Delhi: CPCB.

Ehrenfeld J and Nicholas G. 1997. **Industrial ecology in practice: the evolution of interdependence at Kalundborg.** *Journal of Industrial Ecology* **1**(1): 67–79

Homchean K. 1996. *UNEP Industry and Environment Review* **19**(4): 22.

Hwa T J. 2002. **Asian regional study identifies GP as important strategy for hazardous waste Management.** *International Green Productivity Association Newsletter* 4(4).

MoEF (Ministry of Environment and Forests). 2008. **The Hazardous Wastes (Management, Handling and Transboundary) rules.** New Delhi: MoEF, Government of India.

MoF (Ministry of Finance). 2002. ***Annual Report 2001/02.*** New Delhi: MoF, Government of India

Pappu A, Saxena M, and Asolekar S R. 2007. **Solid waste generation in India and their recycling potential in building materials.** *Building and Environment* **42**(6):2311–2320.

UNEP/IEAT (United Nations Environment Programme/Industrial Estate Authority). 1997. Final Report of the *Regional training workshop on environmental management of industrial estates.*

Bibliography

Côté R P, Dalhousie University, and Eco-Efficiency Centre. Case study presented at *International Conference and Workshop on Industrial Park Management.* Manila, April 2001.

Côté R P, Ellison R, Grant J, Hall J, Klynstra P, Martin M, and Wade P. ***Designing and Operating Industrial Parks as Ecosystems Nova Scotia.*** Canada: School for Resource and Environmental Studies, Dalhousie University, Halifax, 1994.

Freeman H M. ed. ***Standard Handbook of Hazardous Waste Treatment and Disposal.*** New York: McGraw-Hill, 1988.

Mohanlal R. **Industrial Symbiosis through Reuse of Industrial Wastes: a case of Naroda Industrial Estate, Ahmedabad.** Masters Dissertation, School of Planning, Centre for Environmental Planning and Technology, Ahmedabad, 2000.

Noronha J. 1999. **Scavengers and Decomposers in an Industrial Park System: a case Study of Burnside Industrial Park.** MES Thesis. Canada: Dalhousie University, Halifax.

Pangotra P. 1999. **The dyestuffs industry in Gujarat.** *Proceedings of a Workshop on Industry and Environment*, Ahmedabad, February 1999.

Patel R, Modi B, Patwari S, Gopichandran R, Wilderer M. 2001. Aspects of the eco-industrial networking exercise at the Naroda Industrial Estate, Ahmedabad, Gujarat, India: foundations, present status and future directions. Presented as a case study at the *International Conference and Workshop on Industrial Park Management.* Manila, April 2001.

Wilderer M Z. 2001. Principles for developing eco-industrial networks. Paper presented at the *International Conference and Workshop on Industrial Park Management.* Manila, April 2001.

Coal-bed methane

M Lavania
Associate Fellow, Microbial Biotechnology
The Energy and Resources Institute, New Delhi, India
and
B Lal
Director, Environmental and Industrial Biotechnology Division
The Energy and Resources Institute, New Delhi, India

Introduction

Methane (CH_4) is a major component (95%) of natural gas. It is a greenhouse gas that remains in the atmosphere for up to 15 years and is also a relatively inexpensive, clean burning fuel. It can be produced by various ways: natural coalification process, when organic matter, such as trees or vegetation, is quickly buried and then heated; farming and ranching activities; and as a by-product of industrial processes.

Coal-bed methane (CBM) refers to CH_4 trapped in coal beds, and it represents a large unconventional source of natural gas (Beaton, Langenberg, and Pana 2006). Commercial CBM production operations have begun in Canada and are expected to become an important source of natural gas, as conventional sources of natural gas are depleting fast. CBM production involves drilling down into a coal seam and depressurizing the seam by either creating fractures or pumping out the water usually associated with a seam, thus allowing CH_4 to be released/adsorbed from the coal and collected.

Stable isotopes, gas dryness (that is, ratio of CH_4 to the sum of ethane and propane), and geochemical analyses in various geological settings indicate that CH_4 within subsurface formations is of biogenic rather than of thermogenic origin (Ahmed and Smith 2001; Flores, Rice, Stricker, *et al.* 2008). Coal is an extremely complex and heterogeneous material, the structure of which changes with its rank (Levine, Schlosberg, and Silbernagel 1982). Bituminous coal contains nuclei of aromatic and naphthenic rings, coupled to one another by bridges of aliphatic chains, and a significant number of hydroxyl groups that contribute to the integrity of the coal structure through electrostatic binding.

Coal comprises 85%–95% (wt/wt dry coal) of organic materials called macerals, which reflect the nature of the precursor plant material and make up the balance, usually as aluminosilicate sand pyrites. The cellulose- and lignin-based plant material in the coal includes oxygen-containing moieties such as carboxyl, hydroxyl, or ketone functional groups that can be fermented by bacteria (Strąpoć, Picardal, Turich, *et al.* 2008). Because methanogens (all of which are archaea) use a narrow range of 1 and 2 carbon substrates, they require bacterial partners to depolymerize and oxidize complex organic molecules, such as those found in coal, into simple fermentation products such as acetate, methanol, formate, and/or CO_2 and H_2. These products can then be used by methanogens to produce CH_4 in deep coal beds. The two common methanogenic pathways are acetoclastic methanogenesis, in which acetate is converted into CH_4 and CO_2, and CO_2-reduction methanogenesis, in which CO_2 is reduced using H_2 to produce CH_4 and water. A recent in-depth study of coal-bed gas in the Powder River Basin (Green, Flanegan, and Gilcrease 2008) determined that methanogenesis has been occurring for the past 65 million years or so, predominantly through CO_2 reduction (Flores, Rice, Stricker, *et al.* 2008). Whether this is still an ongoing process is not clear, yet "younger" gas generation by acetoclastic methanogenesis due to groundwater recharge events has occurred till recent times (before 1952). Formolo, Martini, and Petsch (2008), Ulrich and Bower (2008), and Strąpoć, Picardal, Turich, *et al.* (2008) have all emphasized on the impact of groundwater influxes on coal methanogenesis, bringing in nutrients, certain essential minerals, and microbes. Depending on the stratigraphy as well as the fracture and cleat systems, there may be localized groundwater flow through the seam, creating zones of biodegradation and thus methanogenesis (Formolo, Martini, and Petsch 2008). Thus, the potential exists to exploit indigenous coal-bed microbial consortia to produce CH_4 in real time and enhance the CH_4 yield of CBM wells (Jones, Voytek, Warwick, *et al.* 2008; Scott 1999; Ulrich and Bower 2008).

The rate of methanogenesis in coal beds, however, appears to be decreasing in situ and needs to be increased for CH_4 recovery (Fakoussa and Hofrichter 1999). Alberta Innovates–Technology Futures (formerly the Alberta Research Council) has been investigating processes to enhance microbial activity and thus increase CBM production.

It is important to understand the microbial ecology and methanogenesis processes in coal seams so that in situ conditions can be modified to enhance microbial CBM production to economic levels.

Concerns related to produced water

In a CBM well, water is produced in large volumes, especially in the early stages of production; as the amount of water in the coal decreases, gas production increases (Figures 1 and 2). The water must be disposed of safely.

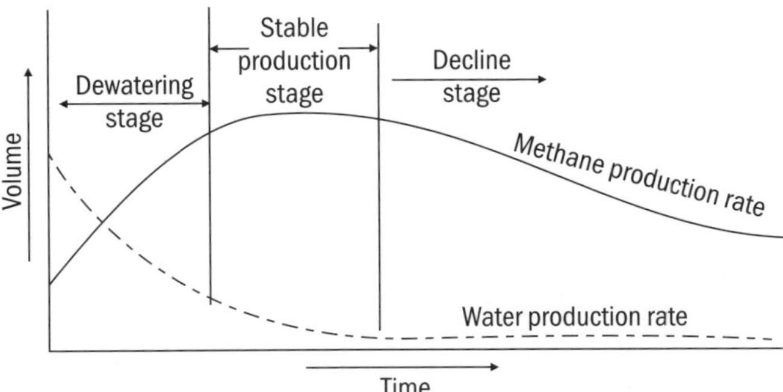

Figure 1 Typical production curves for coal-bed methane well showing relative volumes of methane and water through time, as described by Kuuskraa and Brandenberg (1989)

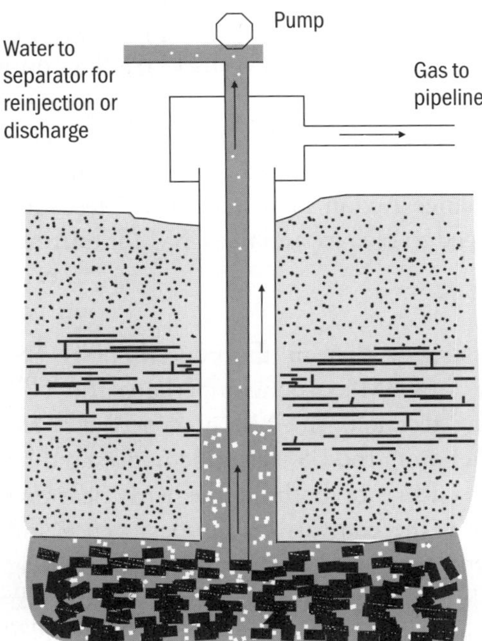

Figure 2 Production scheme of gas and water for a typical coal-bed methane, as described by Rice (1997) in the US Geological Survey

Most frequently, water is reinjected into subsurface rock formations. In some cases, the water is allowed to flow into surficial drainages or is collected in evaporation ponds (Figure 3). In cold regions, it is possible to freeze the water in the winter, collect the salts that separate out, and dispose of or use them independently of the water, which can be discharged.

Figure 3 Production-water evaporation pond, as reported by Rice (1997)

Methane in the atmosphere

CH_4 in the atmosphere has been increasing at a rate of 1% per year over the past 15 years (Rice 1997). In all, 40% of the total CH_4 released in the atmosphere comes from natural systems such as wetlands, swamps, decomposing materials, and forests, 10% from human activities, and the remainder from other activities such as rice cultivation, livestock, landfills, and biomass burning.

Production of CH_4 from coal beds may actually reduce CH_4 emissions to the atmosphere by removing the gas that is otherwise released during coal mining. As part of the President's Climate Change Action Plan, established in 1993, coal-mine operators are being encouraged to recover and use CBM for energy.

International scenario

Conservative estimates (Rice 1997) suggest that in the conterminous US, more than 700 trillion cubic feet (TCF) of CBM exists, with perhaps 100 TCF economically recoverable with existing technology. CBM now accounts for about 7.5% of total natural gas production in the US. Scientific understanding of, and production experience with, CBM are both in the early learning stages. Much is yet to be learned about the controls on the occurrence and recoverability of CBM and about the environmental implications of developing the resource. The CBM industry is still relatively young, and few studies exist on the development and evolution of individual CBM; thus, only few models are available for planning the development of CBM resources on a broader scale. Studies now under way under the aegis of the US Geological Survey (USGS) are designed to develop such models, which will help to further our ability to assess accurately the potential US CBM resource (Figure 4).

Biogenic methane: Indian scenario

India's tremendous economic growth in recent years has intensified the demand for energy sources and has spurred interest in CBM development. India's annual gross domestic product (GDP) has been growing by at least 6% per year since 1991. Two-thirds of India's population still lives in rural areas, with approximately 260 million people living on less than $1 per day.

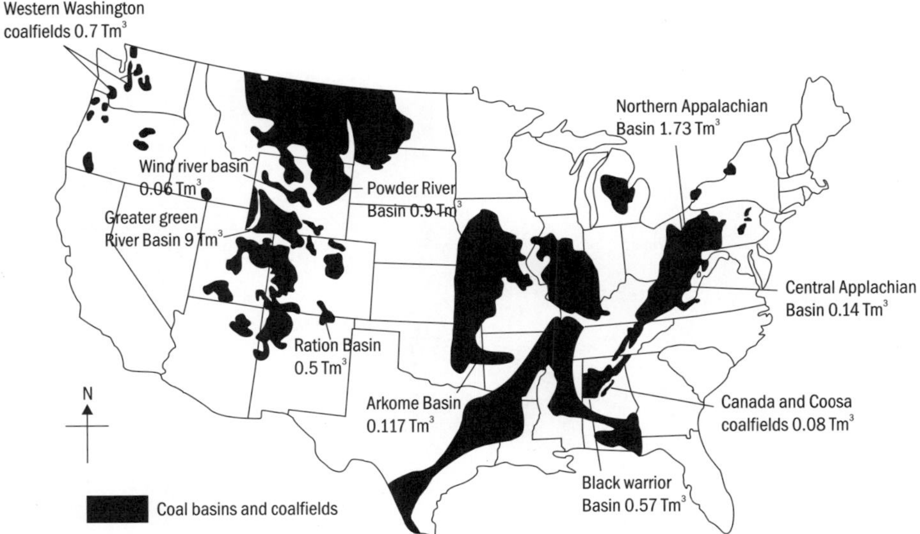

Figure 4 A map showing coal basin scrotal fields in the US and estimates of in-place coal-bed methane resource

Source Bibler, Marshall, and Pilcher (1998); Tyler, Scott, and Kaiser (1997)

Despite India's rapid growth, its GDP per capita is approximately half that of China ($3300 vs $6800). As a result, the Indian government has initiated an economic plan to improve the life of the rural poor while boosting the overall economic performance of the country. India's success and future development will largely be a function of its investment in infrastructure and its access to affordable energy.

India is increasingly turning to unconventional sources of energy to fuel its growth. This is illustrated by the growing interest in Indian CBM. CH_4 is seen as an alternative gas for meeting local gas needs. As gas demand rises sharply in India, on the basis of fuel cost, CBM will be able to compete favourably with imported coal, gas or liquefied natural gas (LNG) for power generation. India's heavy reliance on coal, much of it of low quality, is a major cause of the country's air pollution and relatively high carbon intensity level.

The Indian government has offered 10 blocks for exploration and production of CBM, with the latest round completed on 30 June 2009 (Figure 5), covering an area of approximately 5800 km^2 and estimated to contain 635 billion cubic metres (BCM) of CBM resources.

In India, activities surrounding CBM not only fit into India's strategy to attract investment in infrastructure, but also contribute to country's long-term strategy by providing rural areas with a reliable source of energy for future economic growth.

Production technology

Many ways exist to develop CBM (Figure 6). Production technologies include conventional drilling, drilling before mining, and horizontal drilling.

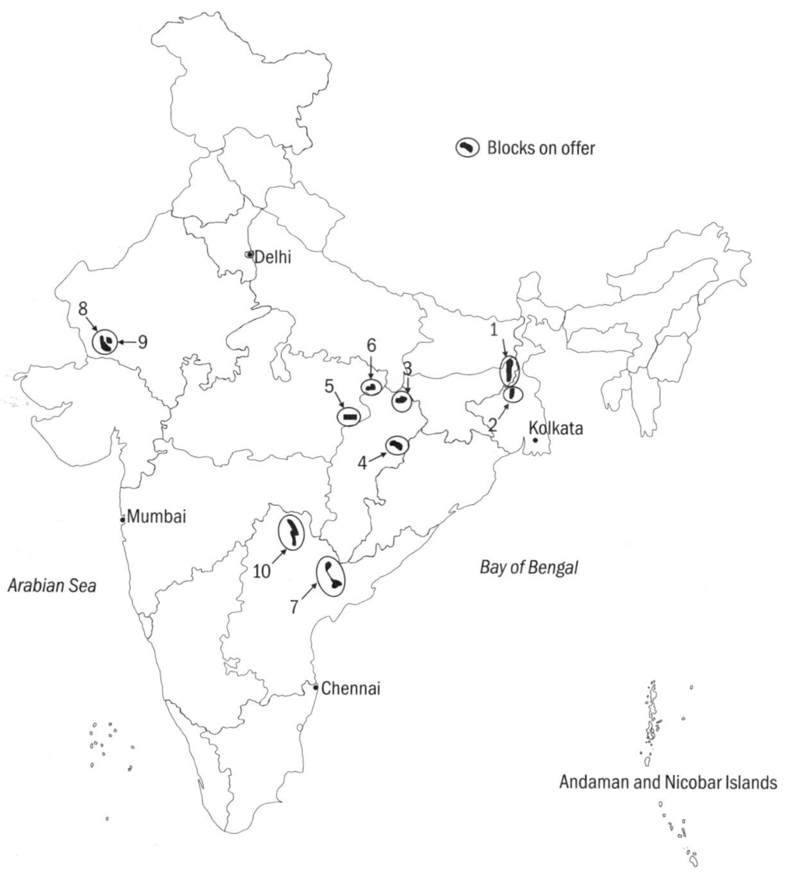

No.	Field name	Block name	State	Area (km^2)	Coal-bed methane resources (BCM)
1	Raj Mahal	RM-CBM-2005/III	Jharkhand	469	158
2	Birbhum	BB-CBM-2005/III	West Bengal	248	50
3	Sohagpur	SP(North)-CBM-2005/III	Madhya Pradesh	609	17
4	Singrauli	SR-CBM-2005	Madhya Pradesh	330	31
5	Tatapani-Ramkola	TR-CBM-2005/III	Chhattisgarh	458	54
6	Mand-Raigarh	MR-CBM-2005/III	Chhattisgarh	634	119
7	Kothagudem	KG(East)-CBM-2005/III	Andhra Pradesh	750	57
8	Barmer	BS(4)-CBM-2005/III	Rajasthan	1168	82
9	Barmer	BS(5)-CBM-2005/III	Rajasthan	739	38
10	Godavari valley	GV(North)-2005/III	Andhra Pradesh	386	30

Figure 5 Coal-bed methane blocks in India.

Sources Data was collected from the website of the Ministry of Petroleum and Natural Gas, Government of India; Ministry of Petroleum and Natural Gas, Government of India

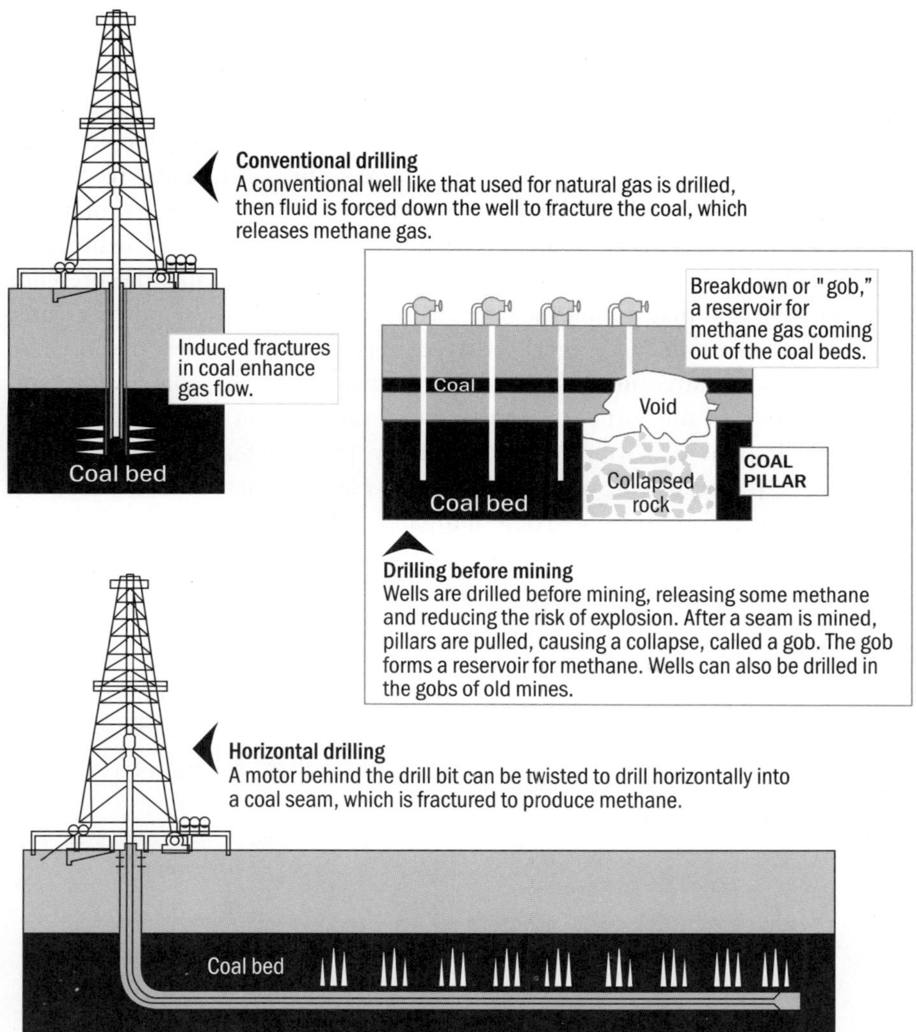

Figure 6 Coal-bed methane can be extracted in different ways, as described by Cobe and Drahovzal (2003)

One method is to simply drill a well to the coal seam, fracture the coal, pump out the water, and connect the well to a gas pipeline. Some success has been achieved when drilling is coordinated with mining of the coal wells, and drilling is done down to the coal bed where some gas is released upon drilling. The coal is then mined out and the roof collapses, creating a large void of collapsed rock that may intersect the coal bed above. Another variation of this post-mining gas extraction is the drilling of gas wells into abandoned underground mines. Many of these old underground mines fill with CH_4 and other gases that blend with the air already in the mines. The gas can be produced, but is generally of lower quality because it is mixed with other gases.

Genesis of coal

For a period in the carboniferous age, lasting about 10–25 million years, lateral compressive forces on the crustal plates caused inward buckling to create concavities of large areas of the earth's surface. The organic sediment was submerged beneath stagnant water and other sediments to forego oxidation. When the driving lateral forces were finally removed, mountains were formed. The resultant folding caused the steep inclines of many coal seams encountered in current CBM exploration. It also explains the presence of higher rank and maturation in some coals that would have required greater temperatures than their current burial depths justify. Besides the organic matter, mineral matter indigenous to the plants enters the coal. Additional mineral matter is air-borne or comes from the erosion of rocks as rock fragments. Inorganic matter to be eventually precipitated also enters with water. Clays, pyrite, calcite, and quartz are thus incorporated in the coal along with other minerals (Table 1). These inorganic substances are integral part of all coals and affect overall chemical and physical properties of the coals.

Table 1 Periods of coal formation, as described by Berkowitz (1979)

	Million of years			Stratigraphic period		
Age	Mean age of coals	Duration of period				Principal species of animals
	—	—			Holocene	Mammalia, man, insects
-1-	<1	1		Quarternary		
	2	11			Pleistocene	
					Pliocene	Domination of Mammalia, fishes, birds
-12						
	15	11			Miocene	
-23-			Cenozoic	Tertiary		
	28	12			Oligocene	
-35-						
-55-	60	20			Eocene	
		15			Paleocene	
-70-						
	120	65		Cretaceous	Upper	Dwindling of Ammonites, Belemnites and Saurja; origin of Mammalia and birds.

Contd...

Table 1 Contd...

Age	Million of years: Mean age of coals	Duration of period	Stratigraphic period			Principal species of animals
					Lower (Weald)	
-135-						
	160	45	Mesozoic	Jurassic	Upper Middle Lower	Domination of Reptilia and Amphibia, (Dinosauria and Ichthyosauria). Mollusca (Ammonites and Bolomnites)
-180-						
	200	40		Triassic	Upper	
					Middle	Crustacea
					Lower	First Reptilia and Mammalia
-220-						
	245	50		Premian		Protozoa Fishes, Mollusca Arthronia (insects) First Amphibia
-270-						
		50			Upper Carbonifer	
	300			Caboniferous		
-320-			Palaeozoic		Lower Carbonifer	
-350-						
	380	50		Devonian		Domination of Coelentrata, Mollusca
-400-						
		90		Silurian		Protozoa First Fishes
-490-						
		110		Gambian		Protozoa, Coelentrata First Mollusca
-600-						
		>1000	Proterozoic	Algonkian		
				Archaean		

Geochemical transformation of biogenic methane

The biogenic CH_4 generated by bacteria in the peatification stage is lost unless burial of the peat is rapid enough and sealing shale lenses are interbedded to form a trap. Later, biogenic CH_4 from other locals may migrate to the developing coal and get adsorbed. Although the largest quantities of CBM are from thermogenic sources, biogenic CH_4 may be retained in quantities of commercial value, as in the thick coal seams of the Fort Union formation of Paleocene Age. These lignite-to-sub-bituminous coals may have a gas content of only 15 SCF/tonne, but seams of the order of 100 feet thick may contain 16 TCF of recoverable CH_4. Therefore, even though the production of biogenic CH_4 occurs from the sub-bituminous coals of the Fort Union formation, it is a combination of thick seams and shallow depths, which makes the production practical.

During the process, CO_2 and water are the volatiles generated. CH_4 generation from bituminous coals becomes appreciable, wherein breakdown of the carbon–carbon bond in the linear-chain components requires high temperature (Figure 7). The rapid generation of CH_4 thereafter, until reaching the anthracite rank, supplies much more gas than can be retained by the coal.

Coal chemistry

Initially and through most of the maturation until the macerals become similar at anthracite, the chemical structure of coal is dependent on depositional environment. The type of vegetation and the chemical constituents of that vegetation provide the starting material in the coalification process that later determines parameters ranging from the amount of gas liberated to the degree of cleating (Figure 8).

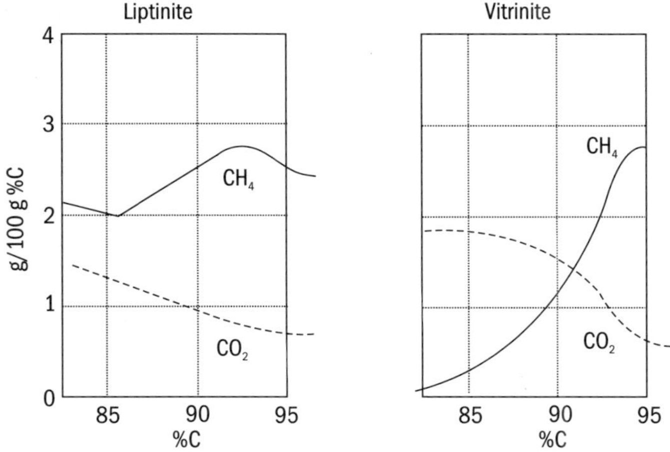

Figure 7 Volatiles of methane (CH_4) and carbon dioxide (CO_2) liberated during the coalification stage, as described by Tissot and Welte (1984)

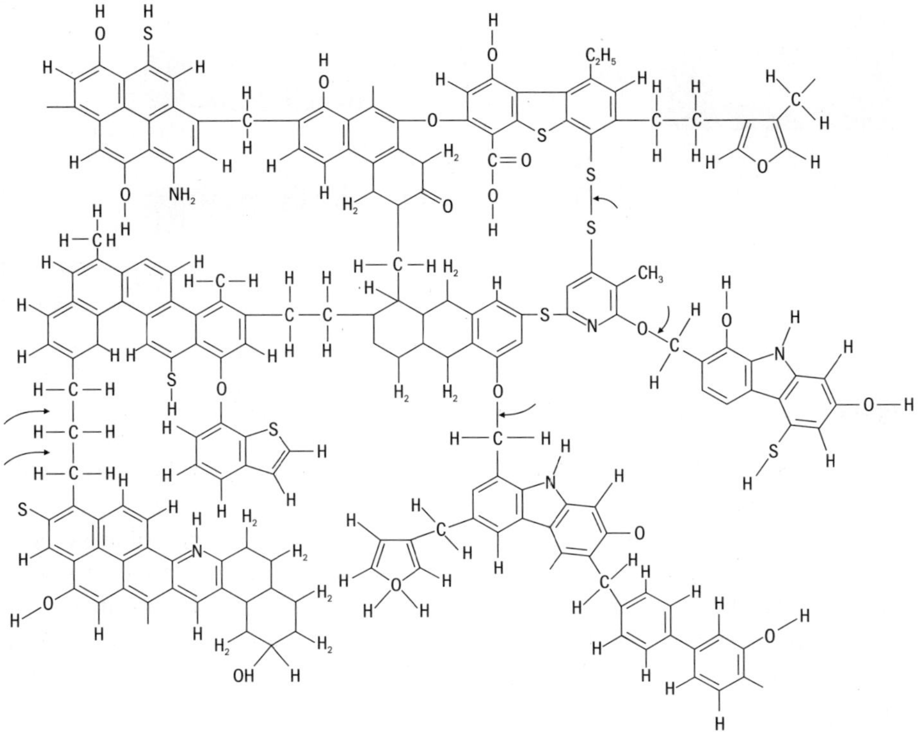

Figure 8 Representative coal molecule
Source Information collected with permission from ACS Publications Division.

Coal-bed methane

Coals have a tremendous amount of surface area and can hold massive quantities of CH_4. As coal beds have large internal surfaces, they can store six to seven times more gas than the equivalent volume of rock in a conventional gas reservoir. Coal varies considerably in terms of its chemical composition, its permeability, and other characteristics. Permeability is a key characteristic, as the coal bed must allow the gas to move once the water pressure is reduced. In higher rank coals, gas is produced, as heat and pressure transform organic material in the coal; gas in low rank coals results from the decomposition of the organic matter by bacteria. Figure 9 provides a simplified view of how CBM is formed. Coal beds are both the source of the gas that is generated and the storage reservoir once it is produced. Gas molecules adhere to the surface of the coal. Most of the CBM is stored within the molecular structure of the coal, and some is stored in the fractures or cleats of the coal or dissolved in the water trapped in the fractures.

Coals generate more gas than they can absorb and conserve. Basins that contain 500–600 SCF of CH_4/tonne are considered to be "very favourable for commercial coal-bed gas production", as long as there is sufficient reservoir permeability and high rate of desorption. Some coals have generated more

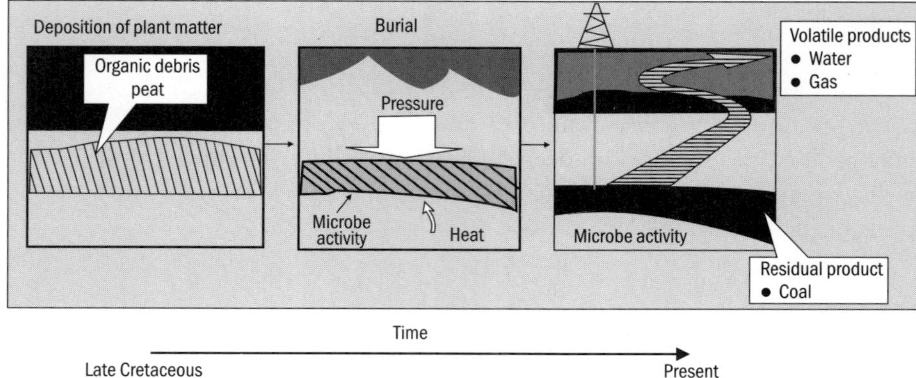

Figure 9 Coalification process, as described by Brown (2002)

than 8000 SCF of CH_4/tonne of coal. The most productive coal beds are highly permeable, saturated with gas, and fractured.

CBM is produced through either chemical reactions or microbial action. Chemical action occurs over time, as heat and pressure are applied to coal. Bacteria that obtain nutrition from coal produce CH_4 as a by-product. CH_4 attaches to the surface areas of coal throughout fractures, and is held in place by water pressure. When the water is released, the gas flows through the fractures into a well bore or migrates to the surface. Figure 10 illustrates the different types of coal and the production of CBM.

Most coals contain CH_4, but it cannot be economically extracted unless there are open fractures that provide the pathway for the desorbed gas to flow

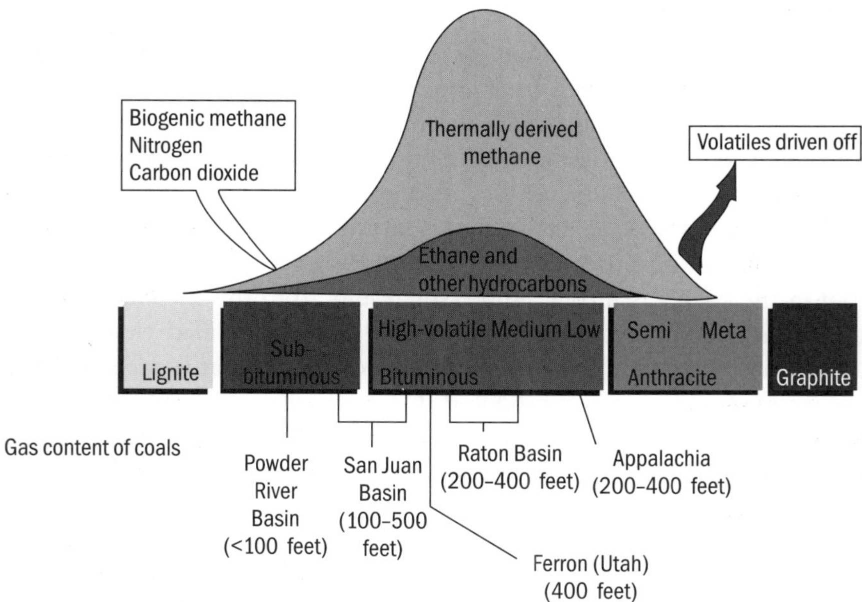

Figure 10 Biogenic methane, nitrogen, and carbon dioxide gas content in different types of coals, as mentioned by Brown (2002)

to the well. CH_4 remains in a coal bed as long as the water level is higher than the coal. These cleats and fractures are typically saturated with water, and the coal must be dewatered (usually pumped out) before the gas flows. Some coals never produce CH_4 if they cannot be dewatered economically. Some coal beds may produce gas but are too deep for feasible drilling to release the gas. CBM wells are typically no more than 5000 feet in depth, although some deeper wells have been drilled to extract the gas.

The deeper the coal bed, lesser the volume of water in the fractures, but more saline it will become. The volume of gas typically increases with coal rank, depth at which the underground coal bed is located, and the reservoir pressure. As the fracture system produces water, the adsorptive capacity of the coal exceeds, pressure falls, and the gas trapped in the coal matrix begins to get adsorbed and move to the empty spaces in the fracture system. The gas remains stored in nearby non-coal reservoirs until it is extracted.

Methanogenesis (methane production)

Methanogenesis, or biomethanation, is the formation of CH_4 by microbes known as methanogens. Organisms capable of producing CH_4 have been identified only from the domain Archaea, a group that is phylogenetically distinct from both eukaryotes and bacteria, although many live in close association with anaerobic bacteria. The production of CH_4 is an important and widespread form of microbial metabolism. In most environments, it is the final step in the decomposition of biomass.

CH_4 production is an anaerobic process and usually occurs at temperatures ranging from 0°C to 100°C. It is considered that the synthesis of adenosine triphosphate (ATP) is linked with methanogenesis by electron transport and proton pumping. CH_4-producing organisms (methanogens) have been isolated from a wide range of habitats where complex organic matter is deposited, including fresh water, marine sediments, rumen, lower intestinal tract, and others. Methanogens are the terminal organisms in microbial food chains comprising three interacting metabolic groups of strictly anaerobic bacteria that together convert complex organic matter into CO_2 and CH_4 (Figure 11). The complex organics are converted to a mixture of volatile fatty acids (VFAs), alcohols, CO_2, and H_2 by fermentative group. The higher fatty acids and alcohols are oxidized to acetic acid at the expense of the reduction of protons to H_2, hence the name obligate proton-reducing acetogenic bacteria. The combined activities of these two groups produce acetic acid, formic acid, and H_2, which are metabolized by the methanogens. The CH_4 produced is inert in anaerobic environments and is oxidized to CO_2 by oxygen-requiring methylotrophic organisms, only after diffusion into aerobic zone. The three orders of methanogens (methanobacterials, methnaococcales, and methnaomicrobials) are characterized by extreme morphological diversity and the ability to proliferate in extreme environments, including saturated brines and temperatures over 90–100°C. The diversity is more evident in

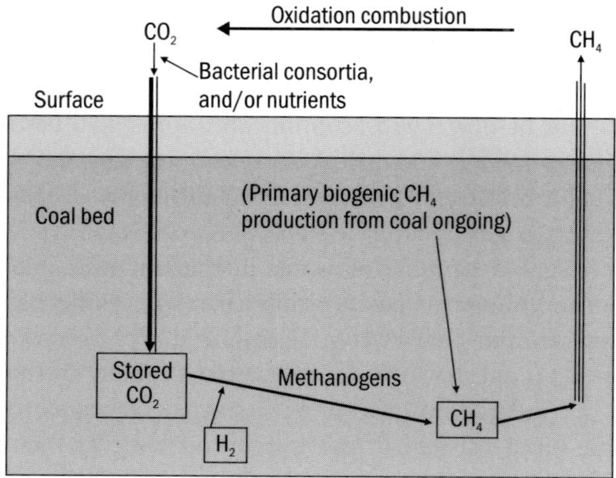

Figure 11 Enhanced coal-bed methane (CBM) followed by microbial regeneration of the CBM reservoir, as described by Gunter (2009)

the cell wall and lipid profile. Cell wall is either pseudomurein protein or heteropolysaccharide. These membrane polar lipids are the derivatives of isoprenyl glycerol ethers, and a variety of neutral lipids constitute a significant portion of the total lipids. These unusual structures of the cell components in methanogens are different from prokaryotic organisms, suggesting a deep phylogenetic division.

Methanogens

Largest group of archaebacteria are characterized as methanogens on the basis of their unique energy metabolism. They convert fermentation products (CO_2, H_2, acetate, formate, methylamine, and methanol) to CH_4, or CH_4 and CO_2. The methanogens are very strict anaerobes (obligate anaerobes) and hence produce CH_4 in oxygen-free as well as low redox potential (less than –333 mV) environment.

Representative genera and their general characteristics

There are 26 known methanogenic genera, mainly *Methanobacterium, Methanothermus, Methanococcus, Methanomicrobium, Methallogenium, Methano spirillum*, and *Methanosarcina.* However, a CH_4-producing strain, *Methanoplasma,* has been isolated; it apparently has no cell wall. Methanogens inhabit anaerobic environments rich in organic matter. They have simple nutritional requirements and use large variety of carbon compounds. Some methanogens can grow as autotrophs and use CO_2 as a carbon source and hydrogen as an energy source, but they may use formate as a carbon source. *Methallothrix* uses acetate and *Methanococcoides* uses methylamine and methanol. A thermophilic strain of *Methanosarcina* uses acetate, methylamines, and methanol. Most known methanogens are mesophilic (requiring optimum temperatures from 21°C to 45°C); some are thermophilic (requiring high temperatures of an optimum between 50°C and 80°C); and so far, there are no reports on psychrophilic methanogens.

Biochemistry of methanogenesis

Methanogenesis is an anaerobic process in bacteria, as they do not use oxygen for their respiration; in fact, oxygen inhibits the growth of methanogens. The terminal electron acceptor in methanogenesis is carbon, not oxygen. Carbon can occur in a small number of organic compounds, all with low molecular weights. The two best-known pathways for CH_4 production are as follows, where CO_2 and acetic acid were used as terminal electron acceptors.

$$CO_2 + 4H_2 \rightarrow CH_4 + 2H_2O$$

$$CH_3COOH \rightarrow CH_4 + CO_2$$

However, methanogenesis has been shown to use carbon from small organic compounds (formic acid [formate], methanol, methylamines, dimethyl sulphide, and methanethiol). The biochemistry of methanogenesis is relatively complex, involving coenzymes and cofactors: F430, coenzyme B, coenzyme M, methanofuran, and methanopterin (Figure 12).

A number of micro-organisms are involved in the process of anaerobic fermentation, including acetic-acid-forming bacteria (acetogens) and CH_4-producing archaea (methanogens). These organisms feed on the initial feedstock and convert it into intermediates (sugars, hydrogen, and acetic acid), before finally being converted to CH_4.

Certain species of bacteria that can grow optimally at temperatures between 35°C and 40°C are called mesophiles, or mesophilic bacteria, and those that can survive at 55–60°C are known as thermophiles, or thermophilic bacteria (Lancaster 1986). Methanogens come from the primitive group of archaea. This family includes species that can grow in the hostile conditions of hydrothermal vents. A property of archaea is unique in bacterial families—that they are more resistant to high temperature and can survive at thermophilic temperatures (Ossmer, Mund, Hartzell, *et al.* 1986).

The oxygen source for these micro-organisms can be the organic material itself or alternatively may be supplied by inorganic oxides. When oxygen source in an anaerobic system is derived from the organic material itself, then the "intermediates" or end products are primarily alcohols, aldehydes, and organic acids plus CO_2. In the presence of methanogens, the intermediates are converted to the end products of CH_4 and CO_2 with trace levels of hydrogen sulphide (H_2S) (Aldrich, Beimborn, Bokranz, *et al.* 1987; Mayer, Rohde, Salzmann, *et al.* 1988). In an anaerobic system, the chemical energy contained within the feedstock is released as CH_4 by methanogens (Figure 13) (Wackett, Hartwieg, King, *et al.* 1987).

$HS—CH_2—CH_2—SO_3^-$
Coenzyme M (HS—CoM)

$CH_3—S—CH_2—CH_2—SO_3^-$
$CH_3—S—CoM$

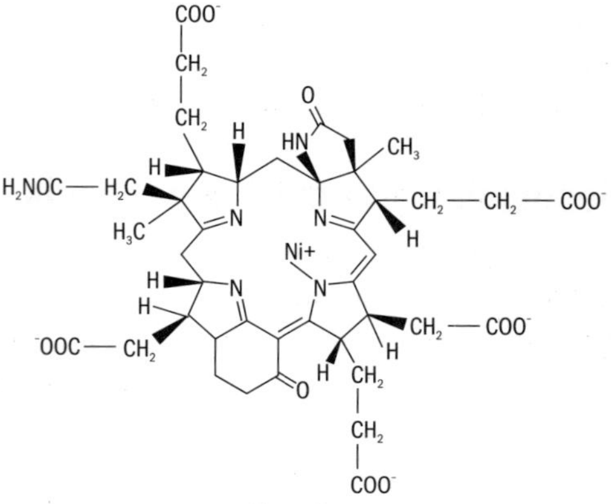

Factor F_{430}

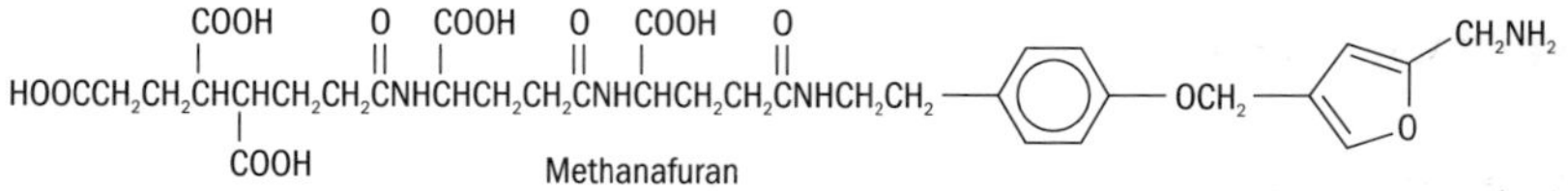

Methanafuran

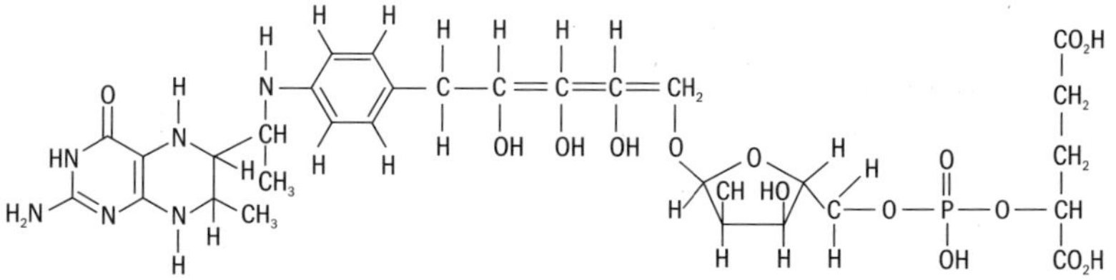

Tetrahydromethonopterin (H_4MPT)

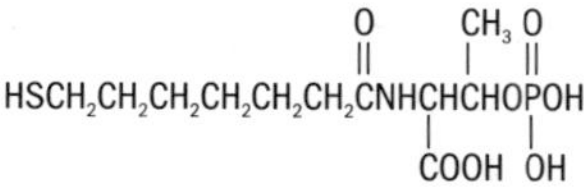

$HSCH_2CH_2CH_2CH_2CH_2CH_2CNHCHCHOPOH$

7-mercaptahepianoylthreonine phosphate
(HS-HTP)

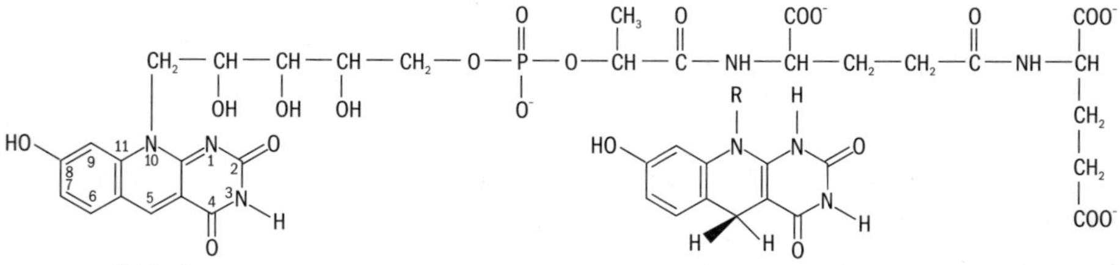

Oxidized

Reduced ($F_{420}H_2$)

Coenzyme F_{420}

Figure 12 Structures of the unusual coenzymes that participate in methanogenesis

Source Rouviere and Ralph (1988)

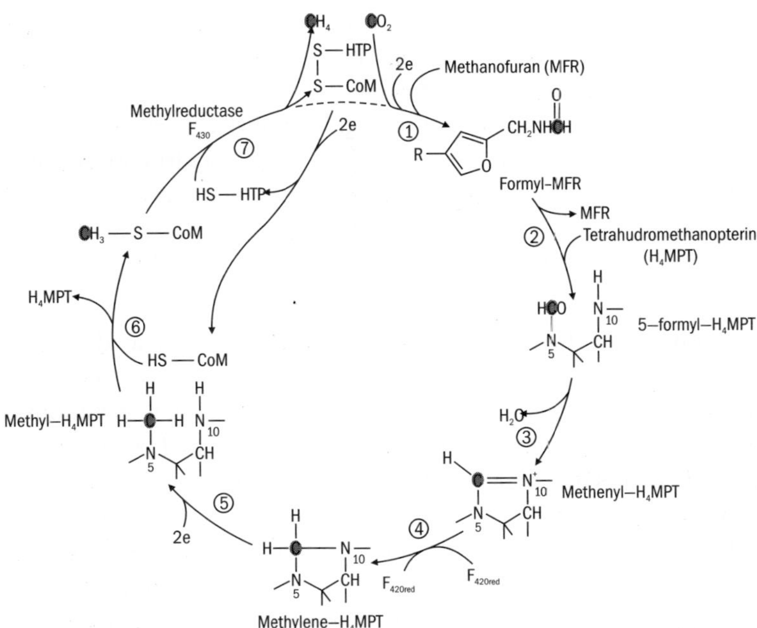

Figure 13 Proposed cycle for the reduction of CO_2 to CH_4. The C1 unit indicated by *grey shade* is sequentially modified, reduced, and transferred in "bucket brigade" fashion bound to coenzymes. Only the portion of the molecule that is involved in C1 attachment is shown for methanofuran and tetrahydromethanopterin

Source Rouviere and Ralph (1988)

In methanogens, major energy for metabolic pathway is associated with the reduction of a methyl group to CH_4. Methanogens are involved in biochemical reactions, although electrons for the reductive step can be obtained from the oxidation of hydrogen, formate, methanol, methylamines, or acetate or ethanol or propanol.

Stages

There are four key biological and chemical stages in anaerobic fermentation of CH_4.

1 Hydrolysis
2 Acidogenesis
3 Acetogenesis
4 Methanogenesis

Hydrolysis

In anaerobic fermentation, to access the energy potential of the feedstock material, these chains must be broken into their smaller constituents. The process of breaking these chains and dissolving the smaller molecules is called hydrolysis. These constituents or monomers (sugars) are readily available in other bacteria. The first fermentative or hydrolytic groups of organisms are able to attack polymeric molecules, despite appearing as solid material (Somayaji, Reddy, Bardiya, *et al.* 1996). These micro-organisms have hydrolytic enzymes, which hydrolyse polymeric substrates into monomers, for example, proteins to amino acids, polysaccharides to oligo and monosaccharides, and lipids to free fatty acids. These hydrolytic bacteria have a wide substrate range and short generation time, along with a diverse substrate-specific nature.

Acidogenesis

In competition with major fast-growing fermentative bacteria, acidogenesis is carried out by some group of bacteria that are unable to use polymeric substances, but are able to utilize smaller molecules for their metabolism and growth. As a result, short-chain fatty acids with two to five (or more) carbon atoms, and similarly, acetic acid, propionic acid, and butyric acid and small quantities of formic, valeric, isovaleric, and caproic acids are produced. Ammonia, sulphate, iso-acids, and certain aromatic compounds are also produced from amino acids (Nyns 1986). The acidogenic population is by far the largest of the trophic groups of bacteria. The population of 10^8 to 10^9 hydrolytic bacteria per millilitre of mesophilic sewage sludge has been documented by several investigators (Somayaji, Reddy, Bardiya, *et al.* 1996). Enumeration studies on hydrolytic bacteria that use specific substrates demonstrated 10^7 proteolytic and 10^5 cellulolytic bacteria in sewage sludge, and 10^4 to 10^5 hemicellulose-degrading bacteria in rumen digestion (Somayaji, Reddy, Bardiya, *et al.* 1996). Another important aspect of hydrolytic fermentation is that many of these bacteria, in the presence of autotrophic methanogens and sulphate-reducing bacteria, produce H_2 for the disposal of excess electrons generated during energy-yielding oxidation of organic material. However, accumulation of H_2 due to the inhibition of autotrophic methanogens results in thermodynamically unfavourable conditions for the disposal of electrons as H molecules. This results in the formation of products such as propionate, butyrate, succinate, and ethanol, which act as the H_2 sink. Unionized organic acids, H ions, and H_2 are inhibitory to the other groups of bacteria, whereas the acidogenic population is much more tolerant to these intermediates (Somayaji, Reddy, Bardiya, *et al.* 1996). Certain polymeric molecules using anaerobic bacteria are *Fibrobacter succinogens, Ruminococcus albus, Ruminococcus flavefaciens,* and *Selenomonas ruminatium* (Somayaji, Reddy, Bardiya, *et al.* 1996).

Acetogenesis

Acetate and H_2 produced in the first stages can be used by methanogens. Other molecules such as VFAs with a chain length greater than acetate must first be catabolized into compounds that can be directly used by methanogens.

The biological process of acetogenesis comprises further breakdown of the remaining components by acetogenic bacteria. Acetogenic bacteria obtain energy from the oxidation of organic acids, alcohols, and VFAs with more than two C atoms. These conversions are extremely sensitive to H_2, and growth of acetogenic bacteria can occur only at a less than 10^{-5} atmosphere partial pressure. Some acetogenic bacteria produce acetate from compounds such as trimethoxy benzoate, trimethoxy cinnamate, and some methoxylated aromatics. Examples of some important acetogenic bacteria are *Acetobacterium woodii* (Bache and Pfenning 1981), *Clostridium aceticum* (Braun, Mayer, and Gottschalk 1981), *Clostridium formicoaceticum*, *Eubacterium limosum*, *Sporomusa termitida* (Breznak, Switzer, and Seitz 1988), *Acetogenium kivui, Acetoanaerobium noterae, Butyribacterium methylotrophicum* (Ziekus, Lynd, Thomson, *et al.* 1980), and *Peptostreptococus productus*. Some acetogens are found to dechlorinate tetrachloromethane through the reductive process. A lot of potential exists for the efficient exploitation of this capability in the degradation of such chemicals in industrial and municipal waste (Somayaji, Reddy, Bardiya, *et al.* 1996). The acetogenic bacteria suffer from a thermodynamic product inhibition by H_2 (Wolin 1976). Hence, the growth rate of slow growing trophic groups is totally dependent on the simultaneous removal of the reduced end-products such as acetate and H_2. For this reason, acetogens are always found in co-cultures with methanogens, which provide them with favourable growth conditions, while their metabolic partners are provided with carbon and energy sources (Somayaji, Reddy, Bardiya, *et al.* 1996).

Methanogenesis

Methanogens are the ultimate trophic group of anaerobic fermentation. They produce the most reduced form of carbon, that is, CH_4. They use terminal end-products of acetogenic fermentation, namely, acetate, formate, methanol, methylamine, H, and CO_2, as carbon and energy source (Ziekus 1980). A few methanogens derive energy from the CO_2 by reduction using H_2. Autotrophic methanogens assimilate CO_2 as a sole carbon source. Other microbial species, which are acetoclastic and chemoorganotrophic archaebacteria, split acetate into CH_4 and CO_2 and obtain energy (Nyns 1986; Somayaji, Reddy, Bardiya, *et al.* 1996). Acetate is an anion and when split into CH_4 and CO_2, it requires a pairing cation. If NH_4^+ or H_3O^+ is used as pairing cations, the end products are bicarbonate and carbonic acids.

During the carbonic acid formation, CO_2 escapes in the evolved CH_4, ultimately resulting in alkalinity or reinforced buffering capacity to combat the acid produced by acetogens (Nyns 1986). In anaerobic fermentation, more than 60% of the biological CH_4 originates from acetate, thus acetoclastic methanogens perform the important function of electron removal while maintaining a neutral pH (Ziekus 1980). Barker (1956) postulated that the pathway of CH_4 generation was initiated by the attachment of CO_2 to a carrier molecule and that all the subsequent reduction steps involved one carbon intermediate attached to a carrier molecule. Its metabolism encompasses nearly the entire spectrum of reactions involved in methanogenesis. In the elucidation of this central pathway, along with conventional co-enzymes, several surprising molecules were discovered.

The enzyme-mediated reactions, which involve six new co-enzymes, are only partially understood in *Methanosaeta soehngenii* by Somayaji, Reddy, Bardiya, *et al.* (1996), and in *Methanothrix* sp., *Methanosarcina* sp., and *Methanosarcina mazei* by Somayaji, Reddy, Bardiya, *et al.* (1996) and Whitman (1985).

In the terminal stage of anaerobic fermentation, methanogens use the intermediate products of the preceding stages and convert them into CH_4, CO_2, and water. Methanogenesis is sensitive to both high and low pH and occurs between pH 6.5 and 8. The remaining non-digestible material, which the microbes cannot feed upon along with any dead bacterial remains, constitutes the digestate.

A simplified generic chemical equation for the overall processes outlined above is as follows.

$$C_6H_{12}O_6 \rightarrow 3CO_2 + 3CH_4$$

Importance in carbon cycle

Methanogenesis is the final step in the decay of organic matter. During the decay process, electron acceptors (such as oxygen, ferric iron, sulphate, nitrate, and manganese) become depleted, whereas H_2, CO_2, and light organics accumulate. During advanced stages of organic decay, all electron acceptors are depleted, except CO_2. CO_2 is a product of most catabolic processes, and so it is not depleted like other potential electron acceptors.

Only methanogenesis and fermentation can occur in the absence of electron acceptors other than carbon. Fermentation only allows the breakdown of larger organic compounds and produces small organic compounds. Methanogenesis effectively removes the intermediates: H_2, small organics, and CO_2. Without methanogenesis, a great deal of carbon (in the form of fermentation products) would accumulate in anaerobic environments.

Methanogenesis in ruminants

Methanogenesis occurs in the guts of humans and animals, especially ruminants. In the rumen, anaerobic organisms, including methanogens, digest cellulose into forms usable by the animal, and without them, livestock such as cattle would not be able to graze grass. The useful products of methanogenesis are absorbed by the gut, but CH_4 is released from the animal mainly by belching (eructation). On average, a cow emits about 250 litres of CH_4 per day.

Methanogenesis in humans

Some humans produce flatus that contains CH_4. In one study of the faeces of nine adults, only five of the samples contained archaea capable of producing CH_4. Similar results are found in samples of gas obtained from within the rectum. Even among humans whose flatus does contain CH_4, the amount is only 0%–10% of the total amount of gas.

Role in global warming

CH_4 in the earth's atmosphere is a greenhouse gas with a global warming potential 21 times greater than CO_2 (averaged over 100 years). Methanogenesis in livestock and the decay of organic material are thus considerable contributors to global warming. Methanogenesis may not be a net contributor in the sense that it works on organic material, which used up atmospheric CO_2 when it was created, but its overall effect is to convert the CO_2 into CH_4, which is a much more potent greenhouse gas.

Methanogenesis can also be beneficially exploited to treat organic waste and to produce useful compounds, and the CH_4 can be collected and used as a fuel. It is the primary pathway, whereby most organic matter disposed of through landfill is broken down.

Factors affecting methane production

Besides the microbial population, environmental factors that govern the process of methanogenesis are anaerobiosis, temperature, pH, substrate composition, carbon:nitrogen (C:N) ratio, micronutrients, and accumulation of inhibitory by-products or intermediates of the process.

Methanogenesis is a strict anaerobic microbial process. The majority of methanogens in this process die in the presence of O_2. Although the major microbial activity in methanogenesis is anaerobic, many workers have reported the presence of aerobic and facultative anaerobic bacteria (Hobson, Bousfield, and Summers 1981; Somayaji, Reddy, Bardiya, *et al.* 1996). They do not have a role in the main degradable reactions but probably play a role in the anaerobic fermentation as scavengers of O_2, rendering the system suitable for growth of the methanogens (Hobson, Bousfield, and Summers 1981). Therefore, the requirement for anaerobic process can be formulated by minimizing the diffusion of air so that the microflora is not exposed to the oxygen tension (Nyns 1986).

Temperature

Methanogenesis occurs practically in an optimal way in two ranges of temperatures: mesophilic at 28–37°C and thermophilic at 50–70°C. Adaptation of a mesophilic to thermophilic conditions is possible (in about 10–20 days), as it always contains approximately 10% of the thermophilic micro-organisms (Nyns 1986). Thermophilic conditions have a number of advantages. In addition to the increased rates of CH_4 production, which is reported to be 1.5 times faster than the mesophilic condition, thermophilic process has improved de-water ability and has enhanced the removal of pathogenic organisms (Buhr and Andrews 1977). However, changes in the overall process efficiency due to increased metabolic activity are balanced by a corresponding increase in the rates of microbial inactivation (Somayaji, Reddy, Bardiya, *et al.* 1996). For example, a higher conversion of proteins was observed at 30–50°C, and conversely, lower temperatures were found suitable for lipid digestion (Somayaji, Reddy, Bardiya, *et al.* 1996).

pH

A pH of 6.5–7.7 has been found to be optimum for *Methanobacterium ruminantium* seen in rumen samples. Maintenance of a near-neutral pH is due to the conversion of acid intermediates to CH_4 by the combined effect of acetogenic and methanogenic bacteria. It should be recalled that the methanogenic process is alkalizing (Capri and Marais 1975). The major controlling buffer in this pH range is the carbonate–bicarbonate. Methanogenesis beyond the range of 6.5–8.0 pH has been found to be less yielding (Nyns 1986).

Substrate composition

The biomass substrate used in methanogenesis comprises various carbon sources such as cellulose, hemicellulose, lipid, lignin, pectin, protein, starch, and other organic materials at varied concentrations. For a pure substrate, it is possible to calculate the amount of gas that may be produced on the basis of a simple carbon balance.

Carbon:nitrogen ratio

Extra attention has to be paid to the N content of the substrate biomass for two reasons. First, N is required for cell growth, a phenomenon necessarily linked to CH_4 production; second, the nitrogenous compounds contribute to the buffering of the medium by releasing ammonium cation. Investigations show that a C:N ratio of about 20 is required for maximum CH_4 production (Figure 14). A decrease in CH_4 yields has also been reported when the C:N ratio increased above 20, probably because of the low N availability (Somayaji, Reddy, Bardiya, *et al.* 1996).

Micronutrients

Methanogens also require the phosphorus (P) and metal ions for optimum growth and CH_4 production. Somayaji, Reddy, Bardiya, *et al.* (1996) have

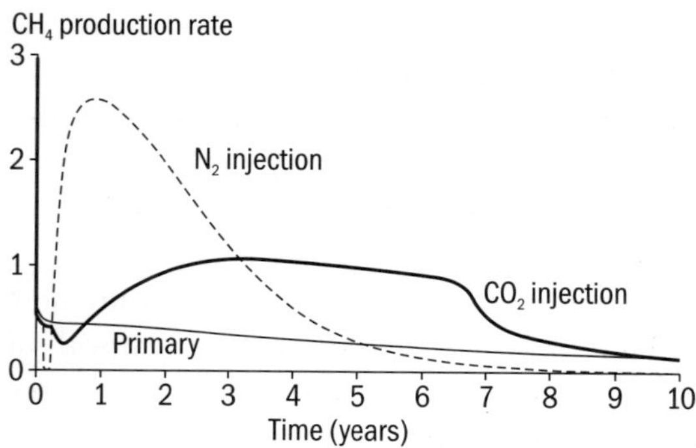

Figure 14 Primary, nitrogen, and carbon dioxide injection scenarios for enhanced coal-bed methane recovery at Fenn-Big Valley. Injection is at constant flow rate. Production rate is normalized, as described by Gunter (2009)

reported that nickel and cobalt ions are necessary at concentrations of 100 nM and 50 nM, respectively. Molybdenum ions at a concentration of 50 nM may enhance the joint effect of nickel and cobalt. Iron cations up to 2 nM and copper ions at a concentration of 4 nM are necessary to enhance the performance of acetoclastic methanogenic archaebacteria. An increase of up to three-fold can be achieved by incorporating the above minerals into a fermentor deficient in these components (Nyns 1986; Somayaji, Reddy, Bardiya, *et al.* 1996). The requirement of P has also been studied, and a C:P ratio of 100:200 is said to be optimum. The required optimum C:N:P ratio for the enhanced yield of CH_4 was found to be 100:2.5:0.5 (Somayaji 1992).

Intermediates

Isobutyric acid, in particular, is inhibitory at low concentrations (50 ppm) (Somayaji, Reddy, Bardiya, *et al.* 1996). Toxic effect of the VFAs at high concentrations is attributed either to the toxicity of the VFAs themselves or to the decrease in pH brought about by them.

Conclusion

It is believed, based on the analyses, that CO_2 reduction is the dominant methanogenic pathway in subsurface coal-associated environments (Ahmed and Smith 2001; Rice and Claypool 1981). Microbial oxidation and decarboxylation reactions generate CO_2, which is then reduced using H_2 to CH_4 in methanogenic zones within the coal seam (Ahmed and Smith 2001). Thus, it is not surprising that methanogens capable of carbonate reduction were detected in the coal-enrichment cultures, as described. The acetoclastic methanogenesis pathway may also occur here to a lesser degree or at basin margins, as suggested by Flores, Rice, Stricker, *et al.* (2008). However, the availability of labile sources of acetate, as with the introduction of a fermentable nutrient, should result in a larger contribution of acetoclastic methanogenesis to the CH_4 yield. This may explain the dominance in the coal enrichment cultures of *Methanosarcina* spp., which can use both acetate and H_2/CO_2. Consortia of methanogens consuming different fermentative waste have been found in hydrocarbon-degrading cultures (Chang, Um, Hoffman, *et al.* 2005; Chang, Um, and Pulliam Holoman 2005), ensuring that thermodynamically favourable conditions exist for the 540 bacterial species to metabolize the organic compounds (Kotsyurbenko, Chin, Glagolev, *et al.* 2004). Changes in the composition of complex nutrients do not directly affect the final steps in the conversion of organic matter to CH_4, as methanogens are restricted to a narrow substrate range for growth. Instead, it is the bacteria providing the precursors for methanogenesis, which are directly affected by the available nutrients and growth conditions. This is clearly reflected by the differences in the composition of the bacterial libraries constructed from uncultured coal samples and from short- and long-term enrichment cultures. The microbial populations shifted from large proportions of oligotrophic and chemolithotrophic members of the proteobacteria in uncultured coal to heterotrophic fermentative *Clostridia* and *Bacteroides* in the mesophilic enrichment cultures. The effect of nutrient type on methanogenesis from coal is being investigated further to determine whether certain nutrient supplements are preferred.

Information obtained from the molecular identification of microbes eventually may be used to modify a target coal seam to make its environment more conducive to enhanced CH_4 production. To regenerate depleted or under-producing CBM wells using microbial metabolic activities, an effective approach would be to identify whether appropriate bacteria are present and add suitable nutrient(s) to selectively stimulate their metabolism and produce substrates supporting methanogenesis. Larger and geologically diverse samples, including groundwater associated with coal beds, should be collected for future investigations to link types of bacteria and archaea to coal rank formation, hydrology (salinities, composition), and depths (Lemay and Konhauser 2006) and to aid in the selection of coal beds appropriate for enhanced methanogenesis.

References

Ahmed M and Smith J W. 2001. **Biogenic methane generation in the degradation of eastern Australian Permian coals.** *Organic Geochemistry* **32**: 809–816

Aldrich H C, Beimborn D B, Bokranz M, Schonheit P. 1987. *Archives of Microbiology* **148**: 193–201

Bache R and Pfenning N. 1981. **Selective isolation of *Acetobacterium woodii* on methoxylated aromatic acids and determination of growth yields.** *Archives of Microbiology* **130**: 255–261

Barker H A. 1956. ***Bacterial Fermentations.*** New York: John Wiley and Sons

Beaton A, Langenberg W, and Pana C. 2006. **Coalbed methane resources and reservoir characteristics from the Alberta Plains, Canada.** *International Journal of Coal Geology* **65**: 93–113.

Berkowitz N. 1979. ***An Introduction to Coal Technology.*** New York: Academic Press, p. 345

Bibler C J, Marshall J S, and Pilcher R C. 1998. **Status of worldwide coal mine methane emissions and use.** In *Coalbed Methane: from coal-mine outbursts to a gas resource*, edited by R M Flores. *International Journal of Coal Geology* **35**: 283–310

Braun M, Mayer F, and Gottschalk G. 1981. ***Clostridium aceticum* (Wieringa), a microorganism producing acetic acid from molecular hydrogen and carbon dioxide.** *Archives of Microbiology* **128**: 288–293

Breznak J A, Switzer J M, and Seitz H J. 1988. ***Sporomusa termitida* sp. nov., an H_2/CO_2-utilizing acetogen isolated from termites.** *Archives of Microbiology* **150**: 282–288

Brown W T. 2002. NRLC Coalbed Methane Conference, 4–5 April 2002

Buhr H O and Andrews J F. 1977. *Water Research* **11**: 129

Capri M G and Marais G R. 1975. **pH adjustment in anaerobic digestion.** *Water Research* **9**: 307–595

Chang W, Um Y, and Pulliam Holoman T R. 2005. **Molecular characterization of anaerobic microbial communities from benzene-degrading sediments under methanogenic conditions.** *Biotechnology Progress* **21**: 1789–1794

Chang W, Um Y, Hoffman B, Pulliam Holoman T R. 2005. **Molecular characterization of polycyclic aromatic hydrocarbon (PAH) degrading methanogenic communities.** *Biotechnology Progress* **21**: 682–688

Cobe J C and Drahovzal A. 2003. ***Coal Bed Methane.*** Fact sheet no. 3

Fakoussa R M and Hofrichter M. 1999. **Biotechnology and microbiology of coal degradation.** *Applied Microbiology and Biotechnology* **52**: 25–40

Flores R M, Rice C A, Stricker G D, Warden A, Ellis M S. 2008. **Methanogenic pathways of coalbed gas in the Powder River Basin, United States: the geologic factor.** *International Journal of Coal Geology* **76** (1–2): 52–75

Formolo M, Martini A, and Petsch S. 2008. **Biodegradation of sedimentary organic matter associated with coalbed methane in the Powder River and San Juan Basins, U.S.A.** *International Journal of Coal Geology* **76** (1–2): 86–97

Green M S, Flanegan K C, and Gilcrease P C. 2008. **Characterization of a methanogenic consortium enriched from a coalbed methane well in the Powder River Basin, U.S.A.** *International Journal of Coal Geology* **76** (1–2): 34–45

Gunter W D. 2009. **Coalbed Methane, A Fossil Fuel Resource with the Potential for Zero Greenhouse Gas Emissions – The Alberta, Canada Program 1996–2009: A Summary.** Alberta CO_2 – ECBM research and field pilots summary October, 2009 Alberta Research Council

Hobson P N, Bousfield S, and Summers R. 1981. ***Methane Production from Agricultural and Domestic Wastes.*** London: Elsevier Applied Science Publishers

Jones E J P, Voytek M A, Warwick P D, Corum M D, Cohn A, Bunnell J E, Clark A C, Orem W H. 2008. **Bioassay for estimating the biogenic methane-generating potential of coal samples.** *International Journal of Coal Geology* **76** (1–2): 138–150

Kotsyurbenko O · R, Chin K J, Glagolev M V, Stubner S, Simankova M V, Nozhevnikova A N, Conrad R. 2004. **Acetoclastic and hydrogenotrophic methane production and methanogenic populations in an acidic West-Siberian peat bog.** *Environmental Microbiology* **6** (11): 1159–1173

Kuuskraa V A and Brandenberg C F. 1989. **Coalbed methane sparks a new energy industry.** *Oil and Gas Journal* 87 (41): 49–56

Lancaster J R Jr. 1986. *FEBS Letters* **199:** 12–18

Lemay T G and Konhauser K O. 2006. **Water chemistry of coalbed methane reservoirs.** Alberta Energy and Utilities Board, EUB/AGS Special Report 081, 372 pp.

Levine D G, Schlosberg R H, and Silbernagel, B G. 1982. **Understanding the chemistry and physics of 655 coal structure (a review).** *Proceedings of the National Academy of Sciences of the United States of America* **79**: 3365–3370

Mayer F, Rohde M, Salzmann M, Jussofie A, Gottachalk G. 1988. *Archives of Microbiology* **147**: 190–194

Nyns E J. 1986. **Biomethanation processes.** In *Biotechnology*, vol. 8, p. 207, edited by J Rehn and G Road. Weinheim, Germany: VCH Publishers Inc.

Ossmer R, Mund T, Hartzell P L, Konheiser U, Kohring G W, Klein A, Wolfe R S, Gottschalk G, Mayer F. 1986. **Immunocytochemical localization of component C of the methylreductase system in *Methanococcus voltae* and *Methanobacterium thermoautotrophicum*.** *Proceedings of the National Academy Sciences of the United States of America* **83** (16): 5789–5792

Rice D D. 1997. **Coalbed methane – an untapped energy resource and an environmental concern: U.S. Geological Survey Fact Sheet FS–019–97.** Details available at http://energy.usgs.gov/factsheets/Coalbed/coalmeth.html

Rice D D and Claypool G E. 1981. **Generation, accumulation and resource potential of biogenic gas.** *AAPG Bulletin* **65**: 5–25

Rouviere P E and Ralph S. 1988. **Wolfe novel biochemistry of methanogenesis.** *Journal of Biological Chemistry* **263** (17): 7913–7916

Scott A R. 1999. **Improving coal gas recovery with microbially enhanced coalbed methane.** In *Coalbed Methane: scientific, environmental, and economic evaluations*, pp. 89–110, edited by M Mastalerz, M Glikson, and S D Golding. The Netherlands: Kluwer Academic Publishers

Somayaji D. 1992. **Microbiological studies on biomethanation of mango peel with special reference to methanogens.** University of Gulbarga [PhD Thesis]

Somayaji D, Reddy M R V P, Bardiya N, Mohan K, Khanna S. 1996. **Methanogenesis.** In *Wealth from Waste*, edited by S Khanna and K Mohan. New Delhi: Tata Energy Research Institute

Strąpoć D, Picardal F W, Turich C, Schaperdoth I, Macalady J L, Lipp J S, Lin Y-S, Ertefai T F, Schubotz F, Hinrichs K-U, Mastalerz M, Schimmelmann A. 2008. **Methane producing microbial community in a coal bed of the Illinois Basin.** *Applied and Environmental Microbiology* **74** (8): 2424–2432

Tissot B P and Welte D H. 1984. ***Petroleum Formation and Occurrence***, 2nd edn, p. 699. New York: Springer and Verlag

Tyler R, Scott A R, and Kaiser W R. 1997. **Defining coalbed gas exploration fairways in low permeability, hydrocarbon over pressured basins: an example from the Piceance basin, northwest Colorado.** Tuscaloosa, AL: University of Alabama, pp. 527–533. [Proceedings of the 1997 International Coalbed Methane Symposium, Tuscaloosa, Alabama, 12–17 May 1997]

Ulrich G and Bower S. 2008. **Active methanogenesis and acetate utilization in Powder River Basin coals, United States.** *International Journal of Coal Geology* **76** (1–2): 25–33

Wackett L P, Hartwieg E A, King J A, Orme-Johnson W H, Walsh C T. 1987. **Electron microscopy of nickel-containing methanogenic enzymes: methyl reductase and F420-reducing hydrogenase.** *Journal of Bacteriology* **169**: 718–727

Whitman W B. 1985. **Methanogenic bacteria in book.** In *The Bacteria*, vol. 8, pp. 3–84, edited by C R Woese and R S Wolfe

Wolin M J. 1976. ***Microbial Production and Utilization of Gases,*** edited by H G Schlegel and G Gottschalk

Ziekus J G. 1980. ***Anaerobic Digestion***, edited by D A Stafford, B L Wheatley, and D E Hughes. London: Elsevier Applied Science Publishers. 61 pp

Ziekus J G, Lynd L G, Thomson T E, Krzycki J A, Weimer P J, Hegge P W. 1980. **Isolation and characterization of a new methylotrophic acidogenic anaerobe, the Marburg strain.** *Current Microbiology* **3**: 381–386.

Bibliography

Anderson G K, Kasapgil B, and Ince O. 1994. **Microbiological studies of two phase anaerobic digestion start-up.** *Water Research* **28**: 2383–2392

Burford J L and Varani F T. 1976. **Energy potential through bioconversion of agricultural wastes.** Final reports to the Four Corner Regional Commission by Biogas of Colarado Inc. and Colarado Energy Research Institute USGS Fact Sheet FS-156-00, November 2000

Bioreactors for waste water treatment

K Ramasamy
Vice Chancellor, Karpagam University, Coimbatore, Tamil Nadu, India
and
P Doraisamy
Professor and Head, Department of Environmental Sciences, Tamil Nadu Agricultural University, Coimbatore ,Tamil Nadu, India

Introduction

Industrialization of developing countries like India has led to the generation of a huge amount of waste water. Indiscriminate disposal of this waste water into rivers, lakes, streams, and other waterbodies and on land has resulted in the pollution of these resources. To reduce the pollution, various treatment technologies have been developed over time. Waste water treatment practice started in the beginning of the 20th century. Among the various treatment systems, the biological treatment system is widely adopted at present, which offers the most viable treatment possibilities owing to the extensive degrading capabilities of the micro-organisms. A biological process often presents a good treatment alternative in terms of costs and environmental benefits.

Biological processes involved in waste water treatment

The biological waste water treatment processes provide a remarkably complex environment for different types of organisms. Although virtually nearly all types of micro-organisms and to some extent macro-organisms may be found in these environments, the most common are usually from three groups: prokaryote bacteria, protozoa, and micro-invertebrates. These organisms occupy different niches, yet are involved in several complex interactions that are important to the treatment process. The bacteria generally occupy the basic trophic level as decomposers and degrade different components in the waste water. The protozoa and the micro-invertebrates, on the other hand, are in most cases consumers, that is, they feed on the bacteria or on other protozoa and metazoa (Hammer and Hammer Jr 2003).

Biofilms in waste water treatment

Biofilms play an important role in waste water treatment, and these are essentially a film of micro-organisms growing on some kind of support material. The biofilm is exposed to waste water, so that the waste material in question (organic compounds, nitrogen, phosphorus, and so on) is transformed when it comes in contact with the biofilm. The support material may comprise different materials, such as gravels, stones, plastic, sand, and activated carbon particles. Consequently, various design possibilities and applications of biofilm processes exist for different circumstances and requirements. The most common designs in practice are trickling filters, rotating biological contactors (RBCs), submerged filters, and suspended carrier systems (Jogdand 1995).

Biofilm processes offer several advantages, such as allowing the development of slow growing micro-organisms, so that a higher degradation potential may be achieved. Furthermore, biofilm processes are generally less sensitive to disturbances, and toxic compounds are relatively easy to manage and usually require small areas compared to the activated sludge process. However, several problems, partly due to the malfunction of the biological processes in different types of biofilm processes, have also been observed. The biofilm may suddenly detach extensively (sloughing) or diverse problems due to certain micro-organisms (fungi, filamentous bacteria, predators, and so on) may occur.

Biological treatment

Biological processes for waste water treatment are generally classified as aerobic processes and anaerobic processes (Barnes and Fitzgerald 1987). The choice between aerobic and anaerobic processes favourably inclined towards the former in the past because these systems were considered to be more reliable, more stable, and better understood in spite of the positive energy recovery aspect of the latter. However, with the advent of anaerobic high-rate processes, the waste water treatment scenario has witnessed a tremendous shift in favour of anaerobic processes (Lettinga 1984). The anaerobic processes are a promising technology to treat industrial waste water, because of the relatively low capital expenditures required and the potential to generate biogas, which can be used as a source of energy. Anaerobic systems generate less waste sludge than aerobic systems, can handle high organic loads, and result in increased profits.

Aerobic biological treatment

In the aerobic process, the organic carbon present in the effluent is used up by the mixed aerobic microbial consortium, as its carbon and energy source and the complex organics finally get converted to microbial biomass (sludge) and CO_2. These treatment methods are based on the systems of sewage treatment and are meant to handle easily biodegradable organic load. When applying to industrial effluents, a careful treatability study is needed to determine the design and working parameters. The basic reaction in aerobic reactor is as follows.

Organic material + O_2	$\longrightarrow$ Aerobic microbes + nutrients	$CO_2 + H_2O$ + New cells

Microbial cells undergo progressive autoxidation of the cell mass.

Cells + O_2	$\longrightarrow$	$CO_2 + H_2O + NH_3$

Anaerobic biological treatment

Anaerobic digestion is a bacterial process that breaks down organic materials present in the waste in the absence of oxygen (Kamaraj, Doraisamy, John Gunasekar, *et al.* 2001). Under anaerobic conditions, the complex organics are first broken down into simple monomers and subsequently to a mixture of volatile fatty acids (VFAs), mostly, acetic, propionic, and butyric acids. This is achieved by hydrolysers and "acidogens", a consortium of hydrolytic and acidogenic bacteria (Gottschalk 1979). The VFAs and $CO_2 + H_2$ are in turn converted to CO_2 and methane (CH_4) by methanogenic bacteria (Zehnder, Ingvorsen, and Marti 1982).

It is generally run in closed tanks. The generalized equation for anaerobic sludge digestion is as follows.

Organic matter + combined oxygen	Anaerobic microbes $\longrightarrow$	New cells + energy for life processes + CH_4 + CO_2 + other gases

The combined oxygen consists of CO_3^{-2}, SO_4^{-2}, NO_3^-, and PO_4^{-3}. Thus, while both aerobic as well as anaerobic degradation routes can equally remove complex organics from the effluents, the anaerobic route has the obvious advantage in that it produces CH_4, a combustible gas with a reasonably good calorific value (24 MJ/m^3). The production of CH_4, a renewable energy source, is alone a strong ecological reason for employing anaerobic treatment, or "biomethanation", and preferring it wherever applicable.

Bioreactors in waste water treatment

In biological treatment processes, special reactors have been designed in which the micro-organisms are maintained under controlled conditions. Due to the versatile physiological abilities of the micro-organisms, in particular, the bacteria, various microbiological reactions can be utilized to remove many types of waste materials, such as different types of biodegradable as wells as recalcitrant organic matter (for example, aliphatic and aromatic compounds, carbohydrates, proteins, fats and oils), nutrients, and many other chemical elements such as sulphur and metals. These reactions may be carried out under aerobic (DO, or dissolved oxygen, present), anoxic (DO absent, nitrate present) or anaerobic (DO and nitrate absent) conditions, depending on which kind of degradation pathway is used.

The biological treatment of industrial effluent has to be decided considering the following criteria.

- Eco-friendly nature of the process
- Flexibility of the process to varying loads and characteristics
- Operator-friendly nature, that is, requiring less personal attention
- Low capital costs and operating costs
- Less energy spent and, on the contrary, generation of energy, if possible
- Possibility of continuous operation
- Easy automation
- Capability to achieve the targeted degradation
- Occupation of minimum land
- Avoidance of dilution, as water is a costly resource
- Production of less sludge, as sludge disposal involves considerable costs (Jogdand 1995).

Different types of bioreactors are often used separately or in combination to treat waste water, both aerobically and anaerobically. The reactors employ both aerobic and anaerobic microflora, which are involved in the effluent treatment (Bitton 1994). These reactors are as follows.

- Aerobic continuous plugflow reactor
- Continuously stirred tank reactor (CSTR)
- Activated sludge
- Anaerobic filter
- Upflow anaerobic sludge blanket (UASB) reactor
- Contact reactor
- Anaerobic attached-film expanded bed reactor (EBR)
- Anaerobic fluidized-bed reactor (AFBR)
- RBC
- Carrier-assisted sludge bed reactor

The types of treatment plant will depend on the complexity and nature of pollutants. Non-biodegradables have to be tackled by physical/chemical methods. The biodegradable pollutants differ widely in their characteristics. Generally, pollutants can be broadly classified as follows:

- Readily degradable organics (for example, food industry)
- Complex organics (for example, organochemical industry)
- Reactable organics (for example, heavy industry and plating)
- Inert organics (for example, coal mining and quarrying)

Aerobic continuous flow reactors

Plug-flow reactor

In the plug-flow reactor, fluid particles travel through the vessel without mixing, and therefore, are discharged in the same sequence in which they enter (Figure 1). If a continuous tracer is introduced starting at time t = 0 (concentration of tracer in the influent being C_0), no tracer appears in the effluent until a time t_h, which is the hydraulic residence time of the fluid in the vessel, has elapsed. Then the concentration of the tracer in the effluent jumps from a zero value to the value C_0 and remains at that value as long as the continuous injection of the tracer is maintained.

Continuous fed stirred tank reactor

In this reactor, immediate dispersion of tracer particles takes place as they enter the vessel (Figure 1). For a continuous tracer, its concentration in the effluent as a function of time is determined by the following material balance equation for the tracer around the reactor (Ramalho 1983).

$$\begin{array}{c}\text{Rate of change in amount}\\\text{of tracer in reactor}\end{array} = \begin{array}{c}\text{Rate of input of}\\\text{tracer to reactor}\end{array} - \begin{array}{c}\text{Rate of output of}\\\text{tracer from reactor}\end{array}$$

Arbitrary flow reactor

The arbitrary flow reactor corresponds to an intermediate amount of dispersion between plug-flow and complete mix types. Mathematical analysis of this type of reactor is considerably more complicated than plug-flow or complete mix types (Figure 1).

Activated sludge process

The activated sludge process is the most widely used biological waste water treatment system today, for both municipal and industrial waste water. It is a suspended growth process, which principally includes an aeration tank and a sedimentation tank. In the aeration tank, influent waste water is mixed with micro-organisms, which oxidize the organic matter (Figure 2). The aeration serves two functions: supplying oxygen to the micro-organisms and maintaining them in suspension so that the maximum contact between the activated sludge flocs and the waste water is achieved. In the sedimentation

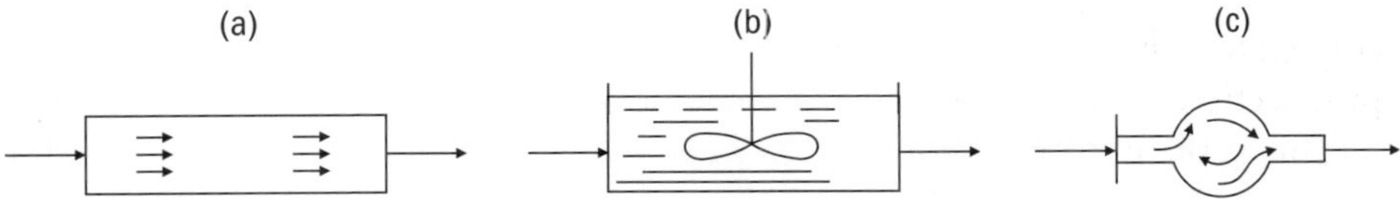

Figure 1 (a) Plug-flow reactor, (b) continuous fed stirred tank reactor, and (c) arbitrary flow reactor

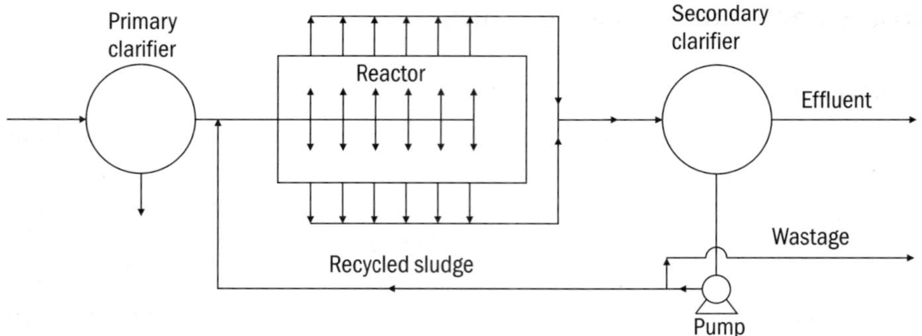

Figure 2 Activated sludge process

tank, the treated water is separated from the activated sludge for further (tertiary) treatment. One part of the separated sludge is re-circulated back to the aeration tank so that a relatively constant amount of biomass can be maintained. Another part is removed as excess sludge and is further treated for final disposal. Depending on the treatment requirements, various modifications of the tank, or a combination of several tanks, may be introduced. Further modifications in the choice of aeration configuration or in the distribution mode of the waste water to the tank are also possible. There are several advantages in using the activated sludge process, the main benefit being the possibility to adapt the process relatively easily to fulfil many requirements. However, several drawbacks are also there, which may make the process difficult to operate. The most common operational problem is the separation in the sedimentation stage. This leads to decreased treatment efficiency due to the loss of biomass. Furthermore, it may also cause problems in consecutive treatment stages such as dewatering and anaerobic digestion.

Fixed-film reactors

Trickling filter

The trickling filter (percolating or biological filter) is a fixed-film system, with micro-organisms growing on the surface of a solid medium. Various biologically inert materials can be used as biofilm support media, and the main consideration is that there be a large specific surface area per unit volume to support biofilm growth. Traditionally, blast-furnace slag, coke or gravel with a diameter of 25–100 mm have been used. However, plastic media have been developed, which are much lighter and allow less sturdy construction techniques to be used. Plastic media comes in a wide variety, such as blocks of corrugated channels, while other "random" media vary in shape, from short pipe sections to weird designs of snowflake complexity.

This is the most common biofilm treatment system, which has been used for sewage treatment since 1900. Trickling filters (Figure 3) consist of circular or rectangular beds of media 1.5–2.5 m in depth. The waste water is distributed

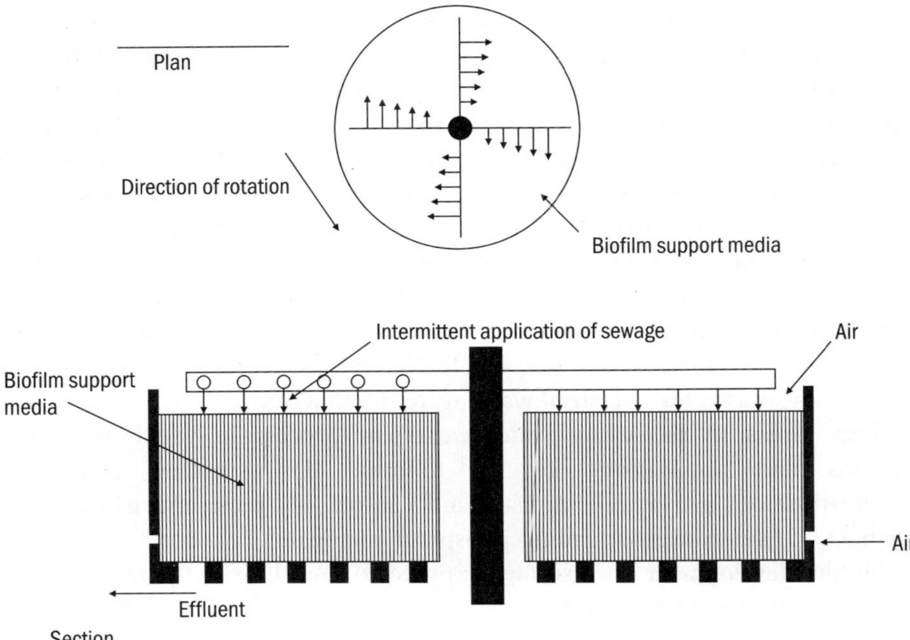

Figure 3 Layout of a circular trickling filter

over the surface of the media by moving arms. This intermittent operation means that the air enters the top of the system and the sewage passes as a film over the surface of the biofilm. Air is also drawn into the system through air bricks set at the sides and through the free draining base. Trickling filters can be operated at organic loadings of 0.3–0.5 kg of biochemical oxygen demand (BOD)/m^3/day.

Biofilms are held together by extracellular polysaccharides excreted by bacteria and enable high numbers of micro-organisms to be maintained in the systems. Organic matter is adsorbed onto the surface of the biofilm and degraded by aerobic bacteria. Microbial activity in biofilms is determined by the concentration gradients created between the internal conditions and the bulk water environment. In aerated aquatic environments, diffusion of oxygen is limited to the surface layers, and progressively more reduced conditions occur with increased depth. Thus, a sequence of oxidation and reduction reactions can occur at different depths within micro-environments. Concentration gradients can create interactions of conditions that favour a variety of microbial processes and enhance nutrient cycling within solid phases.

Rotating biological contactors

RBC is one of the principal types of fixed-film moving medium systems used as digesters. Biological growth is established on the disc surfaces made of polystyrene, polyethylene, polypropylene, stainless steel, cement, aluminium, glass, polyvinyl chloride (PVC), rubber, teflon, wood, wire screens, and so on.

The discs are 2–3 m in diameter and 10–20 mm wide and are mounted on a horizontal shaft. The distance between adjacent discs is 20 mm. The discs are partly submerged (40% area in the medium). They are rotated at 1–7 revolutions per minute. Air is sparged to reduce the risk of anaerobiosis in multiple unit systems. The retention time of the medium is comparable with that of the percolating bed, that is, 7 minutes for a 50-disc array (Figure 4). The biomass on the medium (discs) is of the order of 200 g dry weight per square metre of disc surface, which is equivalent to 40–60 kg/m^3 mixed liquor volatile solids in the activated sludge system. The biological growth is 2–4 mm thick. The standard loading allowed is 6–20 g BOD per square metre of disc per day for domestic sewage, and 80%–90% BOD reduction occurs. The system is also used for industrial waste water treatment.

Some areas in the waste water treatment process in which the RBC systems find applications are as follows: (1) secondary treatment, (2) nitrification, (3) landfill leachate and run-off, (4) pretreatment of water supplies, (5) dentrification, and (6) phosphate removal.

Biofilm development on a surface exposed to fluid flow is the net result of several physical, chemical, and biological processes, which are as follows.

- Transport or adsorption of organic molecules to the surface
- Transport of microbial cells to the surface
- Micro-organisms attached to the surface.
- Microbial transformations (growth and exopolymer production) resulting in biofilm
- Partial detachment of biofilm caused by the fluid shear stress

Fluidized-bed reactors

Fluidized-bed reactors (FBRs) are also referred to as fluidized-bed biofilm reactors (FBBRs) and were introduced in 1980. The reactor is a combination of attached growth (percolating filter) and suspended growth (activated sludge) systems. Biological slime film is developed and maintained on a solid support medium consisting of particles small enough to be maintained in suspension by the upward flow of the liquid being treated. The support medium particles neither sink nor flow out. The reactors are generally cylindrical

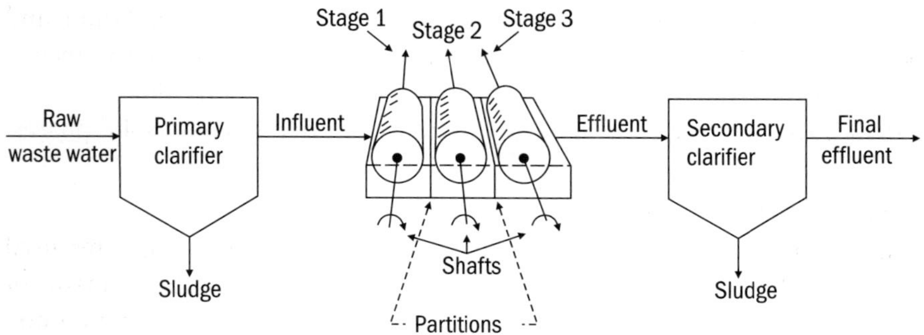

Figure 4 Rotating biological contactor reactor

with perforated distribution plates and tapered or conical entry sections. Fluidization of support particles is allowed, but clumping is prevented. The biofilm has gelatinous matrix composed of polysaccharides. It is a non-viable excretion or secretion. It acts as a binder for micro-organisms (Jogdand 1995).

Expanded bed reactor

The basic mode of operation of an expanded bed reactor (EBR) is similar to that of the packed filters and fluidized-bed processes. Fluidized-bed processes were developed earlier and have diverse applications. The expanded bed system uses the operating mode of the FBR.

Many consider the EBR and FBR to be identical, but the EBR is a better bioreactor because of (1) velocities to maintain the delicate attached living film; (2) separation, retardation, and biocoagulation of fine suspended solids; and (3) achievement of maximum biomass concentrations. Shearing of the microbial film is minimum in the EBR. The comparative efficiencies of the various reactors is summarized in Table 1.

Table 1 Comparison of performance of different treatment systems

Particulars	Trickling filter	Activated sludge	RBC	FBBR
Biomass concentration MLVSS	2000–7000 mg/l	700–2500 mg/l	10 000–20 000 mg/l	10 000–50 000 mg/l
Surface area m^2/m^3 reactor volume	12–30	—	40–50	800–1200
Process loading rate as BOD removal per m^3 of reactor, volume per day	—	0.5–1.2	—	8–16

MLVS – mixed liquour volatile suspended solids; BOD – biochemical oxygen demand; RBC – rotating biological contactor; FBBR – fluidized-bed biofilm reactor

Aerated lagoons

Aerated lagoons are basins having depths varying from 4 feet to 12 feet, in which oxygenation of waste water is accomplished by aeration units of either surface, turbine or diffuser types. The fundamental difference between aerated lagoons and the activated sludge system is that recycling of the sludge is provided in the latter as a means of controlling the amount of biological sludge in the aerator. Aerated lagoons are flow-through devices, that is, no recycle of sludge is provided. Solid concentration in the lagoon is a function of waste water characteristics and detention time. It is usually between 80 mg/l and 200 mg/l, which is much lower than that for conventional activated sludge units (2000–3000 mg/l).

Based on the turbulence level in lagoons, they are classified into two types: completely mixed lagoons and facultative lagoons (Figure 5). The important operational difference between completely mixed and facultative

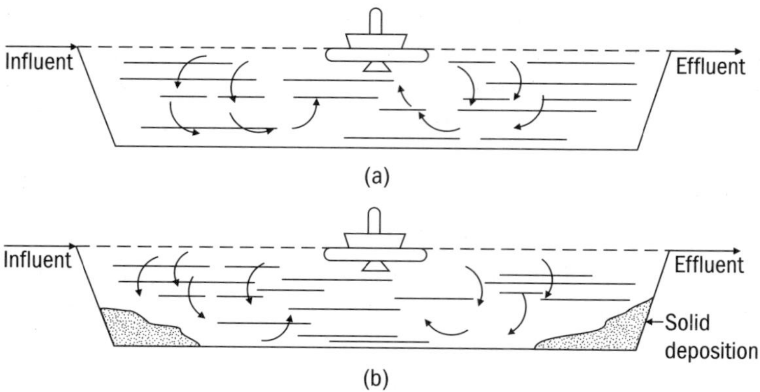

Figure 5 Aerated lagoons: (a) completely mixed lagoon, (b) facultative lagoon

lagoons pertains to the concentration of suspended solids in the effluent. For completely mixed lagoons, all solids are maintained in suspension, and the effluent will contain a much higher suspended solid concentration than the effluent from a facultative lagoon, and consequently, a high value of the insoluble BOD is achieved. If a high-quality effluent is desired, it may be necessary to follow the treatment in the completely mixed lagoon by a sedimentation basin in order to remove most of this insoluble BOD.

Waste water stabilization ponds

Waste water treatment with stabilization ponds is feasible when large land areas are available at low cost, and high-quality effluent is not required. The dissolved oxygen (DO) concentration in the waste water drops to a very low level, and anaerobic degradation prevails if the BOD of the influent is high and oxygen demand is above that provided by photosynthesis and natural surface aeration. The terminal products for the anaerobic degradation of carbonaceous matter are CH_4 and H_2O, in contrast to CO_2 and H_2O obtained during aerobic degradation. Ponds in which the upper layers are aerobic and the lower are anaerobic are referred to as facultative ponds (Figure 6). Most stabilization ponds fall under this category.

When several ponds are operating in series, the first one receiving raw waste water discharge is anaerobic and the second, which receives partially

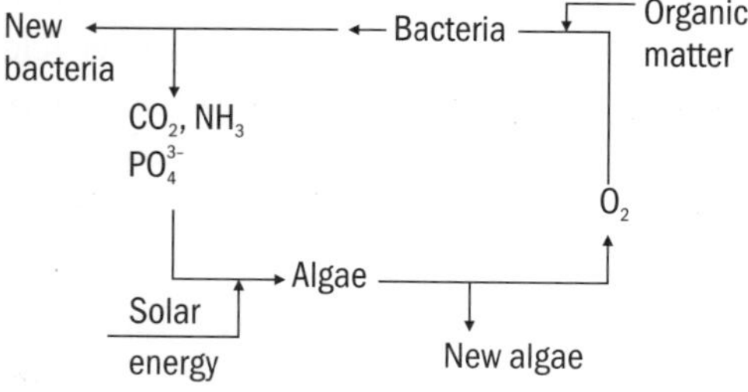

Figure 6 Reactions in a waste water stabilization pond

stabilized waste water from the first, may be facultative. The last one receiving relatively low-BOD waste water discharge from the preceding one might function as an aerobic pond. Because of the high detention time, frequently about two months, removal of refractory organic materials, which cannot be accomplished by activated sludge or aerated lagoon processes, may become possible in stabilization ponds. Thus, a convenient arrangement may be to provide stabilization ponds following an activated sludge (or aerated lagoon) unit to complete the stabilization (Curds and Hawkes 1983).

Anaerobic bioreactors

Anaerobic treatment is applied for waste water as well as for the digestion of sludge. The end products of anaerobic degradation are gases, mainly CH_4, CO_2, and small quantities of H_2S as well as H_2. Some of the advantages of anaerobic treatment over aerobic are as follows.

- Very low sludge production
- Lower consumption of energy
- Production of CH_4, which is of high calorific value
- Possibility of operation at high organic loading rates
- Absence of environmental nuisance of odour or aerosol as in the case with aerobic treatments
- The advantage that micro-organisms can remain dormant for several months and become operational within a week of start-up. This is suitable for seasonally produced waste water.

However, anaerobic treatment systems have some disadvantages also over aerobic treatments, such as high retention time, production of bad odour from gases like H_2S, mercaptans, and so on, and high cost of establishment.

Development of anaerobic technologies for waste water treatment

Anaerobic treatment systems were not considered until recently as viable technologies for treating waste water from pulp and paper mill, sago industries, food processing industries, and so on, as they were mostly considered too dilute for carrying out feasible anaerobic treatment.

These processes were also previously regarded as being too sensitive to inhibitory compounds. The recent advances in the identification of inhibitory substances in the effluents, as well as increasing insight into the biodegradable capacity and the toxicity tolerance of anaerobic micro-organisms, have helped to demonstrate that anaerobic treatment of various inhibitory waste water is feasible.

Single stage anaerobic reactors

The contact process

The contact process, developed in the 1950s, is basically the anaerobic version of the aerobic activated sludge process, employing external settlers and sludge return (Figure 7). The contact or recycled floc process comprises a continuously fed, completely mixed reactor followed by solid/liquid separation. In practice, the main difficulty appeared to be the separation of the sludge from the treated water. This difficulty was largely due to the existing concept that excessive agitation in the bioreactor was necessary. The idea was that the more intensive the mixing, the better would be the contact between the sludge and the waste water.

It is now clear that intensive mixing is also highly undesirable in that respect since these balanced micro-ecosystems are very important for the rapid degradation of intermediates. Sludge concentration in the contact process rarely exceeds 5–10 g VSS/l, and the superficial liquid velocity in the settler should not be greater than about 1 m/h to allow sufficient settling of the flocculated sludge.

Contact digesters are currently used for treating waste from sugar industries and distilleries, citric acid and yeast production factories, production of saurkaraut, canned vegetables, pectin, starch and meat products, and so on. Contact digesters are not much affected by the suspended solids in the feed as this process takes place in retained biomass types of digesters. In food processing waste treatment, the sludge retention time (SRT) is 4 days at 35°C with 95% chemical oxygen demand (COD) removal. In starch manufacturing units employing waste water treatment by contact digesters, 97% COD removal occurs from COD 10^4 g/m^3 at 2.4 kg COD/m^3 day feed. In yeast production units' waste water (COD 45 000 g/m^3), 80% COD removal was achieved with 4 kg/m^3 day at 35°C.

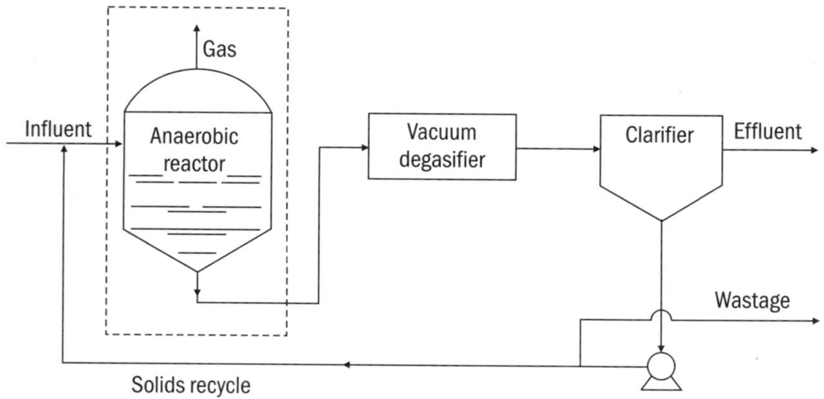

Figure 7 Anaerobic contact process

Continuously stirred tank reactor

The anaerobic CSTR is one of the oldest processes used for the stabilization of sludge. The major application has been, and remains to date, in the stabilization of concentrated sludge produced from the treatment of waste water (Figure 8). To maintain an anaerobic treatment system that will stabilize an organic waste efficiently, the non-methanogenic and methanogenic bacteria must be in a state of dynamic equilibrium. The key factors that govern bacterial growth are as follows.

- Retention time
- Bacterial–substrate contact (mixing)
- Suitable pH
- Favourable temperature
- Adequate concentration of proper nutrients
- Absence (or assimilation) of toxic materials
- Sludge organic composition and concentration

The disadvantage of the anaerobic CSTR treatment is the slow growth of methanogenic bacteria, which results in a longer detention time in the digester for adequate waste treatment.

Anaerobic filter

Anaerobic filters, also referred to as submerged filters, are systems designed to perform anaerobic treatment employing attached growth biomass (Figure 9).

The waste water in the anaerobic filter (fixed-bed) system passes the reactor usually with vertical flow, either upflow or downflow. Various media types have been employed in both upflow and downflow anaerobic filters, of average diameter 0.2–6.0 mm, and are made up of both porous and nonporous material such as stone, plastic, sand, and so on. The micro-organisms in the anaerobic filter reactors get attached to an inert medium in two stages:

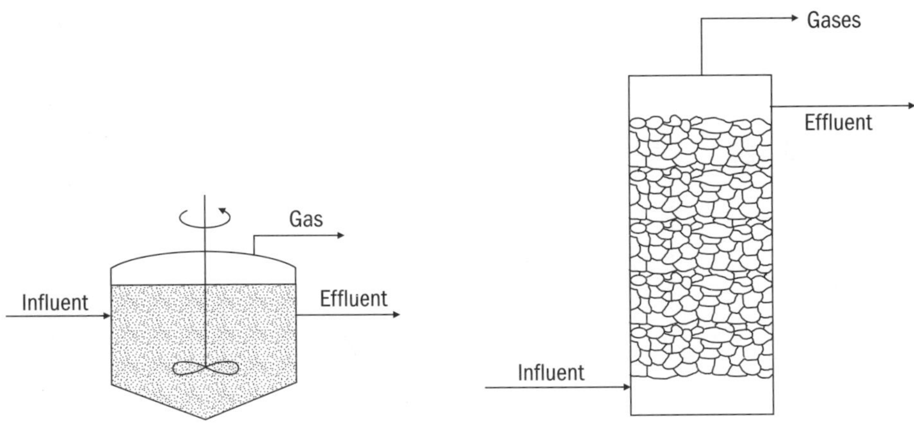

Figure 8 Continuously stirred tank reactor

Figure 9 Anaerobic filter

1 An initial electrostatic attraction to the surface roughness providing shelter against liquid shear forces.
2 Permanent binding by the extracellular secretions.

Waste water is passed upwards or downwards at low velocities through the packing material, providing contact of waste water with attached and entrapped anaerobic biomass, thus allowing degradation of the influent organic matter. Anaerobic filters operate at an HRT in the range of 1–10 days and a COD loading of 4–16 kg/m^3/day. Anaerobic filters have achieved COD removal efficiency of 60%–70%. Their drawback is the substantial volume of packing material that occupies the reactor. Besides these, upflow filters are easily clogged by the accumulated biomass and the suspended solid content of the influent. Clogging of filters leads to channelling, resulting in poor performance of the reactor.

The anaerobic filter constitutes an effective system for anaerobic treatment of soluble waste. No recycling is required since the biomass remains attached to the filter packing and therefore is not lost in the effluent. It is possible to operate anaerobic filters at temperatures lower than those used in the operation of the anaerobic contact process, owing to the high biomass concentration present in the filter. Treatment of waste water containing high concentration of suspended solids is not possible in this system because of clogging problems.

Packed bed reactors—packed column reactors

In packed column reactors (PCRs), the micro-organisms are contained within the packing medium in an enclosed vessel, and the liquid waste passes upwards. The organisms do not form a slime on the packing. Regular backwashing will prevent clogging. This will also prevent high concentration of suspended solids sloughing continuously into the effluent. The treatment rate and efficiency are directly related to surface area of the packing (Figure 10). PCRs are 10–15 times and sometimes even 30 times more efficient than CSTRs.

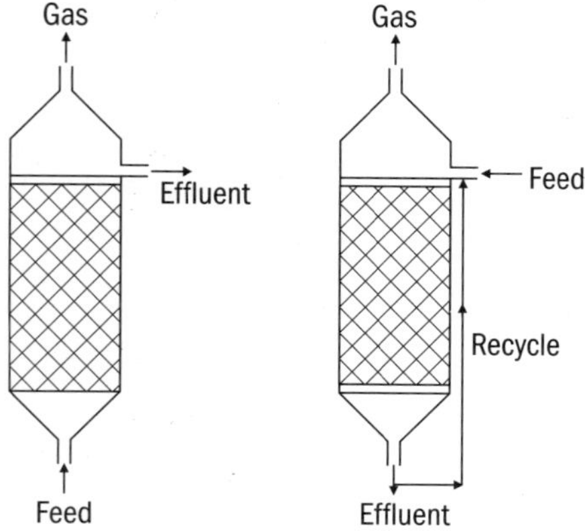

Figure 10 Upflow (left) and downflow (right) packed bed reactor

Anaerobic filters (columns) packed with anthracite coal particles were used for the denitrification of uranium recovery waste water and were effective for effluents with concentrations of 500–5000 mg NO_3-N per litre. Polystyrene spheres in an anaerobic filter or the Neptune-Microfloc filter containing sand, silica, and anthracite coal are used.

Anaerobic baffled reactors

Anaerobic baffled reactors (ABRs) have walls across the tank, which is built from the top to the bottom, so that waste water flowing along the tank has to go alternately under and over the baffles. The baffles tend to keep the bacteria in the tank and also help to prevent problems with the floating solids. ABR is a hybrid anaerobic reactor (Figure 11), and this design offers the advantages of the reactors in series, in the form of high efficiency, and low bypass and resistance to shock loading, together with high biomass retention capacity and high specific activity of the acetoclastic methanogenic biomass in the first chamber of the reactor where high acetate concentration can be found. A COD removal of 95% is achieved with an HRT of 20 h at 20 kg COD/m^3/day.

Fixed-film units

Fixed-film reactors are used for tertiary treatment of industrial and sewage waste water. Typically, these reactors contain rock media or a plastic saddle, which provide points of attachment for the bacteria. The waste stream enters the reactors from the base and passes upwards through the media. Contaminants are trapped on the media and become available for treatment. The media also serve as a biofilter for gases and suspended solids. This system can retain the biomass for long periods but potentially has a propensity for plugging.

Anaerobic fixed film reactor

The anaerobic fixed film (AFF) process is also characterized by the presence of a stationary packing material in the reactor, but here the aim of the design is to avoid the entrapment of suspended solids, thus making the process feasible for pretreatment/treatment of wastes containing a considerable amount of solids

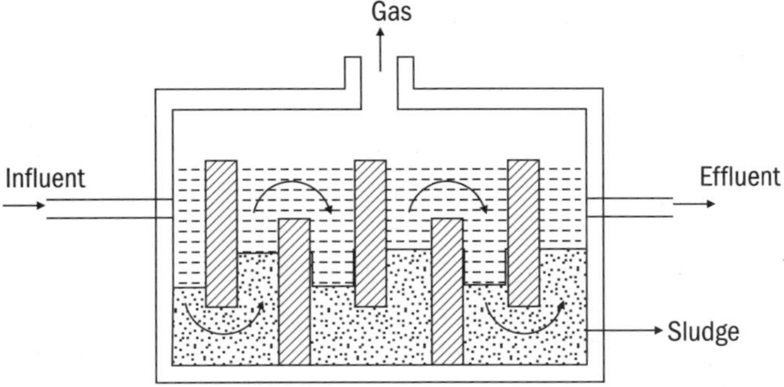

Figure 11 Anaerobic baffled reactors

such as screened manure. In order to prevent the accumulation of solids in the reactor, the AFF process is mostly operated in a downflow mode.

The fixed-film reactor is called a downflow stationary fixed film (DSFF) reactor. DSFF is a true biofilm process in which a high degree of mixing is concentrated in the top, where the activity is higher. In the DSFF reactor, the biomass is immobilized because of the attachment of the biofilm to the support media.

Anaerobic fluidized-bed reactor

In an AFBR, the biomass grows as a biolayer around heavy, small particles. These biolayer-covered particles are maintained in a fluidized stage by an upward-directed flow of water. This configuration has many advantages over other processes and was first recognized for aerobic treatment of waste water. In the early 1970s, the principle of expanded and fluidized beds was applied to anaerobic systems. Systems with a low bed expansion (20%) are generally called EBRs, while in common fluidized-bed reactors, the bed expansions are in the range of 30%–100% or even more.

The AFBR also is an upflow reactor and contains an approximately uniformly sized support medium on which the biofilms grow. The medium is expanded, or fluidized, by the velocity of the influent plus a recycle loop. Owing to the heavy particles (for example, sand) with a high settling velocity (around 50 m/h), high liquid velocities in the reactor (10–30 m/h) can be employed. Because of this high liquid velocity, the inert sediment present in the waste water does not accumulate in the reactor, and therefore, the sludge activity is high. Typical biomass concentrations are about 40 kg/m^3 owing to the very high attachment surface on the small carriers (having a typical diameter of 0.2–0.8 mm) of more than 2000 m^2/m^3. The substrate present in the liquid phase diffuses into the biofilm and gets converted to VFAs and ultimately to CH_4. These products then diffuse out through the biofilm into the bulk liquid (Aivasidis and Wandrey 1988). These reactors have typical loading rates of 25 kg COD/m^3 per day. However, as the biofilm grows, the film-covered particles increase in size, which is accompanied by a decrease in their composite density. This causes the particle to move up in the bed, ultimately resulting in its leaving the reactor and leading to a reduction in the carrier particle concentration inside the reactor (Figure 12). This problem can be overcome by removing the biofilm from the carrier particle that has exited the reactor and then recycling the carrier particle (minus the biofilm) to the reactor. Because of the high biomass concentration and activity, a very high treatment capacity is obtained. Fluidized-bed reactors are, therefore, very compact and require very little space. Compared with other high-rate anaerobic reactors, for example, upflow filter reactors and UASB reactors, the fluidized-bed system is claimed to have the potential advantages in the form of higher purification capacity, no clogging of the reactor (as in filters), no problems of sludge retention (as in UASB systems), and small volume

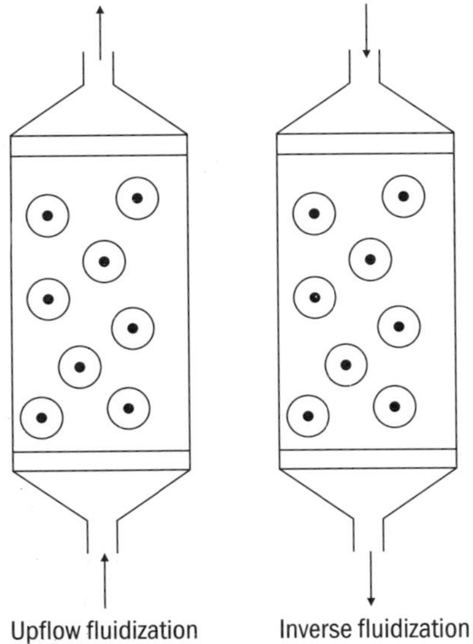

Figure 12 Fluidized-bed reactor

requirement. The major disadvantages of EBRs/fluidized-bed reactors are that they must be highly engineered systems and also that fluidization requires power.

Anaerobic sludge bed reactors

The anaerobic sludge bed (ASB) reactors undoubtedly are by far the most popular anaerobic treatment system so far. The sludge retention of the ASB reactor is based on the formation of easily settling sludge aggregates (flocs or granules) and on the application of an internal gas/sludge/liquid separation system (GSS) device.

Upflow anaerobic sludge blanket reactor

The UASB, which was developed in the Netherlands in the early 1970s (Lettinga, van Velsen, De Zeeuw, *et al.* 1980), is by far the best known reactor-based ASB concept (Figure 13). The UASB reactor is a proven technology, which represents the most used system worldwide (Frankin 2001). More than 1000 full-scale UASB installations have been put in operation until 1998. Most of these full-scale reactors are used for treating agro-industrial waste water, but their application for waste water from chemical industries and sewage is increasing.

The UASB digester requires active bacteria in the form of a high-density granular sludge, which is retained in the digester tank despite gassing and upflow velocity of waste water. Initially, 10%–15% inoculum of granular sludge is required. Hydrodynamics of the digester created by the feed distribution and the shape of digester are important in retaining the correct granular form.

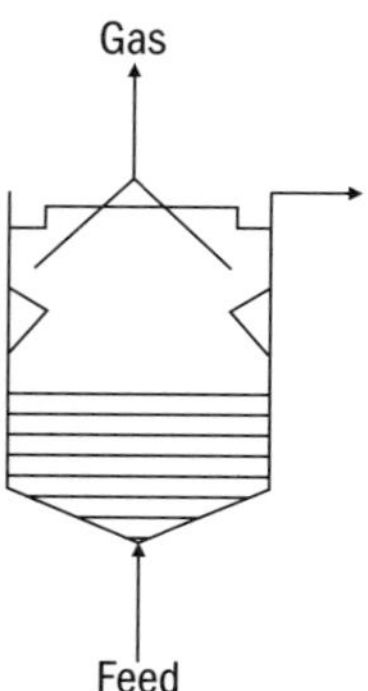

Figure 13 Upflow anaerobic sludge blanket reactor

The main operating problems with the UASB digesters are start-up and maintenance of the flocculated granules. Wastes with a high proportion of insoluble COD (>15%) can give rise to poor granulation. Granulation is a natural process that depends on microbial morphology, nature of the substrate, accumulation of inorganic salts, and so on. Formation of rapidly settling granules from the inoculum of ordinary biomass sludge from stirred tank or other digesters may take only 50 days. The key elements in the feed substrate for successful granule formation are calcium, phosphorus, magnesium, ammonia, aluminium, and silicon. A large population of filamentous methanogens is also essential. Granular sludge develops with the dissolved waste. Baffles in the digester, along with the shape of digester and upward liquid velocity, promote gas–solid separation. Baffles provide the surface area that reduces the upflow velocity and promotes biomass flocculation. Long-chain fatty acids prove toxic to methanogenic bacteria, and addition of calcium chloride can reduce this toxicity and can help in better granulation (Kamaraj, Doraisamy, Gunasekar *et al.* 2001).

The following general conditions favour the successful operation of UASB reactors.

- Loadings of 12–15 kg COD/m^3/day for waste of COD > 1000 mg/l.
- Feed strength not exceeding 10 000 mg/l COD; for high strength waste effluent recycle may be employed.
- A liquid upflow velocity of about 1 m/h.
- Gas loadings of 70–200 m^3/m^2/day to prevent biomass washout.
- Preliminary treatments like rough screening (before the pump) and medium screening (7–12 mm); and grit, oil, and grease removal (to reduce scum build-up).

In UASB reactors, the biomass retention is possible within the reactor without the supply of a carrier material, thereby improving the economics of the process. Moreover, the granular aggregates offer considerable advantages over both flocculent and surface-attached biomass (Sallis and Uyanik 2003).

In UASB reactors, the unique spatial arrangement of bacteria within an anaerobic granule and the phenomenon of granulation offer a number of advantages (McHugh, Carton, Mahony, *et al.* 2003).

- More efficient microbial proliferation.
- Access to resources and niches that cannot be utilized by isolated cells.
- Internal physico-chemical gradients within the aggregates.
- Collective defence against antagonists that eliminate isolated cells.
- Optimization of population survival by differentiation into distinct cell types.
- Continuous operation of reactors beyond normal washout rates.
- Generation of a reactor effluent with low suspended solids.
- Manipulation of biomass in a single phase.
- Manipulation of growth rates independent of the dilution rate.

The formation and stability of the granulation are essential for successful operation. The reactor contains three compartmental units, that is, sludge bed, sludge blanket, and separation zones. Sludge bed is at the base of the reactors. The wastes are passed through this bed under minimal agitation, which ensures uniform distribution of sludge wastes and maximum contact between bacteria, waste, and nutrient utility. The second compartment is the sludge blanket, which occupies about 60% of the reactor volume. It contains flocculated sludge and minute gas bubbles, which aid in achieving ideal mixing of the remaining wastes. The third unit of the separation zone takes up the balance of the volume within the reactor. This collects the gas and releases the biomass that may be attached to the gas bubbles.

The waste water passes upwards through the granular sludge bed. Here, the organic compounds are biologically degraded and biogas is produced. In the top and just above the granular sludge bed, a fluidized zone develops as a result of the production of biogas. In this zone, further biological degradation can take place. The biogas is separated from the liquid in the gas sludge liquid (GSL) separator. Granules with good settling abilities settle back through the fluidized zone to the granular sludge bed, while flocculated and dispersed bacteria are washed out of the reactor with the effluent. The formation of granular sludge, which is of major importance for the achievement of an efficient high-rate reactor, depends on the type of substrate used. Compared to the other advanced anaerobic filters and FBRs, the UASB process is able to retain a high concentration of biomass with high specific activity and handle high organic rates with good COD removal. The capital cost for the UASB process is lower than that for other anaerobic processes, since the separation of gas, liquid, and solids takes place entirely in the reactor, and no support media for bacterial attachment are required. In addition, the process has neither the clogging problem nor the high energy requirements by FBRs. Therefore, its use in anaerobic cellulosic waste water treatment becomes very significant.

Upflow anaerobic hybrid reactors

Hybrid anaerobic digester was a new design developed by Guiot and Van den Berg (1984). The upflow anaerobic hybrid reactor (UAHR) or the upflow sludge bed–filter reactor combines the UASB concept and random packing of a matrix on the top portion of the reactor, which minimizes channelling problem associated with upflow anaerobic filter (UAF) and loss of biomass due to floatation associated with poorly performing UASB reactors (Henry, Donlon, Lens, *et al.* 1996). Additionally, the packing material enhances the development of granular sludge, which favours the methanogenic population, increasing the capacity of the reactor to resist load shocks, the changes in the substrate features, and the presence of inhibiting compounds. The reactor is efficient in the treatment of dilute to high-strength waste water at high organic loading rate (OLR) and short HRT. The UAHR reactor achieved a COD removal efficiency of 86% with the HRT of 5.6 h and OLR of 2.35 kg COD/m^3/day in treating sago waste water (Rajesh Banu, Kaliappan, and Diter 2006). The hybrid anaerobic reactor also operated with distillery spentwash at the organic loading rate of 23.25 kg COD/m^3/day, leading to 64% COD reduction (Acharya, Mohana, and Madamwar 2008)

Expanded granular sludge bed reactor

The expanded granular sludge bed (EGSB) systems employ granular sludge, which has exceptionally good settling characteristics unless very high sludge loads are applied or the settleability deteriorates for some reason. Because of the high settleability of the sludge, superficial liquid velocities exceeding 6 m/h, sometimes even higher than 12.5 m/h (McHugh, Carton, Mahony, *et al.* 2004), can be applied. These high liquid velocities, together with the lifting action of the gas evolved in the bed, lead to a slight expansion of the sludge bed. As a result, an excellent contact between sludge and waste water prevails in the system, leading to significantly higher loading potentials compared to conventional UASB installations.

The EGSB reactor has the potential for the treatment of low-strength waste water (McHugh, Carton, Mahony *et al.* 2004). It also has the potential for treating wastes containing highly toxic and/or recalcitrant compounds. A full-scale version (275 m^3 and 550 m^3) of EGSB has successfully treated waste water containing formaldehyde at a loading rate of 10 kg COD/m^3/day, achieving 90%–98% COD removal (van Lier, Tilche, Ahring, *et al.* 2001).

A special version of the EGSB concept is the so-called internal circulation (IC) reactor, in which the biogas formed is separated halfway up the reactor by means of partitions and then led upwards through a pipe. The lifting forces of the collected biogas are used to bring about the recirculation of granular sludge over the lower part of the reactor, which results in improved contact between sludge and waste water. The biogas is separated in an expansion device. The well-settleable solids (granular sludge) then easily separate and drop back to the bottom of the reactor via another pipe. The extent of liquid/sludge recirculation depends on the gas production. A disadvantage of the IC–UASB concept might be that the construction is rather complex, but the big advantage is that the reactors occupy little space because they are tall, that is, up to approximately 20 m.

Among the anaerobic waste water treatment technologies, the UASB/UAHR systems are the most widely accepted ones, which have found successful application for a vast number of industrial effluents. The use of these reactors, with modifications in the design for the treatment of industrial effluents, is summarized in Table 2.

Treatment of industrial effluents using high-rate reactors

Cassava starch factory effluent

The cassava starch production in our country is mainly concentrated in small to medium scale factories, which generate 30 000 l to 40 000 l of effluent per tonne of sago produced (Figure 14). There are about 1050 industries

Table 2 Examples of the upflow anaerobic sludge blanket (UASB) and upflow anaerobic hybrid reactor (UAHR) applications

Reactor	Substrate	Organic loading rate (kg/ m^3/day)	COD removal efficiency (%)	Reference
Two-stage UASB reactor	Wine vinasse	0.3–7.3	15–92	Kalyuzhnyi, Gladchenko, Sklyar, *et al.* (2001)
UASB reactor	Municipal waste water	—	70–90	Singh and Viraraghaven (2001)
Two-stage UASB reactor	Soyabean curd	50	90	Tawaga, Takahasti, Sekiguchi, *et al.* (2002)
UASB	Fish processing waste water	5–8	78–92	Palenzuela-Rollon, Zeeman, Lubberding, *et al.* (2001)
Anaerobic hybrid reactor	Purified terephthalic acid	9	90	Young, Kim, Page, *et al.* (2000)
Anaerobic hybrid reactor	Baker's yeast factor effluent	9.0	78	Buyukkamaci (2002)
Anaerobic hybrid reactor	Industrial cluster waste water	2.08	94	Kumar, Yadav, Sreekrishnan, *et al.* (2008)
Anaerobic hybrid reactor	Domestic waste water	3.5	80	Jianlong, Hanchang, Yi, *et al.* (2000)
Anaerobic filter + anaerobic hybrid reactor	Sewage	0.9–1.1	70	Elmitwalli (2000)

COD – chemical oxygen demand

UAHR of 225 l capacity installed at sago industries reduced the biochemical oxygen demand (BOD) of the effluent by 99% with 15 days hydraulic retention time (HRT) with the gas production of 909 l/kg of total solids reduced at HRT of 15 days.

Figure 14 Field-scale upflow anaerobic hybrid reactors (UAHR) for cassava starch factory effluent

producing starch/sago from cassava in five districts of Tamil Nadu alone. The effluents from these industries are acidic and highly organic in nature with a COD of 5000–7000 mg/l during the season and 1000–5000 mg/l during off-season.

Field-scale UAHRs of 225-l capacity were fabricated for biomethanation of cassava starch factory effluent (CSFE). Pre-processed coconut shells and polyvinyl chloride (PVC) pall rings were used as media for the micro-organisms.

The reactors were initially loaded with raw effluent mixed with the inocula (the supernatant from an anaerobic lagoon treating CSFE) in the ratio 2:1. The raw effluent had a total solids of 2230 mg/l with a pH of 4.5. The supernatant from the anaerobic lagoon (inoculum) had total solids (TS) of 1827 mg/l with a pH of 6.6. The resulting pH after the mixing of effluents from these sources was 4.7. The reactors were operated in the batch mode, and the variations in the pH were closely monitored.

The reactors took almost five weeks for start-up. The start-up HRT was 15 days. After running the reactor for nine weeks at this HRT, the hydraulic loading rates (HLRs) were successively increased by reducing the HRT gradually to 11, 8, 6, 4, 3, 2, and 1 days, allowing stabilization periods in between.

During the entire period of investigation, the pH of the influent varied from 4.7 to 5.3, while the effluent pH remained almost neutral (pH 6.8–7.4). This is an indication of the stable operation of the reactors throughout the period of operation. Up to six days of HRT, both the reactors exhibited more or less a steady performance, irrespective of the influent concentrations, and thereafter, a decreasing trend was observed in BOD and COD reduction, specific gas production, CH_4 content, and so on due to the increased loading rates. The change was sharp between the 4-day HRT and the 2.5-day HRT because of the drastic change in HLR (60% increase). Very high BOD reductions of 99% for both reactors were obtained at the longest HRT of 15 days (Table 3). The BOD reductions went down with the decrease in HRT, but the decrease in reduction was slow up to the 6-day HRT. Thereafter, the decrease was rather sharp owing to the steep hike in the HLR. The lowest reduction of 78.9% and 77.4% BOD occurred at 1-day HRT.

Gas production rates are the most important indicators of performance for anaerobic reactors. The mean daily gas productions of reactors registered an increase of 7.8 to 8.2 times when the HRT was reduced from 15 days to 1 day, while the specific gas production (l/kg TS) decreased from 909 l (at 15-day HRT) to 574 l (at 1-day HRT). The decrease in TS was 36.8%. A sharp increase in volumetric gas production with reduction in HRT was evident for both the reactors, as the biogas productivity got reduced as the HRT was shortened. The CH_4 content of biogas was in the range 64%–74% for both reactors.

Coffee processing waste water

Coffee is a major plantation crop grown worldwide and is one of the most popular beverages consumed throughout the world. India ranks sixth in the world in coffee production. The average annual production of coffee berries is 0.30 million tonnes from an area of 0.35 million hectares. Both arabica and robusta are cultivated mainly in the hilly tracts of South India and north-eastern states. Coffee is processed either by wet or by dry method. Wet method of processing results in a coffee of superior quality compared to the dry method. Presently in India, about 75%–80% of arabica and 15%–20% of robusta are processed by the wet method. Wet processing of coffee consumes a large quantity of water at different stages of its processing. The resultant

Table 3 Performance of a hybrid reactor at different hydraulic retention times (HRTs) during biomethanation of cassava starch factory effluent (CSFE)

	HRT (days)							
Parameter	15	11	8	6	4	2.5	1.67	1
BOD reduction (%)	99.0	98.6	98.8	98.6	95.6	87.5	83.0	78.9
COD reduction (%)	96.2	96.4	96.2	96.1	93.2	86.3	81.8	77.4
Specific gas production (l/kg TS added)	909	903	919	881	822	775	723	574
Biogas productivity (l/l feed)	3.9	3.9	4.24	4.1	3.45	3.02	2.62	2.04
Volumetric gas production (l/m^3 reactor)	260	356	527	683	863	1210	1576	2038
Methane content (%)	70.0	71.0	72.0	72.0	70.5	68.0	66.0	65.0

BOD – biochemical oxygen demand; COD – chemical oxygen demand effluent contains very high quantities of total suspended and dissolved solids and has a COD and BOD of 6420–8480 mg/l and 3800–4750 mg/l, respectively.

Two types of reactors, that is, the UASB and the UAHR, were fabricated with capacities of 16.75 l and 19.25 l, respectively, for the treatment of coffee processing waste water (CPWW). The reactors were seeded with an anaerobic consortium developed in the laboratory from the enrichment of CPWW with cow dung slurry and goat rumen fluid (Figure 15).

The UAHR was operated at different HRTs of 24 h, 18 h, 12 h, and 6 h. Among the different hours, 18 h performed better in the reduction of pollution load with BOD, COD, and TS reduction of 66.0%, 61.0% and 58.0%, respectively (Table 4).

The population of total anaerobes and methanogens was monitored during the anaerobic process, both in UAHR and UASB, to elucidate the process, and was correlated with the reduction in pollution load, gas production, and so on. The results clearly indicate that biomethanation is an amenable tool to manage CPWW.

In order to enhance the efficiency of the reactor in biomethanation of CPWW during winter, cold-tolerant methanogens were developed by the enrichment of the methanogenic consortium from a cow-dung-based biogas plant from Ooty, Tamil Nadu, where the minimum temperature is usually near-zero, and were acclimatized with CPWW. These cultures produced 16%–18% more gas from CPWW during winter. The effect of different additives in enhancing the biogas production of CPWW was also assessed. Among the different additives, urea at the rate of 5 g/l performed better in enhancing the biogas production from CPWW.

Rubber latex processing waste water

Hevea brasiliensis is the most important commercial source of natural rubber, a product of vital importance recovered from its latex. During the processing

Table 4 Performance of hybrid reactor at different hydraulic retention times (HRTs) during biomethanation of coffee processing waste water (CPWW)

HRT (h)	COD removal (%)	BOD removal (%)	TS removal (%)	Biogas production (l/kg of COD reduction)	Biogas production (l/kg of BOD reduction)	Biogas production (l/kg of TS reduction)	Methane content (%)
24	70	71	64	280	490	535	61.8
18	61	66	58	430	775	840	60.7
12	52	59	49	390	620	690	59.4
6	46	54	42	264	400	440	50.4

COD – chemical oxygen demand; BOD – biochemical oxygen demand; TS – total solids

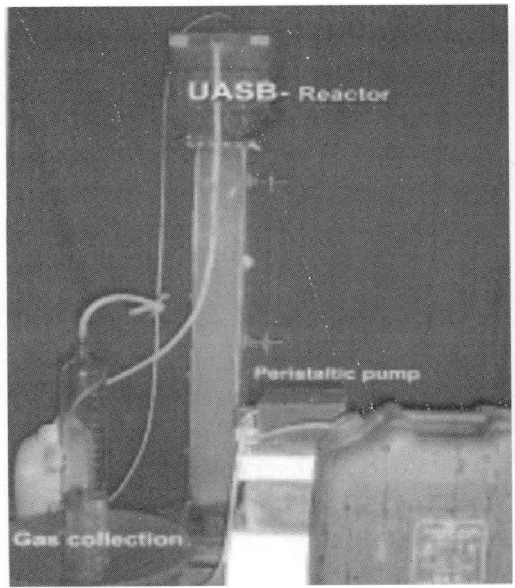

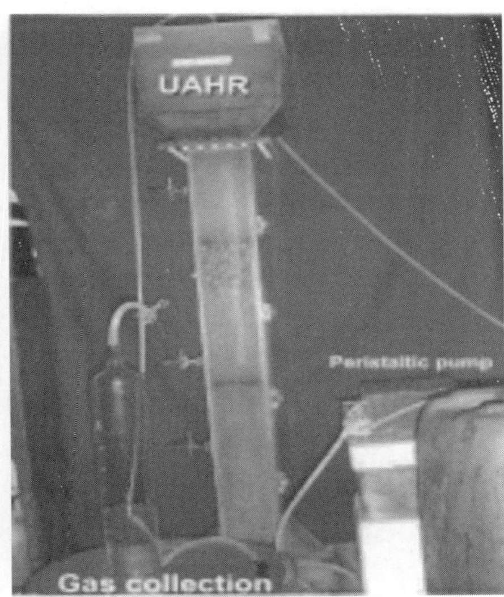

Figure 15 Biomethanation of coffee processing waste water (CPWW) in upflow anaerobic sludge blanket (UASB) reactor
Studies on the biomethanation of CPWW using high rate reactors indicated that upflow anaerobic hybrid reactor (UAHR) was found to be more effective in reducing organic pollution load in terms of total solids (TS), biochemical oxygen demand (BOD), and chemical oxygen demand (COD). A maximum COD reduction of 70.58%, BOD reduction of 71.87%, and total dissolved solid (TDS) reduction of 64.60% were recorded in CPWW because of biomethanation through UAHR. A quantity of 535.7 l, 489.13 l, and 281.25 l of biogas was produced per kilogram of TS, BOD, and COD removed with a methane content of 60.2% in the UAHR.

of rubber latex, large quantity of water is used, and the resulting waste water contains traces of rubber and substances like carbohydrates, proteins, lipids, organic and inorganic salts, and chemicals used in the processing. Various micro-organisms get proliferated using these substances and cause high BOD (4500–6000 mg/l) and COD (13 500–18 500 mg/l), which together result in an objectionable odour (Figure 16).

Biomethanation was carried out at lab and field scale using a high-rate anaerobic reactor, namely the UAHR. The lab-scale reactor was fabricated with a capacity of 16.75 l. The reactor was seeded with an anaerobic consortium developed in the laboratory. The consortium, which was from cow dung slurry and goat rumen fluid, was enriched with rubber latex processing waste water and used as a seed inoculum. After seeding, the reactor was fed with rubber latex processing waste water continuously at 24- and 12-h HRT. Between the two HRTs, the 24-h HRT performed better in reducing pollution load, with BOD, COD and TS reduction of 88.0%, 87.02%, and 68.0%, respectively.

During the start-up period, the pH of the treated effluent ranged from 5.26 to 6.51 in the UAHR reactor. The maximum biogas production was 0.431 l/day, which resulted in the removal of 25.7 kg/day of TS. Based on the results

The upflow anaerobic hybrid reactor (UAHR) installed at Co-operative Rubber Producer's Society, Elavampadam. Maximum biochemical oxygen demand (BOD) and chemical oxygen demand (COD) removal rates of 77.66% and 78.87%, respectively, were recorded with this reactor, and the maximum gas production observed was 3.1 m^3/day, with methane content of 67.8%.

Figure 16 High rate reactor for the treatment of rubber latex processing waste water

obtained from the lab studies, a field-scale UAHR was fabricated and installed at the Co-operative Rubber Producer's Society, Elavampadam, Kerala. The high-rate reactor was fabricated with steel sheets and coated with fibre-reinforced plastic (FRP) material. The high-rate reactor was randomly packed with PVC mesh. This PVC mesh was chosen because it provides the maximum surface area because of its ribs or corrugation, and micro-organisms can attach themselves over this area and remain in the reactor. The volume of the reactor was 1.44 m^3. Based on the results of the lab-scale reactor, for optimization of the process parameters, the field-scale reactor was operated with an HRT of 24 h. The reactor was monitored continuously for 18 months. During the operation period of the reactor, after it attained the steady state, the maximum BOD and COD removal rates were 77.66% and 78.87%, respectively. The maximum gas production of 3.1 m^3/day was achieved with a CH_4 content of 67.8%.

Basic kinetics in waste water treatments

In order to design installations and to predict the effluent quality of a waste water treatment plant, insight into the kinetics of the biological processes is required. For undissolved substrates, the situation is much less satisfactory because of the poor information available so far in the liquefaction (hydrolysis) step, which generally is the rate-limiting step.

The hydrolysis rate strongly depends on the degree to which the substrate to be hydrolysed is accessible to the exo-enzymes and is usually described by means of first order kinetics. This description of hydrolysis by first order kinetics is an empirical relation.

$$dF/dt = -k_h.F \quad ...(1)$$

where,
F = concentration of biodegradable solid substrate (g/l)
k_h = hydrolysis constant (day^{-1})
t = time (day)

After integration, this results in the following equation for batch reactors

$$F=F_0.e^{-k_h t} \quad ...(2)$$

and for a CSTR in steady state,

$$F=F_0/1+k_h.\theta) \quad ...(3)$$

where,
θ = sludge retention time (days)
F_0 = concentration of biodegradable substrate at time t = 0 (g/l)

With the description of hydrolysis by first order kinetics, only the concentration of the substrate is taken into account.

Degradation kinetics of soluble substrates

It is very difficult to mathematically describe the kinetics of anaerobic substrate degradation, because complex substrates and mixed bacterial populations are involved. The basic kinetic equations that are normally used are given below.

Monod kinetics

The best known expression for the specific growth rate of bacteria is the one given by Monod.

$$\mu = \mu_{max} S/K_s + S \quad ...(4)$$

where μ = specific growth rate (day^{-1})
μ_{max} = maximum specific growth rate (day^{-1})
S = concentration of the growth-limiting substrate (g/l)
K_s = affinity coefficient, that is, the substrate concentration at which $\mu = ½ \mu_{max}$ (g/l)

This equation applies only to very specific conditions, namely to suspended growth systems converting unequivocal substrates without nutrients and trace element limitation. In biofilm systems, mass transfer phenomena, for example, diffusion limitation, decrease the conversion rates.

It is convenient to have the kinetic relationships presented in a somewhat different form for water purification system. For example, activities (specific substrate consumption or product yield rate), rather than growth rates, are used. The relationship between the growth rate and the activity of sludge can be deduced as follows, based on the Monod equation. Because this equation does not take the decay rate of micro-organisms into account, it must be combined with

$$\mu_n = \mu - b \quad \text{...(5)}$$

where
μ_n = net (observed) growth rate (day^{-1})
b = decay rate (day^{-1})

which results in the following commonly used equation

$$\mu = Y\, r_{x.max} \,.\, S - b\, /K_s + S \quad \text{...(6)}$$

This equation can be combined with the equation for the maximum specific $r_{x.\,max}$ which is maximum amount of converted substrate per amount of biomass per unit of time.

$$M_{max} = r_{x.max}\, Y \quad \text{...(7)}$$

where
Y = Biomass yield coefficient (g $VSS_{produced}$/g $COD_{removed}$)
$r_{x.max}$ = Specific substrate consumption rate (g$COD_{removed}$/g VSS.d)

which results in the following commonly used equation

$$\mu = Y\, [(r_{x.max} \,.\, S)/(K_s + S)] - b \quad \text{...(8)}$$

These relationships can be worked out for completely mixed fermentation systems, with or without sludge retention or recirculation. The kinetic parameters used in the Monod equation can be determined rather easily; some of them by carrying out batch experiments and others by using a CSTR.

Maximum specific activity: r_x ($r_{x.max}$)

In principle, it is easy to determine the specific activity of anaerobic sludge. The equation describing r_x is as follows.

$$r_x = dS/Xdt \quad \text{...(9)}$$

where
X = biomass concentration (g/l)

The conversion of COD is accompanied by the production of biomass (Y) and methane (1–Y)

$$dM/dT = (1-Y)\, dS/dt \qquad ...(10)$$

where dM = methane production (g COD-CH_4/l)

For anaerobic sludge converting fatty acids, Y amounts to only about 0.015–0.020 kg $VSS_{product}$ /kg $COD_{removed}$. Hence, if one equals the COD consumed to the CH_4 COD formed, underestimation of approximately 2% is achieved.

Refinement of the Monod kinetics—inhibition

During CH_4 fermentation, substrate inhibition often occurs. In order to take this inhibition into account, a term can be included in the Monod equation. An example of this is the Haldane equation.

$$\mu = \mu_{max} \cdot S/[(K_s + S + S^2)/K_1] \qquad ...(11)$$

where K_1 = substrate inhibition constant (g/l)

Contois kinetics

A problem of the Monod kinetics is that the effluent concentration is independent of the influent concentration. This contrasts with the practical experience. This can be overcome using the description of growth rate given by Contois.

$$\mu = \mu_{max}\, S / (BX + S) \qquad ...(12)$$

where
B = Biodegradability constant
X = Biomass concentration (g/l)

The Contois equation has the saturation constant K_{gl} but makes saturation a function of the biomass concentration. This model is used to describe the conversion kinetics systems in which limitation occurs, that is, by mass transport or toxicity during sludge or manure fermentation. This model assumes that both decay and maintenance energy can be neglected.

K is defined as

$$K = BY \qquad ...(13)$$

where
K = kinetic parameter (g/l)
B = yield (g/g)

Then the following equation can be deduced.

$$\mu/\mu_{max} = (S/S_0)/ [\{K + (1 - K) S\}/S_0] \qquad ...(14)$$

Combining the above equation with $\theta = 1/\mu$ results in

$$\theta_x = \mu^{-1}_{max} + (K . \mu^{-1}_{max}) [(S_0 - S)/S] \qquad ...(15)$$

K and μ_{max} can be determined graphically by plotting θ_x against $(S_0 - S)/S$. The intercept of the line with the Y-axis indicates μ_{max}^{-1}, and its gradient is equal to K/μ_{max}. If for a given substrate the values of K and μ_{max} are known, then S can be predicted using the following formula.

$$(S/S_0) = K/ [(\theta_x - 1 + K)/\theta_{xmin}] \qquad ...(16)$$

Equation (16) indicates that the effluent concentration (S) depends on the influent concentration (S_0).

Conclusion

Development in bioreactors for the treatment of industrial waste water has proved that this emerging technology has developed a niche in the waste water treatment sector. High-rate reactors, such as the UASB, anaerobic hybrid reactor, and EGSB, have the potential for the treatment of a wide range of waste, from high strength to very low strength, from relatively tropical to temperate, from simple to complex, and even to those containing toxic or recalcitrant compounds. Even though there remains a constant state, the need for the development of novel processes is being felt, and an enhanced understanding of anaerobic reactors is likely to emerge in the coming decade to further extend the range of applications for waste water treatment.

References

Acharya B K, Mohana S, and Madamwar D. 2008. **Anaerobic treatment of distillery spent wash – a study on upflow anaerobic fixed film bioreactor.** *Bioresource Technology* **99**: 4621–4626

Aivasidis A and Wandrey C. 1988. **Recent developments in process and reactor design for anaerobic wastewater treatment.** *Water Science and Technology* **20**: 211–218

Barnes D and Fitzgerald P A. 1987. **Anaerobic wastewater treatment processes.** In *Environmental Biotechnology*. pp. 57–110. New York: John Wiley and Sons

Bitton G. 1994. ***Wastewater Microbiology***, pp. 139–293. **New York:** John Wiley and Sons

Buyukkamaci N and Filibeli A. 2002. **Concentrated wastewater treatment studies using an anaerobic hybrid reactor.** *Process Biochemistry* **38**: 771–775

Curds C R and Hawkes H A. 1983. ***Ecological Aspects of Used-water Treatment***, pp. 1–11. London: Academic Press.

Elmitwalli T A. 2000. **Anaerobic treatment of domestic sewage at low temperature**, pp. 60–68. Proceedings of the *VI Latin-American Workshop and Seminar on Anaerobic Digestion.*

Frankin R J. 2001. **Full-scale experiences with anaerobic treatment of industrial wastewater.** *Water Science and Technology* **44**: 1–6.

Gottschalk G. 1979. ***Bacterial Metabolism***, pp.167–224. New York: Springer

Guiot S R and Van den Berg. 1984. **Performance and biomass retention of an upflow anaerobic reactor combining a sludge blanket and a filter.** *Biotechnology Letters* **6**: 161–164

Hammer M J and Hammer M J Jr. 2003. ***Water and Wastewater Technology***, pp 82–85. New Delhi: Prentice-Hall of India

Henry M P, Donlon B A, Lens P N, Colleran E. 1996. **Use of anaerobic hybrid reactors for treatment of synthetic pharmaceutical wastewaters containing organic solvents.** *Journal of Chemical Technology and Biotechnology* **66**: 251–264

Jianlong W, Hanchang S, Yi Q. 2000. **Wastewater treatment in a hybrid biological reactor (HBR): effect of organic loading rates.** *Process Biochemistry* **36**: 297–303

Jogdand S N. 1995. ***Environmental Biotechnology***, pp. 24–59. Delhi: Himalaya Publishing House

Kalyuzhnyi S V, Gladchenko M A, Sklyar V I, Kizimenko Y S, Shcherbakov S S. 2001. **Psychrophilic one- and two-step systems for pre-treatment of winery waste water.** *Water Science and Technology* **44**(4): 23–31

Kamaraj S, Doraisamy P, John Gunasekar J, Murugesan V, Sampathrajan A. 2001. ***Anaerobic Processes in Waste Water Management***, pp. 89–95. TNAU Publication

Kumar A, Yadav A K, Sreekrishnan T R, Satya S, Kaushik C P. 2008. **Treatment of low strength industrial cluster wastewater by anaerobic hybrid reactor.** *Bioresource and Technology* **99**: 3123–3129

Lettinga G. 1984. **The prospects of anaerobic wastewater treatment.** In *Anaerobic Digestion and Carbohydrate Hydrolysis of Waste*, pp. 262–273. New York: Elsevier Applied Science Publishers

Lettinga G, van Velsen L, De Zeeuw W, Hobma S W. 1980. ***Anaerobic Digestion***, pp. 167–186. London: Applied Science Publishers

McHugh S, Carton M, Mahony T, O'Flaherty V. 2003. **Methanogenic population structure in a variety of anaerobic bioreactor sludges.** *FEMS Microbiology Letters* **219**: 297–304

McHugh S, Carton M, Collins G, O'Flaherty V. 2004. **Reactor performance and microbial community dynamics during anaerobic biological treatment of wastewaters at 16–37°C.** *FEMS Microbiology Ecology* **48**(3): 369–378

Palenzuela-Rollon A, Zeeman G, Lubberding H J, Lettinga G, Alaerts G J. 2001. **Treatment of fish processing wastewater in a one or two step Upflow anaerobic sludge blanket (UASB) reactor**, pp. 323–328. Proceedings of the *Ninth International Symposium on Anaerobic Digestion*, Antwerp, Belgium

Rajesh Banu J, Kaliappan S, and Diter B. 2006. **High rate anaerobic treatment of Sago wastewater using HUASB with PUF as carrier.** *International Journal of Environmental Science and Technology* **3**(1): 73–81

Ramalho R S. 1983. ***Introduction to Wastewater Treatment Processes***, pp. 211–340. New York: Academic Press

Sallis P J and Uyanik S. 2003. **Granule development in a split-feed anaerobic baffled reactor.** *Bioresource Technology* **89**: 255–265

Singh K S and Viraraghavan T. 2001. **Upflow anaerobic sludge blanket (UASB) reactor treatment of municpal wastewater**, pp, 493–498. Proceedings of the *9th International Symposium on Anaerobic Digestion*, Antwerp, Belgium

Tawaga T, Takahashi H, Sekiguchi Y, Ohashi A, Harada H. 2002. **Pilot-plant study on anaerobic treatment of a lipid and protein-rich food industrial wastewater by a thermophilic multi-staged UASB reactor.** *Water Science and Technology* **45**(10): 225–230

van Lier J B, Tilche A, Ahring B K, Macarie H, Moletta R, Dohanyos M, Hulshoff Pol L W, Lens P, Verstraete W. 2001. **New perspectives in anaerobic digestion.** *Water Science and Technology* **43**(1): 1–18

Young J C, Kim I S, Page I C, Wilson D R, Brown G J, Gocci A A. 2000. **Two stage anaerobic treatment of purified terephthalic acid production wastewaters.** *Water Science and Technology* **42**(5–6): 277–282

Zehnder A J B, Ingvorsen K and Marti T. 1982. ***Anaerobic Digestion,*** pp. 45–68. Amsterdam: Elsevier Biomedical Publications

Ceramic membrane filters from waste fly ash and their applications

M Balakrishnan
Fellow, Energy Environment Technology Development
The Energy and Resources Institute, New Delhi, India
and
V S Batra
Fellow, Energy Environment Technology Development
The Energy and Resources Institute, New Delhi, India

Introduction

In India, various forms of biomass, including firewood, agricultural waste, crop stalks, and agro-residues, are utilized for power generation. The common agricultural residues include bagasse, rice husk, cashewnut shell, arecanut shell, and groundnut shell. These waste products are often used as fuel in many industries, leading to the generation of large amount of residual ash, with associated problems in its disposal. The compounds in biomass fly ash are formed from the inorganic elements present in biomass fuels, such as Ca, K, Na, Si, and P (Table 1). Based on the fuel, the ash could contain high silicon (Si) content (for example, rice husk) or high alkali metal content (for example, wood). Among the agricultural residues, bagasse and rice husk are widely available.

Bagasse

India is one of the largest growers of sugar cane, and sugar industry is the second largest agro-processing industry, employing about 500 000 people in rural areas (Kakati 2006). After the extraction of sugar cane juice, the fibrous residue obtained is called bagasse, which constitutes about 25 wt% to 30 wt% of sugar cane (Babu and Krishna 1998). The bagasse has a calorific value of

Table 1 Composition of fly ash from different biomass

Compound	% Weight Bagasse	Groundnut	Arecanut	Cashew shell	Rice husk
SiO_2	65.03	43.13	42.43	8.16	93.5
Al_2O_3	0.49	10.71	3.85	3.05	0.01
Fe_2O_3	0.49	3.96	4.2	2.05	0.51
TiO_2	0.08	0.55	0.07	0.12	0.04
P_2O_5	1.14	4.19	7.29	14.63	1.06
CaO	2.75	10.81	1.23	7.5	0.68
MgO	3.26	6.1	0.48	10.63	0.47
Na_2O	0.06	5.6	0.2	5.24	0.4
K_2O	1.73	9.61	18.9	21.6	2.4
Cl	0.12	1.44	3.7	1.98	0.11
MnO	0	0	0.02	0.2	–
SO_3	0	0	5.09	4.06	–
Loss on ignition	24.84	3.9	12.5	20.5	0.8

Source Umamaheshwaran and Batra (2008)

8021 kJ/kg and is used as a fuel in boilers to generate steam, which in turn can be used for electricity generation (cogeneration). Sugar cane production in the marketing year 2009/10 was estimated to be about 300 million tonnes (Singh 2009). This corresponds to bagasse generation of 75–90 million tonnes per year. The ash content of bagasse is in the range 2%–22% (Iyer, Rao, and Grover 2002). Assuming an ash content of 4% reported in bagasse from Uttar Pradesh (which is one of the key cane growing states), each year, about 3.0–3.6 million tonnes of bagasse fly ash is likely to be generated in India, if all the bagasse is used as fuel. It is to be noted that this does not include the presence of unburnt carbon, which could be in the range of 16–33 wt% (Umamaheshwaran and Batra 2008; Gupta, Jain, Ali, *et al.* 2003).

At present, bagasse fly ash generated in most Indian sugar mills poses a disposal problem. Landfilling (dumping) is usually practised, thus requiring a large land area. Application of bagasse ash on agricultural land as a soil amender is also practised in some regions. In addition, bagasse fly ash has been examined as an adsorbent as well as an additive in cement and concrete (Srivastava, Mall, and Mishra 2006; Srikanth, Das, Ravikumar, *et al.* 2004; Paya, Monzo, Borrachero, *et al.* 2002).

Rice husk

India is one of the leading countries in rice cultivation, with an average production of about 129 million tonnes per year of paddy over the period 1999–2005 (Gadde, Bonnet, Menke, *et al.* 2009). Rice husk is generated during the milling of rough rice/paddy rice during the production of raw rice or parboiled rice. The amount of rice husk produced is 14%–27% of the weight of paddy (Beagle 1981, cited in Kapur, Kandpal, and Garg 1996). On the basis of a husk-to-paddy ratio of 0.2, the average husk production in India is about 26 million tonnes per year. The ash content of rice husk is in the range 14–22 wt% (Iyer, Rao, and Grover 2002). Assuming an ash content of 18 wt%, the fly ash generation in India is likely to be about 4.6 million tonnes per year, if the entire rice husk is burnt.

Rice husk is primarily used as a fuel, with part of it consumed as boiler feed in rice mills. Because of its high silica content, it has also been tried as an insulator, adsorbent, cement, and concrete additive, and as a substitute for silica (Bronzeoak 2003).

This chapter describes a novel use of biomass fly ash as a precursor for preparing ceramic membrane filters. Select applications of such filters are also described.

Waste-based ceramic membrane filters

Inorganic (ceramic) membrane applications have increased since the early 1980s, especially in areas where organic (polymeric) membranes failed due to their limitations. Ceramic membranes offer high chemical and physical stability, especially at extremes of pH and temperature, and thus, have a relatively longer working life. They have promising applications in biotechnology, environmental technologies (water treatment), and gas cleaning, especially under high temperature conditions. A key limitation of ceramic membranes is the cost, which is approximately 10 times higher than that of polymeric membranes (Mueller and Witte 2006). This can be a deterrent, especially in applications that involve large feed volumes and consequently large filtration area.

Ceramic membranes are usually composed of metal oxides (for example, alumina, titania, zirconia, silica) as well as materials like silicon carbide. These membranes are typically used for microfiltration (0.1–1 μm) and ultrafiltration (0.01–0.1 μm) applications for the separation of particulates, colloids, and dissolved macromolecules. Ceramic nanofiltration (0.001–0.01 μm) membranes for the retention of divalent ions and ceramic gas separation membranes have also been employed.

The support layer in commercial ceramic membranes is prepared using materials such as alumina (Al_2O_3), cordierite ($2MgO{\cdot}2Al_2O_3{\cdot}5SiO_2$), and mullite ($3Al_2O_3{\cdot}2SiO_2$). These are expensive. As a result, there has been an interest in recent years in using naturally occurring materials and waste materials as the starting powders for ceramic membrane preparation. Low-cost natural materials such as kaolin ($Al_2O_3{\cdot}2SiO_2{\cdot}2H_2O$), mineral dolomite ($CaMg\ (CO_3)_2$), and Moroccan clay have been explored (Bouzerara, Harabi, Achour, *et al.* 2006; Seffaj, Alami Younsi, Persin, *et al.* 2005; Belouatek, Benderdouche, Addou, *et al.* 2005). Membranes based on coal fly ash have also been developed (Jedidi, Saidi, Khemakem, *et al.* 2009; Jo, Huchinson, and Raper 1996).

Biomass fly ash based ceramic membranes

The Energy and Resources Institute (TERI) has developed membrane filters in the particulate filtration to ultrafiltration range, using biomass fly ash as the starting material (Batra and Tewari 2006). Of the various types of biomass ash tested, bagasse and rice husk ash were found to be most suitable (TERI 2005). These filters, prepared by tape casting and sintering, have pore sizes in the range 1–10 μm (Figure 1).

These ceramic membrane filters can be used for a variety of solid–gas or solid–liquid separations. Some of the applications that have been successfully tested include the following.

- *Cleaning of hot producer gas from a biomass gasifier* The filters were used for particulate removal from the gas prior to tar removal so that the gas could be used in power generation application (TERI 2005).

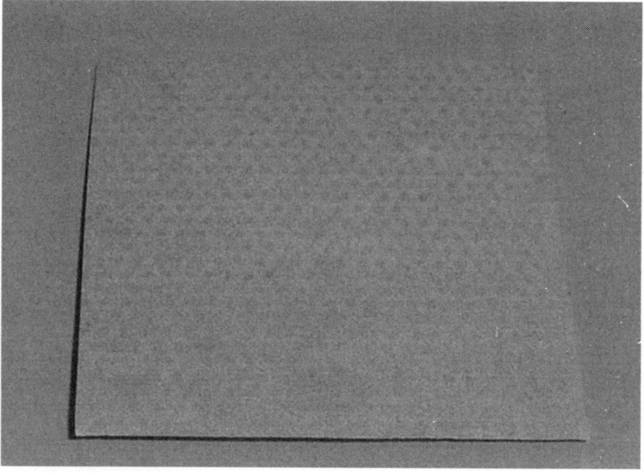

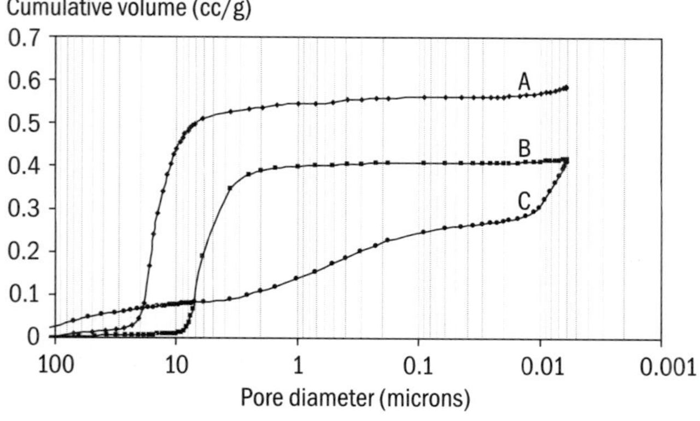

Figure 1 Filter sample and pore size distribution

- *Membrane bioreactor for waste water treatment* The filters were used for sludge retention in a membrane bioreactor (MBR) that integrates biological treatment with membrane separation. Both domestic and industrial (alcohol distillery) waste waters have been tested in a submerged MBR (Tewari, Singh, Batra, *et al.* 2010; Gupta, Satyawali, Batra, *et al.* 2008; Satyawali, Batra, and Balakrishnan 2005).
- *Clarification of sugar cane juice in the sugar manufacturing process* The aim was to remove particulate and colloidal matter from hot limed juice to enhance sugar quality and reduce the loss in molasses (Balakrishnan, Batra, Vashisht, *et al.* 2009).

The details of these applications are presented in the following sections.

Hot producer gas clean-up

Power generation from biomass is an important aspect of sustainable development. The biomass-based power generation potential in India, excluding cogeneration, is estimated at 16 gigawatts (GW), and the country ranks fourth in biomass-based power generation globally (CII Godrej GBC Publication 2004). Among the biomass sources, agricultural waste plays an important role.

One of the key requirements for biomass-based power generation is the physical and chemical clean-up of the producer gas obtained through gasification. The physical clean-up involves the removal of particulate matter comprising ash and unburnt materials. The particulates are typically removed using beds of sand, rice husk, wood pieces, charcoal, and so on or using fabric filters after cooling the gas to low temperatures. Inorganic membrane filters have the advantage of high-temperature particulate removal from producer gas before tar condensation. Thus, if required, they can also be part of a dry gas cleaning system, which eliminates water scrubbing and waste water treatment. In addition, they offer high particulate removal efficiency and ease of cleaning and maintenance. Hot gas cleaning is also required for gasification of coal and other fuels. Other applications of hot gas clean-up using inorganic filters are in incineration, cement manufacture, flue gas clean-up, etc.

TERI has tested the ceramic membrane filters for particulate removal from hot producer gas obtained in charcoal- and wood-fired gasifiers. The pressure drop and particle removal efficiency were initially measured in a laboratory-scale filtration system. Select filters were scaled up and field-tested with TERI's gasifier in Gual Pahari. In the case of charcoal-fired gasifiers, the particulate removal was almost complete (Figure 2), whereas in wood-fired gasifiers it was more than 80%. The pressure drop increased, and the flow rate decreased with time as the dust deposited on the filter. Cleaning by reverse air and heating led to the removal of dust to a certain extent and a lowering of the pressure drop.

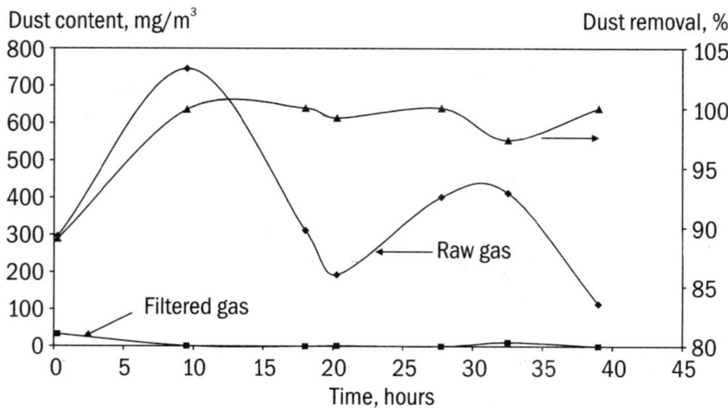

Figure 2 Particulate removal in charcoal-fired gasifier

Sugar cane juice clarification

India is the world's largest producer of "plantation white" or "mill white" sugar obtained by the double sulphitation process (Figure 3). A major challenge in this manufacturing scheme is to ensure that the clarification step consistently results in a juice of high clarity and low colour. However, variations in cane variety, changes in agro-climatic conditions, and fluctuations in the manufacturing process lead to non-uniform juice characteristics.

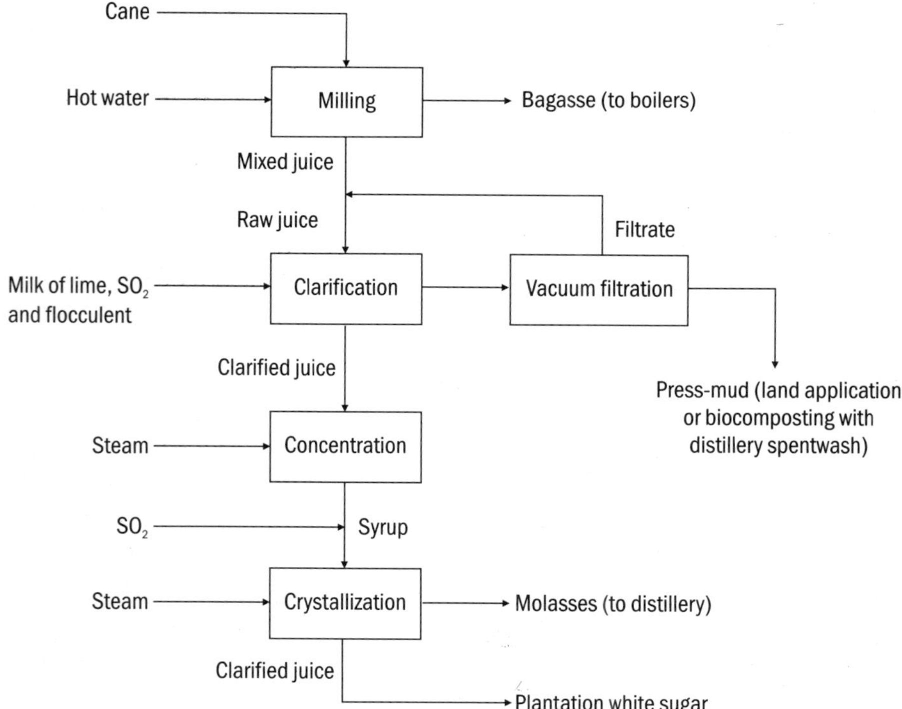

Figure 3 Schematic of plantation white sugar manufacture

Further, the liming–sulfitation process employing conventional clarifier–settlers for removing the precipitated impurities is incapable of eliminating colloidal and dissolved macromolecular substances. For this reason, the clarified juice is typically slightly hazy and has a dark brownish to yellow colour. Thus, the quality of plantation white sugar, characterized by high colour and sulfite content, is noticeably inferior to refined sugar (Table 2).

It is now well established that the filtrate obtained by membrane separation (ultrafiltration or microfiltration) possesses better clarity, lower viscosity, and reduced colour (Chou, Iqbal, Min, *et al.* 2002). Consequent benefits include higher crystal yield, superior sugar quality, energy savings due to reduced evaporator steam consumption, and increased capacity of evaporators, vacuum pans, crystallizers, and centrifuges. Because of their comparatively low cost, polymeric membranes have been tested for this particular application (Eringis and Jaferey 2001; Willet 1997). However, there are concerns about membrane integrity and performance when they are in continuous operation (about 180 days) at elevated temperatures (about 100 °C); further, off-season maintenance of the used polymeric membranes also poses a problem (Ghosh and Balakrishnan 2003).

The bagasse fly ash membrane filters were tested with juice from two different sugar cane crushing seasons for colour and turbidity removal (Balakrishnan, Batra, Vashisht, *et al.* 2009). Dead-end batch filtration was done at ambient temperature using a filtration area of about 18 cm^2. Spectrophotometric technique was used to analyse the colour and turbidity. Turbidity removal was in the range 78%–90%, and colour removal was in the range of 18%–30%. The results were comparable with those obtained using tighter 20-kD polymeric membranes in an Indian sugar mill during earlier pilot trials (Ghosh and Balakrishnan 2003). The ceramic filters were subsequently tested on-site in a sugar mill, and the improvement in juice characteristics was confirmed. This application is particularly attractive since it offers the dual benefits of process improvement in sugar factories combined with innovative value-added utilization of the waste bagasse fly ash generated by this industry.

Waste water treatment

MBRs, which integrate biological treatment with membrane filtration (Figure 4), have emerged as one of the major treatment solutions for domestic and industrial waste water, with over 2200 installations operational worldwide (Yang, Cicek, and Ilg 2006). The development and adoption of

Table 2 Characteristics of various sugars

Property	Raw sugar	Plantation white sugar	Refined sugar
Colour (ICUMSA units)	NA	≤150	≤60
Sulphur (ppm)	≤20	≤70	≤15
Polarization value	NA	≥99.5	≥99.7

Source CODEX Standard for Sugars (1999)

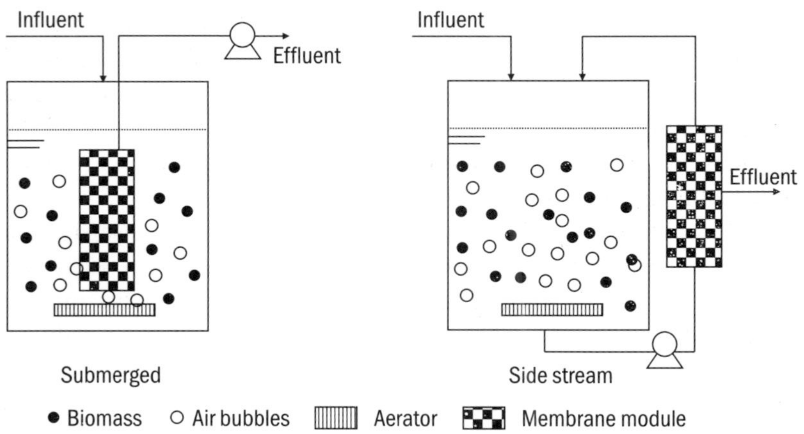

Figure 4 Schematic of membrane bioreactor (MBR) configurations
Source Melin, Jefferson, Bixio, *et al.* (2006)

this technology are being driven by both fresh water shortage and anticipated stricter environmental regulations. MBRs offer the advantages of a compact system with better solids removal, disinfection, and an overall superior quality of treated effluent. The high biomass concentration (typically 10–12 g/l) is conducive to the treatment of high strength waste water. The hydraulic retention time (HRT) is independent of the sludge retention time (SRT), allowing operational flexibility; moreover, degradation of complex compounds is possible at high SRTs. The main limitations are high capital investment as well as membrane fouling, which lowers productivity and requires eventual membrane replacement. Thus, in the context of developing countries, MBRs continue to be an expensive option.

To overcome the limitations of high cost of microfiltration, several workers have explored alternatives such as woven fabric and mesh in submerged MBRs (Chang, Hu, Horng, *et al.* 2007; Wu, Huang, and Zuo 2003; Kiso, Jung, Ichinari, *et al.* 2000). Further, low-cost, clay-based ceramic membranes have also been explored for waste water treatment (Khemakhem, Larbot, and Ben Amar 2006). The biomass fly ash ceramic membranes developed at TERI have been investigated in MBRs for treating both sewage and industrial (distillery) waste water.

An MBR pilot plant for sewage treatment was tested in TERI's residential training complex at Gual Pahari, Haryana. The occupancy of this complex varied from 15 to 150 persons. Raw sewage generated from this complex was screened to remove large particulate matter and then fed into an MBR with submerged membranes. Batch anoxic operation was followed by an aerobic stage. The summary of the MBR performance over 200 days is given in Table 3. The suspended solids retention was consistently high (> 96%). However, the system experienced large fluctuations in influent chemical oxygen

Table 3 Performance summary of sewage treatment in a membrane bioreactor

Parameter	Average value
Flux (L/m^2·h)	10.75 ± 2.48
Transmembrane pressure (mm Hg)	349.22 ± 121.72
Removal (%)	
Suspended solids	95.99 ± 3.51
COD	67.87 ± 16.31
NH_4^+	77.70 ± 18.51
PO_4^{-3}	33.81 ± 24.85

COD – chemical oxygen demand

demand (COD), wide temperature range (9–29°C), and relatively high influent PO_4^{-3} concentration. As result, the COD, total nitrogen (NH_4^+), and total phosphorus (PO_4^{-3}) removals were somewhat lower than those reported in the literature.

Industrial waste water treatment was investigated with distillery spentwash, which is the effluent generated during ethanol manufacture from sugar cane molasses. This waste water is characterized by large volumes (8–15 l/l alcohol), extremely high COD (80 000–100 000 mg/l) and biochemical oxygen demand (BOD) (40 000–50 000 mg/l), high nutrients (nitrogen, phosphorus, and potassium), low pH, strong odour, and dark brown colour. India is the second largest producer of ethanol in Asia, with a projected annual production of about 2300 million litres in 2012 (Market Research Analyst 2008). Moreover, alcohol distilleries are rated as one of the 17 most polluting industries. Anaerobic treatment with biogas recovery is employed as the first treatment step in most Indian distilleries. The diluted, bio-digested spentwash is used for irrigation ("ferti-irrigation"); it is also used for composting with pressmud, which is a solid waste generated in the sugar manufacturing process. Though various physico-chemical and biological options have been explored, a comprehensive treatment scheme for molasses distillery waste water leading to effective removal of both organics and colour is still being explored (Satyawali and Balakrishnan 2008a).

The bagasse fly ash membranes with a median pore size of about 7 µm and porosity of about 30% were tested for distillery waste water treatment in a submerged MBR. The influent was spentwash obtained after anaerobic treatment. The MBR was inoculated with sludge from a distillery unit and operated in the fed-batch mode. The treatment yielded an average of 36% COD removal, with retention of up to 98% suspended solids (Gupta, Satyawali, Batra, *et al.* 2008). It was observed that the lower molecular weight component (< 940 Da) was degraded in the MBR operation, but the higher molecular weight melanoidins fraction, which imparts colour to the solution, remained largely unaffected (Satyawali and Balakrishnan 2008b).

Summary

Commercial ceramic membranes have a broad range of applications in solid–liquid and solid–gas separations in various sectors, but their high cost is a limitation. Fly ash generated by the combustion of agro-residues is a rich source of metal oxides; as such, it is a readily available precursor for the preparation of ceramic membrane filters. TERI has developed ceramic membranes from biomass fly ash and successfully tested their use for hot producer gas clean-up, sugar cane juice clarification, and sludge separation in waste water treatment. These examples, therefore, illustrate the transformation of biomass fly ash waste material into a high-value product with multiple applications.

References

Babu B V and Krishna V R. 1998. **Optimum utilization of waste oil for improved thermal efficiency of bagasse.** *Journal of Indian Association for Environmental Management* **25**: 59–65.

Balakrishnan M, Batra V S, Vashisht P, Singh M. 2009. **Ceramic membranes from waste bagasse fly ash to clarify sugarcane juice.** *International Sugar Journal* **111**(1325): 318–322

Batra V S and Tewari P K. 2006. **A process for the preparation of inorganic membrane filters for solid liquid and solid-gas separations.** Patent application No.1362/Del/2006 dated.7/6/2006

Beagle E C. 1981. ***Rice Husk Conversion to Energy.*** Rome: Food and Agriculture Organization (FAO)

Belouatek A, Benderdouche N, Addou A, Ouagued A, Bettahar N. 2005. **Preparation of inorganic supports for liquid waste treatment.** *Microporous and Mesoporous Materials* **85**: 163–168

Bouzerara F, Harabi A, Achour S, Larbot A. 2006. **Porous ceramic supports for membranes prepared from kaolin and doloma mixtures.** *Journal of the European Ceramic Society* **26**(9): 1663–1671

Bronzeoak. 2003. **Rice husk ash market study.** Report No. ETSU U/00/00061/REP DTI/Pub URN 03/668. Details available at <**http://www.berr.gov.uk/files/file15138.pdf**>, last accessed on 31 May 2010

Chang W K, Hu A Y J, Horng R Y, Tzou W Y. 2007. **Membrane bioreactor with nonwoven fabrics as solid–liquid separation media for wastewater treatment.** *Desalination* **202**: 122–128

Chou C C, Iqbal K, Min Y G, Gao D W, Duffault E. 2002. **White and refined sugar production from cane sugar factories.** First Biennial World Conference on Recent Development in Sugar Technology, May 16–17, Delrey Beach

CII Godrej GBC Publication. 2004. *Biomass Energy Potential & Prospects.* RES -- Fact Sheet No. 1.

CODEX Standard for Sugars. Codex Alimentarius Commission 1999. Details available at <http://www.codexalimentarius.net/download/standards/338/CXS_212e.pdf>, last accessed on 31 May 2010.

Eringis A and Jaferey I. 2001. **Technical and economical benefits of membrane filtration in cane sugar processing.** Proceedings of the International Society of Sugar Cane Technologists, International Congress **24**: 151–152

Gadde B, Bonnet S, Menke C, Garivait S. 2009. **Air pollutant emissions from rice straw open field burning in India, Thailand and the Philippines.** *Environmental Pollution* **157**: 1554–1558

Ghosh A M and Balakrishnan M. 2003. **Pilot demonstration of sugarcane juice ultrafiltration in an Indian sugar factory.** *Journal of Food Engineering* 58: 143–150

Gupta V K, Jain C K, Ali I, Sharma M, Saini V K. 2003. **Removal of cadmium and nickel from wastewater using bagasse fly ash—a sugar industry waste.** *Water Research* 37: 4038–4044

Gupta R, Satyawali Y, Batra V S, Balakrishnan M. 2008. **Submerged membrane bioreactor using fly ash filters: trials with distillery wastewater.** *Water Science and Technology* 58(6): 1281–1284

Iyer P V R, Rao T R, and Grover P D. 2002. ***Biomass Thermo-chemical Characterization,*** 3rd edn. New Delhi: Indian Institute of Technology

Jedidi I, Saıdi S, Khemakem S, Larbot A, Elloumi-Ammar N, Fourati A, Charfi A, Ben Amar R. 2009. **New ceramic microfiltration membranes from mineral coal fly ash.** *Arabian Journal of Chemistry*. 2: 31–39

Jo Y M, Huchinson R, and Raper J A. 1996. **Preparation of ceramic membrane filters, from waste fly ash, suitable for gas cleaning.** *Waste Management Research* 14: 281–295

Kakati D. 2006. **Sweet power: sugar mill source is a cleaner source of energy.** SPAN March/April 2006 18–21. Details available at <http://span.state.gov/wwwfspmarapr069.pdf>, last accessed on 30 January 2010

Kapur T, Kandpal T C, and Garg H P. 1996. **Electricity generation from rice husk in Indian rice mills: Potential and financial viability.** *Biomass and Bioenergy* 10(5–6): 393–403

Khemakhem S, Larbot A, and Ben Amar R. 2006. **Study of performances of ceramic microfiltration membrane from Tunisian clay applied to cuttlefish effluents treatment.** *Desalination* 200(1-3): 307–309

Kiso Y, Jung Y, Ichinari T, Kitao T, Nishimura K, Min K. 2000. **Wastewater treatment performance of a filtration bio-reactor equipped with a mesh as a filter material.** *Water Research* 34: 4143–4150

Market Research Analyst. 2008. Details available at **<http://www.marketresearchanalyst.com/2008/01/26/world-ethanol-production-forecast-2008-2012>**, last accessed on 30 January 2010

Melin T, Jefferson B, Bixio D, Thoeye C, de Wilde W, de Koning J, Van der Graff J, Wintgens T. 2006. **Membrane bioreactor technology for wastewater treatment and reuse.** *Desalination* 187: 271–282

Mueller U and Witte M. 2006. Details available at <http://www.techneau.org/fileadmin/files/Publications/Publications/Deliverables/D2.3.3.5b.pdf>, accessed on 30 January 2010

Paya J, Monzo J, Borrachero M V, Dýaz-Pinzon L, Ordonez L M. 2002. **Sugarcane bagasse ash (SCBA): studies on its properties for reusing in concrete production.** *Journal of Chemical Technology and Biotechnology* 77: 321–325

Satyawali Y and Balakrishnan M. 2008a. **Wastewater treatment in molasses based alcohol distilleries for COD and color removal: a review.** *Journal of Environmental Management* 86: 481–497

Satyawali Y and Balakrishnan M. 2008b. **Treatment of distillery effluent in a membrane bioreactor (MBR) equipped with mesh filter.** *Separation and Purification Technology* 63: 278–286

Satyawali Y, Batra V S, and Balakrishnan M. 2005. **Ash filters for membrane bioreactor applications.** *Proceedings of ENVIROVISION 2005.* September 22–24, Vadodara

Seffaj N, Alami Younsi S, Persin M, Cretin M, Albizane A, Larbot A. 2005. **Processing and characterization of $TiO_2/ZnAl_2O_4$ ultrafiltration membranes deposited on tubular support prepared from Moroccan clay.** *Ceramics International* 31: 205–210

Singh S. 2009. *India SUGAR ANNUAL 2009*. Global Agricultural Information Network (GAIN) Report No. IN9049. Details available at <http://gain.fas.usda.gov/Recent%20GAIN%20Publications/Commodity%20Report_SUGAR%20ANNUAL_New%20Delhi_India_4-8-2009.pdf>, last accessed on 30 January 2010

Srikanth S, Das S K, Ravikumar B, Rao D S, Nandakumar K, Vijayan P. 2004. **Nature of fireside deposits in a bagasse and groundnut shell fired 20 MW thermal boiler.** *Biomass and Bioenergy* **27**: 375–384

Srivastava V C, Mall I D, and Mishra I M. 2006. **Equilibrium modelling of single and binary adsorption of cadmium and nickel onto bagasse fly ash.** *Chemical Engineering Journal* **117**: 79–91

TERI (The Energy and Resources Institute) Report. 2005. **Development and testing of low cost membrane filters from biomass ash for clean up of producer gas for gasifier based power plants.** [Report] Contract No. MNES 03/01/05/2002/BM

Tewari P K, Singh R K, Batra V S, Balakrishnan M. 2010. **Membrane bioreactor (MBR) for wastewater treatment: filtration performance evaluation of low cost polymeric and ceramic membranes.** *Separation and Purification Technology* **71**: 200–204

Umamaheshwaran K and Batra V S. 2008. **Physico-chemical characterization of Indian biomass ashes.** *Fuel* **87**(6): 628–638

Willet C C. 1997. **The A.B.C process. Direct production of refined quality sugar from cane mixed juice.** *International Sugar Journal* **99**: 7–8

Wu Y, Huang X, and Zuo W. 2003. **Effect of mesh pore size on performance of a self-forming dynamic membrane coupled bioreactor for domestic wastewater treatment.** *5th International Membrane Science and Technology Conference*, November 11–13, Sydney

Yang W, Cicek N, and Ilg J. 2006. **State-of-the-art of membrane bioreactors: Worldwide research and commercial applications in North America.** *Journal of Membrane Science* **270**: 201–211

Technological options for municipal solid waste management

R Johri
Senior Fellow, Water Resources Policy and Management
The Energy and Resources Institute, New Delhi, India
and
K V Rajeshwari
The Energy and Resources Institute, New Delhi, India
and
A N Mullick
Research Associate, TERI-BCSD
The Energy and Resources Institute, New Delhi, India

The disposal of municipal solid waste (MSW) has gained immense importance for urban local bodies and waste management experts. The current method of disposal—that is, landfill—is becoming economically and environmentally unacceptable. It is increasingly being felt that the energy present in its organic fraction can be recovered for gainful utilization through the adoption of suitable technologies. Recovery of energy from waste has additional benefits by way of an overall reduction in the (1) volume of waste, (2) demand for land required for landfill, (3) cost of transportation, and (4) environmental pollution.

Solid waste comprises different materials, such as food waste; packaging in the form of paper, metals, plastic, or glass; discarded clothing and furnishing; garden waste; and hazardous and radioactive waste. Domestic waste has three main characteristics: weight generated, density, and constituents; these vary from place to place depending on the level of industrial development and a number of other factors. Currently, most of the MSW in the country is disposed of unscientifically, that is, no "sanitary landfill" exists (Pachauri and Sridharan 1998). This has an adverse impact not only on the ecosystem but also on the human environment. Unscientific disposal practices of leaving the waste unattended at the disposal sites allow birds, rodents, fleas, and so on to access the waste, creating unhygienic conditions (malodour, breeding of air-borne pathogens, and so on). Plastic content of the municipal waste is collected by rag-pickers for recycling, either at the primary collection centres or at the dump sites. Recycling of plastic waste is mostly done in small units, which do not have adequate technologies to process plastic in a safe manner, thus exposing workers to toxic fumes and unhygienic conditions.

In addition, biomedical waste (BMW) generated from different clinics, hospitals, nursing homes, pathological laboratories, blood banks, and veterinary centres is being disposed of along with municipal waste at the dump sites. This is due to inadequate facilities and lack of awareness about the impact of the waste on the environment and health. Like other Asian countries, in India too, most of the waste is landfilled. The management of BMW is not as per the Biomedical Waste (Management & Handling) Rules, 1998. Waste is largely dumped in the low-lying areas prone to flooding, thus increasing the risk of infection during the rainy season. Groundwater pollution, though largely not assessed, is another threat posed by the dumping of waste. Now, the situation with respect to the construction of sanitary landfills and disposal of BMW is gradually improving.

Electronic waste (e-waste), a new entrant among more traditional waste such as plastic and industrial scrap, promises higher return for the recyclers. E-waste is a collective name for discarded electronic devices such as computer, cell phone, refrigerator, television, and many other electrical and electronic appliances.

Environmentally sound management of waste must go beyond merely its safe disposal and recovery, and seek to address the root cause of the problem by attempting to change unsustainable patterns of production and consumption. Accordingly, the framework for requisite action should be based on a hierarchy and focus on four major waste-related programmes, given below.

1 Minimizing waste
2 Maximizing environmentally sound reuse and recycling
3 Promoting environmentally sound waste treatment (converting it into energy) and disposal
4 Extending waste service coverage

This article provides an overview of the status of waste generated in important cities and discusses its composition, characteristics, and management, including segregation, storage, transportation, and disposal. The technology options available for converting waste to energy are also covered. It also highlights the technologies of foreign companies collaborating with India and presents calculations for energy recovery from the various technologies. The merits/demerits of the technologies suitable for India are also given. The biochemical process for degradation of solid waste is covered in detail along with the evaluation of various MSW technologies. It also covers metal recovery from the scrap, which has a lot of potential.

Current scenario of solid waste management

Solid waste generation

Solid waste management has become a major environmental issue in India. Generation of solid waste depends on many factors such as culture and nature of the people, socio-economic conditions, commercial importance of the waste, and its industrial base. Table 1 indicates the solid waste generation in major metropolitan cities in India. It is observed that the total MSW generated by the major metropolitan cities is 27 167 metric tonnes per day (MTPD). The national average for class-I cities is 0.376 kg/person/day. The MSW generated daily in India ranges from about 69 MTPD in smaller cities (Kochi) to 6000 MTPD in bigger cities/towns (Delhi) (Table 1).

Table 1 Solid waste generation in major metropolitan cities in India

State/Union Territory	City	Urban population (2001)	Municipal solid waste generated (metric tonnes/day)
Andhra Pradesh	Hyderabad	3 829 753	957
Andhra Pradesh	Visakhapatnam	982 904	246
Bihar	Patna	1 961 532	588
Delhi	New Delhi	350 000	272
Delhi	Delhi	13 363 471	6000
Gujarat	Ahmedabad	4 215 497	1265
Gujarat	Surat City	2 433 835	730
Gujarat	Vadodara	1 491 045	447
Karnataka	Bengaluru	1 304 008	326
Kerala	Kochi	275 225	69
Maharashtra	Mumbai	11 914 398	7500
Maharashtra	Nagpur	2 040 175	700
Maharashtra	Pune	2 540 000	1000
Madhya Pradesh	Bhopal	1 482 718	445
Madhya Pradesh	Indore	1 550 880	465
Punjab	Ludhiana	1 429 709	500
Rajasthan	Jaipur	1 870 771	561
Tamil Nadu	Chennai	4 343 645	1086
Tamil Nadu	Coimbatore	1 501 373	375
Tamil Nadu	Madurai	1 233 083	308
Uttar Pradesh	Kanpur	2 725 207	954
Uttar Pradesh	Lucknow	2 262 369	792
Uttar Pradesh	Varanasi	1 250 039	438
West Bengal	Kolkata	4 572 876	1143
Grand total		70 924 513	27 167

Source National Solid Waste Association of India (2007)

A survey conducted by the Central Pollution Control Board (CPCB) puts the total municipal waste generated from class I and II cities at about 18 million tonnes (MT) in 1997 (CPCB 2000a, b), whereas Table 1 reveals that the total solid (TS) waste generated only in major metropolitan cities in India is about 70 MT (in 2007) (National Solid Waste Association of India 2007). The present annual solid waste generated in the Indian cities has increased from 6 MT in 1947 to 48 MT in 1997 and is expected to increase to 300 MT/year by 2047 (CPCB 2000a, b) (Figure 1).

Solid waste generation is directly related to the economy of a country. As the society becomes richer, waste generation increases (0.2 >> 1.5 kg/capacity/ day), whereas the traditional recycling practices tend to decline. The US, the European Union (EU), and Japan are by far the world's largest producers of solid waste, with the US producing about 14 times more waste than Japan and the EU countries (US: 200 MT, Japan: 50 MT).

Composition of waste

Physical characteristics of waste

The characteristics of the MSW collected from any area depend on a number of factors, such as the food habits, cultural tradition, lifestyle, and climate. Table 2 shows waste characterization in major metropolitan cities in India.

Table 3 highlights the changes in the relative shares of different constituents of waste in the past several decades. A notable rise in paper, rags, and plastic components and a decline in inert components used for land levelling clearly show that the proportion of recyclable waste is increasing in the total municipal waste. This has led to increased interest in recovering valuables from the waste. Table 3 clearly shows that there is a substantial increase in plastic and paper waste.

Chemical characteristics of waste

The chemical characteristics of waste could be divided into three components: (1) chemical, (2) biochemical, and (3) toxic. Sometimes waste may have fat content (lipid).

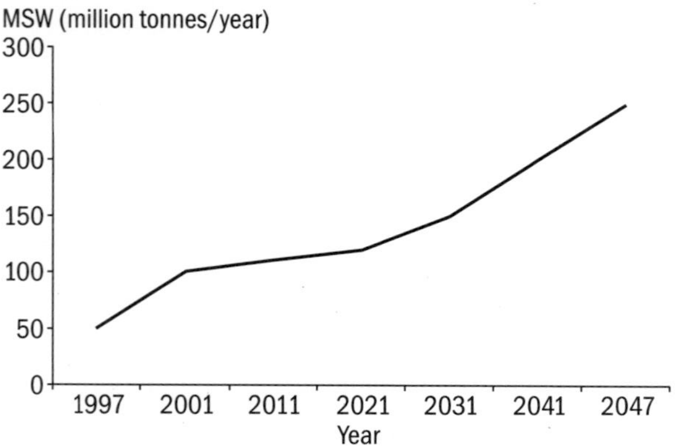

Figure 1 Annual municipal solid waste generated in Indian cities

Table 2 Waste characterization in Indian metropolitan cities

City	Compostables (%)	Recyclables (%)	C:N ratio	HCV (kcal/kg)	Moisture
Hyderabad	54.2	21.60	25.90	1969	46
Visakhapatnam	45.96	24.20	41.70	1602	53
Patna	51.96	12.57	18.62	819	36
Delhi	54.42	15.52	34.87	1802	49
Ahmedabad	40.81	11.65	29.64	1180	32
Surat	56.87	11.21	42.16	990	51
Vadodara	47.43	14.50	40.34	1781	25
Bengaluru	51.84	22.43	35.12	2386	55
Kochi	57.24	19.36	18.22	591	50
Greater Mumbai	62.44	16.66	39.04	1786	54
Nagpur	47.41	15.53	26.37	2632	41
Pune	62.44	16.66	35.54	2531	63
Bhopal	52.44	22.33	21.58	1421	43
Indore	48.97	12.57	29.30	1437	31
Ludhiana	49.80	19.32	52.17	2559	65
Jaipur	45.50	12.10	43.29	834	21
Chennai	41.34	16.34	29.25	2594	47
Coimbatore	50.06	15.52	45.83	2381	54
Madurai	55.32	17.25	32.69	1813	46
Kanpur	47.52	11.93	27.64	1571	46
Lucknow	47.41	15.53	21.41	1557	60
Varanasi	45.18	17.23	19.40	804	44
Kolkata	50.56	11.48	31.81	1201	46

Source Ministry of Statistics and Programme Implementation (2004/05)

Table 3 Comparison, over time, of physical characteristics of municipal solid waste

Component	Composition (%)	
	1996	2005
Biodegradable	42.21	47.430
Paper	3.63	8.130
Rags	—	4.490
Plastic	0.60	9.220
Leathers and rubbers	—	—
Glass	0.60	1.010
Metal	0.49	0.500
Others	—	4.016
Inerts	45.13	25.160

Source Zhu, Asnani, and Zurbrugg (2007)

- *Chemical* This characteristic includes the nitrogen–phosphorus–potassium (N–P–K), pH, total carbon, carbon:nitrogen (C:N) ratio, and calorific value.
- *Biochemical* It includes carbohydrates, proteins, natural fibre, and biodegradable factor.
- *Toxic* Toxicity characteristics include heavy metals, pesticides, insecticides, and toxicity test for leachates.

An analysis of the Indian waste carried out by the National Environmental Engineering Research Institute (NEERI), India, and other organizations indicates that the N content of the waste varies from 0.56% to 0.71%, P varies from 0.52% to 0.82%, K varies from 0.52% to 0.83%, and the C:N ratio varies between 21 and 31.

Effect on health

A rising urban population growth, dwindling municipal resources, and the complexity of MSW management have complicated the relationship between environmental management and public health. The combined effect of uncollected waste, poor handling, and inadequate disposal safeguards for the MSW has always had implications on public health. Public health impacts of MSW can occur along all stages of waste handling due to various factors. Health risks associated with uncollected solid waste are as follows.

- Intoxication or infection due to exposure to uncollected solid waste in streets or at unofficial dump sites.
- Uncollected organic waste ferment creates conditions favourable for the survival and growth of microbial pathogens.
- Uncollected solid waste obstructs storm water run-off, resulting in flooding or creation of stagnant waterbodies, which become habitats and breeding grounds for water-borne vectors of tropical diseases.

Health risks associated with improper disposal of solid waste

- Groundwater used for drinking purposes can become chemically and microbiologically polluted if waste is disposed of near water sources. Direct dumping of untreated solid waste in rivers, lakes or seas can also result in the accumulation of toxic substances in the food chain due to its uptake by plants and animals.
- Disposal of health-care waste requires special attention as it can create major health hazards, example being the transmission of viral infections.

Health risks associated with recycling

- Workers dealing with recycling of waste that has a high metal or chemical content may experience toxic pressure. Scavengers who scour waste sites may sustain injuries and develop infection from vectors such as rats and flies.
- Scavengers and their families are at a risk because they often build their homes very close to landfill sites.

Municipal solid waste management

It is estimated that the solid waste generated by 300 million people living in urban areas is about 38 MT/year. Proper disposal of urban waste is not only necessary for safeguarding the environment and improvement of health, but it also creates immense potential for the recovery of resources. It is estimated that about 100 000 tonnes of MSW is generated daily in India. The per capita generation in major cities is roughly 0.20–0.60 kg. Generally, the collection efficiency is 70%–90% in the metropolitan cities, whereas in smaller cities, the collection efficiency is below 50%. It is also estimated that the urban local bodies spend about Rs 500–1500 per tonne on solid waste for its collection, transportation, treatment, and disposal. Due to the lack of proper disposal sites, even the collection efficiency gets affected.

Functional elements of municipal solid waste management

- Waste generation leads to the generation of material having no value in the present form.
- Waste handling, sorting, storage, and processing at the source are the activities involved till the waste is placed in a container for collection. On-site storage is of primary importance because of the public health concern and aesthetic considerations.
- Collection includes not only the gathering of solid waste and recyclable material but also its transportation to the processing/disposal site.
- Sorting, processing, and transformation of solid waste include separation of bulky items, separation according to size by using screens, manual separation, and separation of ferrous and non-ferrous metals. Waste transformation is undertaken to reduce the volume, weight, and size or the toxicity of waste without the resource recovery.
- Transfer and transportation entail two steps: transfer of the waste from a smaller collection vehicle to a larger vehicle, and subsequently, transportation over long distances to a processing or disposal site.
- Disposal in the landfills or uncontrolled dumping is the practice followed by most municipal bodies. A well-managed landfill is an engineered facility used for the disposal of solid waste. Increase in solid waste generation will have a significant impact on the land area required for disposing waste as well as on the methane (CH_4) emissions.

Other types of wastes in urban areas

Other types of waste found in urban areas include biomedical waste, industrial waste, thermal power plant waste, e-waste, and so on.

Regulations governing solid waste management practices

Following are the regulations that govern the solid waste management practices.

- The Municipal Solid Wastes (Management and Handling) Rules, 2000

- Manual on MSW Management, Central Public Health and Environmental Engineering Organization (CPHEEO), Ministry of Urban Development (MoUD), 2000
- Biomedical Waste (Management and Handling) (Amendment) Rules, 2003, Ministry of Environment and Forests (MoEF)
- Hazardous Waste (Management, Handling, and Transboundary Movement) Rules, 2009, MoEF
- Special notification for other wastes from time to time, by the MoEF
- CPCB guidelines on e-waste management
- Draft regulation on e-waste is already notified for suggestions and objection by the public (May 2010).

Land requirement

An increase in the solid waste generation impacts the cumulative requirement of land (base year 1997) for the disposal of MSW, estimated to cumulate to about 1400 km^2 by 2047 (Figure 2).

Estimates under the business-as-usual (BAU) scenario are made considering the average collection efficiency of waste as 72.5%, average depth of landfill site as 4 m, and the average waste density as 0.9 tonne/m^2 (NIUA 1989). Diversion of land for waste disposal is physically impossible as the area with the largest concentration of solid waste is also the area with a serious scarcity of vacant land. Therefore, if the current methods of solid waste disposal persist, the waste would have to be carried over long distances, requiring improved transport facilities and infrastructure, involving additional finances.

Methane emission

Emissions from landfills, which comprise 50%–60% CH_4, contribute significantly to global warming. In 1997, landfill released about 7 MT of CH_4 into the atmosphere, which is projected to increase to 39 MT by 2047 under the BAU scenario (Figure 3).

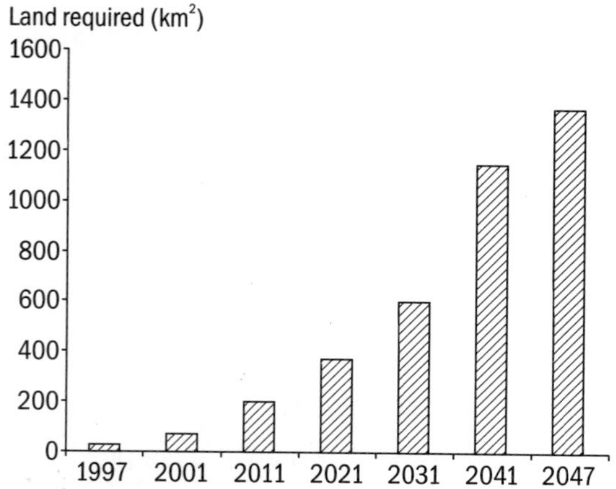

Figure 2 Land requirement for solid waste management
Source CPCB (2000b)

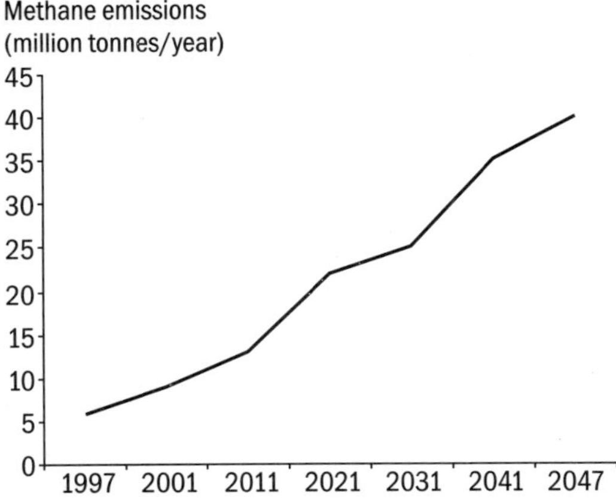

Figure 3 Emission of methane from landfills

Emissions have been calculated using the approach given by Bingemer and Crutzen (1987), which assumes that 50% of the carbon emissions in landfills is transformed into CH_4.

Calculation for energy recovery

The main parameters that determine the potential of recovering energy from the MSW are quantity of waste and physical and chemical characteristics of waste.

The production of energy will depend on the specific treatment processes, seasonal variations, and so on. Important physical parameters are the size of the constituents and the density and moisture content of the waste.

Table 4 Important waste parameters for technical viability of energy recovery.

Waste treatment method	Basic principle	Important waste parameters	Desirable range
Thermo-chemical conversion	Decomposition of organic matter by action of heat	Moisture content, organic/ volatile matter	< 45%, > 40%
Incineration	Decomposition of organic matter by action of heat	Fixed carbon	< 15%
Pyrolysis	Decomposition of organic matter by action of heat	Total inert	< 35%
Gasification	Decomposition of organic matter by action of heat	Calorific value	> 1200 kcal/kg
Biochemical conversion, anaerobic digestion	Decomposition of organic matter by microbial action	Moisture content Organic/volatile matter C: N ratio	> 50% > 40% 25–30

Source CPHEEO (2000)

The important chemical parameters are fixed carbon content, calorific value, volatile solid (VS) contents, C:N ratio, toxicity, and inert fraction present in waste. Table 4 indicates waste parameters for energy recovery.

Assessment of energy recovery potential

Energy can be recovered from biodegradable and non-biodegradable matter through the thermo-chemical and biochemical conversions.

The thermo-chemical conversion

Thermal decomposition of the organic matter results in the production of thermal energy in the form of either fuel oil or gas. The process is suitable for high percentage organic non-biodegradable matter with a low moisture content. Incineration and pyrolysis are more suitable options under this technology (Figures 4 and 5).

Assessment of energy potential through thermo-chemical conversion

Energy recovery potential (kWh) = NCV × W × 1000/860 = 1.16 × NCV × W,

where W = total waste quantity (tonnes); NCV = net calorific value (kcal/kg).

Power generation potential (kW) = 1.16 × NCV × W/24 = 0.048 × NCV × W (assuming a conversion efficiency of 25%)

Net power generation potential (kW) = 0.012 × NCV × W

If NCV = 1200 kcal/kg, then net power generation potential (kW) = 14.4 × W

Biochemical conversion

The process of biochemical conversion is based on the enzymatic decomposition of organic matter by microbial action to produce CH_4 or alcohol. This technology is suited for waste having a high percentage of biodegradable matter with a high moisture content. The options include fermentation and anaerobic digestion. The potential of energy recovery from MSW depends on the calorific value and organic fraction.

Assessment of energy potential through biochemical conversion

- Total organic solids/VS = 50%
- Calorific value of biogas = 5000 kcal/m^3
- Digestion efficiency = 60%
- Organic biodegradable fraction is 66% of the VS
- Biogas yield: B (m^3) = 80 m^3/kg of the VS destroyed
- Biogas yield (m^3) = 0.80 × 0.60 × 0.33 × W × 1000 = 158.4 W, where W is the amount of waste processed in tonnes.

Hence, the energy recovery potential (kWh) = B × 5000 = 921 × W/860; power generation potential (kW) = 921 × W = 38.4 × W/24; net power generation potential (kW) = 11.5 × W (typical conversion efficiency 30%).

In general, 100 tonnes of raw MSW with 50%–60% organic matter can generate about 1–1.5 megawatt (MW) power, depending on the waste characteristics (CPHEEO 2000).

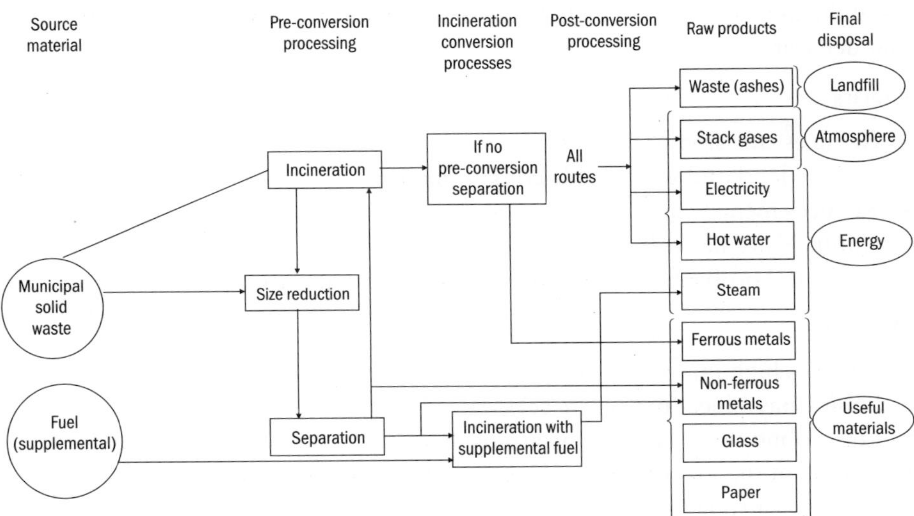

Figure 4 Refuse incineration routes with energy recovery
Source Domino (1979)

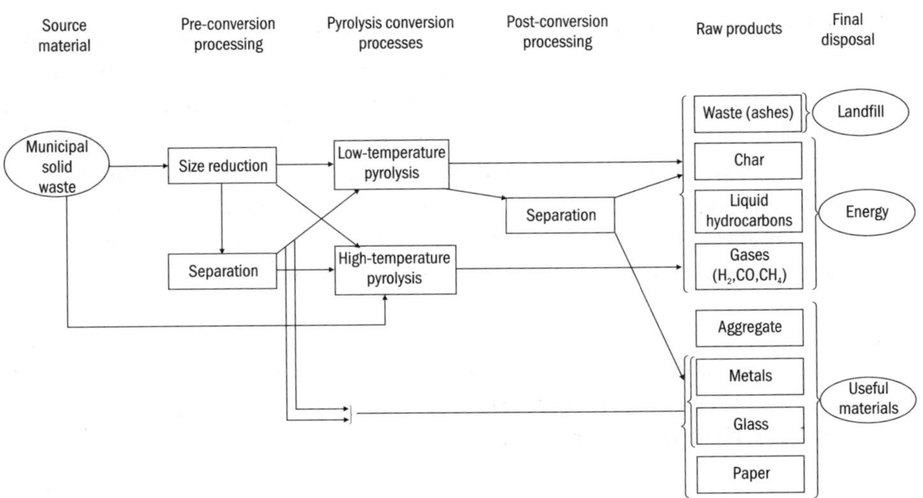

Figure 5 Pyrolysis of refuse
Source Domino (1979)

Technology options

The MSW contains organic as well as inorganic matter. The latent energy of its organic fraction can be recovered for gainful utilization through the adoption of suitable waste processing and treatment technologies. A few such options are as follows: sanitary landfill, incineration, gasification, biodegradation processes, anaerobic digestion, and so on.

Sanitary landfill

Landfill, implying the final disposal of MSW on the land, has several merits and demerits.

Merits

- Reduces emission of CH_4 and toxics.
- Lowers land and water table contamination.
- Reduces menace of birds and rodents.
- Controls fire hazards.
- Curbs the problem of bad odour in the inhabited area.

Demerits

- Involves high initial cost of design and commitment.
- Public may object to the site selected.
- Emits obnoxious gases in the atmosphere.
- Results in the loss of real estate value of the nearby areas.
- Problem of rodents and birds remains.
- Site selection depends on the availability of land at economically transportable distances.

Incineration

Incineration is a process of direct burning of waste in the presence of excess air (O_2), at temperatures of about 800°C and above, liberating heat energy, inert gases, and ash. The net energy yield depends on the density and composition of waste, relative percentage of moisture and inert material (adding to the heat loss), ignition temperature, size and shape of the constituents, design of the combustion system (fixed bed/fluidized bed), and so on. In practice, about 65%–80% of the energy content of the organic matter can be recovered as heat energy, which can be used either for direct thermal application or for generating power using steam turbine (with conversion efficiency of about 30%).

Merits

- Reduction in the size of the waste material reduces the problems of its dumping.
- Complete sterilization prevents the outbreak of epidemics.
- Biologically hazardous waste is reduced to ash.

Demerits

- Air pollution cannot be avoided even in highly sophisticated plants.
- It is a high-investment system.

- Additional cost inputs are required for complete pollution control.
- Dioxins and furan emissions are hazardous.
- Treatment of ash for removing pollutants is expensive.
- There is a high maintenance cost involved.

Gasification

Gasification involves the thermal decomposition of organic matter at high temperatures in the presence of a limited amount of air/O_2, producing a mixture of combustible and non-combustible gases (CO and CO_2).

Merits

- Handling of gas is easier than solid fuel.
- Reduction in volume reduces pressure on landfill dumping.
- Less ash is produced as compared to incineration.
- There is no emission of hazardous gases.

Demerits

- Plant requires regular maintenance.
- Operation and maintenance are costly.
- It releases tar and volatile poisons into the environment.

Biodegradation process

Biodegradation can be defined as the reduction of refuse through the use of an organic methodology (Figure 6). Organic methods are divided into two general

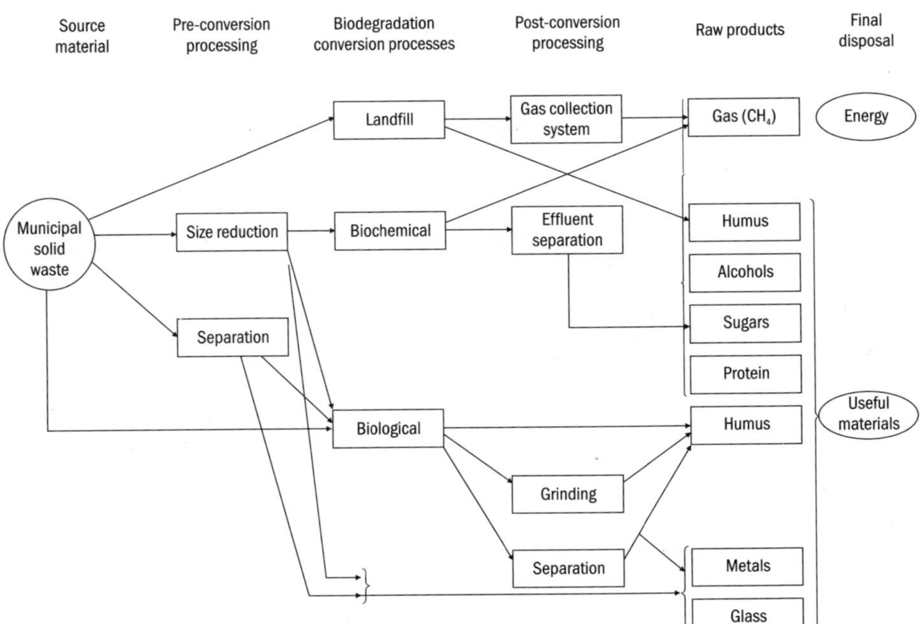

Figure 6 Biodegradation of refuse
Source Domino (1995)

categories. First is the direct reduction of refuse by biological organisms, including aerobic and anaerobic conversion. Second is the reduction of refuse by biochemical methods, including chemical processing, and/or the selected extraction from specific species of protozoa or fungi. In the aerobic degradation process, organic material is oxidized to give a humus product commonly called compost, which can be used as a fertilizer.

As the aerobic degradation process involves decay of the organic material, such as garbage, leaves, and manure, the process takes considerable time. Anaerobic digestion leads to a highly marketable product called CH_4. The first step is to break down the complex organic materials present in the refuse into organic acid and CO_2. The second step involves the action of bacteria known as CH_4 formers on the organic acid to produce CH_4 and CO_2. Biochemical conversion leads to either a reduction in refuse or the conversion of cellulose, and in many applications both aspects may exist.

Composting

Composting involves the breakdown of organic waste in the presence of micro-organisms, heat, and moisture. This can be carried out on a small scale in households or on a large scale, depending on the quantity of waste to be processed. Three types of micro-organisms are involved in the process of composting: bacteria, fungi, and actinomycetes; these micro-organisms act on the waste to convert it into sugars, starch, and organic acids. These, in turn, are acted upon by high-temperature bacteria, which prevail in the compost heap and generate stabilized compost (Figure 7).

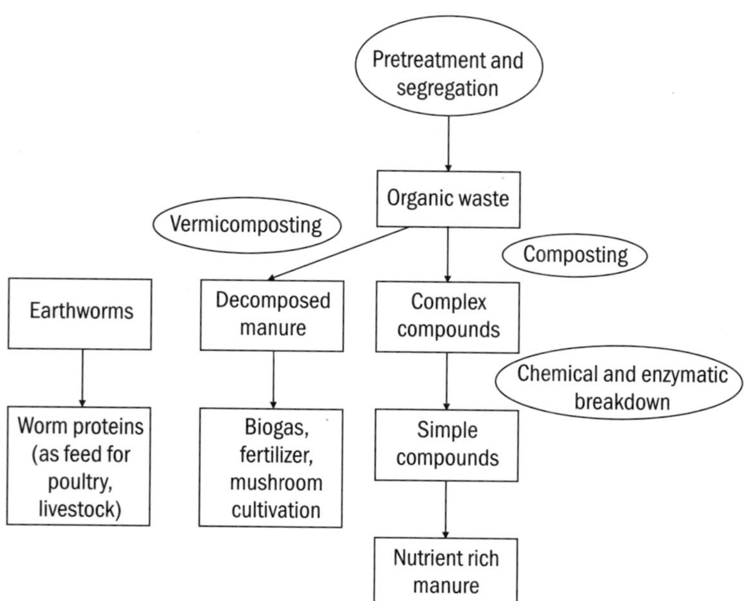

Figure 7 Process flow diagram of composting and vermicomposting of waste

Composting has the following advantages.

- Recycling of waste by the generation of useful manure, which is organic in nature.
- Reduction in volume of waste to be disposed of on land.
- No requirement of any high-end technical expertise.

Optimum conditions for composting

- *Carbon:nitrogen ratio* The ideal ratio of C:N is 30:1 for optimum growth of bacteria.
- *Particle size* Feedstocks should be grounded to increase the surface area of these materials. The optimum size is 1/8–2 inches (0.32–5.08 cm).
- *Oxygen* Compost should contain 15%–20% O_2 for optimum aerobic composting.
- *Temperature* Optimum temperature lies between 131°F and 150°F, which kills most weeds, seeds, animals, and plant pathogens.
- *Moisture* A moisture content of 40%–60% is required to support the growth of micro-organisms for conversion of organic substances into waste.
- *pH* The optimum pH for composting lies between 6.5 and 7.5.

The final compost is characterized by the absence of the smell of ammonia, completion of the reaction, and stabilization, as indicated by low temperature; dark colour of the compost; and presence of at least three species of arthropods, namely, the sow and pill bug, ground beetle, and centipede.

Compost characteristics

The quality of compost depends on the substrate processed. Most of the organic materials processed have high N, P, and K contents, which form enriched manure after undergoing composting. The average composition of the various constituents in compost is given in Table 5. In addition to the presence of various constituents in the compost, there is a need for the stabilized compost to contain heavy metals within the prescribed limits, before mixing it with soil. The United States Environmental Protection Agency (USEPA) limits for heavy metals in the compost are shown in Table 6.

Table 5 Characteristics of compost

Parameter	Concentration
Total nitrogen	1.3%
Total phosphorus	0.2%–0.5%
Total potassium	0.5%
Organic phosphorus	0.054%
pH	8.6
Moisture	45%–50%
Organic matter	30%–70%

Table 6 Standards for compost (parts per million)

Element	Concentration (parts per million)
Chromium	1000
Cadmium	10
Copper	500
Lead	500
Nickel	100
Zinc	1000

Types of composting technologies

There are three different types of composting technologies, namely, windrow, aerated static pile, and in-vessel composting.

Windrow composting

Triangular piles of waste are created to allow the diffusion of O_2 and retention of heat. The piles are occasionally turned by means of specific equipment consisting of paddles to increase porosity and facilitate the diffusion of air. Waste should preferably be piled under a roof to prevent its exposure to rain, which can otherwise form a part of a run-off.

Aerated static pile composting

Mechanical aeration of waste piles is done by placing them over a network of pipes connected to a blower. The blower supplies air for composting and hence creates a positive as well as a negative pressure. Air circulation provides O_2 and prevents heat build-up. Optimum temperature and moisture are maintained for microbial action. As the piles are not turned during the process, in order to achieve complete destruction of the pathogens, a layer of stabilized compost is placed over the pile to maintain the desired temperature. The retention time for the completion of composting is 6–12 weeks.

In-vessel composting

This is a controlled process of composting, wherein aeration, moisture, and temperature requirements for composting are maintained in a chamber or a vessel. The time taken for composting is one to four weeks. The advantages of this system over others are that the environmental conditions for rapid composting are under control, and the problems of malodour and leachate generation are minimal.

Vermicomposting

Vermicomposting is a process in which food material and kitchen waste, including vegetable and fruit peelings, papers, and so on, can be converted into compost through the natural action of worms. An aerobic condition is created by exposing organic waste to air. Vermicomposting has been extensively used in Canada, California, and Florida for the utilization of MSW. Many Asian countries are also adopting the process for waste disposal. Thousands of species of earthworms, such as *Eisenia foetida* and *Eisenia Andrei*, are actively involved in the decomposition of organic waste. Certain biochemical changes in the earthworm's intestine result in the excretion of cocoons and undigested food known as vermicasting, which is an excellent manure, containing vitamins, enzymes, nitrates, phosphates, and potash. The enzymes produced also facilitate the degradation of various biomolecules present in the solid waste into simple compounds for utilization by micro-organisms (Figure 7).

Because of the following characteristics, earthworms are used for composting waste.

- Ability to live in compost
- Fast multiplication, incubation, and maturation
- Complete digestion of the organics
- Adaptability to a wide variety of organic material

Effective micro-organisms for improving compost quality

Effective micro-organisms (EM) are present in a mixture of different beneficial, naturally occurring micro-organisms of three main genera, namely, phototrophic bacteria, lactic acid bacteria, and yeast. In the presence of organic material, these EM, consisting of aerobic and anaerobic bacteria, produce various useful products such as vitamins, organic acids, chelated minerals, and antioxidants. The anaerobic micro-organisms include *Lactobacillus bifidis*, zymogens, sulphur/sulphate-reducing bacteria, chlorobacteria, and brown-green photosynthetic bacteria. The aerobic bacteria consist of blue-green algae, azotobacters, and *Bacillus* sp. (*Bacillus subtilis*) acetobacters.

Application of EM technology results in the degradation of various pollutants, including heavy metals. The process occurs due to the action of two kinds of micro-organisms: zymogenic micro-organisms and synthesizing micro-organisms. Zymogenic micro-organisms reduce organic matter to soluble nutrients, which are readily available for consumption by bacteria. The process produces large quantities and a wide variety of organic acids, amino acids, and several antioxidizing enzymes, which facilitate the breakdown and decomposition of pollutants by EM. The EM technology can be applied to the compost to completely remove pathogens that exist even after three months of composting. Although various harmful bacteria, insects, and other pathogens that cause diseases are killed at a high temperature of 60–70°C in the compost heap, some survive this temperature. This technology was developed by Prof. (Dr) Teruo Higa, University of Ryukyus, Okinawa, Japan.

Quality of the compost improves due to an increase in the population of useful micro-organisms as well as the macro- and micro-nutrients (N, P, K, S, Ca, Mg, Fe, Mn, Zn, Cu, Mo, and B). One of the important benefits of using the EM technology is the absence of odour, which exists in a compost plant due to the emission of various gases such as ammonia and hydrogen sulphide.

The technology has been tested in several countries such as Japan, Pakistan, Switzerland, and United Arab Emirates for converting the city's solid waste and kitchen garbage, leather and oily sludge of petroleum industry, and the sludge containing chromium into bio-sludge and finally into bio-fertilizer. The bio-fertilizers have been successfully tested in rice and onion crops.

Status of composting technology in India

Composting in India has been used at both the household level as well as for handling waste at the city level. Non-governmental organizations (NGOs) and residents welfare associations have been instrumental in promoting this technology on a small scale. In India, two methods of composting are popularly used. The Indore process involves packing of alternate layers of vegetable waste and night soil of 3-inch (7.62 cm) thickness up to a depth of 1.5 m in a trench. This is an aerobic process and is completed within four months. The Bangalore process is an anaerobic process, which does not require occasional mixing of the heap of waste. The time period for the completion of the process is about five months. The compost quality is high, with N, P, and K content of 1.5%, 1.1%, and 1.5%, respectively.

Highly mechanized composting processes involve a range of techniques varying from sieves, hammer mills, and pulverizers, to rotating drums for the separation of the non-biodegradable materials and decomposition of organic waste.

Excel Industries in India has set up composting plants in various cities such as Ahmedabad, Bhopal, Delhi, Kanpur, Kolkata, and Mumbai. The operating parameters of the plants are given in Table 7. However, due to the problem associated with the marketing of the product, the plant is not economical. Privatization has been effective in the revival of closed-down plants and conversion of the waste in the dump sites into compost in Ahmedabad, Bengaluru, and Mumbai.

Table 7 Operating parameters of the compost plant by Excel Industries

Parameter	Value
Weight of garbage	300 tonnes/day
Retention time	45 days
Quantity of compost	100 tonnes/day
Cost of plant	Rs 25 million
Value of product	Rs 1500/tonne

Anaerobic digestion

Various types of organic waste, such as animal dung, sewage sludge, and organic fraction of the MSW, can be anaerobically digested to give biogas and manure. Anaerobic digestion involves the breakdown of organic compounds by micro-organisms in the absence of O_2 to produce biogas, which is a mixture of CH_4 and CO_2. The optimum temperature for anaerobic digestion is 37°C, and the pH is 7. In addition to treating waste, anaerobic degradation generates clean fuel, which can be used for various thermal applications for power generation and for use of digested sludge as manure. The feasibility of biogas generation from the organic fraction of the MSW was known as early as the 1950s. Since then, research efforts were directed towards enhancing the biogas yield by the adoption of various mechanisms, including modification of digester designs and use of thermophilic conditions. Research on the degradation of municipal waste and biogas generation is at different levels and on different types of digestion processes. The various types of anaerobic digestion processes include conventional slurry digestion, dry digestion, and two-phase digestion. Segregation and pre-processing for the separation of biodegradable and non-biodegradable fractions are the foremost steps in almost all categories of the anaerobic digestion processes.

Conventional digestion

The process is popular for biogas generation from animal dung and sewage sludge at a TS concentration of 3%–10%. The homogeneous slurry prepared by mixing the solid with water is introduced into the digester, and an equivalent amount of digested slurry is withdrawn, the volume of which is dependent on the retention time for digestion. Application of this design for MSW does not appear to be feasible due to the mixing of very large volumes of water to attain the desired TS content. Moreover, pre-processing is required to make homogeneous slurry. In addition to this, mechanical mixing of the contents may be required during digestion to prevent the floating of partially digested material and the trapping of biogas. Some of the processes based on continuous digestion have been used in many large-scale demonstration facilities for treating waste in other countries.

Refuse conversion to methane process

Refuse conversion to methane (RefCom) process was demonstrated by the Waste Management, Inc. at Florida, for the conversion of 50–100 TPD of municipal waste and sewage sludge to biogas. In this process, after mechanical segregation of the biodegradable organic material from waste, the waste was mixed with sewage sludge and water obtained from the dewatering of the digested sludge. The mixture was fed into parallel digesters of 1300 m^3 capacity. The system had an operational problem due to the inefficient separation and multiple components in the separation process. However, with the modified screen separation, the process was successful in thermophilic conditions of 58–60°C, and could operate at a maximum TS concentration of 4.8% with the agitation. The monitoring of the system performance indicated an optimal retention time of 10–15 days with a gas yield of 0.34 m^3/kg volatile solid (VS) added. It was necessary to maintain a pH of over 7 to prevent the souring of the digester.

WMC process

The WMC process was developed by the WMC Resources Recovery Ltd, UK, which developed a plant of 5–7 TPD for treating the municipal refuse. After the initial screening and shredding for the separation of non-biodegradable components, the waste is treated under mesophilic conditions at a TS concentration of 10% with a retention time of 15 days. The gas is used for mixing the contents. In addition to biogas, the digested sludge, after dewatering, is converted into manure, following the stabilization process. Alternatively, the digested solids are blended with coal dust and are used as fuel pellets for various industrial and commercial applications.

Cal recovery process

The Cal recovery process was again based on the digestion of segregated biodegradable waste component of municipal waste at Los Angeles. The maximum organic loading rate at a retention time of 15 days was 4 kg VS/m^3/day. The major operational problem experienced during this process was the formation of a scum layer due to insufficient mixing of the digester contents.

Waste biogas process

In this process, the waste biogas, following the shredding and separation processes, is digested under mesophilic conditions at a concentration of 5%–10% after mixing with the sewage sludge. At an average TS concentration of 7%, and loading rate of 1.5 kg VS/m^3/day, and a retention time of 28 days, the CH_4 yield was 0.16 m^3/kg TS added.

The digested sludge after drying can be used as a soil conditioner.

Dry anaerobic digestion process

Dry anaerobic digestion is a novel process for the anaerobic degradation of solid waste at higher TS concentration. The features of this process are (1) TS concentration of 30%–35%, (2) no requirement of mechanical agitation, and (3) no scum formation.

Three different types of processes are popularly used under dry anaerobic conditions: dry anaerobic composting (DRANCO) process, VALOGRA process, and IBVL process.

Dry anaerobic composting process

DRANCO involves the pretreatment segregation step for the isolation of biodegradable components, which are anaerobically digested under thermophilic conditions. After the process of digestion at a TS concentration of 30%–35%, further stabilization occurs in two to three days. The overall retention time for the process is 21 days, and the biogas yield is 125–180 m^3/tonne of input. As in the continuous process, sludge dewatering provides water for making up the initial TS concentration, and the dried digested sludge is used as a soil conditioner after suitable processing to reduce the size. The compost quality is high due to the absence of pathogens as a result of the processing at thermophilic conditions and a high C:N ratio of 15:1.

VALOGRA process

This process can handle the TS content of the slurry in the range of 35%–40%, with the agitation of the reactor contents by occasional supply of biogas under a pressure of 6–7 bars. Biogas production in this case is 4 m^3/m^3 of reactor/day, as compared to the conventional process of 1 m^3/m^3 reactor/day at a retention time of 15 days under the mesophilic conditions. The conversion efficiency is 50%, and the digested sludge is used as a combustion material after further processing.

BIOCEL process

A relatively cheaper and low maintenance dry anaerobic system is the BIOCEL process, wherein the organic fraction of solid waste in the presence of methanogenic bacteria is allowed to remain under a plastic cover for six to eight weeks, resulting in CH_4 yield of 37–50 m^3/tonne of organic waste. Preliminary experiments and pilot plant studies have been carried out by adopting this process.

Two-phase digestion

The two-phase digestion process involves the separation of the hydrolysis, acidification, and methanation phases that occur during the anaerobic digestion of waste and provides the optimal conditions for maximum efficiency in both phases. There are three types of two-phase processes that have been developed for solid waste.

The Hitachi process

The Hitachi process was developed by Hitachi Ltd. It involves a thermo-chemical pretreatment process under alkaline conditions at 60°C, with a pH of 9.8 for 3 hours. The liquefaction step involves a high temperature of 60°C and an alkaline pH of 7.5–8.2. The total retention time for the process is six to eight days.

The IBVL process

The IBVL process was developed for treating organic solid waste, particularly agro-based products by the Institute for Storage and Processing of Agricultural Produce, the Netherlands. The system involves liquefaction and acidification reactor and a methanogenic reactor, namely, the upflow anaerobic sludge blanket (UASB) reactor. The reaction occurs under mesophilic conditions. The digested effluent from the methanogenic reactor is recirculated into the liquefaction/acidification reactor, resulting in the conversion of the volatile fatty acids (VFAs) into CH_4 due to the presence of CH_4-producing microbes in the effluent of the UASB. During this period, the methanogenic reactor handles the acids being produced from a just-started liquefaction/acidification reactor. The rate of degradation becomes constant after a period of 10–14 days. The digested residue is subjected to aerobic composting before being used as a soil conditioner.

Status of application of anaerobic digestion technology for treating municipal solid waste in India

Biomethanation is considered to be the most technically viable option for the Indian MSW due to the presence of high organic and moisture content and the dual benefits associated with the process. In addition to this, the temperature conditions are more suitable for the optimum performance of the digester. In spite of this, there are no commercial plants for biomethanation of solid waste in India. Indigenous technologies are being developed by various organizations, and small-scale prototypes have been found functional. However, these are not commercialized as yet. Several institutes have been carrying out research studies associated with the biomethanation of solid waste for improving the digestibility and biogas yield by providing optimal conditions, improved microbial strains, and digester designs. Table 9 provides a summary of the various parameters involved for the technological options—high rate biomethanation and producing refuse-derived fuel (RDF).

The Centre for Application of Science and Technology to Rural Areas (ASTRA) at the Indian Institute of Science, Bengaluru, has developed a plugflow digester to dispose of the leaves and other urban waste. TERI (2008) has developed a biphasic process for treating different types of organic solid waste. The process called TERI's Enhanced Acidification and Methanation (TEAM) is a two-stage anaerobic digestion process designed specially for biomethanation of organic solid waste that is fibrous and has light floating materials (Rajeshwari, Lata, Pant, *et al.* 2001). Operating between 35°C and 40°C, the first stage of the process extracts the organic content from the solid waste, while the second stage generates biogas and enriched manure from the liquefied waste. The system has six acidification reactors operating in a series and a single UASB for CH_4 production from the VFAs. The total retention time is seven days, and the process does not involve any agitation mechanism, requiring low maintenance. The digested sludge has a high N, P, and K content, and the treated effluent from the UASB is reused for the extraction of the organic contents in the acidification phase (Figure 8). Recently, a 5-MW waste-to-energy (WTE) plant has been commissioned in Lucknow, which is based on the technology by Entec, Austria, for biomethanation of solid waste. The biogas-induced mixing arrangement (BIMA) is a technology developed

Figure 8 TEAM digesters for solid waste treatment

by Entec. The technology is specifically suitable for high biochemical oxygen demand (BOD) waste materials. The digester converts the waste material into biogas and sludge and is adopted for the biomethanation of different types of solid waste such as animal dung, vegetable processing waste, slaughterhouse waste, fish-processing waste, waste water, sewage sludge, and household waste all over the world. The features of the BIMA digester are as follows: (1) no moving parts and hence reduced installation and maintenance costs; (2) high TS content of 12%, which can be treated at a short hydraulic retention time of 10–20 days; (3) high efficiency with the VS reduction of 50%–60%; and (4) no scum formation, with the gas produced inside being used for mixing contents.

Cost of power generation from municipal solid waste to energy projects

A comparative evaluation of different WTE technologies for India is given in Table 8, and the suitability of different MSW WTE technologies for India is given in Table 9.

Thus, the following points emerge from the previous discussion.

- Producing RDF from source-segregated waste and its use in already existing boiler or dedicated boiler hold promise, and the technologies are commercially available.
- High-rate biomethanation must be given high priority for WTE projects in India due to the combination of cost, technology, effectiveness, and environmental benefits.

Table 8 Evaluation of municipal solid waste (MSW) technologies

Description	Plasma gasification vitrification (PGV)	Biomethanation	Pelletization-based combustion	Pyrolysis
Type of process	PGV	Biogas-induced mixing arrangements (BIMA)	Combustion	Solid waste-to-energy recycling facility (SWERF)
MSW treatment capacity (tonnes per day)	500	300	500	500
Quantity of final MSW treated (tonnes/day)	165	—	160	500
Pre-processing of the MSW	Needed	Needed	Needed (segregation, drying, and fluffing)	Not needed
Land requirement	10 hectares (22.7 megawatt [MW])	—	9–10 acres (1 MW)	10 acres (11 MW)
Technology status	Found to be successful abroad mostly for medical and hazardous wastes	Under demonstration in India and abroad	Under demonstration in India	Successful for the MSW in Wollongong, New South Wales
Type of feeding	Mixed with biomass residue in densified form	Slurry preparation	Conveyor system; no mixing of external feed for desired properties	No shredding and drying of feed. For pretreatment, autoclave, magnetic, and eddy current separators are used
Level of automation	High	Moderate	Moderate	Moderate
Power generation capacity (MW/100 tonnes/day)	22.7 MW (4.5 MW)	5.6 MW (1.9 MW)	6 MW (1.2 MW)	11 MW (2 MW)
Plant load factor (%)	90	—	70	90
Auxiliary power consumption (MW/%)	7.7 MW (34.2%)	0.5 MW (8.9%)	10%	14%
By-product	7 tonnes/day of slag	75 tonnes/day of bio-fertilizer	—	Char
Plants in operation in India	None	Not for the MSW but for slaughterhouse waste and tannery waste in Hyderabad	Only MSW pelletization plant is in operation in Hyderabad	Various stages of commissioning

Contd...

Table 8 Contd...

Description	Plasma gasification vitrification (PGV)	Biomethanation	Pelletization-based combustion	Pyrolysis
Capital cost (million rupees/ MW)	187 (82.3)	76 (135.7)	25 (41.6)	11
Cost of power generation (rupees/ kWh)	4.11	—	2.6–2.8	—
Cost of by-product	Rs 8000/tonne	—	—	Market yet to be explored
Post-landfill requirements	Required for inert	Required for inert	Required for inert	Required for inert and char
Manufacturers in India	—	Imported (some of the plant and machinery can be indigenized)	Available	Some of the plant and machinery can be indigenized

Source UPERC (2003)

Table 9 Suitability of different municipal solid waste (MSW)-based waste-to-energy technologies for India

Technology	Advantages	Disadvantages	Applicability for India
Landfill with landfill gas (LFG) extraction	▪ Low-cost energy recovery because infrastructure exists ▪ Controlled LFG recovery results in high gas yield	▪ Uncontrolled methane can escape easily ▪ Requires significant landfill infrastructure ▪ Requires high volume of waste and large land area	Not recommended as per MSW Rules of 2000, which state that organic waste cannot be sent to the landfill
Incineration	▪ Most suitable for high calorific value waste such as paper, plastics, and wood ▪ Relatively free from noise and odour, and requires smaller land area	▪ Highly capital intensive ▪ Low heating value waste not ideal for energy recovery ▪ Toxic metals lead to air pollutant emissions ▪ Requires high operation and maintenance cost and high pollution control cost	Not recommended unless waste stream is supplemented by high calorific value waste free from chlorinated plastics
Biomethanation	▪ Can be small scale, requiring no external power source ▪ Totally enclosed system; modular construction possible ▪ Odour and visible pollution reduced	▪ Highly capital intensive ▪ Requires high degree of biodegradable material	Recommended only with segregated MSW

Contd...

Table 9 Contd....

Technology	Advantages	Disadvantages	Applicability for India
Refuse-derived fuel (RDF) burning	▪ Fuel produced has high energy content ▪ Fuel is convenient to store and transport	▪ High energy consumption for crushing and drying of waste ▪ High inorganic content reduces the effectiveness of fuel ▪ Favourable in areas having lower rainfall ▪ May require gas clean up to avoid toxic emissions	Recommended for Indian conditions. Indian projects are presently using MSW fluff (step before pelletization), thus reducing the cost of production. Solar drying is also an option to reduce pre-processing cost
Gasification/pyrolysis	▪ Can be small scale with modular design ▪ Closed system with no smell and odour ▪ Less expensive gas clean up system than incinerators	▪ Net energy recovery might be affected by excessive moisture ▪ No commercial model on MSW as yet	Technology adaptation for MSW feedstock required, therefore not recommended
Plasma arc pyrolysis	▪ Efficient treatment of any type of waste ▪ End product is vitrified slag, which does not require landfilling	▪ Highly expensive ▪ No commercial model on MSW as yet	Recommended only for very toxic or radioactive waste where cost of treatment is not a concern

Source TERI (2008)

- Gasification/pyrolysis has a distinct potential, and although there are limitations to its uptake, these can be circumvented as the technology matures.
- The present trend is in favour of material recovery facilities and a shift away from landfills for MSW disposal. Landfill with gas recovery system (landfill gas, LFG) has a long track record, but it is no longer pursued abroad. It is also excluded as a potential technological option in this study in view of Point 6 (Disposal of municipal solid wastes), Schedule II (Management of the solid waste) of the Municipal Solid Wastes (Management and Handling) Rules, 2000.

WTE projects provide for a beneficial way for disposing of waste. Many technologies such as incineration, biomethanation, RDF method, gasification, and pyrolysis can be used to derive energy from MSW. Of the above technologies, only biomethanation and RDF projects have proved to be commercially viable in both international and national markets. Hence, for the sake of assessing the cost of power generation from such projects, we need to consider only two types of MSW for energy technologies, that is, biomethanation and RDF.

In the case of biomethanation, three cases are developed, depending on the capacity of plant to process the waste. In case the waste generated in the city is less than 300 TPD, then a small capacity plant would be sufficient. On the other hand, if the waste generated in the city is approximately 1000 TPD as in large cities like Delhi, then a higher capacity plant would be needed. Thus, three cases are developed, that is, 3-MW plant that can process up to 300 TPD of waste, 5-MW plant that can process up to 500 TPD, and 10 MW plant that can process 1000 TPD of waste.

In the case of RDF-based plant, it is necessary to operate the plant at a certain level, and it would not be feasible to operate it in a city that generates only about 300 TPD of waste. It would be economical only in a city that generates at least 500–700 TPD of waste, such as Hyderabad. Thus, only 6.5 MW capacity RDF plant is considered, which can process up to 700 TPD of waste.

Technical parameters

The technical inputs and assumptions used for estimating the levelized cost of power generation using biomethanation and RDF-based WTE plants are summarized in Table 10.

Domestic companies involved in municipal solid waste treatment and disposal

The Ministry of New and Renewable Energy, Government of India, is promoting all the technology options available for setting up projects for the recovery of energy from urban waste. Table 11 provides a list of WTE technology providers/suppliers in India.

- Zen Global Finance Ltd has indigenously developed a technology for the conversion of MSW into a clean fuel. The RDF technology can cost-effectively replace domestic fuels such as firewood and kerosene. Currently, Zen Global has set up a plant in Andhra Pradesh, which has a capacity of 200 TPD with an expected fuel output of 60 TPD. An RDF-based plant is also being retrofitted in Delhi, whereas proposals for Kanpur and Agra are under different stages of consideration.
- EnKem Engineers Ltd is promoting a biomethanation process in technical collaboration with Entec of Austria. EnKem has designed a BIMA digester for the treatment of MSW, resulting in the power generation. The ongoing projects include two projects at Lucknow and Kanpur for treating of 300 TPD and 600 TPD of MSW for the generation of electricity.
- Future Fuel Engineering (India) Pvt. Ltd is promoting anaerobic biodigestion technology in collaboration with ECOTEC of Finland and is presently in the process of installing a plant at Kalyan, near Mumbai. Global Environmental Engineering Ltd is promoting anaerobic biodigestion

Table 10 Technical inputs and assumptions

Particulars	Biomethanation			Refuse-derived fuel
Plant capacity (MW)	3	5	10	6.5[a]
Waste processed (TPD)	300[b]	500[b]	1000[b]	700[a]
Life of plant[b] (year)	15	15	15	15
Capital cost (million)	402.5[b]	573[b]	1020[b]	600[a]
Land cost[a] (million)	3	5	6	4
Operation and maintenance (O&M) cost (million)	27.2[b]	41.3[b]	82.5[b]	30[a]
Annual escalation in O&M cost[c] (%)	4.83	4.83	4.83	4.83
Debt–equity ratio[d]	70:30	70:30	70:30	70:30
Interest rate[d] (%)	12.00	12.00	12.00	12.00
Repayment period (including 1-year moratorium)[d] (years)	10	10	10	10
Return on equity[e] (%)	14.00	14.00	14.00	14.00
Discount rate (weighted average cost of capital, that is, WACC) (%)	12.60	12.60	12.60	12.60
Capital recovery factor (%)	15.16	15.16	15.16	15.16

[a]As per discussion with stakeholders.

[b]Based on National Master Plan for Development of Waste-to-Energy in India, Ministry of New and Renewable Energy (MNRE), Government of India; details available at <www.mnes.nic.in>, last accessed on 25 July 2008.

[c]Based on average wholesale price index for the last three years, that is, 2005/06, 2006/07, and 2007/08; details available at <https://reservebank.org.in/cdbmsi/servlet/login/>, last accessed on 25 July 2008.

[d]Based on financing norms as given by the Indian Renewable Energy Development Agency (IREDA).

[e]Based on the Central Electricity Regulatory Commission (CERC) norms for return on equity for power generating plants.

technology in collaboration with PAQUES Pvt. of the Netherlands. Its current project includes a pilot project in Pune.

- Thermax Ltd, the leader among Indian companies, is offering environment-friendly technologies. The waste management division of Thermax offers incineration plants for solid and hazardous waste. Its collaborators include ACWA of the UK, Danskrodzone of Denmark, and Thermal Process of the US.
- Excel Industries is the most successful organization in the field of MSW disposal in India. It has implemented many projects for producing bio-fertilizer (compost) from solid waste. The process adopted by Excel Industries involves four stages: sanitation of waste by chemical, segregation of organic waste, reduction of volume of waste, and conversion of waste into biological products using bacteria. It has already installed seven plants all over India and is marketing its product at Rs 1500/tonne.

Table 11 Waste-to-energy technology providers/suppliers in India

Biomethanation technology	Incineration/combustion technology
ENKEM Engineers Pvt. Ltd	SELCO International
Chemtrols Engineering Ltd	Shriram Energy Systems Ltd
Mailhem Engineers Pvt. Ltd	Pyrolysis/gasification technology
Degrimont India Ltd	Shri Damodar Synthetics Ltd
REVA Enviro Systems Pvt. Ltd	Batliboi Environmental Engineering Ltd
Global Environmental Engineering Ltd	
Linde Process Technologies India Ltd	
UEM India Ltd	
Hydroair Tectonics Pvt. Ltd	
Plasma arc technology	**Biogas engine supplier**
Terrasafe Technologies (P) Ltd	Green Power International (P) Ltd
Global Enviro Plasma Technologies	Guascor S.A. (Spain)
	APE (India) Ltd
	Cogen India Engineering Pvt. Ltd
	Greaves Ltd

Source Ministry of New and Renewable Energy, Government of India 2010; Recent developments in municipal solid waste management. Details available at http://mnes.nic.in/energy-uwaste.htm#.

Planned projects are located at Vijayawada in Andhra Pradesh, Amboli in Mumbai, and Ahmedabad in Gujarat. A plant in Jaipur, costing Rs 56 million ($1.14 million) is under implementation. Bioconversion units have been commissioned at Bhopal, Kolkata, Malad, Mumbai, and Thane (Mumbai), along with the commissioning of the largest MSW dumping ground at Deonar in Mumbai.

- EDL Power (India) Ltd is the only company in India to provide sanitary landfill technology in collaboration with an Australian company. The company has not been able to make any major headway so far.

The MSWTE projects based on biomethanation technology are under various stages of implementation in India. Most of these designs are imported from Austria, Australia, Germany, and many other countries. These projects fall under the demonstration schemes of the Ministry of New and Reneweable Energy, Government of India, for generating energy from urban and industrial waste. The only MSW-based incineration facility implemented in New Delhi is under revival. The technology based on the RDF, the first of its kind in the country, has been commissioned (6.6 MW) on the outskirts of the Hyderabad city, and the electricity so generated is fed to the grid. Technology Information, Forecasting and Assessment Council (TIFAC), Department of Science and Technology, Government of India; and Technology Development Board have jointly extended financial and techno-managerial support to

SELCO International Ltd, Hyderabad, for implementing the project. Shriram Energy System Ltd has recently commissioned another commercial plant for generating electricity from waste in Vijayawada. It is being synchronized with the APTransco grid. The major foreign companies servicing the Indian market for waste are Entec, Austria; TBW, Germany; BTA, Germany; Lunde, Germany; Nellemen, Neilsen; Rauscven berger, Denmark; and EISU, UK.

International scenario and practices

Germany

Germany promotes recycling by adopting the world's most stringent return requirements for packing material and many other goods, including automobiles. Material management is the responsibility of the manufacturer during the end of product life, including ultimate disposal or reuse of the materials from which it is made. This has led to high recycling rates and heavy monetary costs, prompting the ongoing controversy in Germany and elsewhere. Each year, Germany generates about 30 MT of MSW, and the landfills emit 1.2–1.9 tonnes of CH_4, accounting for 25%–35% of Germany's CH_4 emission and about 3%–7% of the national gross world product (GWP) to meet the provisions of a 1993 law requiring that by 2005, all wastes disposed in landfill will have to have a total organic carbon content of less than 5%. Under the law, CH_4 emissions are projected to drop by 67% by 2005 and 80% by 2015.

United States

The US produces about 200 MT of MSW each year. In 1997, 55% was landfilled, 28% recycled or composted, and 17% incinerated. The 11.6 tonnes of CH_4 emitted by landfills accounted for 37% of the anthropogenic CH_4 emission or about 4% of the national greenhouse gases (GHGs). The US regulation now requires the largest landfills to collect and combust the LFG, which is projected to reduce emissions to 9.1 tonnes in 2010. There are more than 150 LFG projects in operation; 200 more are under development and are being promoted by the government through technological support and tax incentives. If all the materials currently being recycled in the US were to be landfilled, the national GHG emissions would increase by 2%, even with the new LFG regulation. More than 9000 municipal organizations collect household material, and numerous commercial enterprises also recycle material. Many innovative uses of recycled material are reducing the emissions during manufacturing; for example, remanufacture of commercial carpets from recovered fibres lowers the energy input by more than 90%, and some products are now said to have a zero net GHG impact.

Japan

With a large waste stream and a limited land area, Japan relies heavily on both recycling and incineration as an alternative to landfilling. The widespread participation in recycling recovers not only the easily recycled material, such as metal and glass, but also large quantities of unconventional recycled material, such as aseptic packaging (juice boxes). Japan has approximately 1900 waste incinerator facilities, of which 17% generate electric power with a capacity of 710 MW. A major new commitment to convert highly efficient WTE facilities has been announced by the Japanese government. In 1998, a corrosion-resistant, high-temperature, fluidized-bed WTE facility achieved 30% conversion efficiency (to electricity) with low dioxin and stack gas emission. The facility can accept mixed municipal and industrial waste, including plastic, and recover ash for road foundation and recyclable metals.

Best available techniques for recovery of metals from e-scrap

The Tenth Plan document of India estimates that approximately 7.2 MT of hazardous waste is being generated in India per year. Of this, the most significant growth is experienced in the generation of waste electrical and electronic equipment (WEEE), especially waste computers, TVs, printers, mobile phones, and so on. The e-waste inventory for 2005 has been estimated to be 0.15 MT and is expected to exceed 0.8 MT by 2012, according to the CPCB estimates. Of the total e-waste generated in the country, western India accounts for the largest proportion at 35%, followed by the South India at 30%. North India and eastern India account for 21% and 14% of e-waste generation, respectively. Although North India is not a leading generator, it happens to be the leading processing centre of e-waste in the country. Currently, of the 14 registered e-waste re-processors in India, there are three formal recyclers operating in the North, and there are no formal recyclers in the East. A graphical representation of the regional spread of e-waste generation in India is given in Figure 9.

WEEE, commonly known as e-waste, comprises waste generated from used electronic devices and household appliances that are not fit for their original intended use and are passed through the chain of recovery, recycling, or disposal. Such waste items encompass wide range of electrical and electronic devices such as computers, hand-held cellular phones, personal stereos, and large household appliances such as refrigerators and air conditioners. E-waste contains over 1000 different substances, many of which are toxic and potentially hazardous to the environment and human health, if not handled in an environmentally sound manner. However, e-waste contains several useful materials that can be processed to obtain precious metals such as gold and platinum, including base metals, glass, and plastics.

For example, a mobile phone can contain over 40 elements, including base metals such as copper (Cu) and tin (Sn), special metals such as cobalt (Co), indium (In), and antimony (Sb), and precious metals such as silver (Ag), gold

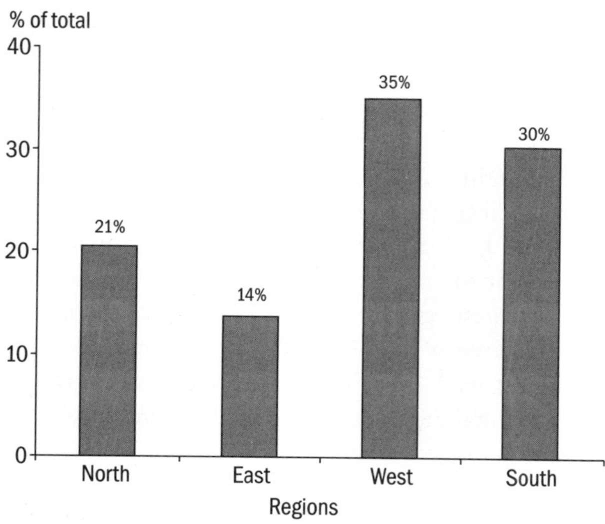

Figure 9 Regional spread of e-waste generation in India

Table 12 Recovery of resources from mobile phones

Components	Composition by weight (%)
Acrylonitrite butadiene styrene–polycarbonate alloy	29
Ceramics	16
Copper and compounds	15
Silicon plastics	10
Epoxy	9
Other plastics	8
Iron	3
Pre-painted steel (PPS)	2
Flame retardant	1
Nickel and compounds	1
Zinc and compounds	1
Silver and compounds	1
Al, Sn, Pb, Au, Pd, Mn, etc.	<1

Al – aluminium; Sn – tin; Pb – lead; Au – gold; Pd – palladium; Mn – manganese

Source <www.envocare.co.uk/mobile_phones.htm>

(Au), and palladium (Pd), as shown in Table 12. Metals represent, on average, 23% of the weight of a phone, the majority being Cu, while the remainder is plastic and ceramic material. Looking at 1 tonne of phone handsets (without battery), this would be 3.5 kg Ag, 340 g Au, 140 g Pd, and 130 kg Cu. For a single unit, the precious metal content is generally in the order of milligrams only: 250 mg Ag, 24 mg Au, 9 mg Pd, 9 g Cu are present on average. Furthermore, the Li ion battery of a phone contains about 3.5 g Co. At first sight, this appears to be very little, but taking into account the figure of 1.2 billion mobile phones sold globally in 2007, this leads to a significant metal demand.

As an example, on average, a computer comprises 23% plastic, 32% ferrous metals, 18% non-ferrous metals (lead, cadmium, antimony, beryllium, chromium, and mercury), 12% electronic boards (gold, palladium, silver, and platinum), and 15% glass (Figure 10). Only about 50% of the computer is recycled, and the rest is dumped. The toxicity of the waste is generally due to the presence of lead, mercury, and cadmium—non-recyclable components of a single computer may contain almost 2 kg of lead. Much of the plastic used contains flame-retardants, which makes it difficult to recycle this plastic.

The e-waste consists of diverse items such as ferrous and non-ferrous metals, glass, plastic, electronic components, and other items, and also hazardous elements. Therefore, the major approach to treat e-waste is to reduce the concentration of these hazardous chemicals and elements through recycle and recovery. In the process of recycling or recovery, certain e-waste fractions act as secondary raw material for the recovery of valuable items. The recycle and recovery include the following unit operations (Figure 11).

1 *Dismantling* Removal of parts containing dangerous substances (chlorofluorocarbons [CFCs], mercury [Hg] switches, polychlorinated biphenyl [PCB]); removal of easily accessible parts containing valuable substances (cable containing copper, steel, iron, precious metal containing parts, for example, contacts).
2 *Segregation of ferrous metal, non-ferrous metal, and plastic* This separation is normally done in a shredder process.
3 *Refurbishment and reuse* Refurbishment and reuse of e-waste have potential for those used electrical and electronic equipment that can be easily refurbished to put to their original use.

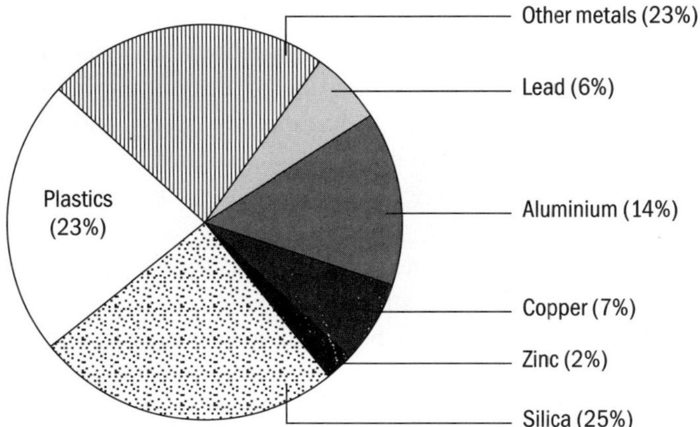

Figure 10 Resource recovery potential from computer waste (based on a typical desktop computer, weighing approximately 27 kg)
Source MCC (1996)

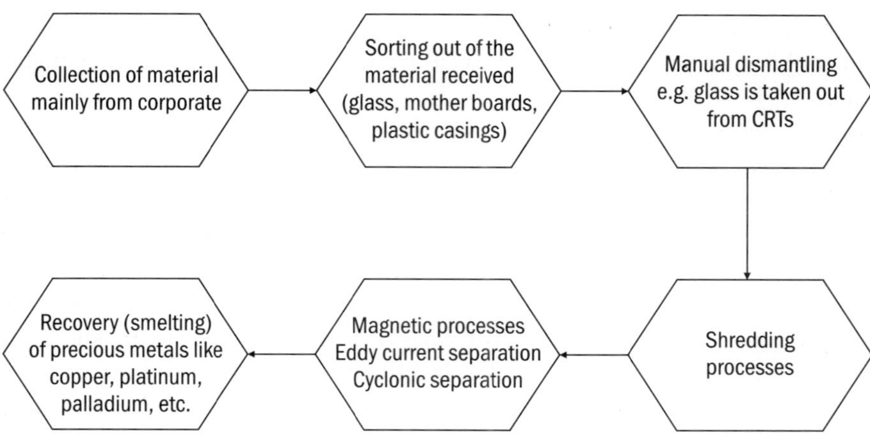

Figure 11 Process followed by formal of e-waste recyclers in India

4 *Recycling/recovery of valuable materials* Ferrous metals in electric arc furnaces, non-ferrous metals in smelting plants, and precious metals in separating works are recovered.
5 *Treatment/disposal of dangerous materials and waste* Shredder light fraction is disposed of in landfill sites or sometimes incinerated (expensive), CFCs are treated thermally, PCB is incinerated or disposed of in underground storages, and Hg is often recycled or disposed of in underground landfill sites.

Smelters across the developed countries use state-of-the-art process for material recovery and safe disposal of the waste materials, with minimal loss and environmental impact. The processes followed in these units include mechanical and metallurgical processes. In the mechanical process, appropriate techniques for dismantling, shredding, and sorting are adopted. The process is based on the physical and optical characteristics of the materials and fractions that are processed. Some of the commonly used methods are magnetic separation of ferrous fractions, eddy current separation of aluminium, manual and optical sorting, and gravity separation. The respective metal and non-metal fractions are then sent for further processing. The magnetic fraction is sent to the steel plant for further processing, and the aluminium and copper fractions are sent to the respective smelters.

The plastic and glass fractions are either further processed in recycling units or disposed of after treatment in landfills. The metallurgical processes in a modern smelter result in the recovery of base and special metals such as lead, copper, nickel, tin, antimony, and bismuth, and precious metals such as silver, gold, platinum, palladium, and rhodium. The processes are based on chemical and metallurgical properties of the materials and consist of three broad steps of extraction, separation, and purification. The recovery yield of various metals is high due to several intermediate steps. The processes are accompanied by appropriate waste water and solid waste treatment and off-gas purification processes, and efforts are made to reuse the treated water and waste slag as construction material and recover chemicals and acids. In addition to glass and various metal fractions, plastics form a major component of e-waste. The huge c-volume and weight of the plastic materials recovered from the waste components have made it necessary for the recyclers to explore economical and environment-friendly solutions for regenerating plastic materials without altering their properties.

The recovery techniques in this particular area are still emerging. A methodology, CreaSolv process (Maurer, Schlummer, and Altanu 2004), has been developed for the recovery of polymer fraction from the waste plastics of mobile phones. The process is economical, safe, and environment-friendly and results in high-quality recycled polymer. The process involves dissolving the main polymers—acrylonitrilite butadiene styrene (ABS) or polycarbonate (PC)—in organic solvents. The dissolved components are then separated from the insoluble materials (metals), generating particles of recycled polymer. The polymer can be further extruded in injection moulding processes, resulting in pure recycled polymers with properties similar to virgin material. The process is economical as the solvent can also be recovered, which can be further reused for the dissolution of plastics. The process is efficient as it does not require any further mechanical processing such as grinding, but can be applied directly on plastic materials after dismantling.

Gartner Inc. (2007) estimates that the Indian e-waste market was worth Rs 510 million (€7.3 million) in 2008 and is predicted to reach Rs 1.5 billion (€21.6 million) by 2013. This is based only on the revenues of authorized e-waste recyclers, which make up just a fraction of the market as a whole. There are 12 authorized recyclers in India [licensed under Schedule IV of Hazardous Waste (Management, Handling and Transboundary Movement) Rules, 2008 (Table 13)]. India's authorized recyclers have capacities ranging from 120 tonnes to 12 000 tonnes per year—yet 3 MT of e-waste was generated in India in 2007 alone. The remainder is processed by India's vast informal e-waste sector, which makes the potential total value of the market much greater.

Strategies for sustainable solid waste management

In order to achieve an efficient and sustainable system of solid waste management, a proper planning and technology selection procedure needs to be carried out, and before implementation, the present and future ways to manage solid waste stream need to be considered. Presently, the responsibility of enforcing the MSW rules lies with the secretaries of the urban development ministry of the concerned states. It is necessary to extend this responsibility to experts with both technical and administrative experience so as to guide the local bodies for effective implementation of the rules. Barring a few municipal corporations in Bengaluru, Delhi,

Table 13 List of the units registered with MoEF/CPCB as recyclers/re-processors having environmentally sound management facilities—e-waste re-processors

Name of the unit	Quantity of waste allowed (MTPA)
Andhra Pradesh	
M/s Earth Sense Recycle Pvt. Ltd, Plot No. 37, APIIC Industrial Park, Mankhal, Maheshwaram Mandal, Rangareddy district, Andhra Pradesh	1800
Haryana	
Earth Sense Recycle Pvt. Ltd, Plot No.225, Sector-VI, IMT Manesar, Gurgaon, Haryana	1200
Karnataka	
K G Nandani Enterprises, #46/4, 46/5, Billakempanahalli Village, Bidadi Hobli, Ramnagar (taluk and district), Karnataka	7200
E-Parisara Pvt. Ltd, Plot No. 30-P3, KIADB Industrial Area, Dabaspet, Nelamangala Taluk, Bangalore Rural district	1800
Ash Recyclers, Unit-II, Shed No. B-3, KSSIDC Industrial Estate, Hoskote, Bengaluru	120
New Port Computer Services (India) Pvt. Ltd, Shed No. B-29, KSSIDC Industrial Estate, Bommasandra, Bengaluru	500
Maharashtra	
Eco Recycling Ltd, 26, Chandivali Village Road, Off. Saki Vihar Road, Andheri(E), Mumbai, Maharashtra	3600
Earth Sense Recycle Pvt. Ltd, A-7, Gala No. 1, 2, and 3, Ground Floor, Prerana Complex, Anjur Phata, Dapoda Road, Val Village, Bhivandi Tal., Thane district, Maharashtra	360
Tamil Nadu	
Trishyiraya Recycling India Pvt. Ltd, Plot No. A-7, Phase-I, MEPZ-SEZ, Tambaram, Chennai, Tamil Nadu	740
TES AMM Pvt. Ltd, Plot No. A-18, SIPCOT Industrial Growth, Centre Oragadam, Panruti 'A' Village, Sriperumbudur, Kanchipuram district, Tamil Nadu	30 000
Uttar Pradesh	
TIC Group India Pvt. Ltd, J-2 and J-6, SDF Block-J, Noida Special Economic Zone (NSEZ), Noida Dadri Road, Noida, Uttar Pradesh	1000
Uttarakhand	
Attero Recycling Pvt. Ltd, 173, Village Bhagwanpur, Raipur Industrial Area, Roorkee district, Haridwar, Uttarakhand	12 000

MoEF – Ministry of Environment and Forests; CPCB – Central Pollution Control Board
Source CPCB; Details available at <http://www.cpcb.nic.in/divisionsofheadoffice/hwmd/e-Waste.pdf>, last accessed on 22 March 2010

Hyderabad, Mumbai, and so on, most local bodies find it financially difficult to incur the heavy capital expenditure involved in improving the collection, transportation, and disposal systems. Therefore, either they seek financial support from the central or state governments, or the resources have to be generated through the privatization of the select services. In order to promote privatization of the services, financial institutions will have to liberalize the procedures for financing private companies. The stipulated demands for guarantees and mortgage by the state government or contracting authorities pose an obstacle in resource generation for private companies. Considering the high cost involved in waste management, the first priority of local bodies, even in the case of privatization, should be waste minimization at the source. Active participation of citizens is a prerequisite for waste minimization.

The municipal waste rules do not put any responsibility on citizens to segregate waste into wet and dry, which is absolutely essential for effective WTE applications. The rules should be amended accordingly. The package material in use must be recyclable, and for that, the packaging rules must be modified. Effective waste minimization can be achieved only through segregation of waste and diversion at the source of recyclable material to recycling centres through scrap-dealers or rag-pickers. Initially, some infrastructural support can be provided to such groups to ensure that environmentally sound recycling is practised. Door-to-door waste collection is absolutely essential for improving the waste-collection efficiency. This responsibility can be given to the private operators to reduce the staff for waste collection, as this is the most labour-intensive job in the waste management cycle. The debris generated during building repairs and other construction and demolition waste must be collected by the local bodies against payment, as this is not a part of the regular municipal waste. This also applies to waste from health care facilities, as this waste too can generate some revenue for the local bodies. Development of landfill sites in a scientific manner as contemplated in the rules will require more time because most municipalities have neither the finances nor the identified land for the purpose; this activity can also be given to the private sector operators. However, it is absolutely necessary to recover the user charges from waste generators. Most of the WTE technologies presently used for utilizing the MSW in India, namely, biomethanation, RDF combustion, incineration, and gasification, though feasible, are yet to be proven for their successful regular operation and replicability. Most plants demonstrated on a pilot scale have been tested for their feasibility but are yet to be established for assessing and prioritizing the technologies under the techno-commercial parameters.

References

Bingemer H G and Crutzen P J. 1987. **The production of methane from solid waste.** *Journal of Geo-physical Research* **92**(D2): 2181–2187

CPCB (Central Pollution Control Board). 2000a. ***Control of Urban Pollution.*** [Series CUPS/48/1999-2000]. New Delhi: CPCB

CPCB (Central Pollution Control Board). 2000b. ***Management of Municipal Solid Waste.*** New Delhi: CPCB

CPHEEO (Central Public Health and Environmental Engineering Organization). 2000. ***Bio Medical Waste (Management & Handling) Rules, 1998. Manual on Municipal***

Solid Waste Management, pp. 264–265. New Delhi: CPHEEO, Ministry of Urban Development, Government of India

Domino F A. 1979. ***Energy from Solid Waste, Recent Development,*** pp. 5–6. [NASA CR-2526] *Energy Technology Review* No. 42, *Pollution Technology Review* No. 56. New Jersey: Noyes Data Corporation

Gartner Inc. 2007. **Gartner says worldwide mobile phone sales increased 16 per cent in 2007.** Press release 27 February 2007

MCC (Microelectronics and Computer Technology Corporation). 1996. ***Electronics Industry Environmental Roadmap.*** Austin, TX: MCC

National Solid Waste Association of India, Mumbai. 2007. *Proceedings of the International Conference on Sustainable Solid Waste Management,* pp. 327–333, 5–7 September 2007, Chennai, India. Mumbai: National Solid Waste Association of India

NIUA (National Institute of Urban Affairs). 1989. ***Upgrading Municipal Services: Norms and Financial Implications.*** New Delhi: NIUA. 197 pp.

Pachauri R K and Sridharan P V (eds). 1998. ***Looking Back to Think Ahead: GREEN India 2047. Growth with Resource Enhancement of Environment and Nature.*** New Delhi: Tata Energy Research Institute. 346 pp.

Rajeshwari K V, Lata K, Pant D C, Kishore V V N. 2001. ***Directions, Innovations, and Strategies for Harnessing Action for Sustainable Development.*** New Delhi: Tata Energy Research Institute. 179 pp.

TERI (The Energy and Resources Institute). 2008. ***Market Development for Clean Energy Sector: focus on municipal solid waste.*** Report for Department of Economic Affairs, Ministry of Finance, New Delhi, and Asian Development Bank, Manila

UPERC (Uttar Pradesh Electricity Regulatory Commission). 2003. ***Review Report on MSW to Energy Project in Allahabad.*** Uttar Pradesh: UPERC

Zhu D, Asnani P U, and Zurbrugg C. 2007. **Improving Municipal Solid Waste Management in India.** Report for the World Bank, 2007

Bibliography

Bhide A D and Sundaresan B B. 1983. ***Solid Waste Management in Developing Countries.*** New Delhi: Indian National Scientific Documentation Centre. 222 pp.

Cheremisinoff N P. 2003. ***Handbook of Solid Waste Management and Waste Minimization Technologies.*** USA: Elsevier Science

Comptroller and Auditor General of India. 2008. **Performance audit on *Management of Waste in India.*** Details available at <http://www.cag.gov.in/html/reports/civil/2008_PA14_SD_civil/chap_1.pdf>

EPTRI (Environmental Protection Training Research Institute). 1995. ***Status of Solid Waste Disposal in Metropolis Hyderabad.*** Hyderabad: EPTRI. 46 pp.

Hagelüken C and Meskers C E M. 2008. **Mining our computers—opportunities and challenges to recover scarce and valuable metals from end-of-life electronic devices.** In *Electronics Goes Green 2008+*, pp. 623–628, edited by H Reichl, N F Nissen, J Müller, and O Deubzer, Stuttgart: Fraunhofer IRB Verlag.

Johri R (ed.). 2009. ***E-Waste – implications, regulations, and management in India and current global best practices.*** **New Delhi: The Energy and Resources Institute**

Pachauri R K and Batra R K (eds). 2001. **A novel process using enhanced acidification and a UASB reactor forbiomethanation of vegetable market waste.** *Waste Management and Research* **19**: 292–300

Domino FA.1979. **Enenrgy from solid waste, recent development [NASA CR-2526], pp. 5–6.** *Energy Technology Review No. 42, Pollution Technology Review No. 56.* New Jersey: Noyes Data Corporation

13

Technology trends and opportunities in cogeneration

P K Dadhich

Senior Fellow, Centre For Research on Energy Security
The Energy and Resources Institute, New Delhi, India

The primary objective of the Electricity Act (EA), 2003, is to improve the technical and economic performance of the electricity sector in the country. This harmonizes and rationalizes the provisions in the Indian EA, 1910; the Electricity (Supply) Act, 1948; and the Electricity Regulatory Commissions Act, 1998. The EA, 2003, encourages private sector participation in generation, transmission, and distribution, with the objective of dispensing the regulatory responsibilities from the state governments to the regulatory commissions. The key features in the EA, 2003, which facilitate distributed generation are as follows.

- Generation is de-licensed and captive generation is freely permitted.
- There is provision for private transmission licensees.
- There is open access in transmission from the outset.
- Distribution licensees are free to undertake generation, and generating companies are free to take up licences for distribution.
- For rural and remote areas, stand-alone systems for generation and distribution are permitted.
- For rural and remote areas, decentralized management of distribution through panchayats, users' associations, cooperatives, or franchisees is permitted.
- Trading as a distinct activity is recognized, with the regulatory commissions being authorized to fix ceilings on trading margins, if necessary.
- Where there is direct commercial relationship between the consumer and the generating company or a trader, the price of power is not regulated and only the transmission and wheeling charges with surcharge are regulated.

- The act specifically mentions cogeneration and also encourages its use as it increases energy efficiency.

Cogeneration technologies

Cogeneration is an energy conservation process wherein heat from a fuel is simultaneously converted into useful thermal energy (such as process steam) and electrical energy. The need for either form can be the primary incentive for cogeneration, but there must be an opportunity for economic captive use or sale of the other. In a processing plant, the need for process and other heating systems is likely to be primary; in a public utility plant, electricity is the usual primary product.

Cogeneration is also called combined heat power (CHP). Cogeneration systems can be designed in the following ways.

- Heat-oriented design and operation mode
- Power-oriented design and operation mode
- Cost-oriented design and operation mode

Heat-oriented design and operation mode

In the heat-operated application of a cogeneration plant, the heat requirements are determined, and the whole plant is aligned accordingly. The surplus power produced is fed to the grid.

Power-oriented design and operation mode

In the power-operated application of cogeneration, focus is on power requirements. The utilized heat varies, depending on the power production.

Cost-oriented design and operation mode

The lowest operating costs of a cogeneration plant are achieved using the cost-operated design and operation mode. The annual total costs (capital costs and running costs) are calculated for different design variants, and the design with the lowest total costs is implemented. The grid and the heat network evens out the surplus or the need for heat and power.

A cogeneration system, thus, is designed from one of the two perspectives: it may be designed to a size to meet the needs for process heat and steam of a plant or community of industrial or institutional users, such that the electric power is treated as a by-product, which should either be used on-site or sold; or it may be designed to a size to meet the electric power demand, and the rejected heat is used to meet the needs for supply at or near the site. The latter approach is the likely one if it is a utility-owned system; it is the former if a processing plant is the owner. A cogeneration system will not match the varying power and heat demands at all times for most applications. Thus, separate on-site generation of heat or the purchase of utility-supplied electric power has to supplement the output of an industrial cogeneration system. If the on-site electric power demand is relatively low, an alternative option is to match the cogeneration system to the heat load, and contract for the sale of excess electricity is given to the local utility grid. Fuel saving is the major incentive for cogeneration. Since all heat- or electric-based power systems emit heat into the environment, the emitted heat can frequently be used to meet all or a part of the local thermal energy needs. The use of the emitted heat usually has no effect on the amount of primary fuel used for the thermal power process. Heat engines also require a high-temperature thermal input, usually receiving the working fluid directly from the source of heat, but, in some cases, they can obtain the input thermal energy from heat emitted from a higher temperature process. In the former case, the cogeneration process employs a heat-engine topping cycle; in the latter, a bottoming cycle is used.

In the topping cycle, fuel is used for power generation in a steam boiler or gas turbine cycle combustor, and the waste heat from the power generation cycle supports an industrial process (Figure 1). The most common topping cycle is one in which a boiler generates steam at a pressure higher than that needed for the process or space conditioning application. The high-pressure steam is then expanded in a turbine to a pressure that is appropriate for the application, generating electricity in the expansion process. Steam turbines, gas turbines, and reciprocating engines are commonly used in topping cycles.

A bottoming cycle is most commonly associated with the recovery or waste heat boiler (Figure 2). In the bottoming cycle, fuel is not supplied directly to the power-generating cycle. Rather, steam is generated from a waste heat source and then expanded in a turbine to do work or to generate electricity. Steam is frequently used in the bottoming cycle because of its ability to condense at low temperatures in the closed Rankine cycle.

Cogeneration systems

All cogeneration systems involve the operation of a heat engine for doing mechanical work, which is used to drive an electric generator (Figure 3).

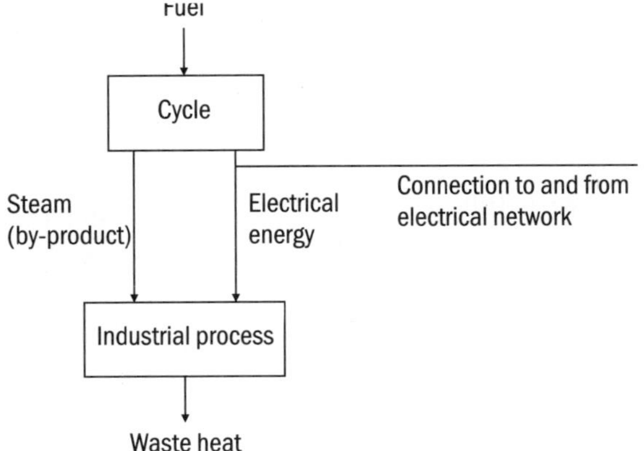

Figure 1 Topping cycle in a cogeneration process

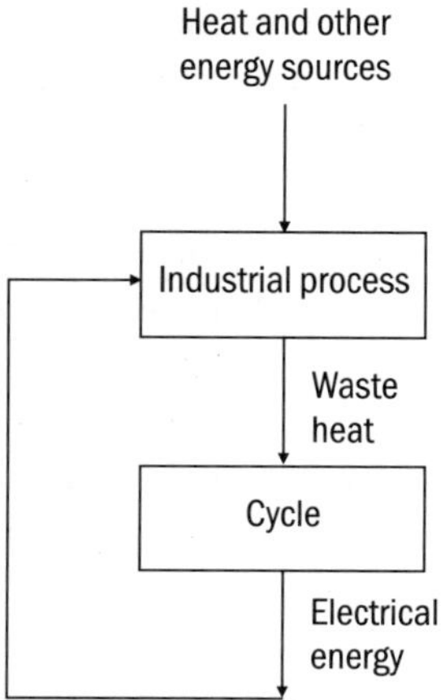

Figure 2 Bottoming cycle in a cogeneration process

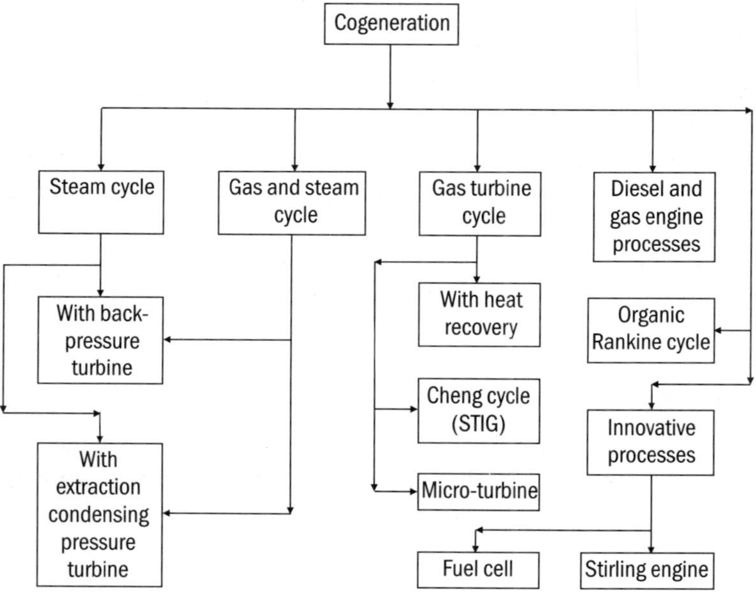

Figure 3 Overview of different cogeneration concepts

The most commonly used systems are as follows.

- Steam turbines (back-pressure and extraction configurations)
- Open-cycle (combustion) gas turbines
- Combined-cycle gas turbine systems + steam turbine
- Diesel engines/gas engines
- Innovative processes

Each heat-engine type has unique characteristics (Table 1), hence, an appropriate choice for specific application is important because of the following features.

- Power-to-heat ratio at design point
- Efficiency at design point
- Capacity range
- Power-to-heat ratio variability
- Off-design (part-load) efficiency
- Multi-fuel capability

Steam turbine cycle

Basic process and components

The Rankine cycle is the thermodynamic cycle for the steam turbine. The cycle is the basis for conventional power-generating stations and comprises a heat source (boiler) that converts water into high-pressure steam. In the steam cycle, water is first pumped from medium to high pressure. It is then heated to the boiling temperature corresponding to the pressure, boiled (heated from liquid to vapour), and then superheated (heated to a temperature above boiling). A multi-stage turbine expands the pressurized steam to lower pressure, and the steam is then exhausted, either into a condenser at vacuum conditions or into an intermediate temperature steam distribution system that delivers the steam to the industrial or commercial application. The condensate from the condenser or from the steam utilization system returns to the feed-water pump for continuation of the cycle.

Table 1 Some characteristic heat engines used in cogeneration systems

Engine type	Efficiency design at point (%)	Part-load efficiency	Multi-fuel capabilities	Maximum temperature of recoverable heat (°C)	Typical heat to power ratio
Steam turbine extraction condensing	25–30	Fair	Excellent	93–315	0.1–0.3
Steam turbine back-pressure	2–25	Fair	Excellent	93–315	0.05–0.2
Combustion gas turbine	2–30	Poor	Poor	538–649	0.3–0.45
Indirectly fired gas turbine open cycle	25–30	Poor	Good	371–482	0.4–1.0
Indirectly fired gas turbine closed cycle	25–30	Excellent	Good	371–482	0.4–1.0
Diesel engines	35–40	Good	Fair-to-poor	260–371	0.6–0.85

Source Perry and Green (1997)

The steam turbine itself consists of a set of stationary blades (called nozzles) and a moving set of adjacent blades (called buckets or rotor blades) installed within a casing. The two sets of blades work together such that the steam turns the shaft of the turbine and the connected load. The stationary nozzles accelerate the steam to a high velocity by expanding it to lower pressure. A rotating bladed disc changes the direction of the steam flow, thereby creating a force on the blades, which because of the wheeled geometry, manifests itself as torque on the shaft on which the bladed wheel is mounted. The combination of torque and speed is the output power of the turbine.

Details of stage design

To maximize turbine efficiency, steam is expanded to work in a number of steps or stages. These stages are characterized by the manner in which energy is extracted from steam. Turbine stages are known as impulse or reaction, although most combine elements of the two. In an impulse turbine, often compared to a water wheel, impulse nozzles orient steam so that it flows in well-formed, high-speed jets (Figure 4). These jets contain kinetic energy,

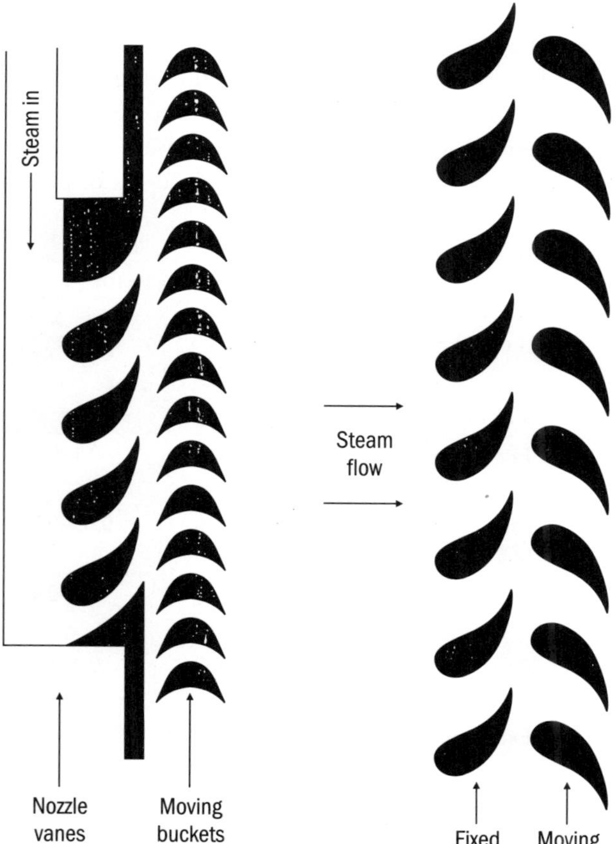

Figure 4 Sample impulse and reaction stage

which the moving blades, or buckets, convert into shaft rotation (mechanical work) as the steam jet changes direction. The pressure drop occurs only across the stationary blades.

Reaction turbines make use of the reaction force created by the blades as steam accelerates through the nozzles. Steam enters a fixed vane or nozzle passage and leaves as a jet that fills the entire circumference of the motor (Figure 5). It then flows through the moving blades, which are shaped to act like moving nozzles. Here, it changes direction and drops in pressure, and its speed increases relative to that of the blades.

Modern steam turbines normally combine impulse and reaction. For example, many impulse turbines use 10%–50% reaction in their designs. All control stages are of the impulse type, while most low-pressure section blades are of the reaction type.

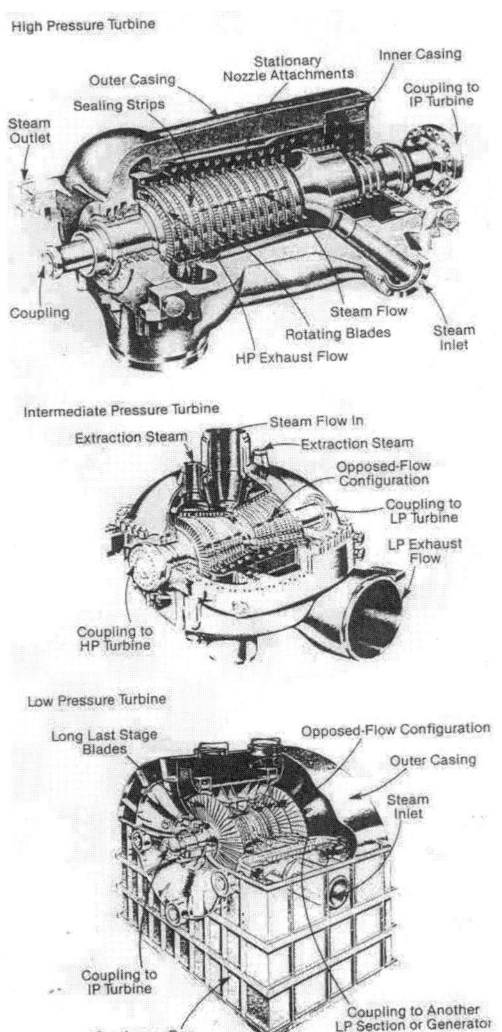

Figure 5 Typical high-, intermediate-, and low-pressure turbine schematics

The connection between the steam supply and power generation is the steam and return feed-water lines. There are numerous options in the steam supply, pressure, temperature, and extent, if any, for partially reheating the expanded steam. Steam systems vary from low-pressure lines used primarily for space heating and food preparation, to medium-pressure lines used in industrial processes and cogeneration, to high-pressure lines used in utility power generation. Generally, as the system gets larger, the economics favour higher pressures and temperatures with their associated heavier walled boiler tubes and more expensive alloys.

In general, utility applications involve raising the steam for the exclusive purpose of power generation. Such systems also use a water-cooled condenser to exhaust the steam from the turbine at the lowest practical pressure. Some utility turbines have dual uses: power generation and steam delivery at higher pressure into district heating systems or to neighbouring industrial plants at the required pressure and, consequently, they do not have condensers. These plants are actually large cogeneration/CHP plants.

Boilers

For cogeneration with plants up to (approximately) 40-megawatt (MW) capacity, horizontal industrial boilers are built. This enables rail shipping, with considerable cost savings and improved quality, as factory labour is usually lower in cost and greater in quality than field labour. Large, shop-assembled boilers are typically capable of firing only gas or distillate oil, as there is inadequate residence time for the complete combustion of most solid and residual fuels in such designs. Large, field-erected industrial boilers firing solid and residual fuels bear a resemblance to utility boilers except for the actual solid fuel injection. Large boilers usually burn pulverized coal; however, intermediate and small boilers burning coal or solid fuel employ various types of solid feeders.

Types of steam turbines

Cogeneration applications use two types of steam turbines: non-condensing and extraction.

Non-condensing (back-pressure) turbine

The non-condensing turbine (also referred to as a back-pressure turbine) exhausts its entire flow of steam into the industrial process or facility steam mains at conditions close to the process heat requirements (Figure 6).

Usually, temperature of the steam sent into the mains is not much above the saturation temperature. The term "back-pressure" refers to turbines that exhaust steam at atmospheric pressures and above. The specific CHP application establishes that the discharge pressure levels 3, 10, and 17 atmosphere gauge (atm [g]) are the most typical pressure levels for steam distribution systems. District heating systems in Europe and North America

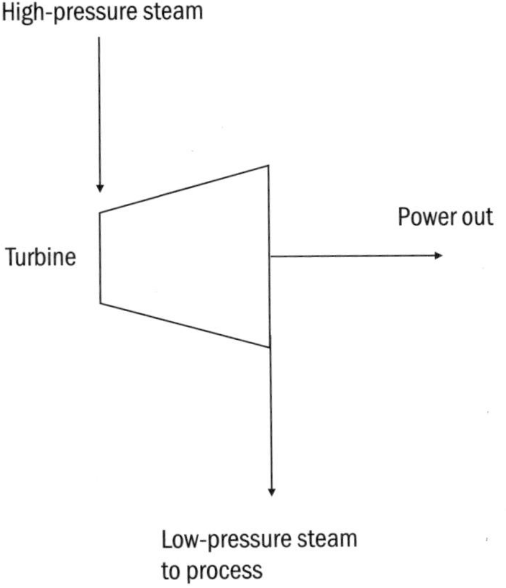

Figure 6 Non-condensing (back-pressure) steam turbine

most often use lower pressures, while industrial processes use higher pressures. Industrial processes often include further expansion for mechanical drives, using small steam turbines for driving heavy equipment that runs continuously for long periods. The power generation capability reduces significantly when steam is used at appreciable pressure rather than being expanded to create vacuum in a condenser. Discharging steam into a steam distribution system at 10 atm (g) can sacrifice slightly more than half the power that could be generated when the inlet steam conditions are 51 atm (g) and 427 °C, typical of small-steam turbine systems.

Extraction turbine

The extraction turbine has opening(s) in its casing for the extraction of a portion of the steam at some intermediate pressure before condensing the remaining steam. Figure 7 illustrates the extracted steam that may be used for processing in a CHP facility or for feed-water heating, as is the case in most utility power plants.

The steam extraction pressure may or may not be automatically regulated. Regulated extraction permits more steam to flow through the turbine to generate additional electricity during periods of low thermal demand by the CHP system. In utility-type steam turbines, there may be several extraction points, each at a different pressure corresponding to a different temperature. The facility's specific needs for steam and power over a period of time, determine the extent to which steam in an extraction turbine is extracted for use in the process. In large, often complex, industrial plants, additional steam to the steam turbine may be admitted (flowing into the casing and increasing the flow in the steam path). Often this happens when multiple boilers are used at different pressures, because of their historical existence. These steam turbines are referred to as admission

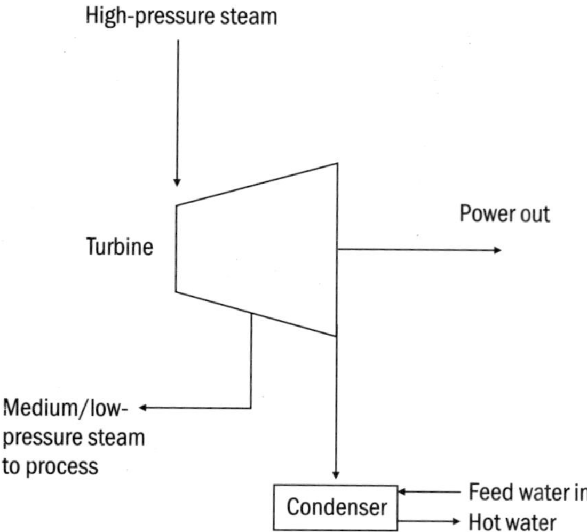

Figure 7 Extraction steam turbine

turbines. At steam extraction and admission locations, there are usually steam flow control valves that add to the steam and control system cost.

Numerous mechanical design features increase efficiency, providing for operation over a range of conditions, simplifying manufacture and repair, and achieving other practical purposes. The long history of steam turbine use has resulted in a large inventory of steam turbine stage designs. For example, the division of steam acceleration and the consequent change in the direction of flow vary between competing turbine manufacturers under the impulse and reaction designs. Manufacturers tailor clients' design requests by varying the flow area in the stages and the extent to which steam is extracted (removed from the flow path between stages) to accommodate the specification of the client.

When the steam expands through a high-pressure ratio, as in utility and large industrial steam systems, it can begin to condense in the turbine when its temperature drops below the saturation temperature at that pressure. If water drops form in the turbine, blade erosion occurs from the drops, impacting the blades. At this point in expansion, steam sometimes returns to the boiler and is reheated to a high temperature before returning to the turbine for further (safe) expansion. In a few large, high-pressure utility steam systems, double reheat systems are installed.

With these choices, "the designer" of the steam supply system and the steam turbine faces the challenge of creating a system design that delivers (seasonally varying) power and steam, which together present the most favourable business opportunity to plant owners. Between the power (only) output of a condensing steam turbine and the power and steam combination of a back-pressure steam turbine, essentially any ratio of power to heat output can be supplied. Back-pressure steam turbines can be obtained with a variety of back pressures, further increasing the variability of the power-to-heat ratio.

Design characteristics

Custom design

Steam turbines are designed to match the CHP design pressure and temperature requirements and maximize electrical efficiency while providing the desired thermal output.

Thermal output

Steam turbines are capable of operating over a broad range of steam pressures. Utility steam turbines operate with inlet steam pressures up to 238 atm (g) and exhaust vacuum conditions as low as 25 mm Hg (absolute). Steam turbines are custom designed to deliver the thermal requirements of the CHP applications through the use of back-pressure or extraction steam at appropriate pressures and temperatures.

Fuel flexibility

Steam turbines offer a wide range of fuel flexibilities using a variety of fuel sources in the associated boiler or other heat sources, including coal, oil, natural gas, wood, and waste products.

Reliability and life

The life of the steam turbine is extremely long. When properly operated and maintained (including proper control of boiler water chemistry), steam turbines are extremely reliable, requiring overhauls only after several years. They require controlled thermal transients to minimize differential expansion of the parts as the massive casing slowly heats up.

Size range

Steam turbines are available in sizes varying from 100 kW to over 250 MW. In the multi-megawatt industrial and utility steam turbine, designations merge with the same turbine (high-pressure section), serving both industrial and small utility applications.

Emissions

Emissions are dependent on the fuel used by the boiler or other steam sources, boiler furnace combustion section design and operation, and built-in and add-on boiler exhaust clean-up systems.

Performance characteristics

Electrical efficiency

The electricity generating efficiency of standard steam-turbine power plants varies from a high of 37% higher heating value (HHV) for large electric utility plants designed for the highest practical annual capacity factor, to under 10% HHV for small simple plants, which make electricity as a by-product of the process delivering steam to processes or district heating systems. Steam-turbine thermodynamic efficiency (isentropic efficiency) refers to the ratio of power actually generated from the turbine to what would be generated by a perfect turbine with no internal losses, using steam at the same inlet conditions and discharging to the same downstream pressure (actual enthalpy drop divided by the isentropic enthalpy drop). Turbine thermodynamic efficiency is not to be confused with electrical generating efficiency, which is the ratio of net power generated to the total fuel input to the cycle. Steam-turbine thermodynamic efficiency measures the efficiency of the turbine in extracting power from steam. Multi-stage (moderate-to-high pressure ratio) steam turbines have thermodynamic efficiencies that vary from 65% for small (under 1000-kW) units to over 90% for large industrial and utility-sized units. Small single-stage steam turbines can have efficiencies as low as 50%. When a steam turbine exhausts are routed to a CHP application, the turbine efficiency is not as critical as in a power-only condensing mode. Majority of the energy not extracted by the steam turbine satisfies the thermal load. Power-only applications waste the exhaust turbine-steam energy in condensers.

CHP system efficiency

Steam-turbine CHP systems generally have low power-to-heat ratios, in the range 0.05–0.2. This is because electricity is a by-product of heat generation, with the system optimized for steam production. Hence, steam-turbine CHP system electrical efficiency may seem to be low because the primary objective is to produce large amounts of steam. The effective electrical efficiency of steam-turbine systems, however, is generally high because almost all the energy difference between the high-pressure boiler output and the lower-pressure turbine output is converted into electricity. This means that efficiency of the total CHP system is generally high and approaches the boiler efficiency level. Steam boiler efficiency ranges from 70% to 85% HHV, depending on boiler type, age, fuel, duty cycle, application, and steam conditions.

Performance and efficiency enhancements

In industrial steam-turbine systems, business conditions determine the requirements and relative values of electric power and steam. Plant system engineers then decide the extent of efficiency-enhancing options to incorporate, in terms of their incremental effects on performance and plant cost, and select appropriate steam-turbine inlet and exhaust conditions. Often the steam turbine is installed into a system that already exists and is being modified so that a number of steam-system-design parameters are already determined by previous decisions, which exist as system hardware characteristics. As the stack temperature of the boiler-exhaust combustion products still contains some heat, trade-offs occur on the extent of investment in heat reclamation equipment for the sake of efficiency improvement. Often the stack-exhaust temperature is set at a level where further heat recovery results in the condensation of corrosive chemical species in the stack, with consequential deleterious effects on stack life and safety.

Steam reheat

Higher pressures and steam reheat increase power generation efficiency in large industrial (and utility) systems. The higher the pressure ratio (the ratio of the steam inlet pressure to the steam exit pressure) across the steam turbine and the higher the steam inlet temperature, more power it will produce per unit of mass flow. To avoid condensation, the inlet steam temperature is increased to the economic practical limit of materials. This limit is now generally 427–482°C for small industrial steam turbines. When the economically practical limit of temperature is reached, the expanding steam can reach a condition of temperature and pressure where condensation to (liquid) water begins. Small amounts of water droplets can be tolerated in the last stage of a steam turbine, provided the droplets are not too large or too numerous. At pressures higher than this point, steam is returned to the boiler and reheated before being returned to the expansion steam turbine for further expansion. When returned to the next stage of the turbine, the steam expands without condensation.

Combustion air preheating

In large industrial systems, air pre-heaters recover heat from the boiler-exhaust gas stream, and use it to preheat the combustion air, thereby reducing fuel consumption. Boiler combustion air pre-heaters are large versions of the heat wheels used for the same purpose in industrial furnaces.

Capital cost

A steam-turbine-based CHP plant is a complex process with many interrelated subsystems that must generally be custom designed. A typical breakdown of installed costs for a steam-turbine CHP plant is 25% for boiler; 25% for fuel-handling, storage, and preparation system; 20% for stack gas clean-up and pollution controls; 15% for steam-turbine generator; and 20% for field construction and plant engineering. Boiler costs are highly competitive. Typically, the only way by which significant cost reductions can be made when designing a system is by fuel-handling/storage/preparation. In a steam-turbine cogeneration plant, especially one burning solid fuel such as biomass, the turbine accounts for a much smaller portion of the total system-installed costs than those associated with internal combustion engines and industrial gas turbines.

Emissions

Emissions associated with a steam turbine are dependent on the source of steam. Steam turbines can be used with a boiler firing any one or a combination of a large variety of fuel sources, or they can be used with a gas turbine in a combined cycle configuration. Boiler emissions vary depending on the fuel type and environmental conditions. Boiler emissions include oxides of nitrogen (NO_x), oxides of sulphur (SO_x), particulate matter (PM), carbon monoxide (CO), and carbon dioxide (CO_2).

Typical emissions

Table 2 illustrates typical emissions of NO_x, PM, and CO for boilers, by the size of the steam-turbine system and by fuel type.

Gas turbine cycle

Basic process and components

Gas-turbine systems operate on the thermodynamic cycle known as the Brayton cycle, in which atmospheric air is compressed, heated, and then expanded, with excess of power produced by the expander (also called the turbine) over that consumed by the compressor being used for power generation. Power produced by an expansion turbine and consumed by a compressor is proportional to the absolute temperature of the gas passing through the device. Consequently, it is advantageous to operate the expansion turbine at the highest practical temperature consistent with economic materials and internal blade cooling technology, and to operate the compressor with inlet air flow at as low a temperature as possible. As technology advances permit higher turbine-inlet temperature, the optimum pressure ratio also increases. Higher temperature and pressure ratios result in higher efficiency and specific power. Thus, the general trend in gas-turbine advancement has been towards a combination of higher temperatures and pressures. While such advancements increase the manufacturing cost of the machine, the higher value, in terms of greater power output and higher efficiency, provide net

Table 2 Typical boiler emission ranges

	System 1 (500 kW)			Systems 2 and 3 (3 MW/15 MW)		
Boiler fuel	Oxides of nitrogen	Carbon monoxide	Particulate matter	Oxides of nitrogen	Carbon monoxide	Particulate matter
Coal (kg/cal)	—	—	—	0.36–2.24	0.04–1.26	—
Wood (kg/cal)	0.396–0.882	1.08	0.594–1.008	0.396–0.882	0.108	0.594–1.008
Fuel oil (kg/cal)	0.27–0.666	0.054	0.018–0.144	0.126–0.666	0.054–1.944	0.018–0.144
Natural gas (kg/cal)	0.054–0.18	0.144	—	0.18–0.504	0.144	—

Note All emission values are without post combustion treatment.

Source USEPA (1997)

economic benefits. The industrial gas turbine is a balance between performance and cost, which results in the most economic machine for both the user and the manufacturer.

Modes of operation

There are several variations of the Brayton cycle in use today. Fuel consumption may be decreased by preheating the compressed air with heat from the turbine exhaust using a recuperator or regenerator; the compressor work may be reduced and net power increased by using inter-cooling or pre-cooling; and the exhaust may be used to raise steam in a boiler to generate additional power in a combined cycle. Figure 8 shows the primary components in a simple cycle gas turbine.

Gas turbine exhaust is quite hot—up to 427–482°C for smaller industrial turbines and up to 593°C for some new large central station utility machines and aero-derivative turbines. Such high exhaust temperatures permit direct use of the exhaust. With the addition of a heat recovery steam generator, the exhaust heat can produce steam or hot water. A portion or all of the steam generated by the heat recovery steam generation (HRSG) may be used to generate additional electricity through a steam turbine in a combined cycle configuration.

A gas-turbine-based system operates in combined heat and power mode when the waste heat generated by the turbine is applied in the end-use. For example, a simple-cycle gas turbine using the exhaust in a direct heating process is a CHP system, while a system that features all of the turbine exhaust feeding an HRSG and all of the steam output used to produce electricity in a combined-cycle steam turbine is not a CHP system.

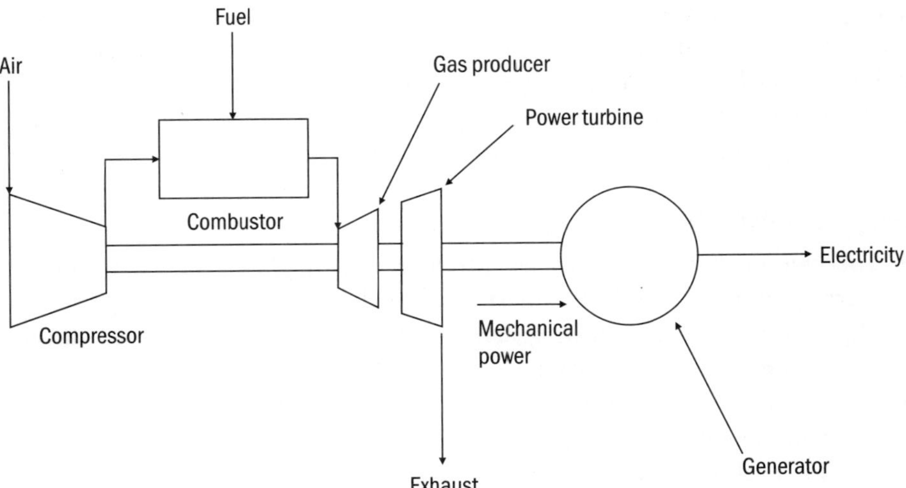

Figure 8 Components of a simple cycle gas turbine

Types of gas turbines

Aero-derivative gas turbines

These are meant for the generation of stationary power and are adapted from their jet and turbo-shaft aircraft engine counterparts. While these turbines are lightweight and thermally efficient, they are usually more expensive than products designed and built exclusively for stationary applications. The largest aero-derivative generation turbines available are of the capacity 40–50 MW. Many aero-derivative gas turbines for stationary use operate with compression ratios in the range of 30:1, requiring a high-pressure external fuel gas compressor. With advanced system developments, larger aero-derivative turbines (>40 MW) are approaching 45% simple-cycle efficiencies on the lower heating value (LHV) basis.

Industrial or frame gas turbines

Industrial and frame gas turbines are meant exclusively for stationary power generation and are available in a capacity range of 1–250 MW. They are generally less expensive, more rugged, can operate longer between overhauls, and are more suited for continuous base-load operation with longer inspection and maintenance intervals than aero-derivative turbines. However, they are less efficient and much heavier. Industrial gas turbines generally have more modest compression ratios (up to 16:1) and often do not require an external fuel gas compressor. Larger industrial gas turbines (>100 MW) are approaching simple-cycle efficiencies of approximately 40% (LHV) and combined-cycle efficiencies of 60% (LHV).

Industry uses gas turbines between 500 kW and 40 MW for on-site power generation and as mechanical drivers. Small gas turbines also drive compressors on long-distance natural gas pipelines. In the petroleum industry, turbines drive gas compressors to maintain well pressures and enable refineries and petrochemical plants to operate at elevated pressures. In the steel industry, turbines drive air compressors used for blast furnaces. In process industries, such as chemicals, refining and paper, and in large commercial and institutional applications, turbines are used in a combined heat and power mode, generating both electricity and steam for use on-site.

Design characteristics

- *Thermal output* Gas turbines produce high-quality (high temperature) thermal output suitable for most combined heat and power applications. High-pressure steam can be generated or the exhaust can be used directly for process drying and heating.
- *Fuel flexibility* Gas turbines operate on natural gas, synthetic gas, landfill gas, and fuel oils. Plants typically operate on gaseous fuel with stored liquid fuel for back-up to obtain the less expensive interruptible rate for natural gas.

- *Reliability and life* Modern gas turbines have proved to be reliable power generators, given proper maintenance is carried out regularly. The overhaul time is typically 25 000–50 000 hours.
- *Size range* Gas turbines are available in sizes from 500 kW to 250 MW.
- *Emissions* Many gas turbines burning gaseous fuels (mainly natural gas) feature lean premixed burners (also called dry low-NO_x combustors) that produce NO_x emissions below 25 parts per million (ppm) with laboratory data down to 9 ppm, and simultaneous low CO emissions in the range 10–50 ppm. The selective catalytic reduction (SCR) or catalytic combustion further reduces NO_x emissions. Many gas turbines sited in locales with stringent emission regulations use the SCR after treatment to achieve single-digit (below 9 ppm) NO_x emissions.
- *Part-load operation* Because gas turbines reduce the power output by reducing the combustion temperature, efficiency at part load can be substantially lowered.

Performance characteristics

Electrical efficiency

Thermal efficiency of the Brayton cycle is a function of the pressure ratio, ambient air temperature, turbine inlet air temperature, efficiency of the compressor and turbine elements, turbine blade cooling requirements, and any performance enhancements (such as recuperation, inter-cooling, inlet air cooling, reheat, steam injection, or combined cycle). All these parameters, along with the gas turbine's internal mechanical design features, have been improving with time. Therefore, newer machines are usually more efficient than older ones of the same size and general type. The performance of a gas turbine is also appreciably influenced by the purpose for which it is intended. Emergency power units generally have lower efficiency and lower capital cost, while turbines intended for prime power, compressor stations, and similar applications with high annual capacity factors have higher efficiency and higher capital costs. Emergency power units are permitted for a maximum number of hours per year and allowed to have considerably higher emissions than turbines permitted for continuous duty.

Table 3 summarizes the performance characteristics of commercially available, typical gas turbine CHP systems over the ones in 1–40 MW size range. The heat rates shown are based on manufacturers' specifications and industry publications. The available thermal energy (steam output) was calculated using the published turbine data on turbine exhaust temperatures and flows. The CHP steam estimates are based on an unfired HRSG with an outlet exhaust temperature of 138°C, producing dry, saturated steam at 10 atm (g). The total efficiency is defined as the sum of the net electricity generated plus steam produced for plant thermal needs divided by the total fuel input to the system. Higher steam pressures can be obtained but at slightly lower

Table 3 Gas turbine combined heat power: typical performance parameters[a]

Cost and performance characteristics	System 1	System 2	System 3	System 4	System 5
Electricity capacity (kW)	1000	5000	10 000	25 000	40 000
Total installed cost ($/kW)	1780	1010	970	860	785
Electricity heat rate (kcal/kWh), higher heating value (HHV)	3926	3173	2965	2506	2323
Electrical efficiency (%), HHV	21.90	27.10	29.00	34.30	37.00
Fuel input (kcal/h)[a] 10^6	3.93	15.85	29.66	62.65	92.94
Required fuel gas pressure (kg/cm^2) g	6.68	11.25	17.58	23.90	30.58
Combined heat power (CHP) characteristics					
Exhaust flow (1000 kg/h)	19.96	73.48	143.33	259.00	432.72
GT exhaust temperature (°C)	510	510	491	510	457
Heat recovery steam generation (HRSG) exhaust temperature (°C)	138	138	138	138	138
Steam output (kcal/h)[a] 10^6	1.79	6.70	12.50	24.09	34.47
Steam output (1000 kg/h)	3.05	11.36	21.18	40.82	58.41
Steam output (kW equivalent)	2080	7800	14 540	28 020	40 100
Total CHP efficiency (%), HHV	68	69	71	73	74
Power/heat ratio	0.48	0.64	0.69	0.89	1.00
Net heat rate (kcal/kWh)	1682	1499	1402	1301	1246
Effective electrical efficiency (%), HHV	51	57	61	66	69

[a] For typical systems commercially available in 2001

total efficiencies. Additional steam can be generated and total efficiency further increased with duct firing in the HRSG. To estimate fuel savings, the effective electrical efficiency is a more useful value than overall efficiency. Effective electric efficiency is calculated by assuming that the useful thermal output from the CHP system would otherwise be generated by an 80% efficient boiler. The theoretical boiler fuel is subtracted from the total fuel input, and the remaining fuel input is used to calculate the effective electric efficiency, which can then be compared to traditional electricity generation.

Table 3 shows that electrical efficiency increases as combustion turbines become larger. As electrical efficiency increases, the absolute quantity of thermal energy available to produce steam decreases per unit of power output, and the ratio of power to heat for economics may affect the decisions that customers make, in terms of CHP acceptance, sizing, and the desirability of selling power.

Fuel supply pressure

Gas turbines need a minimum gas pressure of about 7 atm (g) for the smallest turbines and substantially higher pressures for larger turbines and aero-derivative machines. Depending on the supply pressure of the gas being delivered to the site, the cost and power consumption of the fuel gas compressor can be a significant consideration. Table 4 shows the power required to compress natural gas from supply pressures typical of commercial and industrial service, to the pressures required by typical industrial gas turbines. Required supply pressures generally increase with an increase in the gas turbine size.

Part-load performance

When less than full power is required from a gas turbine, the output is reduced by lowering the turbine inlet temperature. In addition to reducing power, this change in operating conditions also reduces efficiency. Figure 9 shows a typical part-load derate curve. Emissions are generally increased at part-load conditions, especially at half-load and below.

Effects of ambient conditions on performance

The ambient conditions under which a gas turbine operates have a noticeable effect on both power output and efficiency. At elevated inlet air temperatures,

Table 4 Power requirements for natural gas compression

Type of power requirement	System 1	System 2	System 3	System 4	System 5
Turbine electric capacity (kW)	1000	5000	10 000	25 000	40 000
Turbine pressure ratio	6.5	10.9	17.1	23.1	29.6
Required compression power (kW)					
3.52 kg/cm^2 (g) gas-supply pressure	17	125	310	650	1310
10.55 kg/cm^2 (g) gas-supply pressure	—	26	120	300	675
17.58 kg/cm^2 (g) gas-supply pressure	—	—	40	150	380

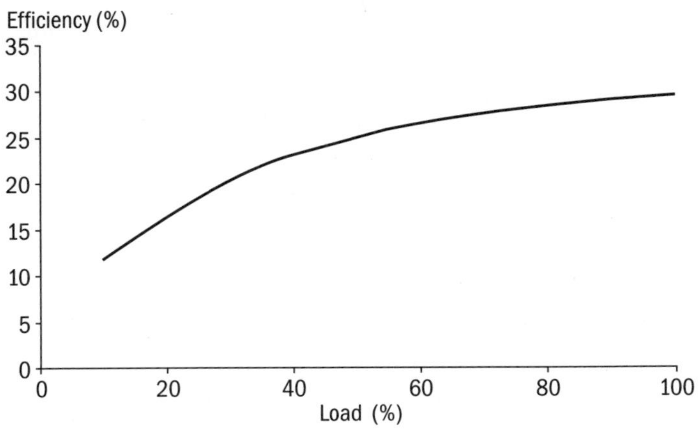

Figure 9 Part-load power performance

both power and efficiency decrease. The power decreases due to the decreased airflow mass rate (the density of air declines as temperature increases), and the efficiency decreases because the compressor requires more power to compress air of higher temperature. Conversely, the power and efficiency increase when the inlet air temperature is reduced. Figure 10 shows the variation in power and efficiency for a gas turbine as a function of ambient temperature compared to the reference ISO condition of sea level at 15°C. At inlet air temperatures of nearly 38°C, the power output can drop to as low as 90% of the ISO-rated power for typical gas turbines. At cooler temperatures of about 4–10°C, power can increase to as high as 105% of the ISO-rated power.

The density of air decreases at altitudes above sea level. Hence, the power output decreases (Figure 11).

Heat recovery

The economics of gas turbines in process applications often depend on the effective use of thermal energy contained in the exhaust gas, which generally

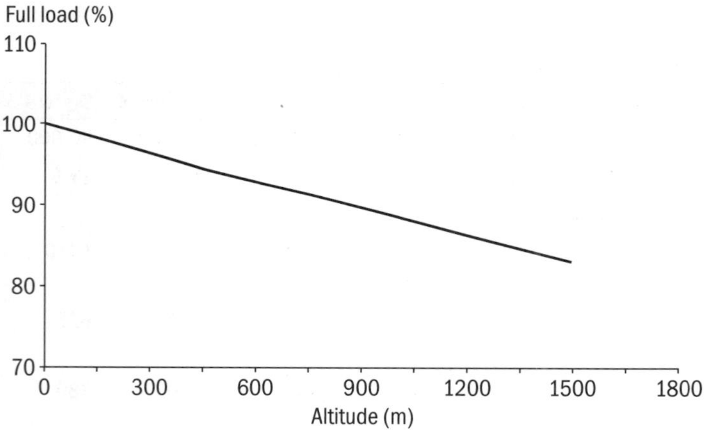

Figure 10 Derate at altitude: ambient temperature effects on performance

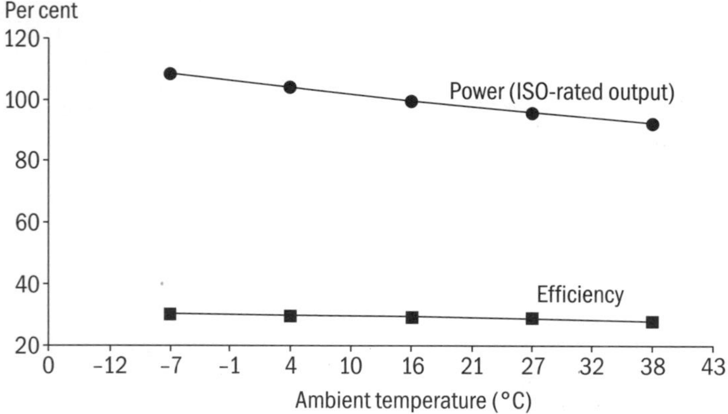

Figure 11 Effect of altitude on performance

represents 60%–70% of the inlet fuel energy. The most common use of this energy is for steam generation in unfired or supplementary-fired heat-recovery steam generators. However, the gas-turbine exhaust gases can also be used as a source of direct process energy for unfired or fired process fluid heaters, or as preheated combustion air for power boilers. Figure 12 shows a typical gas-turbine/HRSG configuration. An unfired HRSG is the simplest steam CHP configuration, and it can generate steam at conditions ranging from 10 atm (g) to 182 atm (g).

CHP system efficiency

The overall or total efficiency of a CHP system is a function of the amount of energy recovered from the turbine exhaust. The two most important factors influencing the amount of energy available for steam generation are gas-turbine exhaust temperature and HRSG stack temperature.

Turbine-firing temperature and turbine-pressure ratio combined determine the exhaust temperature of the gas turbine. Typically aero-derivative gas turbines have higher firing temperatures than industrial gas turbines, but when the higher pressure ratio of aero-derivative gas turbines is recognized, the turbine discharge temperatures of the two turbine types remain somewhat close, at 454–510°C. For the same HRSG exit temperature, a higher turbine exhaust temperature (higher HRSG gas inlet temperature) results in greater availability of thermal energy and increased HRSG output.

Similarly, the lower the HRSG stack temperature, the greater is the amount of energy recovered and the higher is the total system efficiency. HRSG stack temperature is a function of steam conditions and fuel type. Saturated steam temperatures increase with increase in steam pressure. Because of

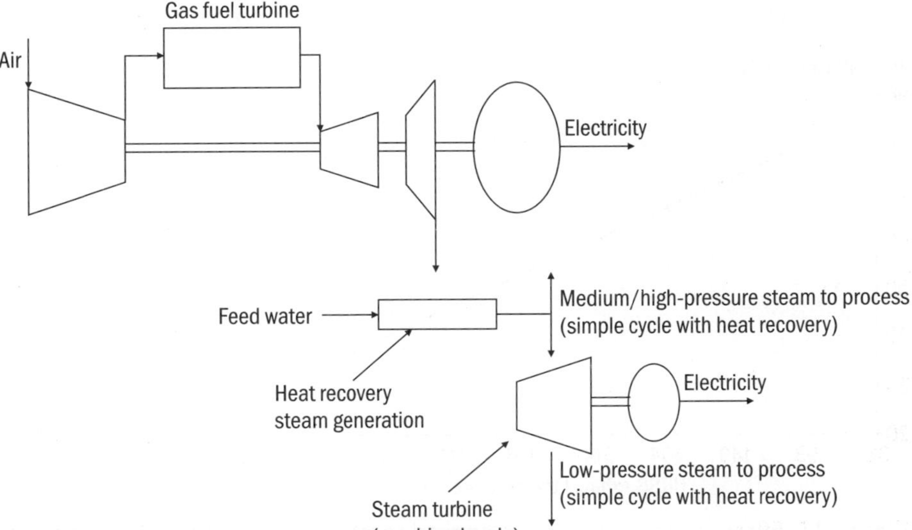

Figure 12 Heat recovery from a gas-turbine system

pinch-point considerations within the HRSG, higher steam pressures result in higher HRSG exhaust stack temperatures, lesser utilization of available thermal energy, and a reduction in total CHP system efficiency. In general, minimum stack temperatures of about 149°C are recommended for sulphur-bearing fuels. Figure 13 illustrates the increase in overall system efficiency as the exhaust temperature decreases through effective heat recovery. Generally, unfired HRSGs can be designed to economically recover approximately 95% of the available energy in the turbine exhaust (the energy released moves from turbine exhaust temperature to the HRSG exhaust temperature).

Overall, the CHP efficiency generally remains high under part-load conditions. The decrease in electric efficiency from the gas turbine under part-load conditions results in a relative increase in the heat available for recovery. This can be a significant operating advantage for applications in which the economics are driven by a high steam demand.

Supplementary firing

As the gas turbine combustion process consumes little of the available oxygen in the turbine airflow, the oxygen content in the gas-turbine exhaust permits supplementary fuel firing ahead of the HRSG to increase the steam production relative to an unfired unit. Supplementary firing can raise the exhaust gas temperature entering the HRSG up to 982°C and increase the amount of steam produced by a factor of two. Moreover, as the turbine exhaust gas is essentially preheated combustion air, fuel consumed in supplementary firing is less than that required for a stand-alone boiler providing the same increment in steam generation. The HHV efficiency of the incremental steam production from supplementary firing is often 85% or more than that of an unfired HRSG when firing natural gas.

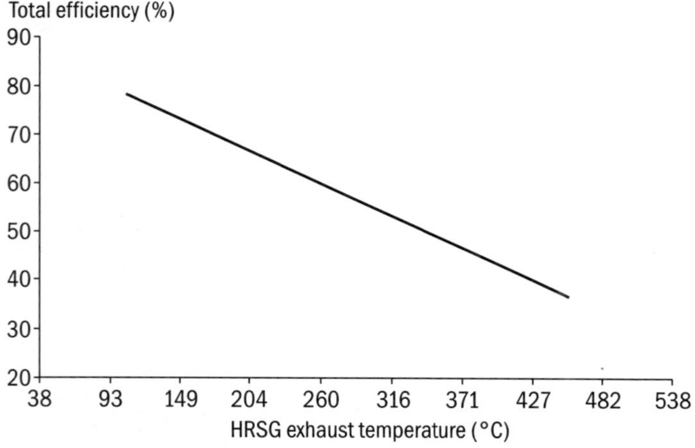

Figure 13 Effect of stack temperature on total combined heat power (CHP) efficiency (based on an LM6000 with unfired heat recovery steam generation [HRSG])

Supplementary firing also increases system flexibility. Unfired HRSGs are typically convective heat exchangers that respond solely to the exhaust conditions of the gas turbine and do not easily allow for steam-flow control. Supplementary firing capability enables the control of steam production within the capacity of the burner system, independent of the normal gas-turbine-operating mode. Low NO_x duct burners with guaranteed emission levels as low as 0.08 lb (pound) NO_x/million British thermal units (MBTU) can be specified to minimize the NO_x contribution of supplemental firing.

Performance and efficiency enhancements

Recuperators

Several technologies that increase the output power and/or efficiency of the gas turbines have been developed and put into limited commercial service. Fuel use can be reduced—hence, efficiency can be improved—by the use of a heat exchanger called a recuperator, which uses the hot-turbine exhaust to preheat the compressed air entering the combustor. Depending on gas-turbine-operating parameters, such a heat exchanger can add up to 10% points in machine efficiency, thereby raising efficiency from 30% to 40%. However, because there is an increased drop in pressure in both the compressed air and turbine exhaust sides of the recuperator, the power output is typically reduced by 10%–15%.

Recuperators are expensive, and their cost can normally be justified only when the gas turbine operates for a large number of full-power hours per year, and the cost of the fuel is relatively high. They also lower the temperature of the gas-turbine exhaust, reducing the turbine's effectiveness in CHP applications. As recuperators are subject to large temperature differences, they have to undergo significant thermal stresses. Cyclic operation, in particular, can fatigue the joints, causing them to develop leaks and lose power and effectiveness. Design and manufacturing advancements have mitigated some of the cost and durability issues, and commercial recuperators have been introduced on micro-turbines and on a 4.2-MW industrial gas turbine (through a project supported by the US Department of Energy).

Intercoolers

Intercoolers are used to increase the gas-turbine power by dividing the compressor into two sections and cooling the compressed air leaving the first section before it enters the second compressor section. Intercoolers reduce power consumption in the second section of the compressor, thereby adding to the net power delivered by the combination of the turbine and compressor. Intercoolers have been used for decades in industrial air compressors and in some reciprocating engine turbochargers.

Intercoolers are generally used where additional capacity is considered particularly valuable. The gas-turbine efficiency does not change significantly with the use of inter-cooling. While intercoolers increase the net output, the reduced power consumption of the second section of the compressor results in a lower temperature for the compressed air entering the combustor, and, consequently, incremental fuel is required.

Inlet air-cooling

As shown in Figure 5, the decreased power and efficiency of gas turbines at high ambient temperatures means that the gas-turbine performance is at its lowest at the times when power is in greatest demand and most valued. Figure 11 also shows that cooling the air entering the turbine at about 10 °C on a hot day can increase the power output by 15%–20%. The decreased power and efficiency resulting from high ambient air temperatures can be mitigated by any of the several approaches to inlet air-cooling, including refrigeration, evaporative cooling, and thermal-energy storage using off-peak cooling.

In refrigeration cooling, either a compression-driven refrigeration cycle or thermally activated (absorption chiller) refrigeration cycle cools the inlet air through a heat exchanger. The heat exchanger in the inlet air stream causes an additional pressure drop in the air entering the compressor, thereby slightly lowering the cycle power and efficiency. However, as the inlet air is now substantially cooler than the ambient air, there is a significant net gain in power and efficiency. Electric motor compression refrigeration requires substantial parasitic power loss. Thermally activated absorption cooling can utilize waste heat from the gas turbine, reducing the direct parasitic loss. However, the complexity and cost of this approach pose potential drawbacks in many applications.

Evaporative cooling, widely used due to its low capital cost, makes use of a spray of water, directed straight into the inlet air stream. The evaporation of water reduces the air temperature. As cooling is limited to the wet bulb air temperature, evaporative cooling is most effective when the wet bulb temperature is appreciably lower than the dry bulb (ordinary) temperature. Evaporative cooling can consume large quantities of water, making it difficult to operate it in arid climates. A few large gas turbines work on evaporative cooling, and this principle is expected to be used more frequently on smaller machines in future.

The use of thermal energy storage systems—typically ice, chilled water, or low-temperature fluids—to cool the inlet air can eliminate most parasitic losses from the augmented power capacity. Thermal energy storage is a viable option if on-peak power pricing occurs only a few hours a day. In that case, the shorter time of energy storage discharge and the longer time for daily charging allow for a smaller and less expensive thermal energy storage system.

Emissions

Gas turbines are among the cleanest fossil-fuelled power generation equipment available commercially. Emission control technologies in gas turbines are continuing to evolve, with the phasing out of older technologies and development and commercialization of new technologies.

The primary pollutants from gas turbines are NO_x, CO, and volatile organic compounds (VOCs). Other pollutants such as SO_x and particulate matter (PM) are primarily dependent on the fuel used. The sulphur content of the fuel determines emissions of sulphur compounds, primarily SO_2. Gas turbines operating on de-sulphized natural gas or distillate oil emit relatively insignificant levels of SO_x. In general, SO_x emissions are greater when heavy oils are fired in the turbine. Control of SO_x is thus an issue of fuel-purchasing rather than that of a gas-turbine technology. PM is a marginally significant pollutant for gas turbines using liquid fuels, and ash and metallic additives in the fuel may contribute to PM in the exhaust.

It is important to note that the gas-turbine-operating load has a significant effect on the emission levels of the primary pollutants of NO_x, CO, and VOCs. Gas turbines typically operate at high loads. Consequently, gas turbines are designed to achieve maximum efficiency and optimum combustion conditions at high loads. Controlling all pollutants simultaneously at all load conditions is difficult. At higher loads, higher NO_x emissions occur due to peak flame temperatures. At lower loads, lower thermal efficiencies and more incomplete combustion occur, resulting in higher emissions of CO and VOCs.

Cheng cycle (steam-injected gas turbine)

The Cheng cycle offers another variant of the gas-turbine cycle with the utilization of waste heat (Figure 14). The steam generated is partly fed

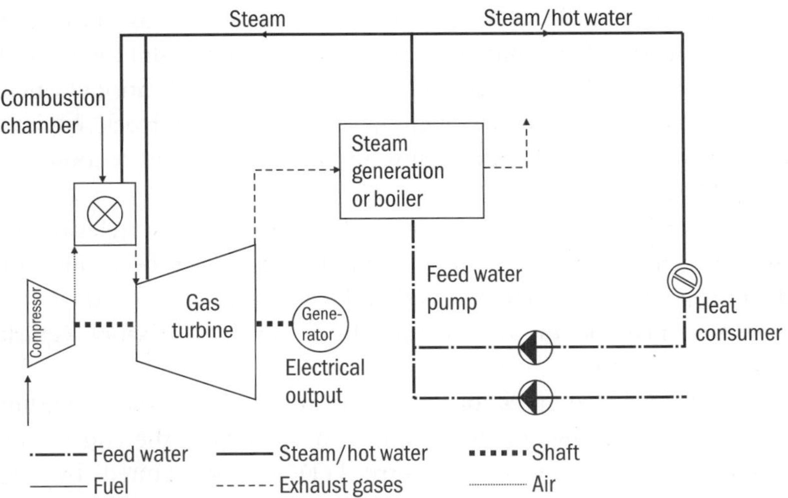

Figure 14 Set-up of the Cheng cycle

into the combustion chamber and the turbine. Thus, output and electrical efficiency are significantly increased. It also allows for an adjustment to the heat requirements of the plant so that steam is fed into the gas turbine as per requirements.

Basic engine processes

There are two primary reciprocating engine designs relevant to stationary power generation applications—the spark-ignition Otto-cycle engine and the compression-ignition diesel-cycle engine. The essential mechanical components of the Otto-cycle and diesel-cycle are the same. Both use a cylindrical combustion chamber in which a close-fitting piston travels the length of the cylinder. The piston connects to a crankshaft that transforms the linear motion of the piston into the rotary motion of the crankshaft. Most engines have multiple cylinders that power a single crankshaft.

The primary difference between the Otto and diesel cycles is the method of igniting fuel. Spark-ignition engines (Otto cycle) use a spark plug to ignite a pre-mixed air–fuel mixture introduced into the cylinder. Compression-ignition engines (diesel-cycle) compress the air introduced into the cylinder to a high pressure, raising its temperature to the auto-ignition temperature of the fuel that is injected at high pressure.

Engines are further categorized by crankshaft speed (revolutions per minute [RPM]), operating cycle (two- or four-stroke), and whether turbo-charging is used. Reciprocating engines are also categorized by their original design purpose—automotive, truck, industrial, locomotive, and marine. Hundreds of small-scale stationary power, CHP, irrigation, and chiller applications use automotive engine models. These are generally low-priced engines due to large production volumes. However, unless conservatively rated, these engines have limited durability. Truck engines have the cost-benefit of production volume and a reasonably long life (for example, one million miles). A number of truck engines are available as stationary engines. Engines intended for industrial use are designed for durability and for a wide range of mechanical drive and electric power applications. Their sizes range from 20 kW to 6 MW, including industrialized truck engines in the 200–600 kW range and industrially applied marine and locomotive engines above 1 MW.

Both the spark ignition and diesel four-stroke engines—most relevant to stationary power generation applications—complete a power cycle in four strokes of the piston within the cylinder, as described below.

1 *Intake stroke* Introduction of air (diesel) or air–fuel mixture (spark ignition) into the cylinder.
2 *Compression stroke* Compression of air or an air–fuel mixture within the cylinder. (In diesel engines, fuel is injected at or near the end of the compression stroke or top dead centre [TDC], and ignited by the elevated temperature of the compressed air in the cylinder. In spark-

ignition engines, the compressed air–fuel mixture is ignited by an ignition source at or near the TDC.)

3 *Power stroke* Acceleration of the piston by the expansion of hot, high pressure combustion gases.
4 *Exhaust stroke* Expulsion of combustion products from the cylinder through the exhaust port.

Types of reciprocating engines

Natural-gas-based spark-ignition engines

Spark-ignition engines use spark plugs with a high-intensity spark of timed duration to ignite a compressed fuel–air mixture within the cylinder. Natural gas is the predominant spark-ignition engine fuel used in electric generation and CHP applications. Other gaseous and volatile liquid fuels, ranging from landfill gas to propane to gasoline, can be used with the proper fuel system, engine compression ratio, and tuning. American manufacturers began to develop large natural gas engines for the burgeoning gas transmission industry after World War II. Smaller engines were developed (or converted from diesel blocks) for collecting gas and for other stationary applications such as developing the infrastructure for natural gas. Natural gas engines for power generation applications are primarily four-stroke engines available in sizes up to about 5 MW.

Depending on engine size, one of the two ignition techniques given below ignites the natural gas.

1 *Open chamber* The spark plug tip is exposed in the combustion chamber of the cylinder, directly igniting the compressed fuel–air mixture. Open chamber ignition is applicable to any engine operating near the stoichiometric air–fuel ratio up to moderately lean mixtures.
2 *Precombustion chamber* It comprises a staged combustion process where the spark plug is housed in a small chamber mounted on the cylinder head. This cylinder charges with a rich mixture of fuel and air, which, upon ignition, shoots into the main combustion chamber in the cylinder as a high-energy torch, providing sufficient ignition energy to light off lean fuel–air mixtures used in large bore engines.

The simplest natural gas engines operate with natural aspiration of air and fuel into the cylinder (via a carburettor or other mixer) by the suction of the intake stroke. High performance natural gas engines are turbocharged to force more air into the cylinders. Natural-gas-based spark-ignition engines operate at modest compression ratios (compared with diesel engines) in the range 9:1–12:1, depending on engine design and turbocharging. Modest compression is required to prevent auto-ignition of the fuel and engine knock, which can cause a serious engine damage.

Using high-energy ignition technology, lean fuel–air mixtures can be burned in natural gas engines, lowering the peak temperatures within the cylinders and resulting in reduced NO_x emissions. The lean-burn approach in reciprocating engines is analogous to dry, low-NO_x combustors in gas turbines. All major natural gas engine manufacturers offer lean-burn, low-emission models and are engaged in the research and development (R&D) to further improve their performance.

Natural-gas-based spark-ignition engine efficiencies are typically lower than those of diesel engines because of their lower compression ratios. However, large, high performance, lean-burn engine efficiencies approach those of diesel engines of the same size. Natural gas engine efficiencies range from about 28% (LHV) for small engines (<50 kW) to 42% (LHV) for the largest high performance, lean-burn engines. Lean-burn engines tuned for maximum efficiency may produce twice the NO_x emissions as the same engine tuned for minimum NO_x. Tuning for low NO_x typically results in a sacrifice of 1%–1.5% in electric generating efficiency from the highest level achievable.

Many natural-gas-based spark-ignition engines are derived from diesel engines, that is, they use the same block, crankshaft, main bearings, camshaft, and connecting rods as the diesel engine. However, natural-gas-based spark-ignition engines operate at lower brake mean effective pressure (BMEP) and peak pressure levels to prevent knock. Due to the derating effects from lower BMEP, the spark-ignition versions of diesel engines often produce only 60%–80% of the power output of the parent diesel.

Manufacturers often enlarge cylinder bores by about 5%–10% to increase the power, but this is only partial compensation for the derated output. Consequently, the dollars-per-kilowatt capital costs of natural gas spark-ignition engines are generally higher than those of diesel engines from which they were derived. However, by operating at lower cylinder pressure and bearing loads as well as in the cleaner combustion environment of natural gas, spark-ignition engines generally offer the benefits of extended component life as compared to their diesel parents.

Diesel engines

Compression-ignition diesel is among the most efficient simple-cycle power generation options in the market. Efficiency levels increase with engine size and range from about 30% (HHV) for small, high-speed diesels and up to 42%–48% (HHV) for the large bore, slow-speed engines. It is expected that efficiencies will improve to a maximum of 52% (HHV). High-speed diesel engines (1200 RPM) up to about 4 MW in size are available. Low-speed diesels (60–275 RPM) are available in sizes as large as 65 MW.

Diesel engines typically require compression ratios of 12:1–17:1 to heat the cylinder air to a temperature at which the injected fuel will ignite. The quality of fuel injection significantly affects the diesel engine's operating characteristics, fuel efficiency, and emissions. Fine atomization and good fuel dispersion by the injectors are essential for rapid ignition, ideal combustion, and emission control. Manufacturers are increasingly moving towards electronically controlled high-pressure injection systems that provide more precise calibration of fuel delivery and accurate injection timing.

Depending on the engine and fuel quality, diesel engines produce 5–20 times the NO_x (on a ppm basis) of a lean-burn natural-gas-based engine. Emergency generators and marine engines often emit over 9 kg NO_x/megawatt hour (MWh), and the present road engines emit less than 6 kg NO_x/MWh. New diesel engines using low-sulphur diesel will achieve rates of approximately 0.29 kg NO_x/MWh. Diesel engines also produce assorted heavy hydrocarbons and lead to particulate emissions. However, they produce significantly less CO than lean-burn gas engines. The NO_x emissions from diesel burning heavy oil are typically 25%–30% higher than the diesel using distillate oil. Common NO_x control techniques include delayed fuel injection, exhaust-gas re-circulation, water injection, fuel water emulsification, inlet air cooling, intake air humidification, and compression ratio and/or turbocharger modifications. In addition, an increasing number of larger diesel engines are equipped with selective catalytic reduction and oxidation catalyst systems for post-combustion emissions reduction.

High-speed diesel engines generally require high-quality fuel oil with good combustion properties: No. 1 and No. 2 distillate oil comprise the standard diesel fuels. Low-sulphur distillate is preferred to minimize SO_2 emissions. High-speed diesel is not suited to burning oil heavier than the distillate. Heavy fuel oil requires more time for combustion, and the combination of high speed and contaminants in lower quality heavy oils causes excessive wear in high-speed diesel engines. Many medium- and low-speed diesel fuels burn heavier oils, including low-grade residual oils or bunker C oils.

Dual-fuel engines

Dual-fuel engines are diesel compression-ignition engines, predominantly fuelled by natural gas with a small percentage of diesel oil as the pilot fuel. The pilot fuel auto-ignites and initiates combustion in the main air–fuel mixture. Pilot fuel percentages can range from 1% to 15% of the total fuel input. Dual-fuel operation is a combination of diesel and Otto cycle operation. With reduction in the percentage of pilot fuel used, it approaches the diesel cycle more closely. Most dual-fuel engines can be switched back and forth on the fly between dual-fuel and 100%-diesel operation. In general, because of lower diesel oil usage, NO_x, smoke, and particulate emissions are lower for dual-fuel engines than for straight diesel operation, particularly at full load. Particulate emissions reduce in line with the percentage reduction in diesel oil consumption, while the level of NO_x reduction depends on combustion characteristics. However, CO and unburned hydrocarbon emissions are often higher, partly because of incomplete combustion.

There are three basic types of dual-fuel engines, as follows.

1 *Conventional, low-pressure gas injection engines* typically require about 5%–10% pilot fuel and may be derated to about 80%–95% of the rated diesel capacity to avoid detonation. The turndown ratio of the diesel fuel injection system sets the minimum pilot fuel requirement. Conventional diesel injectors cannot reliably turn down to less than 5%–6% of the full-load injection rate. Natural gas input is controlled at each cylinder by injecting gas before the air intake valves open. NO_x emissions of conventional dual-fuel engines are generally in the range 5–8 gm/kWh (compared to lean-burn, natural-gas engines with NO_x emissions in the 0.7–2.5 g/kWh range).

2 *High-pressure gas injection engines* attempt to reduce derating by injecting natural gas at high pressures (245–347 atm [g]) into the main combustion chamber as the pilot fuel is injected. However, the parasitic power for gas compression can be as high as 4%–7% of the rated power output, partly offsetting the benefit of reduced derating. This technology has not proved particularly popular because of this issue and the additional equipment costs for gas injection. Pilot fuel consumption is typically 3%–8% and NO_x emissions are generally in the range 5–8 g/kWh.

3 *Micro-pilot pre-chamber engines* are similar to spark-ignition, pre-chamber engines in that the pilot fuel injected into a pre-chamber provides a high-energy torch that ignites the lean, compressed fuel–air mixture in the cylinder. Mixtures leaner than those in spark-ignition engines are achievable as the energy provided by the diesel-fuelled micro-pilot chamber is higher than that obtained with a spark-ignition pre-chamber. Micro-pilot dual-fuel engines with 1% pilot fuel can operate at or close to the diesel engine's compression ratio and BMEP, hence little, if any, derating occurs. In this case, the high-power density and low dollars-per-kilowatt cost advantage of the original diesel engines are retained, and engine efficiency at 75%–100% load is close to that of the 100%-diesel engine. NO_x and other emissions are comparable to those of lean-burn spark-ignition pre-chamber engines (NO_x emissions in the range 0.7–2.5 g/kWh). These engines must be equipped with conventional diesel fuel injectors in order to operate on 100% diesel.

Several independent developers and engine manufacturers are testing and commercializing dual-fuel retrofit kits for converting existing diesel engines to dual-fuel operation. The level of sophistication of these kits varies widely, and some kits require major engine modifications. Derating, efficiencies, and emissions also vary widely and have yet to be fully tested or certified. However, dual-fuel conversions are unlikely to be as low in emissions as dedicated natural-gas-based engines. In addition, manufacturers may not honour warrantees on an engine retrofitted by an independent third party.

Engine speed classifications

Reciprocating engines are classified as high, medium, or low speed. Table 5 presents the standard speed ranges in each class and the types and sizes of engines available. Engine-driven electric generators typically must run at fixed (or synchronous) speeds to maintain a constant 50–60 Hz output, setting the requisite engine speed within the classifications (that is, a 60-Hz generator driven by a high-speed engine would require engine speeds of 1200, 1800 or 3600 RPM versus a 50-Hz generator, which requires engine speeds of 1000, 1500 or 3000 RPM).

Engine power output is proportional to engine speed, affording high-speed engines the highest output per unit of displacement (cylinder size) and the highest power density. Consequently, high-speed engines generally have the lowest dollars-per-kilowatt production costs among the three types. The cost benefits of high-speed engines must be weighed against other factors. Smaller high-speed engines tend to have lower efficiencies than large-bore, lower-speed engines, in part due to the higher surface area to volume ratio for small cylinders, resulting in slightly higher heat losses. In addition, higher-speed engines tend to have higher wear rates, resulting in shorter periods between minor and major overhauls. These factors are often less important than capital costs for limited duty cycle applications.

Medium-speed stationary power engines are largely derived from marine and locomotive engines. Medium-speed engines have high cost, but are generally also better in efficiency than high-speed engines. Because of their massive physical size and speed-related power reduction, low-speed engines are increasingly being displaced by medium- and high-speed engines as the primary choice for stationary power applications.

Design characteristics

The features that have made reciprocating engines a leading prime mover for CHP and other distributed generation applications include the following.

- *Size range* Reciprocating engines are available in the sizes of 5–10 MW.
- *Thermal output* Reciprocating engines can produce hot water and low-pressure steam.

Table 5 Reciprocating engine types by speed (available MW ratings)

Speed classification	Engine speed (RPM)	Stoic/rich burn, spark ignition	Lean burn, spark ignition	Dual fuel	Diesel
High speed	1000–3600	0.01–1.5 MW	0.15–3.0 MW	1.0–3.5 MW	0.01–3.5 MW
Medium speed	275–1000	—	1.0–6.0 MW	1.0–25 MW	0.5–35 MW
Low speed	58–275	—	—	2.0–65 MW	2–65 MW

- *Fast start-up* The fast start-up capability of reciprocating engines allows timely resumption of the system, following a maintenance procedure. In peaking or emergency power applications, reciprocating engines can quickly supply electricity on demand.
- *Black-start capability* In the event of an electric utility outage, the starting reciprocating engines require minimal auxiliary power. Generally, only batteries are required.
- *Availability* Reciprocating engines have typically demonstrated availability in excess of 95% in stationary power generation applications.
- *Part-load operation* The high part-load efficiency of reciprocating engines ensures economical operation in electric load following applications.
- *Reliability and life* Reciprocating engines have proved to be reliable power generators, given proper maintenance.
- *Emissions* Diesel engines have relatively high emission levels of NO_x and particulates. However, natural gas based spark-ignition engines have improved emission profiles and are easier to site.

Performance characteristics

Electrical efficiency

Table 6 summarizes the performance characteristics for typical commercially available natural-gas-based spark-ignition engine CHP systems in the range 100 kW to 5 MW. This range covers the majority of the market applications for engine-driven CHP. Heat rates and efficiencies shown were based on the manufacturers' specifications and industry publications. The available thermal energy was calculated from published engine data on engine exhaust temperatures, and engine jacket and lube system coolant flows. The CHP thermal recovery estimates are based on the production of hot water. As shown in Table 6, 50%–60% of the waste heat from engine systems is recovered from jacket cooling water and lube oil cooling systems at a temperature too low to produce steam. This feature is generally less critical in commercial/institutional applications where it is more common to have hot water thermal loads. Steam can be produced from the exhaust heat if required (maximum pressure of 10 kg/cm^2 [g]), but if no hot water is needed, the amount of heat recovered from the engine is reduced, and the total CHP system efficiency drops accordingly.

Data in Table 6 shows that electrical efficiency increases as engine size becomes larger. Subsequently, the useful thermal energy decreases with the increase in electrical output, while the ratio of power to heat for the CHP system generally increases. A changing ratio of power to heat impacts project economics and may affect the decisions that customers take in terms of CHP acceptance, sizing, and the desirability of selling power.

Table 6 Gas engine CHP: typical performance parameters[a]

Cost and performance characteristics	System 1	System 2	System 3	System 4	System 5
Base-load electricity capacity (kW)	100	300	800	3000	5000
Total installed cost ($/kW)	1515	1200	1000	920	920
Electric heat rate (kg/kWh), higher-heating value (HHV)	2809	2764	2582	2392	2207
Electrical efficiency (%), HHV	30.60	31.10	33.30	36.00	39.00
Engine speed (revolution per minute)	1800	1800	1200	900	720
Fuel input (kcal/h)[a] × 10^6	0.28	0.83	2.07	7.18	11.04
	708	696	651	603	556
Combined heat power (CHP) characteristics					
Exhaust flow (1000 kg/h)	0.45	1.50	4.95	22.00	30.50
Exhaust temperature (°C)	571	575	465	364	370
Heat recovered from exhaust (kcal/h) × 10^6	0.05	0.21	0.53	1.40	1.80
Heat recovered from cooling jacket (kcal/h) × 10^6	0.09	0.17	0.27	1.10	1.58
Heat recovered from lube system (kcal/h) × 10^6	0	0	0.07	0.31	0.49
Total heat recovered (kcal/h) × 10^6	0.14	0.38	0.88	2.80	3.88
Total heat recovered (kW)	167	443	1025	3259	4508
Form of recovered heat	Hot water	Hot water	Hot water	Hot water	Hot water
Total efficiency (%)	81	77	76	75	74
Power/heat ratio	0.60	0.68	0.78	0.92	1.11
Net heat rate (kcal/kWh)	1024	1181	1203	1224	1238
Effective electrical efficiency	0.84	0.73	0.71	0.7	0.69

[a] For typical systems commercially available in 2001

Part-load performance

In power generation and CHP applications, reciprocating engines generally drive synchronous generators at constant speed to produce steady alternating current (AC) power. At reduced loads, the heat rate of spark-ignition engines increases and efficiency decreases. Figure 15 shows the part-load efficiency curve for a typical lean-burn natural-gas-based engine. The efficiency at 50% load is approximately 8%–10% less than full-load efficiency. As the load decreases further, the curve becomes somewhat steeper. While gas engines compare favourably to gas turbines, which typically

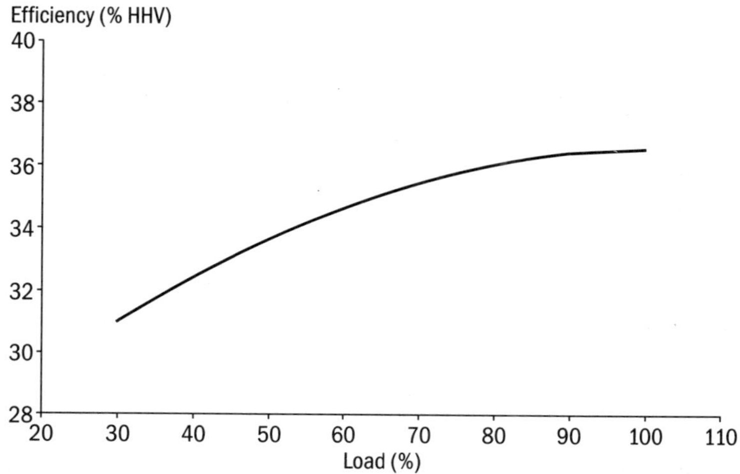

Figure 15 Part-load efficiency performance
Source Caterpillar (2001)

experience efficiency decreases of 15%–25% at half-load conditions, multiple engines may be preferable to a single large unit to avoid efficiency penalties wherein significant load reductions are expected on a regular basis. Diesel engines exhibit even more favourable part-load characteristics than spark ignition engines. The efficiency curve for diesel engines is comparatively flat between 50% and 100% load.

Effects of ambient conditions on performance

Reciprocating engines are generally rated at ISO conditions of 25°C and 0.987 atmospheric (1 bar) pressure. Like gas turbines, reciprocating engine performance—both output and efficiency—degrades as ambient temperature or site elevation increases. While the effect on gas turbines can be significant, it is less so on engines. Reciprocating engine efficiency and power are reduced by approximately 4% per 300 m of altitude above 300 m, and about 1% for every 5.5°C above 25°C.

Heat recovery

The economics of engines in on-site power generation applications often depend on effective use of the thermal energy contained in the exhaust gas and cooling systems, which generally represents 60%–70% of the inlet fuel energy. Most of the waste heat is available in the engine exhaust and jacket coolant, with smaller amounts recoverable from the lube oil cooler and the turbocharger's intercooler and after-cooler (if so equipped). The most common use of this heat is to generate hot water or low-pressure steam for process use or for space heating, process needs, domestic hot water, or absorption cooling. However, the engine exhaust gases can also be used as a source of direct energy for drying or other direct heat processes.

Heat in the engine jacket coolant accounts for up to 30% of the energy input and is capable of producing hot water at temperatures 90–100°C. Some engines, such as those with high pressure or ebullient cooling systems, can operate with water jacket temperatures up to 129°C. Engine exhaust heat represents 30%–50% of the available waste heat. Exhaust temperatures of 455–650°C are typical. By recovering heat in the cooling systems and exhaust, approximately 70%–80% of the energy of the fuel is effectively utilized to produce both power and useful thermal energy.

Closed-loop cooling systems

Figure 16 shows the most common method for recovering engine heat by using closed-loop cooling system. These systems cool the engine by forced circulation of a coolant through engine passages and an external heat exchanger. An external heat exchanger transfers engine heat to a cooling tower or a radiator when excess heat is generated. Closed-loop water cooling systems can operate at coolant temperatures varying from 90°C to 120°C. Depending on the engine and CHP system's requirements, the lube oil cooling and turbocharger after-cooling may be either separate or part of the jacket-cooling system.

Ebullient cooling systems

Ebullient cooling systems cool the engine by natural circulation of a boiling coolant through the engine. This type of cooling system is typically used in conjunction with exhaust heat recovery for the production of low-pressure steam. Cooling water is introduced at the bottom of the engine where the transferred heat brings the coolant to boil, generating a two-phase flow. Formation of bubbles lowers the density of the coolant, causing natural circulation at the top of the engine.

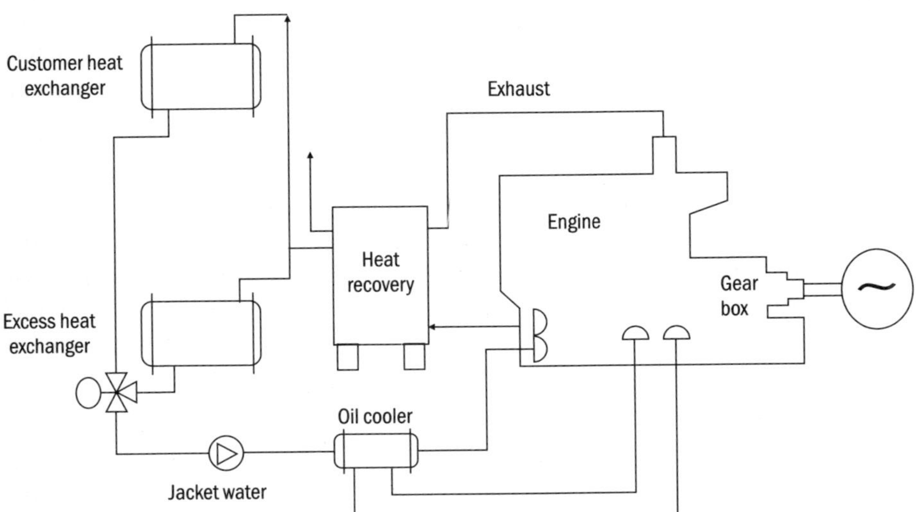

Figure 16 Closed-loop heat recovery system

The coolant at the engine outlet is maintained at saturated steam conditions and is usually limited to 40°C and a maximum of 1 atm (g). Inlet cooling water is also near saturation conditions and is generally 1–2°C below the outlet temperature. The uniform temperature throughout the coolant circuit extends the engine life and contributes to improved combustion efficiencies.

Exhaust heat recovery

Recovered exhaust heat generates hot water at about 100°C or low-pressure steam (up to 10 kg/m^2). To prevent the corrosive effects of condensation in exhaust piping, the exhaust-gas temperatures are generally kept above temperature thresholds, recovering only a portion of the exhaust heat. For this reason, most heat recovery units use a 120–180 °C exhaust outlet temperature.

Exhaust heat recovery can be independent of the engine-cooling system or coupled with it. For example, hot water from the engine cooling can be used as feed-water or feed-water pre-heat to the exhaust recovery unit. In a typical district-heating system, jacket cooling, lube oil cooling, single stage after-cooling, and exhaust gas heat recovery are all integrated for steam production.

Performance and efficiency enhancements

BMEP and engine speed

Engine speed and the BMEP during the power stroke relate to engine power. BMEP is an "average" cylinder pressure on the piston during the power stroke, and is a measure of the effectiveness of engine power output or mechanical efficiency. Engine manufacturers often include BMEP values in their product specifications. Typical BMEP values are as high as 15 atm (g) for large natural-gas-based engines and 25 atm (g) for diesel engines. Corresponding peak combustion pressures are about 120 atm (g) and 180 atm (g), respectively. High BMEP levels increase power output, improve efficiency, and result in lower specific costs (dollars per kilowatt).

BMEP can be increased by raising combustion cylinder air pressure through increased turbocharging, improved after-cooling, and reduced pressure losses through improved air passage design. All these factors increase air-charge density and raise the peak combustion pressures, which translates into higher BMEP levels. However, a higher BMEP increases thermal and pneumatic stresses within the engine, and proper design and testing are required to ensure continued engine durability and reliability.

Turbocharging

Essentially, all modern engines above 300 kW are turbocharged to achieve higher power densities. A turbocharger is a turbine-driven intake air compressor. The hot high-velocity exhaust gases leaving the engine cylinders power the turbine. Large engines are typically equipped with two turbochargers. On a carburetted engine, turbocharging forces more air and fuel into the cylinders, increasing the engine output. On a fuel-injected engine, the mass of fuel injected increases in proportion to the increased air input. Cylinder pressure and temperature normally increase as a result of turbocharging, increasing the tendency for detonation for both spark-ignition and dual-fuel engines and requiring a careful balance between the compression ratio and the turbocharger boost level. Turbochargers normally boost inlet air pressure on a 3:1–4:1 ratio. A wide range of turbocharger designs and models are used. Heat exchangers (called after-coolers or intercoolers) cool the discharge air from the turbocharger to keep the temperature of the air going into the engine under a specified limit.

Availability

Reciprocating engines are maintenance-intensive, but they are widely available, even in high-load factor applications. Natural-gas-based engine availabilities vary with engine type, speed, and fuel quality; Table 7 illustrates typical availability numbers based on a survey of natural gas engine gensets in the CHP applications.

The use of multiple units or back-up units at a site can further increase the availability of the overall facility. Some engine manufacturers offer engine exchange programmes or other maintenance options that increase the ability to promptly deliver and install replacement units on short notice, typically increasing the facility availabilities to greater than 95%.

Emissions

Exhaust emissions from reciprocating engines are of primary environmental concern. The primary pollutants are NO_x, CO, and VOCs—unburned, non-methane hydrocarbons. Emissions of other pollutants such as SO_x and PM are primarily dependent on the fuel used. Sulphur content of the fuel determines the emissions of the sulphur compounds, primarily SO_2. Engines operating on natural gas or de-sulphurized distillate oil emit insignificant levels of SO_x. In general, SO_x emissions are an issue only in large, slow-speed diesel firing heavy oils. PM can be an important pollutant for engines using liquid fuels. Ash and metallic additives in the fuel contribute to PM in the exhaust.

Table 7 Availabilities and outage rates for natural gas engines

Gas engine	80–800 kW	>800 kW
Availability factor (%)	94.5	91.2
Forced outage rate (%)	4.7	6.1
Scheduled outage rate (%)	2	3.5

Source Gas Research Institute (1999)

Micro-turbine cycle

Basic processes

Micro-turbines are small gas turbines, most of which feature an internal heat exchanger called a recuperator. In a micro-turbine, a radial flow (centrifugal) compressor compresses the inlet air, which is then preheated in the recuperator using heat from the turbine exhaust. Next, the heated air from the recuperator mixes with fuel in the combustor, and the hot combustion gas expands through the expansion and power turbines. The expansion turbine turns the compressor and, in single-shaft models, turns the generator as well. Two-shaft models use the compressor drive turbine's exhaust to power a second turbine that drives the generator. Finally, the recuperator uses the exhaust of the power turbine to preheat the air from the compressor.

Single-shaft models generally operate at speeds over 60 000 RPM and generate electrical power of high and variable frequency—AC. This power is rectified to direct current (DC) and then inverted to 50/60 Hz as required for commercial use. In the two-shaft version, the power turbine connects via a gearbox to a generator that produces power at the required frequency.

Thermodynamic cycle

Micro-turbines operate on the same thermodynamic cycle, known as the Brayton cycle, as larger gas turbines. In this cycle, atmospheric air is compressed, heated, and then expanded, with the excess power produced by the expander (also called the turbine) over that consumed by the compressor used for power generation. The power produced by an expansion turbine and consumed by a compressor is proportional to the absolute temperature of the gas passing through those devices. Consequently, it is advantageous to operate the expansion turbine at the highest practical temperature consistent with economic materials and to operate the compressor with inlet airflow at as low a temperature as possible. As advances in technology permit higher turbine inlet temperature, the optimum pressure ratio also increases. Higher temperature and pressure ratios result in higher efficiency and specific power. Thus, the general trend in gas-turbine advancement has been towards a combination of higher temperatures and pressures. However, micro-turbine inlet temperatures are generally limited to 120°C or below to enable the use of relatively inexpensive materials for the turbine wheel, and to maintain comparatively low pressure ratio at 3.5–4.0.

Basic components

Turbo-compressor package

The basic components of a micro-turbine are the compressor, turbine generator, and recuperator (Figure 17). The heart of the micro-turbine is the compressor–turbine package, which is commonly mounted on a single

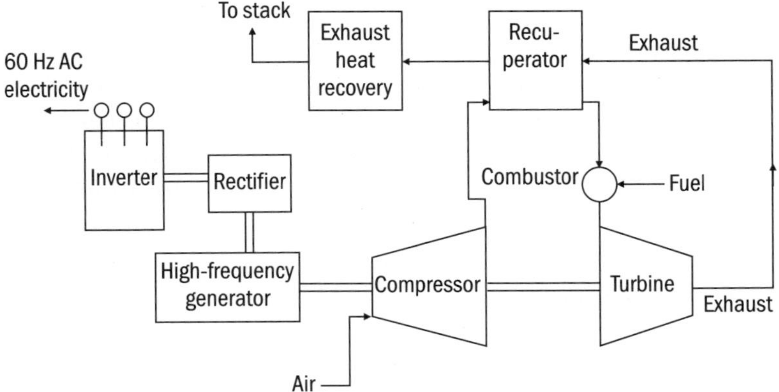

Figure 17 Micro-turbine-based CHP system (single-shaft design)

shaft along with the electric generator. Two bearings support the single shaft. The single moving part of the one-shaft design has the potential for reducing maintenance needs and enhancing overall reliability. There are also two-shaft versions, in which the turbine on the first shaft directly drives the compressor, while a power turbine on the second shaft drives a gearbox and the conventional electrical generator producing 50-Hz power. The two-shaft design features more moving parts, but does not require complicated power electronics to convert high frequency AC power output into 50 Hz.

Moderate-to-large-size gas turbines use multi-stage axial flow turbines and compressors, in which the gas flows along the axis of the shaft and is compressed and expanded in multiple stages. However, micro-turbine turbo machinery is based on single-stage radial-flow compressors and turbines. Radial-flow turbo machinery handles the small volumetric flows of air and combustion products with reasonably high component efficiency. Large-size axial-flow turbines and compressors are typically more efficient than radial-flow components. However, in the size range of micro-turbines—0.2–2 kg/second of air/gas flow—radial-flow components offer minimum surface and end-wall losses to provide the highest efficiency.

In micro-turbines, the turbo-compressor shaft generally turns at high rotational speeds of about 96 000 RPM in the case of a 30-kW machine and about 80 000 RPM in a 75-kW machine; one 45-kW model on the market turns at 116 000 RPM. There is no single rotational speed–power size rule as the specific turbine and compressor design characteristics strongly influence the physical size of components and, consequently, rotational speed. For a specific aerodynamic design, when the power rating decreases, the shaft speed increases, resulting in high shaft speed of the small micro-turbines.

Generator

The micro-turbine produces electrical power via either a high-speed generator turning on the single turbo-compressor shaft or a separate power turbine driving a gearbox and conventional 3600 RPM generator. The high-speed generator of the single-shaft design employs a permanent magnet (typically samarium-cobalt) alternator, and requires that the high frequency AC output (about 1600 Hz for a 30-kW machine) be converted to 50/60 Hz for general use. This power conditioning involves rectifying the high-frequency AC to DC, and then inverting the DC to 50 Hz AC. Power conversion comes with an efficiency penalty (approximately 5%). To start up a single-shaft design, the generator acts as a motor, turning the turbo-compressor shaft until a sufficient RPM rate is reached to start the combustor. Full start-up requires several minutes. If the system is operating independent of the grid (black starting), a power storage unit (typically a battery uninterrupted power supply [UPS]) is used to power the generator for start-up.

Recuperators

Recuperators are heat exchangers that use the hot turbine exhaust gas (about 650°C) to preheat the compressed air (about 150°C) going into the combustor, thereby reducing the fuel needed to heat the compressed air to turbine inlet temperature. Depending on the micro-turbine operating parameters, recuperators can more than double machine efficiency. However, as there is increased pressure drop in both the compressed air and turbine-exhaust sides of the recuperator, power output typically declines by 10%–15% from that attainable without the recuperator. Recuperators also lower the temperature of the micro-turbine exhaust, reducing the micro turbine's effectiveness in the CHP applications.

Power electronics

As discussed, single-shaft micro turbines feature digital power controllers to convert the high frequency AC power produced by the generator into usable electricity. The high-frequency AC is rectified to DC and inverted back to 60- or 50-Hz AC, and then filtered to reduce harmonic distortion. This is a critical component in the single-shaft micro-turbine design, and it represents significant design challenges, specifically in matching turbine output to the required load. To allow for transients and voltage spikes, power electronic designs are generally able to handle seven times the nominal voltage. Most micro-turbine powered electronics generate three-phase electricity.

Electronic components also direct all operating and start-up functions. Micro-turbines are generally equipped with controls that allow the unit to be operated in parallel to or independent of the grid, and internally incorporate many of the grid and system protection features required for the interconnect. The controls also allow for remote monitoring and operation.

CHP operation

In CHP operation, a second heat exchanger, the exhaust gas heat exchanger, transfers the remaining energy from the micro-turbine exhaust to a hot water system. Exhaust heat can be used for a number of different applications, including potable water heating, driving absorption cooling and desiccant dehumidification equipment, space heating, process heating, and other building or site uses. Some micro-turbine-based CHP applications do not use recuperators. With these micro-turbines, the temperature of the exhaust is higher, thus more heat is available for recovery (Figure 17).

Design characteristics

- *Thermal output* Micro-turbines produce thermal output at temperatures in the range 90–160°C, which could meet a variety of thermal needs in buildings.
- *Fuel flexibility* Micro-turbines can operate using a number of different fuels, such as natural gas; sour gases (high sulphur, low Btu content); and liquid fuels (gasoline, kerosene, and diesel fuel/heating oil).
- *Reliability and life* Design life is estimated to be in the range 40 000–80 000 hours. While units have demonstrated reliability, they have not been in commercial service long enough to provide a definitive life data.
- *Size range* Micro-turbines available and under development are sized from 30 kW to 350 kW.
- *Emissions* Low-inlet temperatures and high fuel-to-air ratios result in NO_x emissions of less than 10 ppm when operating on natural gas.
- *Modularity* Units may be connected in parallel to serve larger loads and provide power reliability.
- *Part-load operation* Because micro-turbines reduce power output by reducing mass flow and combustion temperature, efficiency at part load can be below that of full-power efficiency.
- *Dimensions* The dimension could be of about a cubic metre.

Performance characteristics

Micro-turbines are more complex than conventional simple-cycle gas turbines as the addition of the recuperator reduces fuel consumption (thereby substantially increasing efficiency) and introduces additional internal pressure losses that moderately lower efficiency and power. As the recuperator has four connections—to the compressor discharge, the expansion turbine discharge, the combustor inlet, and the system exhaust—it is a challenge to the micro-turbine designer to make all the connections in a manner that minimizes pressure loss, keeps the manufacturing cost low, and entails least compromise of system reliability. Each of the manufacturers' models has evolved in unique ways.

The addition of a recuperator opens numerous design parameters to performance-cost trade-offs. In addition to selecting the pressure ratio for high efficiency and best business opportunity (high power for low price), the recuperator has two performance parameters—effectiveness and pressure drop—that also have to be selected for the combination of efficiency and cost, which creates the best business conditions. Higher effectiveness recuperation requires greater recuperator surface area, which both increases cost and incurs additional pressure drop. Such increased internal pressure drop reduces the net power production and consequently increases micro-turbine cost per kilowatt.

Micro-turbine performance, in terms of both efficiency and specific power, is highly sensitive to small variations in component performance and internal losses. This is because the high efficiency recuperated cycle processes a much larger amount of air and combustion products flow per kilowatt of net power delivered than is the case for high-pressure ratio simple-cycle machines. When the net output is the small difference between two large numbers (the compressor and expansion turbine work per unit of mass flow), small losses in component efficiency, internal pressure losses, and recuperator effectiveness have large impacts on net efficiency and net power per unit of mass flow.

For these reasons, it is advisable to focus on trends and comparisons in considering performance while relying on manufacturers' guarantees for precise values.

Electrical efficiency

Figure 18 shows a recuperated micro-turbine electrical efficiency as a function of micro-turbine compressor ratio for a range of turbine-firing temperatures, that is, from 845°C to 955°C, corresponding to conservative to optimistic turbine material life behaviour. The reported efficiency is the gross generator output (without consideration of parasitic or conversion losses). Often this is at high frequency, so the output must be rectified and inverted to provide 60 Hz AC power. The efficiency loss in such frequency conversion (about 5%, which would lower efficiency from 30% to 28.5%) is not included. Figure 18 shows that a broad optimum of performance exists in the pressure ratio of 3–4.

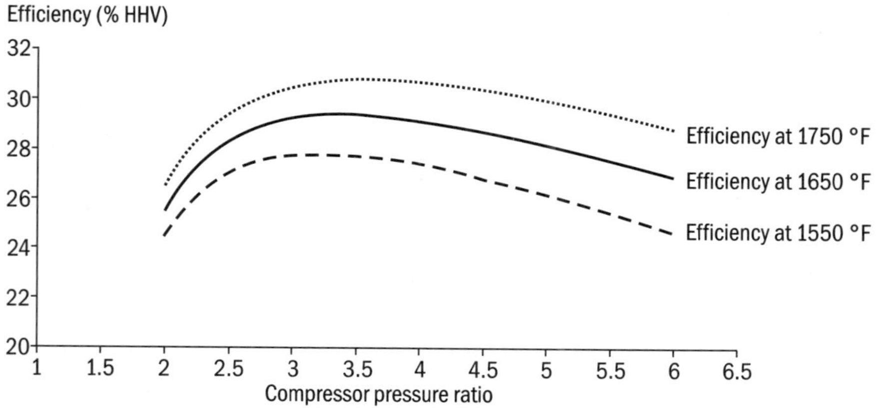

Figure 18 Micro-turbine electrical efficiency as a function of compressor pressure ratio and turbine firing temperature

Figure 19 shows micro-turbine-specific power for the same range of firing temperatures and pressure ratios. Higher pressure ratios result in greater specific power. However, practical considerations limit compressor and turbine component tip speed due to centrifugal forces and allowable stresses in economic materials, resulting in compressor pressure ratio limits of 3.5–5 in the micro turbines currently entering the market.

Part-load performance

When full power from a micro-turbine is not required, the output is reduced by a combination of mass flow reduction (achieved by decreasing the compressor speed) and turbine-inlet temperature reduction. It takes, approximately, 15 seconds for a micro turbine to go from no-load to full-load conditions. Rapid off-loading (removal of load) causes the stored energy in the micro-turbine to over-speed and damage the turbine. In addition to reducing power, this change in operating conditions also reduces efficiency. Figure 20 shows a sample part-load derate curve for a micro-turbine.

Effect of ambient conditions on performance of micro-turbines

The performance of micro-turbines is affected by the ambient conditions that are same as those for gas turbines. These are as discussed in the section pertaining to gas turbines.

Heat recovery

Effective use of the thermal energy contained in exhaust gas improves the micro-turbine system economics. Exhaust heat can be recovered and used in a variety of ways, including water heating, space heating, and driving thermally activated equipment such as an absorption chiller or a desiccant dehumidifier.

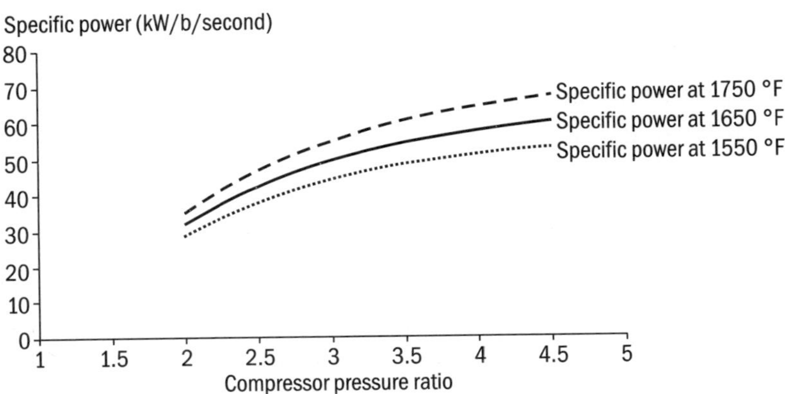

Figure 19 Micro-turbine-specific power as a function of compressor pressure ratio and turbine firing temperature

Source Allied Signal Inc.

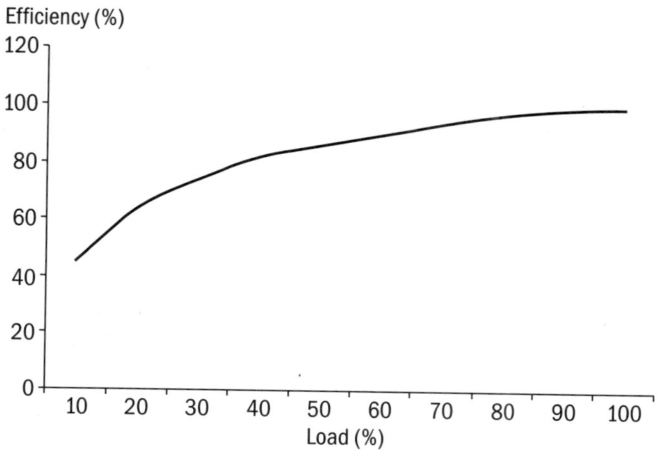

Figure 20 Micro-turbine part-load power performance of a 30-kW micro-turbine

Micro-turbine CHP system efficiency is a function of the exhaust heat temperature. Recuperator effectiveness strongly influences the micro-turbine exhaust temperature.

Consequently, the various micro-turbine CHP systems have a substantially different CHP efficiency and net heat rate chargeable to power. These variations in CHP efficiency and net heat rate are generally due to the mechanical design and manufacturing cost of the recuperators and their resulting impact on system cost, rather than due to differences in system size.

Availability

As only a number of units are in commercial service, it is difficult to come to conclusions about the reliability and availability of micro turbines. The basic design and low number of moving parts hold the potential for the systems of high availability; manufacturers have targeted availabilities of 98%–99%. The use of multiple units or back-up units at a site can further increase the availability of the overall facility.

Micro-turbine emissions characteristics

Table 8 presents typical emissions for micro-turbine systems reflecting the manufacturers' guaranteed levels.

Fuel cells

Fuel cells produce DC electricity through an electrochemical process, much like a standard battery. Unlike a standard battery, a fuel supply continuously replenishes the fuel cell. The reactants, generally hydrogen and oxygen gas, are fed into the fuel cell reactor and power is generated as long as these are supplied. Hydrogen is generally generated from a hydrocarbon fuel such as natural gas or liquefied petroleum gas (LPG), and the oxygen is generated from ambient air.

Table 8 Micro turbine emissions characteristics

Emissions characteristics	System 1	System 2	System 3	System 4
Nominal electricity capacity (kW)	30	70	100	350
Electrical efficiency (higher heating value)	23	25	27	29
NO_x (ppmv)	9	9	15	9
NO_x, (kg/MWh)	0.25	0.23	0.36	0.24
CO (PPMV)	40	9	15	25
CO (kg/MWh)	0.66	0.14	0.22	0.33
THC (PPMV)	<9	<9	<10	<10
THC (kg/MWh)	<0.09	<0.08	<0.09	<0.09
CO_2 (kg/MWh)	876	806	775	695
Carbon (kg/MWh)	239	220	211	190

ppmv – parts per million volume; NO_x – oxides of nitrogen; CO – carbon monoxide; THC – total hydrocarbon; CO_2 – carbon dioxide

Basic processes and components

Fuel cell systems designed for distributed generation (DG) applications are primarily natural-gas- or LPG-fuelled systems. Each fuel cell system consists of three primary subsystems—(1) fuel cell stack that generates DC; (2) fuel processor that converts the natural gas into a hydrogen-rich feed stream; and (3) power conditioner that processes the electric energy into AC or regulated DC.

Figure 21 illustrates the electrochemical process in a single-cell, acid-type fuel cell. A fuel cell consists of a cathode (positively charged electrode), anode (negatively charged electrode), electrolyte, and an external load. The anode provides an interface between the fuel and the electrolyte, catalyses the fuel reaction, and provides a path through which free electrons conduct to the load via the external circuit. The cathode provides interface between oxygen and the electrolyte, catalyses the oxygen reaction, and provides a path

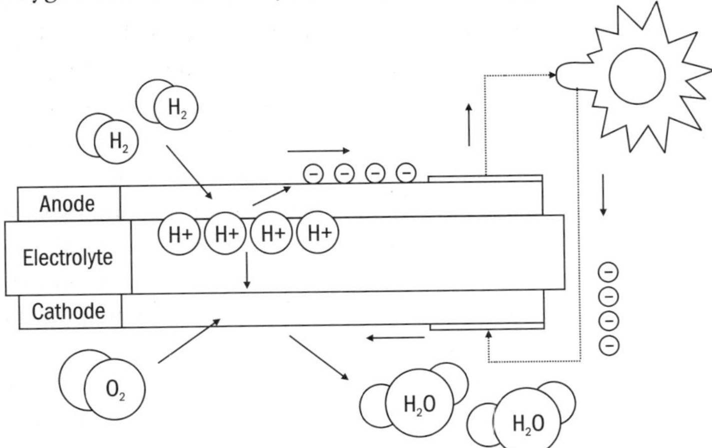

Figure 21 Fuel cell electrochemical process

through which free electrons conduct from the load to the oxygen electrode via the external circuit. The electrolyte, an ionic conductive (non-electrically conductive) medium, acts as the separator between hydrogen and oxygen to prevent mixing and the resultant direct combustion. It completes the electrical circuit of transporting ions between the electrodes.

Hydrogen and oxygen are fed to the anode and cathode, respectively. The hydrogen and oxygen gases do not mix directly, and combustion does not occur. Instead, hydrogen oxidizes one molecule at a time in the presence of a catalyst. Because the reaction is controlled at the molecular level, there is no opportunity for the formation of NO_x and other pollutants. At the anode, the hydrogen gas is electrochemically dissociated (in presence of a catalyst) into hydrogen ions (H^+) and free electrons (e^-) (Equation 1).

$$\text{Anode reaction: } 2H_2 = 4H^+ + 4e^- \quad \dots (1)$$

The electrons flow out of the anode through an external electrical circuit. The hydrogen ions flow into the electrolyte layer and eventually to the cathode, driven by both concentration and potential forces. At the cathode, the oxygen gas is electrochemically combined (in presence of a catalyst) with hydrogen ions and free electrons to generate water (Equation 2).

$$\text{Cathode reaction: } O_2 + 4H^+ + 4e^- = 2H_2O \quad \dots (2)$$

The overall reaction in a fuel cell is as follows (Equation 3).

$$\text{Net fuel cell reaction: } 2H_2 + O_2 = 2H_2O \text{ (vapour)} + \text{energy} \quad \dots (3)$$

The amount of energy released is equal to the difference between the Gibbs free energy of the product and the Gibbs free energy of the reactants. When generating power, electrons flow through the external circuit, ions flow through the electrolyte layer, and chemicals flow into and out of the electrodes. Each process has natural resistances, and overcoming these brings down the operational cell voltage below the theoretical potential. There is also irreversibility that affects actual open circuit potentials. Therefore, some of the chemical potential energy converts into heat. The electrical power generated by the fuel cell is a product of the current measured in amperes and the operational voltage. Based on the application and economics, a typical operating fuel cell will have an operating voltage between 0.55 V and 0.80 V. The ratio of the operating voltage and the theoretical maximum of 1.48 V represent a simplified estimate of the stack electrical efficiency on an HHV basis.

Resistance heat is also generated along with the power. As electric power is the product of the operating voltage and current, the quantity of heat that must be removed from the fuel cell is a product of the current and the difference between the theoretical potential and the operating voltage. In most cases, water produced by the fuel cell reactions releases the fuel cell as vapour and, therefore, the 1.23-V LHV theoretical potential is used to estimate the sensible heat generated by the fuel cell electrochemical process.

The overall electrical efficiency of the cell is the ratio of the power generated and the heating value of the hydrogen consumed. The maximum thermodynamic efficiency of a hydrogen fuel cell is the ratio of the Gibbs free energy and the heating value of the hydrogen. The Gibbs free energy decreases with the increase in temperature because the product water produced at the elevated temperature of the fuel cell includes the sensible heat produced at that temperature, and this energy cannot be converted into electricity without the addition of a thermal energy conversion cycle (such as a steam turbine). Therefore, the maximum efficiency of a pure fuel cell system decreases with increasing temperature. Figure 22 illustrates this characteristic in comparison to the Carnot cycle efficiency limits through a condenser at 50–100°C. This characteristic has led system developers to investigate hybrid fuel-cell-turbine combined cycle systems to achieve system electrical efficiencies in excess of 70% HHV.

Fuel cell stacks

Practical fuel cell systems require voltages higher than 0.55–0.80 V. Combining several cells in electrical series into a fuel cell stack achieves this. Generally, there are several hundred cells in a single cell stack. Increasing the active area of individual cells helps in the current flow. The cell area usually ranges from 100 cm^2 to 1 m^2, depending on the type of fuel cell and application power requirements.

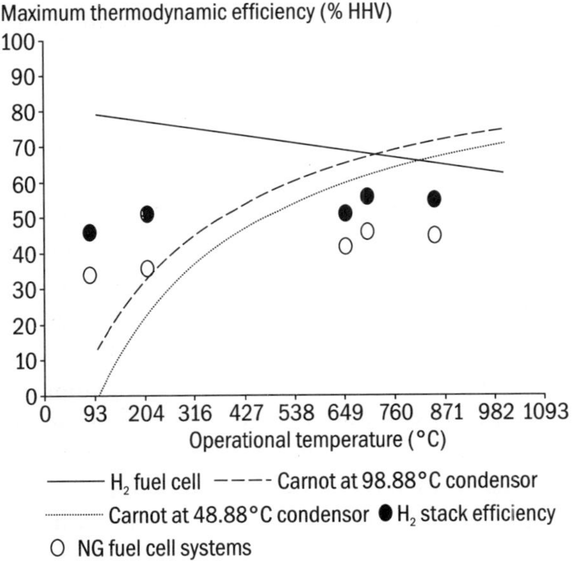

Figure 22 Effect of operating temperature on fuel cell efficiency

Fuel processors

In distributed generation applications, the most viable fuel cell technologies use natural gas as the system's fuel source. To operate on natural gas or other fuels, fuel cells require a fuel processor or a reformer, which is a device that converts fuel into hydrogen-rich gas stream. While adding fuel flexibility to the system, the reformer adds significant cost and complexity. There are three primary types of reformers: (1) steam reformers, (2) auto-thermal reformers, and (3) partial oxidation reformers. The fundamental differences are the sources of oxygen that combine with the carbon within the fuel to release the hydrogen gases and the thermal balance of the chemical process. Steam reformers use steam while partial oxidation units use oxygen gas, and auto-thermal reformers use both steam and oxygen.

Steam reform is highly endothermic and requires a substantial amount of heat input. Auto-thermal reformers operate at or near the thermal neutral point and, therefore, do not generate or consume thermal energy. Partial oxidation units combust a portion of the fuel (that is, partially oxidize it), releasing heat in the process. When integrated into a fuel cell system that allows the use of anode off-gas, a typical natural-gas reformer can achieve conversion efficiencies in the 75%–90% LHV range, with 83%–85% being the expected level of performance. These efficiencies are defined as the LHV of hydrogen generated divided by the LHV of the natural gas consumed by the reformer.

Some fuel cells function as internally steam-reforming fuel cells. As the reformer is an endothermic catalytic converter and the fuel cell is an exothermic catalytic oxidizer, the two combine into one with mutual thermal benefits. More complex than a pure hydrogen fuel cell, these types of fuel cells are more difficult to design and operate. While combining two catalytic processes is difficult to arrange and control, these internally reforming fuel cells are expected to account for a significant market share, as fuel-cell-based DG becomes more common.

Power-conditioning subsystem

The fuel cell generates DC electricity, which requires conditioning before serving a DG application. Depending on the cell area and number of cells, this DC electricity is approximately 200–400 V per stack. If the system is large enough, stacks can operate in series to double or triple individual stack voltages. As the voltage of each individual cell decreases with increase in load or power, the output is considered an unregulated voltage source. The power-conditioning subsystem boosts the output voltage to provide a regulated higher voltage input source to an electronic inverter. The inverter then uses pulse-width modulation technique at high frequencies to generate a simulated AC output. The inverter controls the frequency of the output, which can be adjusted to enhance the power factor characteristics. Because the inverter generates AC within itself, the output power is generally clean and reliable. This characteristic is important to sensitive electronic equipment in premium power applications. Efficiency of the power-conditioning process is typically 92%–96%, and is dependent on system capacity and input voltage–current characteristic.

Types of fuel cells

There are five basic types of fuel cells under consideration for DG applications. The fuel cell's electrolyte or ion conduction material defines the basic type. Two of these fuel cell types—polymer electrolyte membrane fuel cell, or proton exchange membrane fuel cell (PEMFC) and phosphoric acid fuel cell (PAFC)—have acidic electrolytes and rely on the transport of H^+ ions. Two others—alkaline fuel cell (AFC) and molten carbonate fuel cell (MCFC)—have basic electrolytes that rely on the transport of (OH) and $(CO_3)^{-2}$ ions, respectively. The fifth type—solid oxide fuel cell (SOFC)—is based on a solid-state ceramic electrolyte in which oxygen ions (O_2^-) are the conductive transport ion.

Each fuel cell type operates at an optimum temperature, which is a balance between the ionic conductivity and component stability. These temperatures differ significantly among the five basic types, ranging from near ambient to as high as 980°C. The proton-conducting fuel cell type generates water at the cathode, and the anion-conducting fuel cell type generates water at the anode (Table 9).

PEMFC

PEMFC uses a solid polymer electrolyte and operates at low temperatures (about 95°C). Over the past 10 years, the PEMFC has received significant media coverage due to the large auto industry investment in the technology. Due to their modularity and potential for simple manufacturing, reformer/PEMFC systems for residential DG applications have attracted considerable development capital. PEMFCs have high power density and can vary their output quickly to meet demand. This type of fuel cell is highly sensitive to CO poisoning.

AFC

AFC technology uses alkaline potassium hydroxide as the electrolyte. The primary advantages of AFC technology are improved performance (electrical efficiencies above 60% HHV), use of non-precious metal electrodes, and the fact that no unusual materials are needed. The primary disadvantage is the tendency to absorb CO_2, converting the alkaline electrolyte into an aqueous carbonate electrolyte that is less conductive. The attraction of AFC has declined substantially with the interest and improvements in PEMFC technology.

PAFC

PAFC uses phosphoric acid as the electrolyte and is generally considered the most established fuel cell technology. The first PAFC DG system was designed and demonstrated in the early 1970s. PAFCs are capable of fuel-to-electricity

Table 9 Characteristics of major fuel cell types

Characteristics	PEMFC	AFC	PAFC	MCFC	SOFC
Type of electrolyte	H^+ ions (with anions bound in polymer membrane)	OH ions (typically aqueous KOH solution)	H^+ ions (H3PO4)	CO_2^{3-} ions (typically, molten $LiKaCO_3$ eutectics)	O_2 ions (stabilized ceramic matrix with free oxide ions)
Typical construction	Plastic, metal, or carbon	Plastic or metal	Carbon or porous ceramics	High temperature metals or porous ceramic	Ceramic or high-temperature metals
Internal reforming	No	No	No	Yes, good temperature match	Yes, good temperature match
Oxidant	Air to O_2	Purified air to O_2	Enhanced air	Air	Air
Operational	150–180°F	190–500°F	370–410°F	1200–1300°F	1350–1550°F
Temperature	65–85°C	90–260°C	190–210°C	650–700°C	750–1000°C
DG-system-level efficiency, HHV (higher heating value)	25%–35%	32%–40%	35%–45%	40%–50%	45%–55%
Primary contaminate sensitivities	CO, sulphur, and NH_3	CO, CO_2, and sulphur	CO < 1% sulphur	Sulphur	Sulphur

efficiencies of 36% HHV or greater. A 200-kW PAFC has been commercially available since the early 1990s. Over 200 of these commercial units are operating in the US, Europe, and Japan. The current 200-kW product is reported to have a stack lifetime of over 40 000 hours, units with nearly eight years of operation, and commercially based reliabilities in the 90%–95% range. The major market barrier has been the installation cost that has not yet fallen below the $4500–5500/kW range.

MCFC

MCFC uses an alkali metal carbonate (lithium [Li], sodium [Na], and potassium [K]) as the electrolyte and has a developmental history that dates back to the early part of the twentieth century. Due to its operating temperature of 595–760°C, MCFC holds promise in both CHP and DG applications. This type of fuel cell can be internally reformed, can operate at high efficiencies (50% HHV), and is relatively tolerant to fuel impurities. Government/industry R&D programmes during the 1980s and 1990s resulted in several individual pre-prototype system demonstrations. Fuel cell energy is presently the only company with a commercial MCFC. The primary technical issue with MCFC technologies is the degradation of cell components due to the corrosive nature of the electrolyte/operating temperature combination.

SOFC

SOFC uses solid, non-porous metal oxide electrolytes and is generally considered less mature in its development than the MCFC and PAFC technologies. Several SOFC units—up to 100 kW in size and based on a concentric tubular design—have been built and tested. In addition, there are many companies developing planar SOFC designs, which offer higher power densities and lower costs than the tubular design, but these have yet to achieve reliability of the tubular design. Despite relative immaturity, SOFC has several advantages (high efficiency, stability and reliability, and high internal temperatures) that have attracted development support. SOFC has projected service electric efficiencies of 45%–60% and higher for larger hybrid, combined cycle plants. Efficiencies for smaller SOFC DG units are expected to be in the 50% range.

The stability and reliability of SOFC are due to an all-solid-state ceramic construction. Test units have operated in excess of 10 years with acceptable performances. The high internal temperatures of SOFC are both an asset and a liability. As an asset, high temperatures make internal reforming possible; as a liability, these high temperatures add to materials and mechanical design difficulties, which reduce stack life and increase cost. While the SOFC research has been going on for 30 years, costs of these stacks are still comparatively high.

Design characteristics

The features that have the potential to make fuel cell systems a leading prime mover for the CHP and other DG applications include the following.

- *Size range* Fuel cell systems are constructed from individual cells that generate $100/W^2$ kW per cell. This allows the systems to have extreme flexibility in capacity. Systems under development for DG application range in sizes from 5 kW to 2 MW. Multiple systems can operate in parallel at a single site to provide incremental capacity.
- *Thermal output* Fuel cells can achieve overall efficiencies in the 65%–85% range. Waste heat can be used primarily for domestic hot water applications and space heating.
- *Availability* The commercially available 200-kW PC_{25} system fleet (200-plus units) demonstrates greater than 90% availability during the over four-million operating hours. As fuel cell systems mature, their reliability also improves.
- *Part-load operation* Fuel cell stack efficiency improves at lower loads, which results in a system electric efficiency that is relatively steady—down to one-third to one-quarter of rated capacity. This provides the systems with excellent load-following characteristics.

- *Cycling* While part-load efficiencies of fuel cells are generally high, MCFC and SOFC fuel cells require long heat-up and cool-down periods, restricting their ability to operate in many cyclic applications.
- *High quality power* Electrical output is computer-grade power, meeting critical power requirements without interruption. This minimizes loss of productivity, loss of revenues, product loss, and opportunity cost.
- *Reliability and life* As only auxiliary components have moving parts, the reliability of fuel cells is expected to be high. A few of the initial PC_{25} systems have achieved operational lives of 70 000 hours.
- *Emissions* The only combustion within a fuel cell system is the low energy content of hydrogen stream exhausted from the stack. This stream is combusted within the reformer and can achieve emission signatures of < 2 ppmv CO, < 1 ppmv NO_x, and negligible SO_x (on 15% O_2, dry basis).
- *Efficiency* Different types of fuel cells have varied efficiencies. Depending on the type and design of fuel cells, electric efficiency ranges from 30% to 50% HHV.
- *Quiet operation* Noise level of 60 decibels (dBA) at 10-m distance is acceptable for indoor installation.
- *Site* Installation can be carried out both indoor and outdoor.
- *Fuel use* The primary fuel source for the fuel cell is hydrogen, which can be obtained from natural gas, coal gas, methanol, and other fuels containing hydrocarbons.

Performance characteristics

Fuel cell performance is a function of the type of fuel cell and its capacity. As the fuel cell system is a series of chemical, electrochemical, and electronic subsystems, optimization of electric efficiency and performance characteristics can be a challenging engineering task.

Electrical efficiency

As with all generation technologies, electrical efficiency is the ratio of the power generated and the heating value of the fuel consumed. Because the fuel cell system has several subsystems in series, the electrical efficiency of the DG unit is a multiple of the efficiencies of the individual sections. The concept of stack electric efficiency was introduced earlier. Equation (4) calculates the electric efficiency of a fuel cell system.

$$\text{Elec}_{\text{Eff}} = (\text{FPS}_{\text{Eff}} \times \text{H}_2 \text{ utilization} \times \text{Stack}_{\text{Eff}} \times \text{PC}_{\text{Eff}}) \times (\text{HHV/LHV ratio of the fuel}) \quad \ldots (4)$$

where,

FPS_{Eff} = fuel-processing subsystem efficiency; LHV = (LHV of H_2 generated/LHV of fuel consumed); H_2 utilization = % of H_2 actually consumed in the stack; stack $_{Eff}$ = operating voltage/energy potential ~1.23 V; and

PC_{Eff} = AC power delivered/DC power generated (auxilliary loads are assumed as DC loads here).

For example, PAFC = (84% FPS) × (83% utilization) × (0.75 V/1.25 V) × (95% PC) × (0.9 HHV/LHV) = 36% electric efficiency HHV

As the operating temperature range of the fuel cell system increases, the electric efficiency of the system tends to increase. Although the maximum thermodynamic efficiency decreases as shown in Figure 18, improvements in reformer subsystem integration and increases in reactant activity balance out to provide the system-level increase. Advanced high temperature MCFC and SOFC systems are projected to achieve simple cycle efficiencies of 50%–55% HHV, while hybrid combined fuel cell heat engine systems are calculated to achieve efficiencies above 60% in the DG applications.

Part-load performance

In power generation and CHP applications, the fuel cell systems follow either the electric or thermal load of the applications to maximize DG energy economics. Figure 23 shows the part-load efficiency curve for a market-entry PAFC fuel cell in comparison to a typical lean-burn natural-gas engine. The efficiency at 50% load is within 2% of its full load efficiency characteristic. As the load decreases further, the curve becomes somewhat steeper as inefficiencies in air blowers and the fuel processor begin to override the stack efficiency improvement.

Effects of ambient conditions on performance

Fuel cells are rated at ISO conditions of 25°C and 0.987 atm (1 bar) pressure. Fuel cell system performance—both output and efficiency—can degrade when ambient temperature or site elevation increases. Ancillary equipment

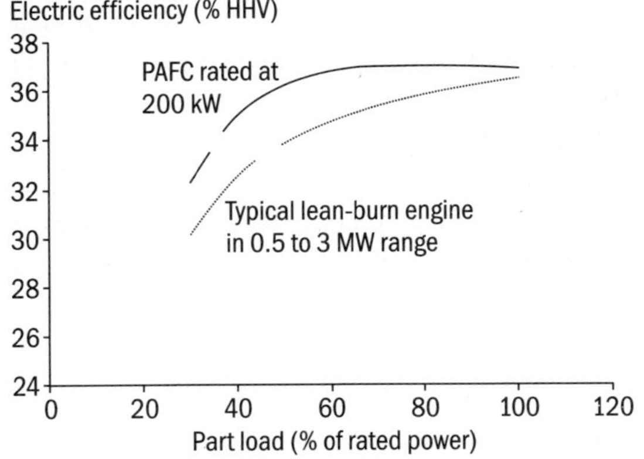

Figure 23 Comparison of part-load efficiency derate

performance—primarily the air-handling blowers or compressors—accounts for the degradation in performance. Performance degradations will be greater for pressurized systems operating with turbochargers or small air compressors as their primary air supply components.

Heat recovery

The economics of the fuel cells in on-site power-generation applications depends lesser on the effective use of the thermal energy recovered than in the case with lower efficiency prime movers; but thermal load displacements improve operating economics as in any CHP application. Generally, the stack and reformer subsystems contain 25% of the inlet fuel energy in the form of higher quality thermal energy. The exhaust gases (including the latent heat of the product water generated in the fuel cell) contain another 25% of recoverable energy. The most common use of this heat is to generate hot water or low-pressure steam for process use or for space heating, process needs, or domestic hot water.

Heat can generally be recovered in the form of hot water or low-pressure steam (< 2 kg/cm^2 [g]), but the quality of heat is dependent on the type of fuel cell and its operating temperature. The one exception to this is PEMFC, which operates at temperatures below 100°C and, therefore, has only low-quality heat.

As an example, there are four primary potential sources of usable waste heat from a fuel cell system: (1) exhaust gas, including water condensation, (2) stack cooling, (3) anode off-gas combustion, and (4) reformer heat. The PAFC system achieves 36% electric efficiency and 72% overall CHP efficiency, which means that it has a 36% thermal efficiency or power to heat ratio of 1. Of the available heat, 25%–45% is recovered from the stack-cooling loop that operates at approximately 205°C and can deliver low- to medium-pressure steam. The exhaust-gas-cooling loop provides the balance of the heat and serves two functions. The first is the condensation of product water, thus rendering the system water self-sufficient, and the second is recovery of by-product heat. As its primary function is water recovery, the balance of heat available from the PAFC is recoverable with 50°C return and 150°C supply temperatures. This tends to limit the application of heat to domestic hot water applications. Maximum system efficiency occurs when all available anode off-gas heat and internal reformer heat are used internally.

In the case of SOFC and MCFC, medium-pressure steam (up to about 10 atm [g]) can be generated from the fuel cell's high-temperature exhaust gas, but the primary use of the hot exhaust gas is in recuperative heat exchange with the inlet process gases. Like engine and turbine systems, the fuel cell exhaust gas can be used directly for process drying.

Emissions

Fuel cell systems generate few emissions as the primary power generation process does not involve combustion. In fact, the fuel processing subsystem is the only significant source of emissions. The anode off-gas that typically consists of 8%–15% hydrogen is combusted in a catalytic or surface burner element to provide heat to the reforming process. The temperature of this lean combustion can be maintained at less than 120 °C, which also prevents the formation of NO_x but is sufficiently high to ensure oxidation of CO and VOCs—unburned, non-methane hydrocarbons. Typically, an absorbed bed before the fuel processor removes and eliminates other pollutants such as SO_x.

CHP on the basis of an organic Rankine cycle process

Basic process

The organic Rankine cycle (ORC) process differs from the steam-turbine cycle in that instead of water an organic working fluid (hydrocarbons like iso-pentane, iso-octane, toluene, or silicone fluid) is used. As this working fluid vaporizes at temperatures lower than water, the process can be better adjusted to a fuel like biomass with its low combustion temperature. In order to delay the ageing process of the working fluid, the permitted temperatures close to the wall may not be exceeded. Therefore, an intermediate thermo-oil cycle allowing better temperature control is necessary. The thermo-oil cycle further allows de-pressurized operation at high temperatures and thus does not require a steam guard.

The exhaust gas resulting from the combustion process in the biomass boiler supplies heat to the thermo-oil cycle. Later, heat is fed to an organic working fluid, which, therefore, vaporizes. The vaporized fluid is expanded in a turbine and the mechanical work obtained is passed to a generator where it is converted into electrical energy. The expanded fluid then enters a condenser where the discharged heat is available at a temperature level that allows operation of a hot-water network for process heat supply. Later, the condensate is brought to operating pressure by the pump and is fed to the evaporator again. In order to increase the electrical output, the working fluid escaping from the turbine can be passed through a recuperator (not included in Figure 24) before it enters the condenser.

Table 10 outlines some data from a plant within a certain range of performance.

Development stage

In the field of geothermal processes, there are already many ORC plants in use. Therefore, the process represents a proven technology. In 1999, Austria's first biomass-fired ORC plant was put in operation by the timber industry in Admont (Styria). More plants are being planned. Presently, the technology is in the demonstration stage and a cost reduction potential exists.

Stirling engine process

Basic process

Exhaust gas resulting from a combustion process enters a boiler-heat exchanger and releases heat to the active gas in the engine (Figure 25).

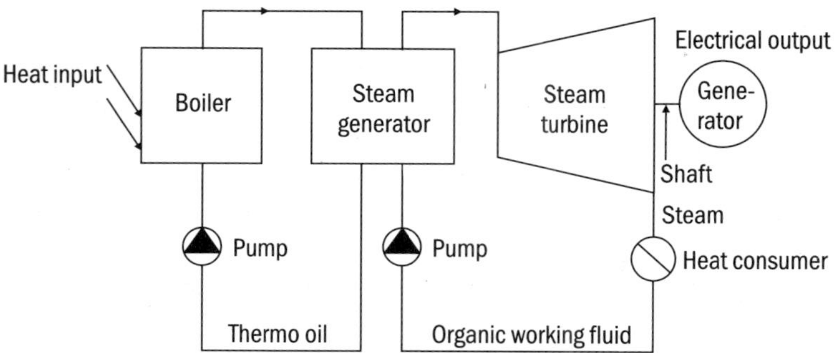

Figure 24 Set-up of an organic Rankine cycle process

Table 10 Data of an organic Rankine cycle process

Plant size ~ 500 kW_{el} (biomass-fired)	Value
Specific investment costs (EUR/kW_{el})	~2.300
Silicon oil (EUR/L)	~22
Specific maintenance costs (EUR/kW_{el})	7
Electrical efficiency el (%)	10–20
Overall efficiency (%)	Up to 85
Emissions (NO_x) (mg/Nm^3)	Depending on fuel ~250–400

kW_{el} – kilowatt electrical; mg/Nm^3 – milligrams per normalized cubic metre; EUR/L – euros per litre

Source Dezentrale Biomasse-Kraft-Wärme- Kopplungstechnologien, Austria

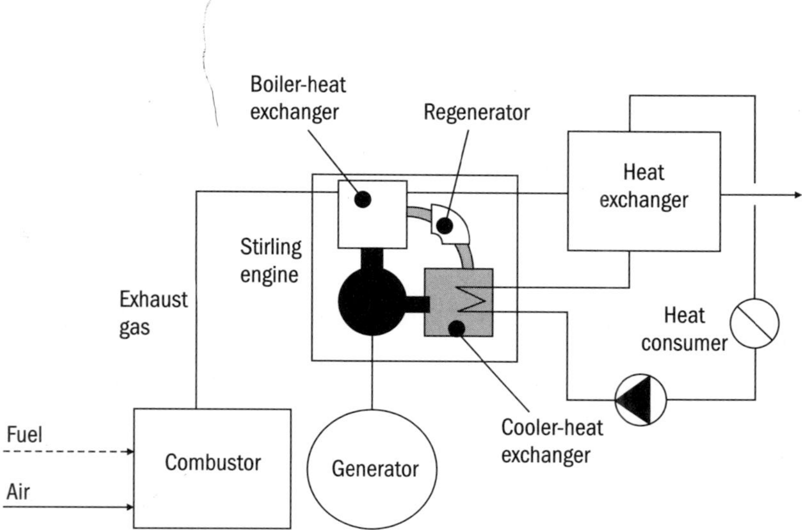

Figure 25 Set-up of a combined heat power (CHP) plant with Stirling engine

Air, nitrogen, helium, and hydrogen serve as possible working fluids. The residual heat of the exhaust gas can be used for supplying heat with the help of an additional heat exchanger. Cooling in the cooler-heat exchanger takes place with the help of the return pipe of the heat supply network. Thus, the heat discharged in the engine can be further utilized for heating purposes (Table 11).

Operating mode of the Stirling engine

In the Stirling engine, the working fluid moves in a closed system between two cylinders. In the working cylinder, heat is supplied for the work output, whereas in the compression cylinder, heat is discharged in order to reduce the compression work. When the working fluid is moved from the working cylinder to the compression cylinder, the residual heat is stored in the regenerator. When the compressed fluid is moved back to the working cylinder, the stored heat can be utilized again.

Design aspects

There are two important designs differing in the arrangement of pistons.

1 The pistons are at a 90° angle to each other.

2 The two pistons—working and compression pistons—are arranged one above the other in the same cylinder. The necessary phase shift of the two pistons is reached with the help of a special linkage.

Control

The Stirling engine is controlled by adjusting the temperature in the boiler-heat exchanger.

Operating state

The operating pressure ranges from 33 bar to 150 bar. The exhaust gas temperature at the boiler-heat exchanger can be used as the operating temperature, which is in the range of 600–1400°C.

Maintenance

The maintenance intervals range from 5000 to 7000 operating hours.

Table 11 Data of a Stirling engine process

Plant size 10–40 kW_{el} (biomass-fired)	Value
Specific investment costs (EUR/kW_el)	~2.400
Specific maintenance costs (EUR/kW_{el})	~0.004–0.011
Electrical efficiency (%)	21–28
Overall efficiency (%)	63–86
Emissions (NO_x) (mg/Nm^3)	~10–15

Source Dezentrale Biomasse-Kraft-Wärme- Kopplungstechnologien, Austria

Development stage

Stirling engines are gaining more and more importance in their application as block heat and power plants. The basis for this is an intensive advancement of the aggregates. Any increase in the working pressure and reduction in the mass of the moved parts can significantly improve the output. CHP plants with Stirling engines are close to market maturity. Intensive research is on at Joanneum Research in Graz in Austria; largely, tests with biomass-firing equipment are being carried out. The technology is in the pilot phase and has a high possibility of cost reduction.

Fuels in different cogeneration systems

The choice of fuel for a cogeneration system is determined by the primary heat-engine cycle. Closed-cycle power systems (steam turbine), which are externally fired, and open-cycle gas turbine and closed-cycle gas turbine systems, which are indirectly fired, can use virtually any fuel that can be burned in a safe and environmentally acceptable manner. Coal, municipal solid waste (MSW), biomass, and industrial waste can be burnt in closed power systems. Internal combustion engines, including open-cycle gas turbines, are restricted to the combustion characteristics specific to the engine type and the yield combustion products clean enough to pass through the engine without damaging it. Natural-gas- and petroleum-derived liquid fuels and coal- and biomass-derived gaseous fuels fall under this category. Figures 26, 27, and 28 show the use of various fuels in different heat engines.

Conclusion

Cogeneration technologies are becoming an increasingly popular means of providing on-site power, meeting the thermal loads, increasing reliability, and improving overall fuel-use efficiency. Advances in conventional equipment and commercialization of emerging generation technologies provide a broad range of equipment types and sizes designed to meet the needs of commercial, institutional, and industrial facilities of all kinds. In addition, these

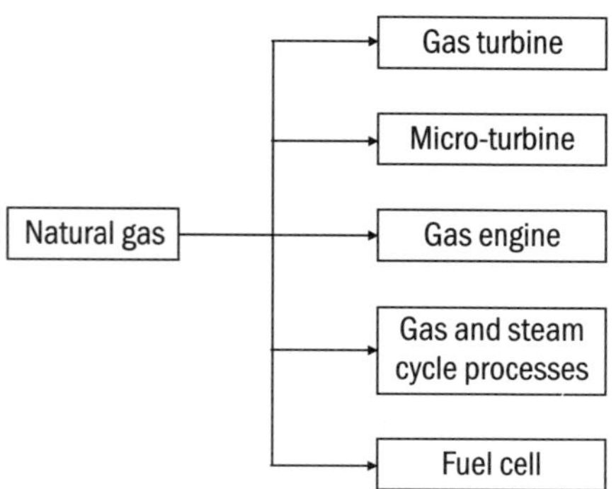

Figure 26 Use of natural gas as fuel in various heat engines

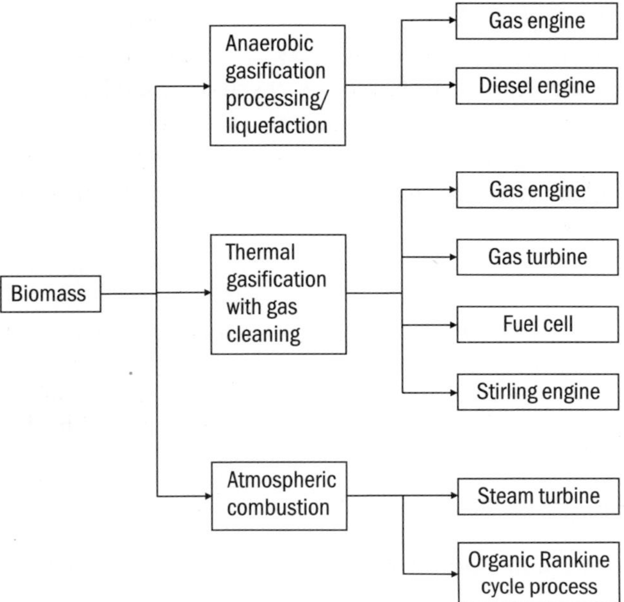

Figure 27 Use of biomass as fuel in various heat engines

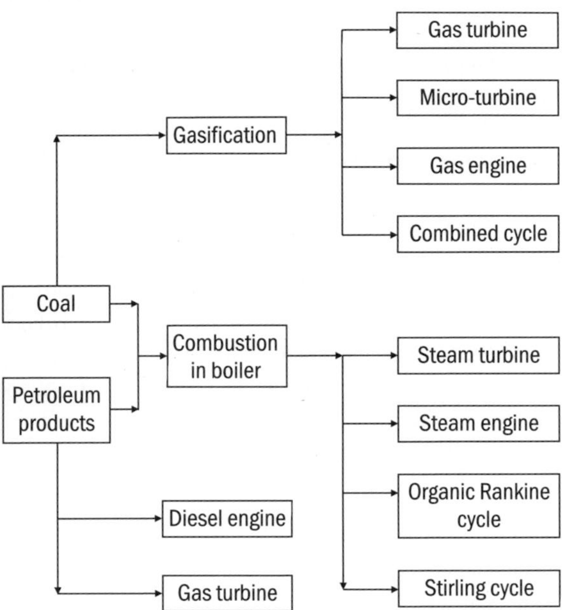

Figure 28 Use of coal and petroleum products as fuels in various heat engines

technologies can also be used to meet decentralized power generation needs in India, especially in locations where a large number of agro-based industries like sugar, rice mills, oil extraction plants, and so on exist. These technologies become even more relevant today, especially when the government is planning to electrify all villages by 2007. Apart from rural electrification, these technologies have a significant role in reducing peak demand for power, in combination with appropriate storage technologies (excluding chilled water storage to meet refrigeration loads) or by enhancing production of power during peak periods. With the EA, 2003, coming into force, most of the institutional barriers should be removed, making way for the adoption of the most competitive technologies.

References

Perry R H and Green D W (eds). 1997. ***Perry's Chemical Engineers' Handbook*** (7th edn). USA: McGraw-Hill Professional Publishing

USEPA (United States Environmental Protection Agency). 1995. ***Compilation of Air Pollutant Emission Factors, Volume I: stationary point and area sources*** (5th edn). [AP-42]. Washington, DC: USEPA

Bibliography

Babcock and Wilcox. 1992. ***Steam: its generation and use*** (40th edn). Ohio: Babcock and Wilcox

Descriptive literature by allied signal power systems. Details available at <http://www.alliedsignal.com>

Descriptive literature by Capstone. Details available at <http://www.capstoneturbine.com>

Energy Technology Austria. 2002. ***Cogeneration (CHP) Technology Portrait: Institute for Thermal Turbomachinery and Machine Dynamics.*** Details available at <http://www.energytech.at>

Arthur D Little Inc. 2000. ***Opportunities for Micro-power and Fuel Cell/Gas Turbine Hybrid Systems in Industrial Applications.*** Cambridge, Massachusetts: Arthur D Little, Inc.

USEPA (United States Environmental Protection Agency). 2002. ***Technology Characterization Report on Steam Turbines.*** Washington, DC: USEPA

USEPA (United States Environmental Protection Agency). 2002. ***Technology Characterization Report on Gas Turbines.*** Washington, DC: USEPA

USEPA (United States Environmental Protection Agency). 2002. ***Technology Characterization Report on Fuel Cells.*** Washington, DC: USEPA

Wilkinson B W and Barnes R W. 1980. ***Cogeneration of Electricity and Useful Heat.*** Florida: CRC Press Inc.

Environmental regulations for waste management in India

Sandeep Shrivastava
Head – Environment, Ambuja Cements Ltd, New Delhi, India

Waste management has become a major challenge these days, with the quantity of waste multiplying at a fast rate and posing a grave threat to human health and environment. Hence, there is a dire need to ensure proper handling and management at every stage of the life cycle of waste, be it collection, segregation, and transportation, as well as treatment/recycling and final disposal.

This article gives an overview of the regulatory framework available in the country for the management of hazardous as well as other industrial and non-industrial waste. It also provides a glimpse on the agencies responsible for the regulation of these wastes, both at the central- and state-government levels.

Evolution of environmental law

The inception of environmental law in India dates back to 1974 when the Water (Prevention and Control) of Pollution Act was promulgated. The United Nations Conference on the Human Environment, held in Stockholm in 1972, brought into sharp focus the environmental problems and paved the way for legislation in India and elsewhere in the world.

This was followed by the Water (Prevention and Control of Pollution) Cess Act, 1977; Air (Prevention and Control of Pollution) Act, 1981, and Environment (Protection) Act, 1986. The Cess Act deals with the charges to be paid for industrial/municipal water use, and the Environment (Protection) Act is a comprehensive legislation that addresses all aspects of the environment.

A snapshot of the status of generation of different types of waste in the country and the pertinent regulations dealing with them is presented in the following sections.

Hazardous waste

Major industrial sectors, which generate hazardous waste in the country, include organic and inorganic chemicals, caustic soda, paints and dyes, asbestos, fertilizers, pesticides, petrochemicals and refinery, and pharmaceuticals. The waste streams emanating from these sectors contain some highly hazardous substances such as heavy metals, cyanides, pesticides, complex aromatic compounds (such as polychlorinated biphenyls), and other chemicals that exhibit inflammable, toxic, corrosive, reactive, and explosive properties.

The Central Pollution Control Board (CPCB) has released the National Inventory of Hazardous Waste Generating Industries and Hazardous Waste Management in India in March 2009. The Inventory was prepared as per the directions of the Supreme Court of India in 2003. The report was published by the CPCB in March 2009, based on the information provided by the state pollution control boards (SPCBs) and the state pollution control committees (SPCCs), pertaining to the period 2007/08.

As per the Inventory, in India, there are 36 165 hazardous waste industries, generating over 6.23 million tonnes per annum (MTPA) of hazardous waste, of which approximately 3.09 MTPA (49.55%) is recyclable, 2.73 MTPA (43.78%) is landfillable, and 0.41 MTPA (6.67%) is incinerable.[1]

Gujarat, Maharashtra, and Andhra Pradesh are the top three hazardous waste generating states. The relative contributions by these states are 28.76%, 25.16%, and 8.93%, respectively. Following these states, Chhattisgarh (4.74%), Rajasthan (4.38%), West Bengal (4.17%), and Tamil Nadu (4.15 %) are found to be major generators of hazardous waste. These seven states together generate over 80% of the country's total hazardous waste.

Hazardous Waste (Management, Handling and Transboundary Movement) Rules, 2008 Amendment July and September 2009

The earlier rules of 1989 got superseded recently in 2008. This set of regulations deals with the generation, collection, treatment, and disposal of hazardous waste. Its important features are as follows.

- As per the 2008 rules, "hazardous waste" means any waste that by reason of its physical, chemical, reactive, toxic, flammable, explosive, or corrosive characteristics causes danger or is likely to cause danger to health or the environment, whether alone or when in contact with other waste or substances and shall include (1) waste specified in schedule-I of the regulation, (2) waste having constituents specified under schedule-II with

[1]CPCB National Inventory of Hazardous Waste Generating Industries and Hazardous Waste Management in India, March 2009.

specified concentration or (3) waste specified in Part A and B of schedule-III with respect to import/export only.

- If the waste generated in an industry is identified to be "hazardous" with reference to the regulation, the generator (referred as "occupier") of the waste-generating unit has to take an "authorization" for the handling of the hazardous waste from the respective SPCB or SPCC.
- The hazardous waste generated in the establishment of the occupier shall be sent or sold to a recycler/reprocessor/reuser authorized under the said rules. The transportation of the same shall be done in accordance with the said rules.
- The occupiers/recyclers/reprocessors/operators of facilities may store hazardous waste for a period not exceeding 90 days and shall maintain records of the same for inspection.
- The rules prescribe the procedure to be followed in the case of import/export of hazardous waste. Its export will be allowed only to an actual user or operator of a disposal facility. Its import and export should strictly be carried out as per Schedule-III of the rules. Hazardous waste can be imported only for recycling/reprocessing and in no case for disposal.
- The rules prescribe packaging, labelling, and transportation conditions and also the liabilities of the occupier, transporter, and operator of the facility and importer.

Recycling of hazardous waste

As per the said rules, all types of hazardous waste recyclers/reprocessors should have consent and authorization, along with the deployment of environmentally sound technologies.

As per the information collected from the CPCB (as of January 2010), a total of 285 units have been granted registration with the Ministry of Environment and Forests (MoEF)/CPCB as recyclers/reprocessors having environmentally sound management facilities for lead waste, and 316 units have been granted registration for other non-ferrous metal waste reprocessors.

With regard to used/waste oil re-processors, as many as 205 units have been registered, while a total of 11 units have been granted registration for e-waste and three units for paint sludge/ink sludge reprocessors.[2]

Disposal of hazardous waste

Facilities for the disposal of hazardous waste in India are developing at a fast pace.

Common treatment, storage, and disposal facilities

Treatment, storage, and disposal facilities (TSDFs) have been developed for the disposal of land-disposable hazardous waste at 22 different locations

[2]CPCB (as of January 2010).

in 10 states, namely, Gujarat (7), Maharashtra (4), Uttar Pradesh (3), Andhra Pradesh (2), Himachal Pradesh (1), Madhya Pradesh (1), Punjab (1), Rajasthan (1), Tamil Nadu (1), and West Bengal (1). The total waste-handling capacity (disposal capacity) of these facilities is approximately 1.5 MTPA, which is much less than the present generation of 2.73 MTPA of land-disposable hazardous waste.[3]

The additional TSDFs with waste-handling capacities to the tune of 1.5 MTPA or so must be developed to accommodate the present and future quantities of land-disposable hazardous waste.

Common hazardous waste incineration

The emission standards for common hazardous waste incinerators were revised with the Environment (Protection) Fifth Amendment Rules, 2008. The discharge of treated flue gas emissions through the stack to the atmosphere shall be according to the parameter-specific emission standards presented in Annexure I.

Used oil versus waste oil

Hazardous Wastes (Management and Handling) Rules Amendment, 2003, differentiated for the first time between "used oil" and "waste oil."

"Used oil" means oil that is (1) derived from crude oil or mixtures containing synthetic oil, including used engine oil, gear oil, hydraulic oil, turbine oil, compressor oil, industrial gear oil, heat transfer oil, transformer oil, spent oil, and their tank bottom sludge, and (2) suitable for reprocessing if it meets the specifications laid down in Part A of Schedule-V but does not include waste oil.

"Waste oil" means oil that includes spills of crude oil, emulsions, tank-bottom sludge, and slop oil generated from petroleum refineries, installations, or ships, and can be used as fuel in furnaces for energy recovery, if it meets specifications laid down in Part B of Schedule-V either as such or after reprocessing.

Coprocessing of hazardous wastes

Coprocessing is the use of waste materials in energy-intensive industries such as cement, lime, steel, glass, and power generation to replace fossil fuels and natural resources. It helps in the upgradation of waste management within the waste hierarchy, reducing global environmental impacts, so as to maintain and improve the industrial competitiveness, decreasing the costs of waste management and optimizing the use of resources. A number of cement companies in India have started coprocessing a variety of hazardous waste in their kilns.

---

[3]CPCB National Inventory of Hazardous Waste Generating Industries and Hazardous Waste Management in India, March 2009.

Hazardous chemicals

A number of hazardous chemicals have been processed/stored by industries in India. The list varies from petrochemicals and refineries to drugs and pharmaceuticals, pesticides, organic and inorganic chemicals, and so on. The CPCB released "Compilation of Material Safety Data Sheet (MSDS) for the 708+ Hazardous, Toxic and/or Flammable Chemicals" in February 2009.

In order to study the inventory of hazardous chemicals being imported by various categories of industrial units in India, the data provided by these industrial units importing hazardous chemicals has been compiled by the CPCB as the *Inventory of Hazardous Chemicals Import in India* (September 2008). The report contains information regarding various chemicals imported, quantity of import, ports used for import, name of importing units, and so on.

Manufacture, Storage and Import of Hazardous Chemicals Rules, 1989: Amendments 1994 and 2000

The aspects of manufacturing, storage, and import of most of these chemicals are covered under the Manufacture, Storage, and Import of the Hazardous Chemicals Rules, 1989: Amendments 1994 and 2000. The regulation provides a comprehensive list of 684 chemicals apart from laying down the criteria pertaining to physical and chemical characteristics of chemicals that need to be provided for proper safeguards that need to be adopted in their handling, processing, and storage. Medium levels of requirements are for 179 chemicals and a high level of safeguards are for 17 chemicals, if their quantities exceed the threshold quantities. "Isolated storages" located away from the processing facilities are governed by a specific schedule of the rules.

Apart from the above, the Public Liability Insurance Act, 1991: Amendment 2008 is a specific legislation dealing with the accidents arising out of handling, storage, manufacturing or importing, and transportation of hazardous chemicals, and is intended to provide an interim financial relief to victims in case of chemical accidents.

Batteries

Incidents of spent lead-acid batteries being reprocessed by unscrupulous backyard smelters to cause severe local environmental impact have led to the promulgation of this regulation so as to ensure that scrap batteries were channellized to the authorized units only for reprocessing.

Batteries (Management and Handling) Rules, 2001

The rules assign responsibilities to all stakeholders, including the manufacturer, importer, assembler and reconditioner, dealer, recycler, consumer or bulk consumer, and auctioneer of the lead-acid batteries.

Biomedical waste

Biomedical waste needs to be stringently managed in order to prevent human health hazards. As per the information available (as of March 2007), as many as 157 common biomedical waste treatment facilities (CBWTF) exist in the country for the management of biomedical waste, of which 149 are currently operational, while eight are yet to be commissioned/are on a trial basis/are under installation.[4]

Biomedical Wastes (Management and Handling) Rules, 1998: Amendment 2003

The rules require that occupants of institutions (hospitals, nursing homes, clinics, dispensaries, veterinary institutions, animal houses, pathological laboratories, blood banks, and so on) generating biomedical waste take all steps to ensure that such waste is handled without any adverse effect on human health or on the environment.

The salient features of the regulation are discussed in the subsequent sections.

Treatment and disposal

The rules require the biomedical facilities, depending on their bed size and location, to install incinerators/autoclaves/microwaves/deep-burial units or ensure the requisite treatment of waste at a common waste treatment facility or any other waste treatment facility. All facilities, of whatever size or location, are required to install the treatment systems as cited above.

All biomedical waste is to be treated and disposed of in accordance with the schedule provided in the rules, which enlist the categories of waste along with the way in which they need to be treated and disposed of. Another schedule specifies the operating parameters and emission and effluent standards of incineration/microwave/autoclave/deep-burial as the treatment options.

Segregation, packaging, transportation, and storage

- Biomedical waste should not be mixed with other types of waste.
- Also, biomedical waste is to be segregated into containers/bags at the point of generation in accordance with the segregation scheme provided in the schedule prior to its storage, transportation, treatment, and disposal. Labelling of containers and the additional information they must carry are also mentioned in the rules.
- The rules forbid storage of an untreated biomedical waste beyond a period of 48 hours. However, this is allowed in case it becomes necessary to store the waste beyond such period, on taking permission from the prescribed

---

[4]CPCB (as in January 2010).

authority and adopting measures to ensure that the waste does not adversely affect human health and the environment.

Prescribed authority

The rules require that every state and union territory establish a prescribed authority with such members, as may be specified, for granting authorization and implementing the rules. In most states/union territories, the SPCBs/SPCCs have been designated as the prescribed authority.

Authorization

The rules make it compulsory for every occupant of an institution generating, collecting, receiving, storing, transporting, treating, disposing, and/or handling biomedical waste in any other manner (except those health-care facilities providing treatment/service to less than 1000 patients per month) to put in an application in the prescribed format to the prescribed authority for the grant of authorization. This authorization is valid for a period of three years, including an initial trial period of one year from the date of issue. Thereafter, an application shall be made by the occupant/operator for renewal. It is mandatory for the prescribed authority to dispose of the application for authorization within 90 days from the date of its receipt.

Municipal solid waste

"Municipal solid waste", or MSW, includes commercial and residential waste generated in municipal or notified areas in either solid or semisolid form, excluding industrial hazardous waste but including treated biomedical waste. Currently, MSW is being processed by composting, biomethanation, converting waste to refuse-derived fuel to be used as fuel, and through projects converting waste to energy.

Municipal Solid Waste (Management and Handling) Rules, 2000

The Municipal Solid Waste (Management and Handling) Rules, enacted in 2000, apply to every municipal authority responsible for collection, segregation, storage, transportation, processing, and disposal of MSW.

In brief, the following requirements are specified for the municipal authorities under the rules.

- Every municipal authority shall, within the territorial area of the municipality, be responsible for the implementation of the provisions of these rules, and for any infrastructure development for collection, storage, segregation, transportation, processing, and disposal of MSW.
- To comply with the implementation programme laid down in the rules, the municipal authority or the operator of a facility needs to apply in a specified format to the SPCB or the SPCC for the grant of an "authorization" for setting up waste processing and disposal facilities, including landfills.

- The rules also lay down the responsibilities of the state governments and union territory administrations, which are as follows.
    - The secretary in-charge of the Department of Urban Development of the state or the union territory concerned, as the case may be, has the overall responsibility for the enforcement of the provisions of these rules in metropolitan cities.
    - The district magistrate or the deputy commissioner of the district concerned has the overall responsibility for the enforcement of provisions of these rules within the territorial limits of his/her jurisdiction.
- Responsibilities of the CPCB and the SPCB/SPCCs are as mentioned below.
    - The SPCB/SPCC has to monitor the compliance of standards regarding groundwater, ambient air, leachate quality, and the compost quality, including incineration standards, as specified in the relevant schedules in the rules.
    - Prior to issuing an authorization, the SPCB/SPCC has to consult other agencies like the state urban development department, the town and country planning department, air port or air base authority, the groundwater board or any other such agency.
    - The SPCB/SPCC has to issue an authorization to the municipal authority or the operator of a facility within 45 days of stipulating the compliance criteria and standards, and any other such conditions as may be necessary.

Plastic waste

As per the CPCB website, plastic waste forms a significant portion of the total MSW. It is estimated that approximately 10 000 tonnes per day (TPD) of plastic waste is generated, that is 9% of 120 000 TPD of MSW in India.

Plastic waste constitutes two major categories: (1) thermoplastics and (2) thermoset plastics. Thermoplastics constitute 80%, and the remaining is thermoset. Thermoplastics are recyclable plastics, which include polyethylene terephthalate (PET), low-density poly ethylene (LDPE), poly vinyl chloride (PVC), high-density poly ethylene (HDPE), polypropylene (PP), polystyrene (PS), and so on; however, thermoset plastics contain alkyd, epoxy, ester, melamine formaldehyde, phenolic formaldehyde, silicon, urea, formaldehyde, polyurethane, metalized and multilayer plastics, and so on[5].

However, waste collection and recycling in the unorganized sector also pose serious environmental and health problems. The central government enacted the Plastics Wastes (Management and Handling) Rules, 1999: Amendment 2003 to apply certain restrictions on the manufacture and use of plastic bags.

---

[5]CPCB (as of January 2010).

Plastic Wastes (Management and Recycling) Rules, 1999: Amendment 2003

The following are the important provisions of the regulation.

Prohibition of the use of carry bags or containers made of recycled plastics

Carry bags and containers should not be used by vendors for storing, carrying, dispensing, or packaging food.

Conditions of manufacture of carry bags and containers made of plastic

- No person shall manufacture, stock, distribute, or sell carry bags made of virgin or recycled plastic bags that are less than 8 × 12 inches in size.
- The minimum thickness of carry bags made of virgin plastics or recycled plastics shall not be less than 20 microns.
- Carry bags and containers made of recycled plastic and used for purposes other than storing and packaging food should be manufactured using pigments and colourants as per IS: 9833: 1981 titled *List of pigments and colourants for use in plastics in contact with foodstuffs, pharmaceuticals, and drinking water.*
- Recycling of plastics shall be undertaken strictly in accordance with the Bureau of Indian Standards (BIS) Specification IS 14 534: 1998 titled *The guidelines for recycling of plastics.*
- Manufacturers of recycled plastic carry bags having printing facilities should code/mark carry bags and containers as per the BIS Specification IS 14 534: 1998 titled *The guidelines for recycling of plastics* and the end-product made of recycled plastics should be marked as "recycled" along with an indication of the percentage of use of recycled material.
- Every occupant manufacturing carry bags or containers of virgin plastics or recycled plastics or both shall make an application in prescribed format for the grant of registration and renewal of registration.

Fly ash

With the country having a coal-based power generation capacity of over 85 000 MW, the amount of fly ash generation has gone up to 125 MTPA.

Fly Ash Notification 1999: Amendment 2003 and 2009

With a notification (September 1999), fly ash utilization of up to 100% has been made mandatory for the coal- and lignite-based power plants within a span of 9–15 years, depending on the plant age. Thereafter, through a notification (August 2003), it was stipulated that construction companies must purchase fly-ash-made products like cement and bricks as per their preferential purchase policy and gradually increase the quantities in subsequent years.

Fly ash notification amendment S.O, 2804 (E) published on November 3 2009 has the following important features.

- Every construction agency engaged in the construction of buildings within a radius of 100 km (by road) from a coal- or lignite-based thermal power plant shall use only fly-ash-based products for construction, such as the following: cement/concrete, fly ash bricks, blocks, tiles, or clay fly ash bricks; blocks, tiles, or cement fly ash bricks; or bricks, blocks, or similar products or a combination or aggregate of them in every construction project.
- Minimum fly ash content for building materials or products to qualify as "fly-ash-based products" category is given in Annexure II.
- No agency, person, or organization shall, within a distance of 100 km (by road) of a thermal power plant, undertake construction or approve design for the construction of roads or flyover embankments with top soil, and not undertake or approve or allow reclamation and compaction of low-lying areas with soil. Only fly ash shall be used for compaction and reclamation.
- No person or agency shall, within 50 km (by road) from a coal- or lignite-based thermal power plant, undertake or approve stowing of mine without using at least 25% of fly ash on weight-to-weight basis, of the total stowing materials used or undertake or approve without using at least 20% of fly ash on a volume-to-volume basis of the total materials used for external dump of overburden and the same percentage in upper benches of back filling of opencast mines, and this shall be done under the guidance of the Director General of Mines Safety (DGMS) or Central Mine Planning and Design Institute Ltd (CMPDIL).
- All coal- or lignite-based thermal power stations would be free to sell fly ash to the user agencies subject to the following conditions.
  - The pond ash and mound ash should be made available free of cost on an "as is where is" basis to the manufacturers of bricks, blocks, or tiles, including clay fly ash product manufacturing unit(s), farmers, the central and the state road construction agencies, public works department, and to agencies engaged in backfilling or stowing of mines.
  - At least 20% of the dry electrostatic precipitator (ESP) fly ash shall be made available free of cost to units manufacturing fly ash or clay-fly ash bricks, blocks, and tiles, on a priority basis. If the demand from such agencies falls short of 20% of the quantity, the balance quantity can be sold or disposed of by the power station as may be possible.

E-waste

Guidelines for Environmentally Sound Management of E-waste, released by the CPCB, March 2008

E-waste comprises waste generated from used electronic devices and household appliances that are not fit for their original intended use and are destined for recovery, recycling, or disposal. Such waste encompasses a wide range of electrical and electronic devises such as computers, hand-held cellular phones, personal stereos, including large household appliances such as refrigerators and air conditioners. E-waste contains over 1000 different substances, many of which are toxic and potentially hazardous to the environment and human health, if these are not handled in an environmentally sound manner.

The growth of e-waste has significant economic and social impacts. The increase in the number of electrical and electronic products, increase in consumption rates, and higher obsolescence rate lead to higher generation of e-waste. The increasing obsolescence rate of electronic products also adds to the huge import of used electronics products. The e-waste inventory based on this obsolescence rate in India for 2005 has been estimated to be 146 180 tonnes, which is expected to exceed 800 000 tonnes by 2012.[6] There is no large-scale organized e-waste-recycling facility in India; there are only two small e-waste-dismantling facilities functioning in Chennai and Bengaluru, while most of the e-waste-recycling units are operating in the unorganized sector.

In India, there are no specific environmental laws or guidelines for e-waste. None of the existing environmental laws have any direct reference to e-waste or refer to its handling as hazardous in nature. However, several provisions of these laws may apply to various aspects of e-waste. As e-waste or its constituents fall under the category of "hazardous" and "nonhazardous" waste, they shall be covered under the purview of "The Hazardous Waste Management Rules, 2008, 2009."

Currently, a total of "11" units have been granted registration by the MoEF/ CPCB as recyclers/reprocessors having environmentally sound management facilities for e-waste, updated up to January 2010.

Administrative mechanism for adjudication of environmental law

There is a two-tier system in the country at the central- and state-government levels for the adjudication of environmental laws in the country. The MoEF is the apex-level central government agency, which frames relevant policies on an ongoing basis. The CPCB assists MoEF in policy-making and developing environmental norms and standards, while the day-to-day implementation of legislation is entrusted to the SPCBs and SPCCs at the state level.

Court decisions

The courts in India have expanded the scope of Article 21, *Right to Life*, in the Constitution of India to the *Right to Clean Environment*. In the *Rural Litigation and Entitlement Kendra vs the State of Uttar Pradesh case*, the Court held that the right to a non-polluted environment and the preservation

[6]Guidelines for Environmentally Sound Management of E-Waste released by CPCB, March 2008

and protection of nature's gifts have also been conceded under Article 21 of the Constitution. Landmark court judgments were made in the area of waste management, which include pollution of Yamuna River, import of hazardous wastes, the Bichheri case, and so on.

Going beyond compliance

Many industries across the world and in India consider waste management as part of their overall management strategy when they plan and implement management systems in their facilities. The Environment Management System (EMS) (as per ISO 14001) and the Occupational Health and Safety Assessment System (OSHAS) (as per BSI 18001) have been increasingly adopted by organizations in India and other countries for taking care of the environment and occupational health and safety issues.

Annexure I Treated flue gas emissions discharge standards for common hazardous waste incinerators

Parameter	Emission standard	Sampling duration
Particulates	50 mg/Nm^3	30 min
HCl	50 mg/Nm^3	30 min
SO_2	200 mg/Nm^3	30 min
CO	100 mg/Nm^3	30 min
	50 mg/Nm^3	24 hours
Total organic carbon	20 mg/Nm^3	30 min
HF	4 mg/Nm^3	30 min
NO_x (NO and NO_2 expressed as NO_2)	400 mg/Nm^3	30 min
Total dioxins and furans	0.1 ng TEQ/Nm^3	8 hours sampling.
Cd + Th + their compounds	0.05 mg/Nm^3	2 hours
Hg and its compounds	0.05 mg/Nm^3	2 hours
Sb + As + Pb + Cr + Co + Cu + Mn + Ni + V + their compounds	0.5 mg/Nm^3	2 hours

TEQ – toxicity equivalent

- All monitored values corrected to 11% oxygen on a dry basis.
- The CO_2 concentration in tail gas shall not be less than 7%.
- In case the halogenated organic waste in input waste is less than 1% by weight, all the facilities in twin chamber incinerators shall be designed to achieve a minimum temperature of 950°C in a secondary combustion chamber, with a gas residence time of not less than 2 s.
- In case the halogenated organic waste is more than 1% by weight in input waste, waste shall be incinerated only in twin-chamber incinerators designed to achieve a minimum temperature of 1100°C in a secondary combustion chamber, with a gas residence time not less than 2 s.
- Incineration plants shall be operated (combustion chambers) with such temperature, retention time, and turbulence, so as to achieve total organic carbon (TOC) content in the slag and bottom ashes less than 3%, or so that their loss on ignition is less than 5% of the dry weight of the material.

Annexure II Minimum fly ash content for building materials or products

Building materials or products	Minimum percentage of fly ash by weight
Fly ash bricks, blocks, tiles, and so on made with fly ash, lime, gypsum, sand, stone dust, cement, and so on (without clay)	50% of total raw material.
Paving blocks, paving tiles, checker tiles, mosaic tiles, roofing sheets, precast elements, and so on wherein cement is used as binder	Usage of Portland-Pozzolana cements (PPC) fly ash or 20% of Ordinary Portland cement content.
Cement	20% of total raw materials.
Clay-based building materials such as bricks, blocks, tiles, and so on.	25% of total raw materials.
Concrete, mortar, and plaster	Usage of PPC (fly ash) or 20% of total raw material.

Index

About the editors

Dr Banwari Lal

Dr Banwari Lal is currently the Director of the Environmental and Industrial Biotechnology Division at TERI, New Delhi, and CEO of ONGC TERI Biotech Ltd. He coordinates the research activities of Microbial Biotechnology and Plant Biotechnology area at TERI. He also coordinates the activities of TERI North East Centre. Currently, his main areas of research include development of clean technology, especially bioremediation of oil spill sites, microbial enhanced oil recovery, and prevention of paraffin deposition in oil well tubing and flow lines. He is a recipient of many awards. He has eight patents and has published 50 research papers in high-rated international journals. He is a member of various advisory and expert committees constituted by the Department of Biotechnology, Government of India, and Department of Science and Technology, Government of India.

Dr Priyangshu M Sarma

Dr Priyangshu M Sarma is associated with TERI for the last ten years. He is currently designated as a Fellow and is also the Convener of the Microbial Biotechnology group of TERI. Before joining TERI for his PhD, Dr Sarma graduated from Pune University in microbiology. He initiated his research career in National Chemical Laboratory, Pune, and continued working on the assessment of genomic and phenotypic diversity of bacterial isolated from petroleum-contaminated sites. With a brief post-doctoral stint in INSA Tolouse, France, he continued his research in the Microbial Biotechnology area of TERI. He is associated with TERI University as an adjunct faculty since 2004.